Probability and Mathematical Statistics (Continued)

ROGERS and WILLIAMS • Diffusions, Markov Processes, and Martingales, Volume II: Îto Calculus
ROHATGI • A Introduction to Probability Theory and Mathematical Statistics
ROSS • Stochastic Processes
RUBINSTEIN • Simulation and the Monte Carlo Method
RUBINSTEIN • Discrete Event Systems: Sensitivity Analysis and Stochastic Optimization by the Score Function Method
RUZSA and SZEKELY • Algebraic Probability Theory
SCHEFFE • The Analysis of Variance
SEBER • Linear Regression Analysis
SEBER • Multivariate Observations
SEBER and WILD • Nonlinear Regression
SERFLING • Approximation Theorems of Mathematical Statistics
SHORACK and WELLNER • Empirical Processes with Applications to Statistics
STAUDTE and SHEATHER • Robust Estimation and Testing
STOYANOV • Counterexamples in Probability
STYAN • The Collected Papers of T. W. Anderson: 1943–1985
WHITTAKER • Graphical Models in Applied Multivariate Statistics
YANG • The Construction Theory of Denumerable Markov Processes

Applied Probability and Statistics

ABRAHAM and LEDOLTER • Statistical Methods for Forecasting
AGRESTI • Analysis of Ordinal Categorical Data
AGRESTI • Categorical Data Analysis
ANDERSON and LOYNES • The Teaching of Practical Statistics
ANDERSON, AUQUIER, HAUCK, OAKES, VANDAELE, and WEISBERG • Statistical Methods for Comparative Studies
ASMUSSEN • Applied Probability and Queues
*BAILEY • The Elements of Stochastic Processes with Applications to the Natural Sciences
BARNETT • Interpreting Multivariate Data
BARNETT and LEWIS • Outliers in Statistical Data, *Second Edition*
BARTHOLOMEW, FORBES, and McLEAN • Statistical Techniques for Manpower Planning, *Second Edition*
BATES and WATTS • Nonlinear Regression Analysis and Its Applications
BELSLEY • Conditioning Diagnostics: Collinearity and Weak Data in Regression
BELSLEY, KUH, and WELSCH • Regression Diagnostics: Identifying Influential Data and Sources of Collinearity
BHAT • Elements of Applied Stochastic Processes, *Second Edition*
BHATTACHARYA and WAYMIRE • Stochastic Processes with Applications
BIEMER, GROVES, LYBERG, MATHIOWETZ, and SUDMAN • Measurement Errors in Surveys
BLOOMFIELD • Fourier Analysis of Time Series: An Introduction
BOLLEN • Structural Equations with Latent Variables
BOX • R. A. Fisher, the Life of a Scientist
BOX and DRAPER • Empirical Model-Building and Response Surfaces
BOX and DRAPER • Evolutionary Operation: A Statistical Method for Process Improvement
BOX, HUNTER, and HUNTER • Statistics for Experimenters: An Introduction to Design, Data Analysis, and Model Building
BROWN and HOLLANDER • Statistics: A Biomedical Introduction
BUCKLEW • Large Deviation Techniques in Decision, Simulation, and Estimation
BUNKE and BUNKE • Nonlinear Regression, Functional Relations and Robust Methods: Statistical Methods of Mod
CHATTERJEE and HADI • Sensitivit
CHATTERJEE and PRICE • Regress ... on
CLARKE and DISNEY • Probability ... rse with Applications, *Second Edition*

... d papers

*Now available in a lower priced paperback edition in the Wiley Classics Library.

Discrete Event Systems

Discrete Event Systems

Sensitivity Analysis and Stochastic Optimization by the Score Function Method

REUVEN Y. RUBINSTEIN
Technion—Israel Institute of Technology, Haifa, Israel

and

ALEXANDER SHAPIRO
Georgia Institute of Technology, Atlanta, Georgia, USA

JOHN WILEY & SONS
Chichester · New York · Brisbane · Toronto · Singapore

Other Wiley Editorial Offices

John Wiley & Sons, Inc., 605 Third Avenue,
New York, NY 10158-0012, USA

Jacaranda Wiley Ltd, G.P.O. Box 859, Brisbane,
Queensland 4001, Australia

John Wiley & Sons (Canada) Ltd, 22 Worcester Road,
Rexdale, Ontario M9W 1L1, Canada

John Wiley & Sons (SEA) Pte Ltd, 37 Jalan Pemimpin #05-04,
Block B, Union Industrial Building, Singapore 2057

Library of Congress Cataloging-in-Publication Data

Rubinstein, Reuven Y.
Discrete event systems : sensitivity analysis and stochastic optimization by the score function method / Reuven Rubinstein, Alexander Shapiro.
p. cm.— (Wiley series in probability and mathematical statistics)
ISBN 0 471 93419 4
1. System analysis. 2. Discrete-time systems. I. Shapiro, Alexander. II. Title . III. Series.
T57.6.R75 1993
003—dc20 92–32372
CIP

British Library Cataloguing in Publication Data

A catalogue record for this book is available from the British Library

ISBN 0 471 93419 4

Produced from camera-ready copy supplied by the authors
Printed and bound in Great Britain by Biddles Ltd, Guildford, Surrey

To our wives
Julia Shapiro and
Rina Rubinstein

Preface

Many modern complex real world systems can be modeled as *discrete-event systems* (DES). Examples are computer-communication networks, flexible manufacturing systems, PERT (project evaluation and review techniques) networks and flow networks. These systems are typically driven by the occurrence of discrete events and their state changes with time. In view of the complex interaction of such discrete events, DES must be studied either via deterministic approximation techniques or via stochastic simulation. There are numerous useful textbooks on performance evaluation of complex DES under both approaches. Simulation is now one of the most widely used techniques in management science and engineering.

In designing, analyzing and operating such complex DES we are interested, however, not only in performance evaluation but in sensitivity analysis and optimization as well. This analysis deals with sensitivities of the performance measure with respect to decision variables, and provides guidance on improving the decisions and identifying the most important ones. With sensitivities (gradients, Hessians, etc.) at our disposal we can optimize the system performance,that is,find the optimal decisions.Consider,for example:

(i) **Stochastic PERT networks.** We might be interested in minimizing the mean shortest path in the network with respect to the parameters of the network links activity durations, subject to some constraints.

(ii) **Queueing networks.** We might be interested in minimizing the steady-state mean waiting time of the customer in such a system, maximizing the throughput, or utilization, with respect to the mean service time and routing probabilities at a particular service station, subject to some constraints.

Until about a decade ago sensitivity analysis and optimization of DES was associated with conventional statistical experimental design, which essentially represents a Crude Monte Carlo (CMC) method. It is a time-consuming and inaccurate procedure. For example, it requires at least $(n+1)$ samples of the output process in order to estimate the gradient, where n is the number of design variables.

In the last decade two new methods for sensitivity analysis and optimization of DES have been developed. They are called *perturbation analysis* (PA) and the *score function* (SF). The SF method is sometimes also called the *likelihood ratio* method.

This book is about the SF method. We shall show that the SF method allows us to evaluate, *simultaneously from a single sample path* (simulation experiment) not only the performance and all its sensitivities (gradient, Hessian, etc.) but to solve an entire optimization problem as well, provided the vector of decision parameters belongs to a parametric family distribution. We shall also show that in order to do so we neither have to know explicitly the sample performance (output process), nor to assume its differentiability; all we need is a single sample from the output process, differentiability of the underlying parametric family of distributions with respect to its vector of parameters, and some mild regularity conditions related to the interchangeability of the operators' expectation and differentiation.

It is well established now that in contrast to the SF approach, the IPA (infinitesimal perturbation analysis) approach requires differentiation of the output process with respect to the parameters of the system. As a result, the interchangeability conditions of the differentiation and expectation operators hold for a restrictive class of problems. This is the main reason why the IPA approach has limited applications.

Today, the SF method allows us to perform sensitivity analysis and optimization in the order of hundreds of decision parameters; **and all this from a single sample path!** SF algorithms and procedures are implemented in a simulation package called QNSO (queueing network stabilizer and optimizer) and they can be readily implemented in any existing discrete-event simulation language, like SLAM, SIMAN and GPSS. The extra computational time required by SF is about 10–50% of the time of the main simulation.

This book is based on a graduate course on performance evaluation, sensitivity analysis and stochastic optimization, which has been taught by Reuven Rubinstein at the Technion—Israel Institute of Technology for the last 5 years, and on some recent advances in perturbation theory and asymptotic analysis of stochastic programs. Although in some sections the treatment is theoretical, emphasis is on concepts rather than on mathematical completeness. We assume that the reader has a basic knowledge of probability, statistics and optimization. More sophisticated concepts will be either explained or referred to the literature.

Chapter 1 presents the necessary background on simulation and modeling, with particular emphasis on the output-analysis and variance-reduction techniques. Chapters 2 and 3 deal with sensitivity analysis and optimization of *discrete event static systems* (DESS) and *discrete event dynamic systems* (DEDS), respectively. The main difference between these two types is that DESS do not evolve with time while DEDS do. An example of DESS is a stochastic-PERT network, and an example of DEDS is a queueing network. In chapter 4 we show how to estimate the standard performance measures (like the steady state waiting time) and rare events in an open non-Markovian queueing network simultaneously from a single simulation experiment. We also present a unified framework for the SF method, importance sampling, rare events and bottleneck networks. Chapter 5 deals with some extensions of the SF method and in particular with autocorrelated input processes. Here we also discuss how to combine efficiently the crude Monte Carlo and the experimental design methods with the SF method. Finally, chapter 6 has a more theoretical orientation. It presents a systematic treatment of perturbation theory of optimization problems and asymptotic analysis of stochastic programs which provides a theoretical basis for chapters 2 and 3. This chapter is almost self-contained and may be useful to the reader interested in sensitivity analysis of deterministic or stochastic programs.

We wish to express our indebtedness to Sòren Asmussen, Herman Chernoff, J. Dai, Alexei Gaivoronski, Paul Glasserman, Dave Goldsman, Vladimir Katkovnik, Yakov Kogan, Joseph Kreimer, Don McLeish, Ben Melamed, Pierre L'Ecuyer, Georg Pflug, Stanislav Uryas'ev, Yorai Wardi, K. Watanabe and Ward Whitt for many valuable suggestions on the earlier draft of this book. Special thanks to Don McLeish for providing us with most of the material for section 3.4 "Conditional Score Function Method", Sòren Asmussen, Paul Glasserman, Vladimir Katkovnik, Pierre L'Ecuyer, Ben Melamed and Ward Whitt for enlightening and stimulating discussions on the SF method.

Many thanks to Eliesar Goldberg from Technion for editing the manuscript, Mirjam Shiran for excellent consulting services with Excel package, Lillian Bluestein, Tsiona Bradu and Eitan Rubinstein for excellent typing of the major part of this book, Tamar Gadrich, Dagan Gilat, Yaron Izhaki, Vladimir Kriman, Alonso Perez-Luna, Idmit Raz and Yoel Shterntal, former and present graduate students of the Technion who collaborated with us on this exciting subject and who made valuable contributions to the entire subject.

A part of this book was written during Reuven Rubinstein's 1990, 1991 and 1992 visits at AT & T, Bell Laboratories, at Ecole Polytechnical Federal Institute (EPFL), Lausanne, and NEC Research Institute. Their hospitality, and especially of G.A. Kochman and C.J. McCallum Jr, from AT&T, T. Liebling from EPFL and B. Melamed and K. Watanabe from NEC USA, Inc. C&C Research Institute is greatly appreciated.

Reuven Rubinstein
Faculty of Industrial Engineering and Management
Technion—Israel Institute of Technology, Haifa, Israel

Alexander Shapiro
School of Industrial and Systems Engineering
Georgia Institute of Technology , Atlanta, Georgia, U.S.A.

List of Acronyms

cdf: cumulative distribution function
CLT: Central Limit Theorem
CMC: Crude Monte Carlo
CSF: Conditional Score Function
$\xrightarrow{\mathcal{D}}$ convergence in distribution
DEDS: Discrete Event Dynamic System
DES: Discrete Event System
DESS: Discrete Event Static System
DSF: Decomposable Score Function
iid: independent identically distributed
IS: Importance Sampling
IPA: Infinitesimal Perturbation Analysis
LCRV: Linear Control Random Variables
LLN: Law of Large Numbers
LR: Likelihood Ratio
MIS: Modified Importance Sampling
NCRV: Nonlinear Control Random Variables
PA: Perturbation Analysis
PERT: Project Evaluation and Review Techniques
pdf: probability distribution function
QNSO: Queueing Network Stabilizer and Optimizer
RBM: Regulated Brownian Motiotion
RSF: Regenerative Score Function
SF: Score Function
TSF: Truncated Score Function
w.p.1: with probability 1

Contents

Chapter 1

Preliminaries

1.1 Introduction

Let

$$\ell(\boldsymbol{v}) = \mathbb{E}_{\boldsymbol{v}}\{L(\boldsymbol{Y})\} \tag{1.1.1}$$

be the expected performance of a stochastic system, where $L(\boldsymbol{Y})$ is the sample performance driven by an input vector $\boldsymbol{Y}$ having a cumulative distribution function (cdf) $F(\boldsymbol{y}, \boldsymbol{v})$. Here $\boldsymbol{v}$ is a vector of parameters lying in a parameter set $V \subset \mathbb{R}^n$ and the subscript $\boldsymbol{v}$ in $\mathbb{E}_{\boldsymbol{v}} L$ means that the expectation is taken with respect to $F(\boldsymbol{y}, \boldsymbol{v})$.

Relevant examples are traffic systems, flexible manufacturing systems, computer-communication systems, production lines, coherent lifetime systems, PERT-project networks and flow networks (see below). Most of these systems operate through the occurrence of discrete events which may change the state of a system and hence are referred to as *discrete event systems* (DES). For most real-world DES, analytical methods are not available and they must be studied numerically, that is, either by standard deterministic numerical techniques or via simulation.

Simulation is now one of the most widely used techniques in management science, business and engineering. Its popularity constantly increases. There are several good commercial computer simulation languages available (e.g. GPSS, SLAM, SIMAN) and numerous useful textbooks on performance evaluation of complex DES such as queueing systems. Discrete-event simulation is a compulsory part of almost all BA and MA engineering and management science curricula.

To estimate the performance $\ell(\boldsymbol{v})$ through simulation (for a fixed $\boldsymbol{v}$) we generate a random vector $\boldsymbol{Y}_1$ from the cdf $F(\boldsymbol{y}, \boldsymbol{v})$ and compute $L(\boldsymbol{Y}_1)$. We next generate a second random vector $\boldsymbol{Y}_2$ (independent of $\boldsymbol{Y}_1$) and compute $L(\boldsymbol{Y}_2)$. Repeating this N times we estimate $\ell(\boldsymbol{v})$ by the sample mean $\hat{\ell}_N(\boldsymbol{v}) = N^{-1} \sum_{i=1}^{N} L(\boldsymbol{Y}_i)$. By the Strong Law of Large Numbers, $\lim_{N\to\infty} \hat{\ell}_N(\boldsymbol{v}) = \mathbb{E}_{\boldsymbol{v}} L(\boldsymbol{Y})$ with probability one. Using standard statistical tools we can derive a Central Limit Theorem (CLT) and confidence intervals for $\ell(\boldsymbol{v})$.

Clearly, we still have the problem of how to generate on the computer a sequence of random vectors $\boldsymbol{Y}_1, \ldots, \boldsymbol{Y}_N$ from a specified joint cdf $F(\boldsymbol{y}, \boldsymbol{v})$. It is common in simulation to generate (approximations of) independent random numbers from a uniform distribution over the interval $(0, 1)$, and transform them appropriately to approximate random variables from the target distribution. The uniform random numbers are produced by a *random number generator*, which is in fact a small deterministic computer program which produces a sequence of the so-called *pseudorandom numbers*. That sequence must resemble a true random sequence in the statistical sense. A good generator should also be fast and be able to produce the billions of random numbers often required in modern simulations without getting into periodic behavior. L'Ecuyer (1990b) surveys the "state-of-the-art" random number generation techniques for simulation. In the past, most of the generators used in practice were simple *linear congruential generators* (Law and Kelton (1991)). But such generators have too small periods for modern applications. Better alternatives, like multiple recursive, non-linear, and combined generators (L'Ecuyer (1990b); L'Ecuyer and Tezuka (1991)), which have much longer periods and better statistical behavior, are now available. L'Ecuyer and Côté (1992) give the code of a random number package that facilitates the implementation of some variance reduction techniques.

With the random number generator in hand, we can generate a sequence of random vectors $\boldsymbol{Y}_1, \ldots, \boldsymbol{Y}_N$ from an arbitrary cdf. Two of the most widely used generation techniques are the *inverse transform method* and the *acceptance/rejection method.* These techniques are treated in most textbooks on simulation. For an excellent treatment of random variable generation, see Devroye (1986) and for an attractive survey, see Schmeiser (1980).

After this brief remark on random number and random variable generation, let us return to problem (1.1.1). There are many good references (e.g. Elishakoff (1983a,b), Kalos and Whitlock (1986), Law and Kelton (1991), and Whitt (1989b)) dealing with performance evaluation, that is with evaluation of $\ell(\boldsymbol{v})$. These studies comprise both analytical and numerical methods.

In designing, analyzing and operating such complex DES we are interested, however, not only in *performance evaluation* but in *sensitivity analysis* and *optimization* as well. The first deals with evaluation of sensitivities (gradients, Hessians, etc.) of the performance measures with respect to the parameters of the system. It provides guidance in improving the decisions and identifying the most important ones, as well as the bottlenecks of the system. The second deals with optimization of an entire DES and with sensitivity analysis of the optimal solution with respect to the parameters of the system.

Consider for example:

(1) **Traffic light system.** Here (i) the performance measure may be a vehicle's average delay as it proceeds from a given point of origin to a given destination, or the average number of vehicles waiting for a green light at a given intersection in the system; (ii) the sensitivity and decision parameters may be the average rate at which the vehicles are arriving at the intersections in the system and the rate at which the light changes from green to red. We might be interested to know:

   (a) What will be the vehicle's average delay in such a system if the interarrival rate at a given intersection increases (decreases), say, by 10–50 percent, or we add one or several traffic lights to the system

   (b) What parameters cause bottlenecks (high congestion in the system) and how can these bottlenecks be prevented

   (c) How can the average delay in such a system be minimized, with allowance for some constraints.

(2) **Manufacturing system.** Here (i) the performance measure may be the average waiting time of an item to be processed at several work stations (robots) according to a given schedule and route; (ii) the sensitivity and decision parameters may be the average rate at which the workstations (robots) process the item. In such a system we might be interested in minimizing the average make-span consisting of the processing time and delay time with allowance for some constraints (e.g. cost).

(3) **A general open k-station queueing network** with performances being, say, the steady -state sojourn time of a customer in the network, the throughput, the utilization and the number of customers in the network. We might be interested to know:

   (a) What will be the expected performance of the system, say, the expected sojourn time if (i) we perturb some parameters of the system like the vector of service rates, routing probabilities, the buffer size at each node by 10–50 percent or (ii) we change the topology of the system by adding, say, an extra node (or several nodes) to the existing system.

(b) What are the sensitivities (gradients, Hessians, etc.) of the expected performance $\ell(\boldsymbol{v})$ with respect to the parameters of the system, like the vector of service rates, the vector of interarrival rates and the routing probabilities.

(c) What are the bottlenecks (highly congested) nodes, and how do we prevent them.

(d) How to solve an entire optimization problem for such a queueing system, where, say, the objective function is the customers' expected delay, the vector of the decision parameters is the vector of service rates and we assume, additionally, that the expected number of customers waiting at each node must not exceed a certain quantity.

(e) Let $\boldsymbol{v}^*$ and $\ell(\boldsymbol{v}^*)$ be an optimal solution and the optimal value, respectively. Let $\boldsymbol{r}$ be a vector of parameters. Assume, for example, that $\boldsymbol{v}^*$ is the vector of the optimal service rates and $\boldsymbol{r}$ is the parameter vector of interarrival rates. Assume further that $\boldsymbol{v}^*(\boldsymbol{r})$ and $\ell(\boldsymbol{v}^*, \boldsymbol{r})$ are known or can be evaluated, provided $\boldsymbol{r}$ is fixed, say, when $\boldsymbol{r}$ equals $\boldsymbol{r}_0$. We would like to know how $\boldsymbol{v}^*(\boldsymbol{r})$ and $\ell(\boldsymbol{v}^*(\boldsymbol{r}), \boldsymbol{r})$ behave when $\boldsymbol{r}$ is close to $\boldsymbol{r}_0$ or, in other words, we would like to perform sensitivity and stability analyses of the optimal solutions $\boldsymbol{v}^*$ and $\ell(\boldsymbol{v}^*)$ with respect to $\boldsymbol{r}$.

(f) How robust is the optimal solution $\boldsymbol{v}^*$ if the underlying cdf $F(\boldsymbol{y}, \boldsymbol{v})$ is contaminated by another cdf, say $H(\boldsymbol{y}, \boldsymbol{v})$. In other words, what will be the optimal solution $\boldsymbol{v}^*$ if we believe that the "true" cdf is F while in "fact" it is $(1-t)F(\boldsymbol{y}, \boldsymbol{v}) + tH(\boldsymbol{y}, \boldsymbol{v})$, where t is the percentage of contamination of $F(\boldsymbol{y}, \boldsymbol{v})$ by $H(\boldsymbol{y}, \boldsymbol{v})$?

(4) **Stochastic network.** We might be interested in sensitivity analysis and minimization of the mean shortest path in a system with respect to the parameters of the network links, with allowance for some constraints. Examples are PERT-project networks and flow networks. In the first case, the input and output variables may represent activity durations and the minimum project duration, respectively; in the second case, they may represent flow capacities and maximal flow capacities.

Let $\boldsymbol{v}$ be the decision vector and consider the following minimization problem:

$$(\mathrm{P}_0) \qquad \begin{array}{lll} \text{minimize} & \ell_0(\boldsymbol{v}) = \mathbb{E}_{\boldsymbol{v}} L_0(\boldsymbol{Y}), & \boldsymbol{v} \in V, \\ \text{subject to} & \ell_j(\boldsymbol{v}) = \mathbb{E}_{\boldsymbol{v}} L_j(\boldsymbol{Y}) \leq 0, & j = 1, \ldots, k, \\ & \ell_j(\boldsymbol{v}) = \mathbb{E}_{\boldsymbol{v}} L_j(\boldsymbol{Y}) = 0, & j = k+1, \ldots, M. \end{array} \qquad (1.1.2)$$

It is important to note that when the objective function $\ell_0(\boldsymbol{v})$ and the constraint functions $\ell_j(\boldsymbol{v}),\quad j = 1, \ldots, M$, are available analytically, (P_0) becomes a standard nonlinear programming problem

$$(\mathrm{P}_0) \qquad \begin{array}{llll} \text{minimize} & \ell_0(\boldsymbol{v}), & \boldsymbol{v} \in V, & \\ \text{subject to} & \ell_j(\boldsymbol{v}) \leq 0, & j = 1, \ldots, k, & \\ & \ell_j(\boldsymbol{v}) = 0, & j = k+1, \ldots, M, & \end{array} \qquad (1.1.3)$$

which can be solved either analytically or numerically by standard methods of nonlinear programming (e.g. Avriel (1976)). As an example of (1.1.3) we can think of optimization as a Markovian queueing system (see Kleinrock (1976)).

As pointed out earlier, we shall further assume that the objective function and some of the constraints in (P_0) are not available analytically (because of the complexity of DES) and we have to resort to stochastic optimization methods, in particular to the Monte Carlo method.

Until about a decade ago, Monte Carlo methods for sensitivity analysis and optimization of DES were associated with conventional statistical experimental design, which is based on finite-difference estimation of the gradient and Hessian and actually represented a Crude Monte Carlo (CMC) method. It is a time-consuming and inaccurate procedure. For example, it requires at least $(n+1)$ simulations to estimate the gradient, where n is the number of design variables. In the late seventies, two new methods were introduced for sensitivity analysis and optimization of DES, called *perturbation analysis* (PA) (see Brémaud (1990), (1991); Brémaud and Vázquez-Abad (1991); Gaivoronski (1990), (1991); Glasserman (1988a,b), (1990a,b,c), (1991a,b,c) (1992a,b); Gong and Ho (1987); Heidelberger et al. (1988); Ho, Eyler and Chien (1979); Ho and Strikland (1990); Ho and Cao (1991); L'Ecuyer (1990a), (1991a); L'Ecuyer and Perron (1992); Suri (1983), (1987), (1989); Suri and Zazanis (1988); Vázquez-Abad and Kushner (1991), (1992); Wardi (1990); Wardi avd Lee (1991); and Wardi et al. (1992)); and *score function* (SF) (see Aleksandrov, Sysoyev and Shemeneva (1968); Arsham (1988); Arsham et al. (1989); Asmussen and Rubinstein (1992a,b); Asmussen, Rubinstein and Wang (1992); Beckman and McKay (1987); Glynn (1990), (1992); Glynn, L'Ecuyer and Ades (1991); L'Ecuyer (1990a), (1991a,b), (1992); Katkovnik and Rubinstein (1992); Kreimer (1984); McLeish and Rollans (1992); Meketon (1987); Mikhailov (1967); Miller (1967); Pflug (1988), (1989), (1990), (1991), (1992); Reiman and Weiss (1989a,b); Rief (1984), (1988); Rubinstein (1969), (1976), (1986a,b), (1989), (1991), (1992a,b,c,d); Rubinstein and Shapiro (1990); and Zhang and Ho (1989)) which is the main theme of this work. Both of them permit estimation of *all* sensitivities (gradients, Hessians, etc.) from a *single simulation run* (experiment) in the course of evaluation of the performance (output) measures. Their approaches are however different. The PA approach entails *perturbation and differentiation of the sample performance with respect to the decision vector* $\boldsymbol{v}$. In these circumstances interchangeability conditions of the differentiation and expectation operators hold for a restrictive class of problems. This is the main reason for limited applications of PA (see Heidelberger et al. (1988), Reiman and Weiss (1989a), and Rubinstein (1989)). By contrast, the SF approach entails *only perturbation and differentiation of the underlying* pdf $f(\boldsymbol{y},\boldsymbol{v}) = \mathrm{d}F(\boldsymbol{y},\boldsymbol{v})/\mathrm{d}\boldsymbol{y}$ (which is assumed to exist) and some mild regularity conditions (see below). It is associated with such classical notions and concepts as *efficient score, Fisher information, Radon-Nikodym derivative, and likelihood ratio* (e.g. Lehmann (1983) and sections 2.3 and 2.7 below). Its main idea consists in replacing the deterministic sensitivities (gradients, Hessians, etc.) of $\ell_j(\boldsymbol{v}), j = 0,1,\ldots,M$, in the deterministic optimization procedures by their stochastic counterparts generated from a *single* simulation experiment. Operating with the stochastic counterpart of (P_0) we shall evaluate its optimal solution. Note that this is due to the assumption that *only* $f(\boldsymbol{y},\boldsymbol{v})$ depends on $\boldsymbol{v}$, while $L_j(\boldsymbol{y})$ do not. Note also that there is no need for analytic expressions of $L_j(\boldsymbol{y})$ which in fact are not available in most practical applications. All we need to have at our disposal is *a sample path (realization)* of the output processes $L_{jt} = L_j(\boldsymbol{Y}_t), j = 0,1,\ldots,M; \;\; t > 0$, and *a sample path (realization)* of what we call the *score function process* (see below). The last is based on the *efficient score* (that is, on $\nabla \log f(\boldsymbol{Y}_t,\boldsymbol{v}), \;\; t \geq 0$) and is defined in chapters 2 and 3 below. Note finally that to the best of our knowledge likelihood ratios were used for derivative estimation in a simulation context for the first time by Mikhailov (1967), Miller (1967), Aleksandrov, Sysoyev and Shemeneva (1968) and Rubinstein (1969).

It is worthwhile mentioning that recourse to stochastic rather than to deterministic derivatives is not new in stochastic optimization. Such derivatives (sensitivities) were extensively used (e.g. Ermoliev (1969), Katkovnik (1976), (1985), Kushner and Clark (1978), Kreimer and Rubinstein (1988), L'Ecuyer, Giroux and Glynn (1991) and Wardi (1990)) in numerical (iterative) procedures of the stochastic-approximation type. What we regard as *innovative* and *advantageous* in the SF method is that in solving (P_0) *it does not require iterative procedures based on a stochastic approximation, hence multiple simulation; in fact, it is able to generate, from a single simulation experiment, a complete stochastic optimization procedure.* In other words, the SF method "mimics" its deterministic counterpart in the sense that the optimization procedures are the same, only that the sensitivities are stochastic rather than deterministic.

It is important to note that in its original form the SF method (e.g. Rubinstein (1976)) has certain limitations, which will be discussed through the book. In particular it is *only* suitable for those sensitivity and optimization problems where the sample performance L *does not depend* on $\boldsymbol{v}$, and $\boldsymbol{v}$ is a parameter vector in a family of distributions $f(\boldsymbol{y}, \boldsymbol{v}), \quad \boldsymbol{v} \in V$, where the efficient score $\nabla \log f(\boldsymbol{Y}, \boldsymbol{v})$ exists and has limited variance. Its modifications, however, the so-called *decomposable score function* (DSF), *conditional score function* (CSF), and *truncated score function* (TSF) (see chapters 3 and 5) can be applied to a rather broad class of problems.

Our purpose in this work is to demonstrate the feasibility and efficiency of the SF method and its modified versions for performance evaluation, sensitivity analysis and optimization of rather complex DES (in particular for complex non-Markovian queueing systems) and to develop foundations (in particular to present a statistical theory describing the properties of the derived estimators and their associated procedures and algorithms). While we are not aware of any alternative optimization method capable of yielding the optimal solution of an entire optimization problem of type (P_0) from a single simulation run, it is not our intention here to compare the SF method with any other stochastic optimization method employed in simulation studies, in particular with:

(1) The infinitesimal perturbation analysis (IPA) method which resembles the SF, in the sense that, subject to limitations (see Heidelberger et al. (1988), L'Ecuyer (1990a) and chapter 5), it permits estimation of the sensitivities $\nabla^k \ell(\boldsymbol{v}), k \geq 1$, from a *single* simulation experiment, but at the same time requires iterative (numerical) procedures of stochastic approximation type (see Gaivoronski (1990), (1991), Ermoliev and Gaivoronski (1992), L'Ecuyer, Giroux and Glynn (1991)).

(2) The experimental design techniques (e.g. Biles and Ozmen (1987) and Kleijnen (1986), (1991)), in which the sensitivities are estimated by using finite differences (e.g. L'Ecuyer and Perron (1992)), and thus by the Crude Monte Carlo method.

Apart from the comparative aspect, if necessary, we shall touch upon these approaches as well.

We shall also consider in chapter 5 the extended version of the program (P_0), namely the case where the sample performance L *depends* on a parameter vector, that is, where $\ell_j(\boldsymbol{v})$ in (1.1.3) can be expressed as

$$\ell_j(\boldsymbol{v}) = \mathbb{E}_{\boldsymbol{v}_1}\{L_j(\boldsymbol{Y}, \boldsymbol{v}_2)\} = \int L_j(\boldsymbol{y}, \boldsymbol{v}_2) \mathrm{d}F(\boldsymbol{y}, \boldsymbol{v}_1), \qquad j = 0, 1, \ldots, M. \tag{1.1.4}$$

Here $\boldsymbol{v} = (\boldsymbol{v}_1, \boldsymbol{v}_2)$ is a parameter vector, and the subscript $\boldsymbol{v}_1$ in $\mathbb{E}_{\boldsymbol{v}_1} L$ denotes that the expectation is taken with respect to the pdf $f(\boldsymbol{y}, \boldsymbol{v}_1)$. Note that we assume that f depends on the parameter vector $\boldsymbol{v}_1$ and not on $\boldsymbol{v}_2$, and vice versa for L.

As we mentioned earlier, the SF approach is usefull for sensitivity analysis and optimization with respect to the parameter vector $\boldsymbol{v}_1$ only. To perform sensitivity analysis and optimization with respect to the entire parameter vector $\boldsymbol{v} = (\boldsymbol{v}_1, \boldsymbol{v}_2)$ we shall present in chapter 5 the so-called direct, "push in" and "push out" techniques.

Moreover, we shall deal in this book with both DEDS (discrete-event dynamic systems) and DESS (discrete-event static systems), which differ mainly in that the first evolve over time, and the second do not. Examples of DEDS are queueing networks, and examples of DESS are reliability systems and PERT networks. Note that DESS require only a constant number of random vectors $\boldsymbol{Y}$, whereas DEDS may require a random number of $\boldsymbol{Y}$'s (e.g. the dimension of $\boldsymbol{Y}$ might be the number of customers served in a busy period).

In the remainder of this chapter we present some background on simulation, modeling and analysis. More specifically, section 1.2 deals with the output analysis of the simulated data, and in particular with estimation of transient and steady-state quantities of queueing models; section 1.3 presents several commonly used variance reduction techniques, and finally section 1.4 presents a rather general framework for the sensitivity analysis of DES.

1.2 Output Analysis of the Simulated Data

The goal of this section is to present a brief treatment of statistical analysis for the output data obtained via simulation.

Let $\{L_t : t > 0\}$ be the stochastic process under consideration and assume that we want to estimate the expected value of $X_t = \varphi(L_t)$, where φ is a real valued measurable function on the state space of L_t. For example, if $\varphi(\boldsymbol{x}) = \boldsymbol{x}^2$, then $\mathbb{E}\{X_t\} = \mathbb{E}\{L_t^2\}$. If $\varphi(\boldsymbol{x}) = I_{\{L<\boldsymbol{x}\}}$ is the indicator function of the interval $(-\infty, a)$, then $\mathbb{E}X_t = P\{L_t < a\}$.

Before proceeding further, we shall distinguish between the *finite-horizon* and *steady -state* simulations (regimes). In the finite-horizon, which is also called the *terminating* simulation, the measurements of system performance are defined relatively to the interval of simulation time $[0, T]$, where T is the instant of time when a specific event, say event A, occurs. Note that the event A, as well as the initial state of the model, must be specified before the simulation begins. Note also that T might be a random variable, say the time when a new busy cycle begins.

In the steady -state, which is also called the long-run simulation, the performance measures are defined as limits at the time, that is, as the length of the simulation goes to infinity. It is important to note that in the steady -state there always exist an event specifying the length of a run. Also care must be taken when making inferences (drawing conclusions) about the performance in the steady -state, that is about the steady -state parameters. The reason is that the output data are typically correlated and therefore the classical statistical analysis, based on independent observations, is not applicable. Note also that for some stochastic models, only the finite-horizon simulation is feasible, since the steady -state either does not exist or the finite-horizon period is so long that the steady -state analysis is meaningless within a given computer time budget (e.g. Law and Kelton (1991)).

To provide a better insight into the meanings of finite-horizon and steady -state simulation, assume that L_t represents the sojourn time of the t-th customer in a stable $GI/G/c/b$ queue, $\mathcal{L}(0)$ is the initial state of the system, say the number of customers in the queue at time $t = 0$. Consider

$$F_{t,m}(x) = P\{\varphi(L_t) \leq x\}, \tag{1.2.1}$$

given the initial state of the system $\mathcal{L}(0) = m$. Here GI, G, b, and c in $GI/G/c/b$ stand for the interarrival time distribution, the service time distribution, the number of servers, and the buffer size, respectively (e.g. GI means that the interarrival time has a general (G) distribution and are independent (I)), $F_{t,m}$ is called the *finite-horizon distribution* of the random variable (rv) $\varphi(L_t)$, provided the initial state of the system $\mathcal{L}(0)$ is m (m customers are present). Note that in general the distribution $F_{t,m}(x)$, will be different for each pair (t, m).

We say that the process $\{L_t\}$ settles down into the steady -state or the steady -state exists if for all m

$$\lim_{t\to\infty} F_{t,m}(x) = F(x) \equiv P\{\varphi(L_\infty) \leq x\}. \tag{1.2.2}$$

In other words, "steady -state" means that as $t \to \infty$ the transient cdf $F_{t,m}(x)$ approaches a steady -state cdf $F(x)$ which *does not depend* on the initial state of the system $\mathcal{L}(0)$. In particular (1.2.2) assumes that the underlying stochastic process L_t converges in distribution to a proper limit L_∞ independently of the event $\{\mathcal{L}(0) = m\}$. Notice that for practical purposes "steady -state" means that after some period of time the transient cdf $F_{t,m}(x)$ of the process $\{\varphi(L_t)\}$ becomes very close to its limiting (steady -state) cdf $F(x)$, and, of course, we do not mean that the realizations of $\{L_t\}$ generated from the simulation run become independent or even constant after some period in time. The exact distributions (transient and steady -state) are available only for simple Markovian models (see, for example, Gross and Harris (1985) and Abate and Whitt (1987a,b,c)). Note also that for queueing models of the $GI/G/1$ type, not only the distributions (transient and steady -state) but the associated moments are likewise not available. For such models one usually resorts to simulation.

In the rest of this section we shall present several simple statistical procedures for estimating system parameters for both (i) finite-horizon and (ii) steady -state simulations.

(i) Let L_t be a discrete time process and suppose that we want to estimate the following expected value (parameter)

$$\ell(k,m) = \mathbb{E}\left\{k^{-1}\sum_{t=1}^{k} L_t\right\} , \tag{1.2.3}$$

given, as before, that the initial state of the system $\mathcal{L}(0) = m$. If L_t is a continuous time process, then the sum $\sum_{t=1}^{k} L_t$ is replaced by $\int_0^k L_t dt$. With such definitions it is readily seen that if $\{L_t\}$ is the delay process in the stable $GI/G/1$ queue, then $\ell(k,m)$ represents the average delay of the first k customers in the system, provided the initial state of the system is m, and if $\{L_t\}$ is the queueing length process, then $\ell(k,m)$ represents the average number of customers in the system during the time interval $(0,k]$, provided again the initial state of the system is m.

Assume now that we have performed N independent replications (simulation runs), each starting at the state m. Then the point estimator and the $(1-\alpha)$ 100% confidence interval for $\ell(k,m)$ can be written as

$$\bar{\ell}_N(k,m) = N^{-1}\sum_{i=1}^{N} X_i \tag{1.2.4}$$

and

$$\left\{\bar{\ell}_N(k,m) \pm z_{1-\alpha/2}\sigma N^{-1/2}\right\} , \tag{1.2.5}$$

respectively, where $X_i = k^{-1}\sum_{t=1}^{k} L_{ti}$, L_{ti} is, say, the delay of the t-th customer at the i-th replication, σ^2 is the variance of X_i, $i = 1,\ldots,k$, and $\Phi(z_{1-\alpha}) = 1-\alpha$, with Φ being the standard normal cdf.

Typically σ^2 is unknown and must be replaced by the sample variance

$$S^2 = \frac{1}{N-1}\sum_{i=1}^{N}(X_i - \bar{\ell}_N)^2 .$$

The algorithm for estimating the finite-horizon performance $\ell(k,m)$ is simple indeed and can be written as

Algorithm 1.2.1 :

1. *Perform N independent replications (simulation runs), each of length k while starting the simulations at the initial state $\mathcal{L}(0) = m$.*

2. *Calculate the point estimator and the confidence interval of $\ell(k, m)$ according to (1.2.4) and (1.2.5), respectively.*

Suppose that the steady -state cdf $F(x)$ of $\varphi(L_t)$ exists and we want to estimate the expected value ℓ of the random variable $\varphi(L_\infty)$ given $\mathcal{L}(0) = m$, say to estimate the expected steady-state delay of a customer in the $GI/G/1$ queue, provided the initial state of the system $\mathcal{L}(0)$ equals m. Taking into account (1.2.2) we can represent ℓ as $\ell = \mathbb{E}\{\varphi(L_\infty)\} = \mathbb{E}\{\varphi(L)\}$. Denote further $X_t = \varphi(L_t)$. Let $X_1, \ldots, X_T$ be a sample taken in the steady -state. Then, depending on whether $\{L_t\}$ is a discrete or continuous time process, we can estimate ℓ as

$$\bar{\ell}_T = T^{-1} \sum_{t=1}^{T} X_t \text{ or as } \bar{\ell}_T = T^{-1} \int_0^T X_t \mathrm{d}t,$$

respectively.

The standard statistical analysis for the steady -state is based on the Central Limit Theorem (CLT) for $\bar{\ell}_T$ as $T \to \infty$; that is on the assumption that $T^{1/2}(\bar{\ell}_T - \ell)$ converges in distribution to a normal distribution with mean 0 and variance σ^2. We assume that the CLT holds, that is

$$T^{1/2}(\bar{\ell}_T - \ell) \xrightarrow{\mathcal{D}} N(0, \sigma^2) \text{ as } T \to \infty,$$

where $\xrightarrow{\mathcal{D}}$ means convergence in distribution. For large T, it is appropriate to regard $\bar{\ell}_T$ as being approximately normally distributed with mean $\ell = \mathbb{E}L$ and variance σ^2/T. We call σ^2 the *asymptotic variance* of the sample mean $\bar{\ell}_T$.

It is well known (e.g. Karlin and Taylor (1975)) that if $\{L_t\}$ is a continuous stationary process then

$$\begin{aligned} \sigma^2 &= \lim_{T\to\infty} T \text{ Var } \bar{\ell}_T = \lim_{T\to\infty} \int_{-T}^{T} \left(1 - \frac{|s|}{T}\right) R(s)\mathrm{d}s \\ &= 2\int_0^\infty R(s)\mathrm{d}s, \end{aligned} \tag{1.2.6}$$

where

$$R(s) = \mathrm{Cov}\{(L_t,\ L_{t+s})\} = \mathbb{E}\{(L_t\ L_{t+s})\} - \mathbb{E}L_t^2$$

is the covariance (autocovariance) function. Similarly, if $\{L_t : t > 0\}$ is a discrete process, we have to replace the integrals in (1.2.6) with the corresponding sums while all other data remain the same. It follows from (1.2.6) that if

L_t and L_{t+s}, $s \neq 0$ are uncorrelated, then $\sigma^2 = \text{Var } L_t = R(0) > 0$; and if L_t and L_{t+s} are "perfectly" correlated, that is $\text{Cov}(L_t, L_{t+s}) = \text{Var } L_t = R(0)$ then $\sigma^2 = \lim_{T\to\infty} T \text{ Var } \bar{L}_T = \infty$.

The most interesting applications fall between these two extreme cases. Note also that for typical steady-state queueing processes like the sojourn time process, the process describing the number of customers in the queue, the throughput and utilization we have that

(i) The autocovariance function $R(s)$ is non-negative and decays in time, provided some regularity conditions hold (e.g. Reynolds (1975), Whitt (1991)).

(ii) The speed of decay of $R(s)$ typically decreases as the traffic intensity ρ increases.

Consider now confidence intervals for $\ell = \mathbb{E}L$. Based on the normal approximation for $\bar{\ell}_T$ a $(1-\alpha)$ 100% confidence interval for $\ell = \mathbb{E}\{\varphi(L)\}$ can be written as

$$\left\{\bar{\ell}_T \pm z_{1-\alpha/2}\frac{\sigma}{T^{1/2}}\right\}, \tag{1.2.7}$$

where as before $\Phi(z_{1-\alpha}) = 1-\alpha$ and Φ denotes the standard normal cumulative distribution function. For any given α and T the width of the confidence interval can be defined as

$$w_a(\alpha, T) = 2z_{1-\alpha/2}\sigma/T^{-1/2}. \tag{1.2.8}$$

We shall also use the following relative width of the confidence interval J:

$$w_r(\alpha, T) = \frac{w_a(\alpha, T)}{\ell}, \tag{1.2.9}$$

provided $\ell = \mathbb{E}L > 0$. We shall call w_a and w_r the *absolute* and the *relative* width of the confidence interval J, respectively.

For specified *absolute width* Δ and level of precision $1-\alpha$, the *required simulation run length* is

$$T_a(\Delta, \alpha) = \frac{4\sigma^2 z^2_{1-\alpha/2}}{\Delta^2}, \tag{1.2.10}$$

Similarly, for specified *relative* width Δ and level of precision $1-\alpha$, the *required simulation run* length is

$$T_r(\Delta, \alpha) = \frac{T_a(\Delta, \alpha)}{\ell^2} = \frac{4\sigma^2 z^2_{1-\alpha/2}}{\ell^2\Delta^2}. \tag{1.2.11}$$

Unless stated otherwise, we shall use the relative width criterion. It follows from (1.2.11) that with the relative width criterion, the required run length is proportional to σ^2/ℓ^2, which we call the (relative width) *run-length ratio* (see Whitt (1989a)). It is often helpful to decompose the run-length ratio σ^2/ℓ^2 as

$$\frac{\sigma^2}{\ell^2} = \frac{\sigma^2}{\mathrm{Var}L} \cdot \frac{\mathrm{Var}L}{\ell^2} \tag{1.2.12}$$

Following the terminology of Whitt (1989a) we shall call $\sigma^2/\mathrm{Var}L$ the *correlation factor*, because it describes the effect of the correlation over time, while we shall call $\mathrm{Var}L/\ell^2$ the *squared coefficient of variation*, because it describes the effect of the variability over time. Note that typically $\sigma^2/\mathrm{Var}L \geq 1$, with strict equality (equal to unity) when L_t and L_{t+s}, $s \neq 0$ are uncorrelated. Note also that $\mathrm{Var}L/\ell^2 \geq 0$, with strict equality (equal to zero) when $L =$ const and $\ell^2 > 0$. It is shown in Asmussen (1992) and Whitt (1989b) that in typical situations both σ^2/ℓ^2 and $\sigma^2/\mathrm{Var}L$ are greater than unity and they increase rather fast in ρ, especially when ρ approaches unity (see, for example, (1.2.13) below and also Whitt (1992)). The reason is that typical output processes are positively correlated and the correlation increases fast in ρ. To demonstrate this phenomenon quantitatively, consider the following two examples borrowed from Whitt (1989a).

Example 1.2.1 $M/M/1$ **queue.** Let L_t be the queue length process (including the customer in the service) in a stable $M/M/1$ queue. Assume for simplicity that the service rate is $\mu = 1$, that is the traffic intensity is $\rho = \lambda/\mu = \lambda$. In this case (see Whitt (1989a))

$$\mathbb{E}L = \frac{\rho}{1-\rho}, \quad \mathrm{Var}L = \frac{\rho}{(1-\rho)^2} \quad \text{and} \quad \sigma^2 = \frac{2\rho(1+\rho)}{(1-\rho)^4}. \tag{1.2.13}$$

For the run-length ratio, the correlation factor and the squared coefficient of variation, we obtain

$$\frac{\sigma^2}{\ell^2} = \frac{2(1+\rho)}{\rho(1-\rho)^2}, \quad \frac{\sigma^2}{\mathrm{Var}L} = \frac{2(1+\rho)}{(1-\rho)^2} \quad \text{and} \quad \frac{\mathrm{Var}L}{\ell^2} = \frac{1}{\rho}$$

respectively.

It follows, from the above, that as $\rho \to 1$ both σ^2/ℓ^2 and $\sigma^2/\mathrm{Var}L$ tend to infinity at a rate of $O(1-\rho)^{-2}$.

Consider now the heavy traffic situation, that is as $\rho \to 1$. In this case, L_t can be approximated by the *regulated Brownian motion* or reflected Brownian motion (RBM), which represents a Brownian motion on the positive real line with constant negative drift a, and constant positive diffusion coefficient b (see Abate and Whitt (1987a)). The stationary distribution of the RBM, say $R(t,a,b)$, is exponential with mean $b/|2a|$. In the particular case of

$a = -1$ and $b = 1$, we have the so-called *canonical* RBM which plays an important role since, under appropriate normalization, it represents a common limit process for rather general queueing models in heavy traffic. To be more specific, consider a family of stationary queueing processes $\{X^\rho(t) : t \geq 0\}$ indexed by the traffic intensity ρ, that is, the ρ-th system has traffic intensity ρ. Let

$$\hat{X}^\rho(t) = (1-\rho)\, X^\rho\,(t(1-\rho)^{-2}), \quad t \geq 0$$

be the normalized process. Note that the process $\hat{X}^\rho(t)$ uses both *space scaling* $(1-\rho)$ and *time scaling* $1/(1-\rho)^2$. Such scaling is consistent with the majority of existing heavy-traffic limit theorems (e.g. Whitt, 1989b) and in particular, it is consistent with the assumption that the process $\hat{X}^\rho(t)$ converges in distribution to the stationary RBM, $R(t,a,b)$, while the so-called rescaled process $\{|a|/b\; \hat{X}^\rho(bt|a^2), \quad t \geq 0\}$ converges in distribution to the canonical RBM, $R(t,-1,+1)$, which has an exponential stationary distribution with mean $1/2$. In this case we have

$$\mathbb{E}L = 1/2,\ \mathrm{Var}L = 1/4,\ \sigma^2 = 1/2,\ \frac{\sigma^2}{\ell^2} = \frac{\sigma^2}{\mathrm{Var}L} = 2,\ \text{and}\ \frac{\mathrm{Var}L}{\ell^2} = 1.$$

Let $L(t)$ be the queueing-length process in the standard $GI/G/c$ queue, where c is the number of servers. It is shown in section 5.1 of Whitt (1989b) that in this case, as $\rho \to 1$, we obtain

$$\Big((1-\rho)/d)\; L(dt/(1-\rho)^2\Big) \xrightarrow{\mathcal{D}} R(t,-1,1)$$

or, equivalently,

$$L(t) \xrightarrow{\mathcal{D}} dR((1-\rho)^2 t/d)/(1-\rho).$$

Here $d = c_{ar}^2 + c_s^2$, c_{ar}^2 and c_s^2 are the squared coefficient of variation of the interarrival and service times rv's. In this case, as $\rho \to 1$, we have

$$\ell = \mathbb{E}L \approx d/2(1-\rho),\ \mathrm{Var}L \approx \ell^2,\ \ \sigma^2\ \approx d^3/2(1-\rho)^4$$

and the run-length ratio is

$$\frac{\sigma^2}{\ell^2} \approx 2d/(1-\rho)^2,$$

that is the run-length ratio is proportional to d, and inversely proportional to $(1-\rho)^2$.

1.2.1 Planning Queueing Simulations

Although in a typical simulation, neither Var$L$ nor $\ell \equiv \mathbb{E}L$ is available we would like, however, to plan our simulation ahead of time, so that we would be able to predict how long the simulation must be run in order to obtain good confidence intervals.

To provide a better insight into this important matter, we consider the $M/M/1$ queue. Substituting (1.2.13) in (1.2.10) and (1.2.11), the estimated lengths of the simulation run while using the criteria of absolute and relative widths are:

$$T_a(\Delta, \alpha) = \frac{8\rho(1+\rho)\, z_{1-\alpha/2}^2}{(1-\rho)^4 \Delta^2} \tag{1.2.14}$$

and

$$T_r(\Delta, \alpha) = \frac{8(1+\rho)\, z_{1-\alpha/2}^2}{\rho(1-\rho)^2 \Delta^2}\ , \tag{1.2.15}$$

respectively. It follows from (1.2.14) and (1.2.15) that as $\rho \to 0$ and $\rho \to 1$ both T_a and T_r approach zero and infinity, respectively.

Let, for example, $\Delta = 0.05$, $z_{1-\alpha/2} = 1.96$ (which corresponds to $1-\alpha = 0.95$), and assume that we want to estimate $\ell = \mathbb{E}L$ for $\rho = 0.8$. In this case (see (1.2.13))

$$\ell = \rho/(1-\rho) = 4, \quad \sigma^2 = \frac{2\rho(1+\rho)}{(1-\rho)^4} = 1800$$

and (see (1.2.14) and (1.2.15))

$$\begin{aligned} T_a &= 7200 \left(\frac{1.96}{0.05}\right)^2 = 1.1 \times 10^7\ , \\ T_r &= 450 \left(\frac{1.96}{0.05}\right)^2 = 6.9 \times 10^5. \end{aligned}$$

To have a width $\Delta = 0.005$ instead of $\Delta = 0.05$, we would have to multiply these T_a and T_r by 100.

Table 1.2.1 displays T_a and T_r as a function of ρ for $\Delta = 0.05$ and $z_{1-\alpha/2} = 1.96$.

Table 1.2.1 The required length of the simulation run for the queue-length process in the $M/M/1$ queue.

ρ	0.3	0.5	0.7	0.8	0.9	0.95	0.99
$(1-\rho)^{-4}$	4.1	16	123	625	10^4	$1.6 \cdot 10^5$	10^8
$(1-\rho)^{-2}$	2.0	4	11.1	25	10^2	4.10^2	10^4
T_a	$2.0 \cdot 10^4$	$1.5 \cdot 10^5$	$1.8 \cdot 10^6$	$1.1 \cdot 10^7$	$2.1 \cdot 10^8$	$3.6 \cdot 10^9$	$2.4 \cdot 10^{13}$
T_r	$1.1 \cdot 10^5$	$1.5 \cdot 10^5$	$3.3 \cdot 10^5$	$6.9 \cdot 10^5$	$2.6 \cdot 10^6$	$1.0 \cdot 10^7$	$2.5 \cdot 10^9$

It follows from the results of table 1.2.1 that the lengths of the simulations T_a and T_r increase very fast in ρ, and as $\rho \geq 0.95$ (heavy traffic) simulation is virtually useless since it requires an enormous amount of time (e.g. $T_a = 3.6 \times 10^9$ and $T_r = 1.0 \times 10^7$ for $\rho = 0.95$). In this case one can use analytic approximations instead, which are based on heavy traffic limit theorems and the theory of Brownian motion. Using this theory Asmussen (1992) and Whitt (1989b) show that for typical output processes $\{L_t\}$, like the waiting time and queueing length processes

$$\sigma^2 = O\left[(1-\rho)^{-4}\right] \text{ and } \frac{\sigma^2}{\ell^2} = O\left[(1-\rho)^{-2}\right], \tag{1.2.16}$$

as $\rho \to 1$, that is in heavy traffic T_a and T_r are proportional to $(1-\rho)^{-4}$ and $(1-\rho)^{-2}$, respectively. Notice that these results match formulas (1.2.14) and (1.2.15) for the $M/M/1$ queue, provided $\rho \to 1$.

The following two subsections deal with the *batch mean* and the *regenerative* methods for estimating steady-state system parameters. Some other methods, like the *replication/deletion* method, the *spectral* method, developed by Heidelberger and Welch (1981), the *standardized times series* method, developed by Schruben (1982), (1983), can be found in Law and Kelton (1991).

1.2.2 Batch Mean Method

We present now the *batch mean* method which is the method most widely used by simulation practitioners. According to the batch mean we make a single run (single simulation) of length M, say. We delete at the beginning K initial observations, corresponding to the finite-horizon simulation, divide the remaining $M-K$ steady-state observations into N batches each of length

$$T = \frac{M-K}{N}$$

and estimate the expected steady-state performance $\ell = \mathbb{E}L$ as

$$\tilde{\ell}_N = N^{-1}T^{-1}\sum_{i=1}^{N}\sum_{t=1}^{T} L_{ti} \,. \tag{1.2.17}$$

Assume for simplicity the L_t is a discrete time process and let $X_i = T^{-1}\sum_{t=1}^{T} L_{ti}$ and $X_j = T^{-1}\sum_{t=1}^{T} L_{tj}$ be the sample means corresponding to the i-th and j-th batches. Then for $i \neq j$ the rv's X_i and X_j are approximately independent, provided the batches are sufficiently large, say $T \approx T_r$, where $T_r = T_r(\rho)$ is given in table 1.2.1. Under these assumptions an approximate confidence interval for ℓ can be written as

$$\left\{\tilde{\ell}_N \pm z_{1-\alpha/2}\frac{S}{N^{1/2}}\right\}, \tag{1.2.18}$$

where

$$S^2 = \frac{1}{N-1}\sum_{i=1}^{N}(X_i - \tilde{\ell}_N)^2.$$

Following Schmeiser (1982) and Whitt (1991) it is desirable to choose the number of batches $N \approx 20 - 30$. (See also the discussion by Glynn and Whitt (1991) for consistently estimating the asymptotic variance with the batch means method as the run length increases).

In order to have an idea how to choose the length of the initial deletion K consider the following simple example. Let L_t be the queue length (not including the customer in service) in a $M/M/1$ queue and assume that we start the simulation at the origin (with an empty queue). It is shown in Abate and Whitt (1987a,b) that in order to be within 1% of the steady-state mean, the length of the initial portion K to be deleted should be of order $8/(1-\rho)^2$ expected service times. For $\rho = 0.5$, 0.8, 0.9, and 0.95, K equals 32, 200, 800, and 3200 expected service times, respectively. From this and table 1.2.1 the ratio of the batch size T_r, while using the criterion of relative width to the initial deletion K equals $T_r/K = 53$; $137,5$; $168,8$; and 185 for $\rho = 0.5$, 0.8, 0.9, and 0.95, respectively. For more details on the transient behavior of queueing models, see Abate and Whitt (1987a,b,c) and Whitt (1991), (1992).

In summary, the batch mean algorithm can be written as follows:

Algorithm 1.2.2 :

1. *Make a single simulation run of length M with the underlying model, and delete K observations corresponding to the finite-horizon simulation.*

2. *Divide the remaining $M - K$ steady-state observations into N batches each of length*

$$T = \frac{M-K}{N}.$$

3. *Calculate the point estimator and the confidence interval for ℓ according to (1.2.17) and (1.2.18), respectively.*

1.2.3 Regenerative Simulation

Roughly speaking, a stochastic process $\{L_t\}$ is called regenerative if there exist certain random times $0 = T_0 < T_1 < T_2, \ldots$ such that at each such time the future of the process becomes a probabilistic replica of the process itself. In other words *regenerative* means that the process $\{L_t\}$ can be split into independent identically distributed epochs, called *regenerative cycles* $\tau_i = T_i - T_{i-1}$, $i = 1, 2, \ldots$.

A classical example is the $GI/G/1$ queue with instances of time $0 = T_0 < T_1 < T_2, \ldots$ corresponding to a customer arriving at an empty system. At each such random time T_i, $i = 0, 1, \ldots$, the queue starts completely from scratch independently of the past, that is, it *regenerates*. We say that a random variable T taking values in $[0, \infty)$ is the *stopping time* for a stochastic process $\{L_t\}$ if for every $t \geq 0$, the occurrence of the event $\{T \leq t\}$ can be determined from the history $\{L_s : s \leq t\}$ of the process up to time t, or in other words, if events of the form $\{T \leq t\}$ depend on the process only up to time t.

Table 1.2.2 and fig. 1.2.1 present a typical realization (sample path) of the process $\{L_t\}$, presenting the number of customers in the $GI/G/1$ system during four regenerative cycles.

Table 1.2.2 A possible realization of the steady-state number of customers L_t in the $GI/G/1$ system as a function of t.

t	0-3	3-5	5-8	8-10	10-11	11-12	12-16	16- 20	20-21	21-25
L_t	1	2	1	0	1	0	1	2	1	0

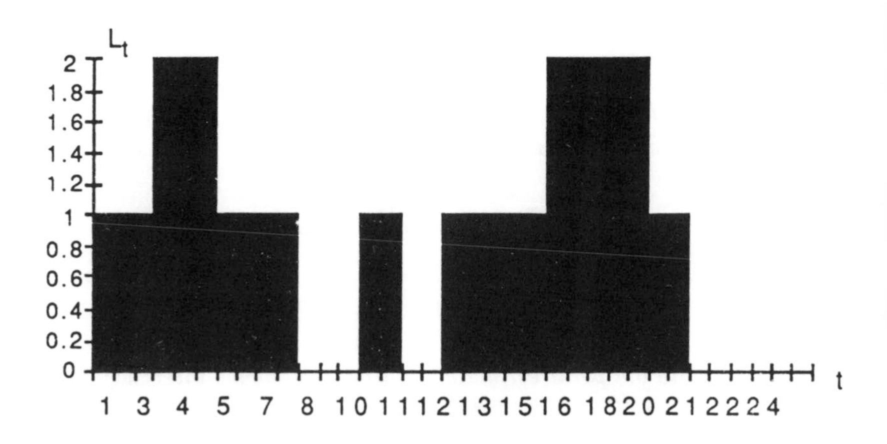

Figure 1.2.1 A possible sample path of the steady-state number of customers in the $GI/G/1$ system.

Another example of a regenerative process is an irreducible and positive recurrent (ergodic) Markov chain $\{L_t\}$ with a countable state space. Let j

be a fixed state. In this case, each time the state j is reached is a time of regeneration.

More formally (see Asmussen (1987)), a stochastic process $\{L_t\}$ is regenerative if there exist a sequence of stopping times $T_0, T_1, T_2, \ldots$ such that:

(i) The sequence of random variables $\{T_i : i \geq 0\}$ represents a renewal process, that is, the random variables $\tau_{i+1} = T_{i+1} - T_i$, $i = 0, 1, \ldots$, $(T_0 = 0)$ are iid.

(ii) For any indices $l, m \in \{0, 1, \ldots\}$ and $t_1, \ldots, t_l > 0$, the random vectors $\{L_{t_1}, \ldots, L_{t_l}\}$ and $\{L_{t_1+T_m}, \ldots, L_{t_l+T_m}\}$ are identically distributed and the process $\{L_t : t < T_m\}$ is independent of the process $\{L_{T_m+t} : t \geq 0\}$.

We shall further need the following properties of the regenerative processes (e.g. Asmussen (1987)).

(a) If $\{L_t\}$ is regenerative, then for any real-valued measurable function φ the process $\{\varphi(L_t)\}$ is regenerative as well.

(b) Under mild regularity conditions the process $\{\varphi(L_t)\}$ has a limiting steady-state distribution in the sense that there exists a random variable L such that

$$\lim_{t \to \infty} P\{\varphi(L_t) \leq x\} = P\{\varphi(L) \leq x\} = F(x),$$

that is $\varphi(L_t) \xrightarrow{\mathcal{D}} \varphi(L)$.

(c) Consider a cycle $L_1, \ldots, L_\tau$ of the process. Assume that $\mathbb{E}\tau < \infty$ and, depending on whether the process $\{L_t\}$ is of discrete or continuous type, define

$$X = \sum_{t=1}^{\tau} \varphi(L_t)$$

or

$$X = \int_0^\tau \varphi(L_t) \mathrm{d}t,$$

respectively. Then under mild conditions the steady-state expected value $\ell = \mathbb{E}\{\varphi(L)\}$ exists and equals (e.g. Asmussen (1987))

$$\ell = \mathbb{E}\{\varphi(L)\} = \frac{\mathbb{E}X}{\mathbb{E}\tau}, \tag{1.2.19}$$

where X can be viewed as the reward obtained during a cycle of length τ, and ℓ as the expected reward $\mathbb{E}X$ obtained during a cycle divided by the expected length of the cycle.

(d) (X_i, τ_i), $i = 1, \ldots$, is a sequence of iid random vectors, which depending on whether the process $\{L_t\}$ is of discrete or continuous time,

$$X_i = \sum_{t=T_{i-1}}^{T_i} \varphi(L_t) \text{ or } X_i = \int_{t=T_{i-1}}^{T_i} \varphi(L_t)\mathrm{d}t,$$

respectively.

Note that property (d) says that the behavior patterns of the system during different cycles are statistically independent and identically distributed.

We now focus our attention on the basic formula (1.2.19). Since for typical non-Markovian queueing models the parameter ℓ, which is called the system steady-state expected performance, is unknown, we must evaluate it either by using analytical approximations (see, for example, Whitt (1983)) or via simulation. We shall use simulation and in particular the methodology for regenerative simulation first introduced by Polyak (1970a,b) and later developed by Crane and Iglehart (1974a,b) and Iglehart (1975), (1976). In particular, given an iid sequence of two-dimensional random vectors $\{X_i, \tau_i\}$, $i = 1, \ldots$, we shall obtain point estimators and confidence intervals for the unknown parameter

$$\ell = \mathbb{E}\{\varphi(L)\} = \frac{\mathbb{E}X}{\mathbb{E}\tau}.$$

To obtain a point estimator of ℓ we generate N regenerative cycles, calculate the iid sequence of two-dimensional random vectors $\{X_i, \tau_i\}$, $i = 1, \ldots, N$, and finally estimate ℓ by

$$\bar{\ell}_N = \frac{\bar{X}}{\bar{\tau}}, \tag{1.2.20}$$

where $\bar{X} = N^{-1}\sum_{i=1}^{N} X_i$ and $\bar{\tau} = N^{-1}\sum_{i=1}^{N} \tau_i$. By the Law of Large Numbers (LLN) $\bar{X}$ and $\bar{\tau}$ are strongly consistent estimators for $\mathbb{E}X$ and $\mathbb{E}\tau$, and hence $\bar{\ell}$ is a strongly consistent estimator for ℓ.

The *advantages* of the regenerative method are:

(a) No initial transient is needed.

(b) No prior parameters, like the number of batches, are needed.

(c) It is asymptotically exact.

(d) It is easy to understand and implement.

The *disadvantages* of the regenerative method are:

(a) For many practical examples the output process L_t is either nonregenerative or it is difficult to identify its regenerative structure.

(b) The estimator $\bar{\ell}_N$ is baised.

(c) The regenerative cycles may be very long.

(d) Convergence in CLT may be rather slow.

We now derive confidence intervals for $\ell(\boldsymbol{v})$. To do so denote

$$\begin{aligned} \chi_1 &= \sum_{t=1}^{\tau} L_t, \\ \chi_2 &= \tau, \\ \widehat{\chi}_1 &= N^{-1}\sum_{i=1}^{N}\sum_{t=1}^{\tau_i} L_{ti}, \\ \widehat{\chi}_2 &= N^{-1}\sum_{i=1}^{N}\sum_{t=1}^{\tau_i} 1, \end{aligned}$$

where τ_i and L_{ti} are the length of the i-th cycle and, say, the waiting time of the t-th customer at the i-th cycle, respectively. Then $(\widehat{\chi}_1, \widehat{\chi}_2)$ is asymptotically distributed as a two-dimensional normal random vector with mean

$$(\mathbb{E}\chi_1, \mathbb{E}\chi_2)$$

and covariance matrix $N^{-1}\boldsymbol{\Sigma}$, where

$$\boldsymbol{\Sigma} = \begin{pmatrix} \mathrm{Var}\{\chi_1\} & \mathrm{Cov}\{\chi_1, \chi_2\} \\ \mathrm{Cov}\{\chi_2, \chi_1\} & \mathrm{Var}\{\chi_2\} \end{pmatrix}. \tag{1.2.21}$$

Considering $\varphi(x_1, x_2) = x_1/x_2$, it follows from the *delta method* (see Chapter 6) that $\bar{\ell}_N$ is asymptotically normally distributed with mean ℓ and variance σ^2/N, where

$$\sigma^2 = (\nabla\varphi)'\boldsymbol{\Sigma}(\nabla\varphi),$$

$$\nabla\varphi = \begin{pmatrix} \partial\varphi/\partial x_1 \\ \partial\varphi/\partial x_2 \end{pmatrix} = \begin{pmatrix} 1/x_2 \\ -x_1/x_2^2 \end{pmatrix},$$

and $(x_1, x_2) = (\mathbb{E}\chi_1, \mathbb{E}\chi_2)$.

Denoting now $s^2 = (\nabla\widehat{\varphi})'\boldsymbol{S}(\nabla\widehat{\varphi})$ as the sample equivalent of σ^2, where

$$\nabla\widehat{\varphi} = \begin{pmatrix} 1/\widehat{\chi}_2 \\ -\widehat{\chi}_1/\widehat{\chi}_2^2, \end{pmatrix}$$

and $\boldsymbol{S}$ is the sample covariance matrix of $\boldsymbol{\Sigma}$, that is

$$\boldsymbol{S} = \begin{pmatrix} S_{11} & S_{12} \\ S_{21} & S_{22} \end{pmatrix}, \tag{1.2.22}$$

$$S_{kj} = \frac{1}{N-1}\sum_{i=1}^{N}(\chi_{ki} - \widehat{\chi}_k)(\chi_{ji} - \widehat{\chi}_j), \quad k, j = 1, 2,$$

$$\begin{aligned} \chi_{1i} &= \sum_{t=1}^{\tau_i} L_{ti}, \\ \chi_{2i} &= \tau_i, \end{aligned}$$

we obtain by straightforward calculation that

$$s^2 = \frac{\widehat{\chi}_1^2}{\widehat{\chi}_2^4} S_{22} - 2\frac{\widehat{\chi}_1}{\widehat{\chi}_2^3} S_{12} + \frac{1}{\widehat{\chi}_2^2} S_{11}. \tag{1.2.23}$$

The asymptotic $(1-\alpha)$ 100% confidence interval for $\ell(\boldsymbol{v})$ can be written finally as

$$J = \{\bar{\ell}_N \pm z_{1-\alpha/2} s/\sqrt{N}\}. \tag{1.2.24}$$

The algorithm for deriving the $(1-\alpha)$ 100% confidence interval for $\ell(\boldsymbol{v})$ can be written as follows:

Algorithm 1.2.3 :

1. *Simulate N regenerative cycles of the process $\{L_t\}$.*
2. *Compute the sequence $\{X_i, \tau_i\}, i = 1, \ldots, N$.*
3. *Calculate the point estimator $\bar{\ell}_N$ and the confidence interval according to (1.2.20) and (1.2.24), respectively.*

Example 1.2.2 $GI/G/1$ **queue.** Consider fig. 1.2.1 which displays a possible sample path of the steady-state number of customers L_t in the $GI/G/1$ system during the period $0 \le t \le 25$

It follows from table 1.2.2 that we have three complete cycles with the following pairs $(X_1, \tau_1) = (10, 10)$; $(X_2, \tau_2) = (1, 2)$; $(X_3, \tau_3) = (13, 13)$.

1.3 Variance Reduction Techniques

In this section we consider importance sampling and control random variables to improve the accuracy of the performance estimators while simulating complex DES. Some other variance reduction techniques such as common and antithetic random variables, stratified sampling, etc. can be found, for example, in Kleijnen (1974), (1986), (1991); Nelson (1987); Rubinstein (1981); Wilson (1984); Wilson and Pritsker (1984a,b).

1.3.1 Importance and Modified Importance Sampling

Let

$$\ell = \mathbb{E}_f\{L(\boldsymbol{Y})\} = \int L(\boldsymbol{y})f(\boldsymbol{y})\mathrm{d}\boldsymbol{y} \tag{1.3.1}$$

be the expected performance measure of a computer simulation model, where $\boldsymbol{Y}$ is the input random variable (vector) having a pdf (probability density function) $f(\boldsymbol{y})$, $L(\boldsymbol{Y})$ is the sample performance measure (output random variable) and the subscript f in $\mathbb{E}_f\{L(\boldsymbol{Y})\}$ means that the expectation is taken with respect to f. The *Crude Monte Carlo* (CMC) estimator of ℓ is

$$\hat{\ell}_f = \frac{1}{N}\sum_{i=1}^{N} L(\boldsymbol{Y}_i), \tag{1.3.2}$$

where $\boldsymbol{Y}_1, \ldots, \boldsymbol{Y}_N$ is a random sample from the pdf $f(\boldsymbol{y})$.

Let $G(\boldsymbol{y})$ be a probability measure (distribution) such that $\mathrm{d}G(\boldsymbol{y}) = g(\boldsymbol{y})\mathrm{d}\boldsymbol{y}$, where $g(\boldsymbol{y})$ is a pdf. Assume that $g(\boldsymbol{y})$ dominates $f(\boldsymbol{y})$ in the absolutely continuous sense, that is

$$\mathrm{supp}\{f(\boldsymbol{y})\} \subset \mathrm{supp}\{g(\boldsymbol{y})\}.$$

In this case we can represent ℓ as

$$\ell = \mathbb{E}_g\left\{L(\boldsymbol{Z})\frac{f(\boldsymbol{Z})}{g(\boldsymbol{Z})}\right\} = \int L(\boldsymbol{z})\frac{f(\boldsymbol{z})}{g(\boldsymbol{z})}g(\boldsymbol{z})\mathrm{d}\boldsymbol{z}, \tag{1.3.3}$$

and an alternative to the CMC estimator $\hat{\ell}_f$ can be written as

$$\bar{\ell}_g = \frac{1}{N}\sum_{i=1}^{N} L(\boldsymbol{Z}_i)W(\boldsymbol{Z}_i)\ , \tag{1.3.4}$$

where $W(\boldsymbol{z}) = f(\boldsymbol{z})/g(\boldsymbol{z})$ and $\boldsymbol{Z}_1, \ldots, \boldsymbol{Z}_N$ is a random sample from $g(\boldsymbol{z})$.

It is well known (e.g. Rubinstein (1981)) that the minimum variance of $\bar{\ell}_g$ is obtained if $g(\boldsymbol{z})$ is chosen as

$$g^*(\boldsymbol{z}) = \frac{\mid L(\boldsymbol{z}) \mid f(\boldsymbol{z})}{\int \mid L(\boldsymbol{z}) \mid f(\boldsymbol{z})\mathrm{d}\boldsymbol{z}}\ . \tag{1.3.5}$$

Note that if $L(\boldsymbol{z}) \geq 0$, then

$$g^*(\boldsymbol{z}) = \frac{L(\boldsymbol{z})f(\boldsymbol{z})}{\ell} \tag{1.3.6}$$

and $\mathrm{Var}\{\bar{\ell}_{g^*}\} = 0$. The density $g^*(\boldsymbol{z})$ as per (1.3.5) and (1.3.6) is called the *importance sampling density.*

Discussions and applications of importance sampling can be found in Asmussen (1985), (1987), (1990); Cottrel, Fort and Malgouyres (1983); Ermakov (1976); Glynn (1992); Glynn and Iglehart (1989); Hammersley and Handscomb (1964); Mikhailov (1974); Siegmund (1976); Walrand (1987) and Wilson (1984).

In general, however, implementation of the importance sampling pdf $g^*(\boldsymbol{z})$ as per (1.3.5) and (1.3.6) is problematic. The main difficulty lies in the fact that knowledge of $g(\boldsymbol{z})$ implies knowledge (availability) of ℓ. But ℓ is precisely the quantity we want to estimate from the simulation! In most simulation studies the situation is even worse since the analytical expression for the sample performance L is unknown in advance. To overcome this difficulty one can make a pilot run with the underlying model, obtain a sample $L(\boldsymbol{Y}_1), \ldots, L(\boldsymbol{Y}_N)$, and then use it to estimate (approximate) the importance sampling pdf $g(\boldsymbol{z})$. It is important to note that sampling from such an artificially constructed pdf $g(\boldsymbol{z})$ might be a very complicated and time-consuming task, especially when $g(\boldsymbol{z})$ is a high-dimensional pdf.

We shall present now an alternative to the importance sampling estimator $\bar{\ell}_{g^*}$ which is called *modified importance sampling* (MIS), and basically represents a generalized version of the *weighted uniform sampling* estimator of Hammersley and Handscomb (1964) and of Powell and Swann (1966). We show that MIS estimators allow us to obtain substantial variance reduction while sampling from the underlying pdf $f(\boldsymbol{y})$. Important applications of the MIS estimators to the so-called "what if" problem will be given in the following chapters.

To introduce MIS estimators we argue as follows. We first rewrite (1.3.1) as

$$\ell = \frac{C \int L(\boldsymbol{y}) f(\boldsymbol{y}) \mathrm{d}\boldsymbol{y}}{\int Q(\boldsymbol{y}) f(\boldsymbol{y}) \mathrm{d}\boldsymbol{y}} = \frac{C \mathbb{E}_f\{L(\boldsymbol{Y})\}}{\mathbb{E}_f\{Q(\boldsymbol{Y})\}}, \tag{1.3.7}$$

where $Q(\boldsymbol{y})$ is an arbitrary integrable function, $C = \int Q(\boldsymbol{y}) f(\boldsymbol{y}) \mathrm{d}\boldsymbol{y}$, and it is assumed that $C \neq 0$. We define next the following estimator (alternative to $\bar{\ell}_g$):

$$\tilde{\ell}_f = \frac{C \sum_{i=1}^{N} L(\boldsymbol{Y}_i)}{\sum_{i=1}^{N} Q(\boldsymbol{Y}_i)} = \frac{C \hat{\ell}_f}{\bar{Q}}, \tag{1.3.8}$$

where $\bar{Q} = 1/N \sum i = 1^N Q(\boldsymbol{Y}_i)$, and $\boldsymbol{Y}_i, \;\; i = 1, \ldots, N$, is a random sample from $f(\boldsymbol{y})$. It is readily seen that $\tilde{\ell}_f$ is a *biased* but *consistent* estimator of ℓ. The consistency follows from $1/N \sum_{i=1}^N L(\boldsymbol{Y}_i) \to \ell$ w.p.1 and $1/N \sum_{i=1}^N Q(\boldsymbol{Y}_i) \to C$ w.p.1 as $N \to \infty$.

Note that $\tilde{\ell}_f$ preserves the main property of the importance sampling (IS) estimator $\bar{\ell}_{g^*}$ in the sense that if $Q(\boldsymbol{y})$ is chosen such that the random variable $L(\boldsymbol{Y}_i)/Q(\boldsymbol{Y}_i)$ varies little, i.e. close to a constant, then the variance of $\tilde{\ell}_f$ will be small, and if $Q(\boldsymbol{y})$ can be chosen proportional to $L(\boldsymbol{y})$, i.e. $Q(\boldsymbol{y}) = C_1 L(\boldsymbol{y})$, then $\mathbb{E}\tilde{\ell}_f = \ell$ and $\mathbb{E}\{(\tilde{\ell}_f - \ell)^2\} = \text{Var}\{\tilde{\ell}_f\} = 0$.

Although, following Glynn and Whitt (1989), $\tilde{\ell}_f$ can be called the *ratio type nonlinear control random variable* for $\bar{\ell}_f$, we call it the modified importance sampling estimator. The reason for that is that the main goal of $Q(\boldsymbol{y})$ is similar to the importance sampling density $g^*(\boldsymbol{z})$, namely to mimic the sample performance $L(\boldsymbol{Y})$. Note that the CMC estimator $\bar{\ell}_f$ is a particular case of the MIS estimator $\tilde{\ell}_f$ with $Q(\boldsymbol{y}) = 1$.

At first glance, one might think that the bias of $\tilde{\ell}_f$ would be a serious disadvantage compared with the unbiased estimator $\bar{\ell}_g$, its counterpart. Fortunately, this is generally not the case, since under some mild regularity conditions (see, for example, Cochran (1977, Section 6), Glynn and Whitt (1989, Theorem 11) and Powell and Swann (1966, Theorem 1)), the confidence intervals (asymptotic efficiency) of the ratio estimators of type (1.3.7) are typically of order $N^{-1/2}$, while the bias $(\mathbb{E}\tilde{\ell}_f - \ell)$ is of order N^{-1}. This actually means that for large N the asymptotic bias is negligible compared with the asymptotic efficiency. (See also our numerical results of chapters 2–4.)

Assume now that $Q(\boldsymbol{y}) \geq 0$. Then clearly $g(\boldsymbol{z}) = C^{-1}Q(\boldsymbol{z})f(\boldsymbol{z})$ is a pdf and we can use $\bar{\ell}_g$ (see (1.3.4)) to estimate the unknown quantity ℓ. In this case the unbiased estimator $\bar{\ell}_g$ can also be written as

$$\bar{\ell}_g = \frac{1}{N} \sum_{i=1}^{N} \frac{L(\boldsymbol{Z}_i)}{C^{-1}Q(\boldsymbol{Z}_i)},$$

where $\boldsymbol{Z}_i \sim g(\boldsymbol{z})$, that is $\boldsymbol{Z}_i$, $i = 1, \ldots, N$, is a random sample from $g(\boldsymbol{z})$. Note that if $Q(\boldsymbol{y}) = L(\boldsymbol{y})$, then $g(\boldsymbol{z})$ coincides with the IS pdf $g^*(\boldsymbol{z})$.

The immediate advantages of $\tilde{\ell}_f$ over $\bar{\ell}_g$ are

(i) We no longer have the restriction of sampling from an importance sampling pdf $g(\boldsymbol{z})$ since we can sample *directly* from the *original* $f(\boldsymbol{z})$.

(ii) We no longer have the restriction that $C^{-1}Q(\boldsymbol{y})f(\boldsymbol{y})$ must be a pdf (it can assume negative values as well). Clearly, because of this we have more freedom to choose $Q(\boldsymbol{y})$. As a result, the estimator $\tilde{\ell}_f$ can be applied to a broader range of practical problems than the IS estimator $\bar{\ell}_g$.

Powell and Swann (1966) demonstrated numerically the high efficiency of both estimators $\bar{\ell}_g$ and $\tilde{\ell}_f$ compared with $\bar{\ell}_f$, assuming that each component of the vector $\boldsymbol{Y}$ is distributed uniformly on the interval [0,1] and thus $g(\boldsymbol{z}) = C^{-1}Q(\boldsymbol{z})$.

Although the estimator $\tilde{\ell}_f$ has certain advantages compared with $\bar{\ell}_g$ (see (i) and (ii) above), its asymptotic efficiency still depends on how well $Q(\boldsymbol{y})$ approximates (mimics) $L(\boldsymbol{y})$. A good approximation of $L(\boldsymbol{y})$ typically requires a large *pilot run* from the underlying model and might be very time-consuming.

We shall introduce now a class of sample functions $L(\boldsymbol{Y})$ where a substantial variance reduction can be achieved without resorting to a pilot run. The class of functions we further deal with assumes that $L(\boldsymbol{y})$ can be represented as $L(\boldsymbol{y}) = R(\boldsymbol{y})Q(\boldsymbol{y})$, where $Q(\boldsymbol{y})$ is analytically available and the constant $C = \int Q(\boldsymbol{y})f(\boldsymbol{y})\mathrm{d}\boldsymbol{y}$ is readily computable.

Clearly, under these assumptions the estimator $\tilde{\ell}_f$ can be written (see (1.3.8)) as

$$\tilde{\ell}_f = \frac{\sum_{i=1}^{N} R(\boldsymbol{Y}_i)Q(\boldsymbol{Y}_i)}{C^{-1}\sum_{i=1}^{N} Q(\boldsymbol{Y}_i)} = \frac{\hat{\ell}_f}{C^{-1}\bar{Q}}, \tag{1.3.9}$$

where $\boldsymbol{Y}_i \sim f(\boldsymbol{y})$. Note again that if $Q(\boldsymbol{y}) \geq 0$, then $g(\boldsymbol{z}) = C^{-1}Q(\boldsymbol{z})f(\boldsymbol{z})$ can be considered to be a pdf. In this case we have $\bar{\ell}_g = 1/N\sum_{i=1}^{N} CR(\boldsymbol{Z}_i)$, where $\boldsymbol{Z}_i \sim g(\boldsymbol{z})$.

Assume without loss of generality that $C = 1$ and define the following random variable $X = RQ - \ell Q$. We have $\mathbb{E}(X) = 0$ and

$$\mathrm{Var}X = \mathrm{Var}\{RQ\} - 2\ell\,\mathrm{Cov}\{RQ, Q\} + \ell^2\mathrm{Var}Q.$$

Clearly, the MIS estimator $\tilde{\ell}_f$ is asymptotically more accurate than the CMC $\bar{\ell}_f$ if

$$2\ell\,\mathrm{Cov}\{RQ, Q\} - \ell^2\mathrm{Var}Q > 0. \tag{1.3.10}$$

Although more applications of MIS estimators of the type (1.3.9) will be given in Chapters 2 below, we present here the following simple example.

Example 1.3.1 Let $L(y) = y^{-3/2}$ and $f(y) = 3y^2,\ \ 0 \leq y \leq 1$. We then have $\ell = \int_0^1 3y^{1/2}\mathrm{d}y = 2$.

Assume $N = 1$. Clearly, in this case the CMC estimator $\bar{\ell}_f = Y^{-3/2}$ and we have $\mathrm{Var}\{\bar{\ell}_f\} = \infty$. Consider the following two cases: (i) $Q_1(y) = y^{-2}$ and (ii) $Q_2(y) = 3y^{-3/2}$. We have

(i) $R_1(y) = y^{1/2},\ \ g_1(y) = 1,\ \ \bar{\ell}_{g1} = 3Y^{1/2}$, and $\mathrm{Var}\{\bar{\ell}_{g1}\} = 0.13775$. It is not difficult to see that in this case $\mathbb{E}(\tilde{\ell}_f - \ell)^2 < \infty$ for $N \geq 1$.

(ii) $R_2(y) = 1/3,\ \ g_2(y) = 3/2\ \ y^{1/2},\ \ \bar{\ell}_{g2} = \tilde{\ell}_f = 2$ and $\mathrm{Var}\{\tilde{\ell}_f\} = 0$.

1.3.2 Control Random Variables

Control random variables is one of the most widely used variance reduction techniques. Its applications can be found in Asmussen and Rubinstein (1992a); Cheng and Feast (1980); Glynn and Whitt (1989); Kleijnen (1974); Lavenberg, Moeller and Welch (1982); Lavenberg and Welch (1981); Rubinstein and Marcus (1985); and Wilson (1984).

First consider the one-dimensional case. Suppose that X is an unbiased estimator of μ, which has to be estimated from a simulation run. A random variable C is called a *control variate* for X if it is correlated with X and its expectation r is known. The control variate C is used to construct an unbiased estimator of μ with a variance smaller than the variance of X. The new estimator $X(\alpha)$ is defined as

$$X(\alpha) = X - \alpha(C - r), \tag{1.3.11}$$

and is called the *linear control random variable.* Here α is a scalar parameter. It follows immediately from (1.3.11) that the variance of $X(\alpha)$ is given by

$$\text{Var}\{X(\alpha)\} = \text{Var}\{X\} - 2\alpha \,\text{Cov}\{X, C\} + \alpha^2 \text{Var}\{C\}.$$

Consequently, the value α^* that minimizes $\text{Var}\{X(\alpha)\}$ is

$$\alpha^* = \frac{\text{Cov}\{X, C\}}{\text{Var}\{C\}} \,. \tag{1.3.12}$$

The corresponding minimal variance equals

$$\text{Var}\{X(\alpha^*)\} = (1 - \rho_{XC}^2)\text{Var}\{X\}, \tag{1.3.13}$$

where ρ_{XC} denotes the correlation coefficient between X and C. Notice that the more C is correlated with X, the greater is the reduction in variance.

Formulas (1.3.11)–(1.3.13) can be easily extended to the case of more than one control variable. Indeed, let $\boldsymbol{C} = (C_1, \ldots, C_m)'$ be a vector of m control variables with known mean vector $\mathbf{r} = \mathbb{E}(\boldsymbol{C})$, $\mathbf{r} = (r_1, \ldots, r_m)'$ and $r_i = E(C_i)$. Then the extended version of (1.3.11) can be written as

$$X(\alpha) = X - \boldsymbol{\alpha}'(\boldsymbol{C} - \mathbf{r}), \tag{1.3.14}$$

where α is an m-dimensional vector of parameters. It is not difficult to see that the value $\boldsymbol{\alpha}^*$ that minimizes $\text{Var}\{X(\boldsymbol{\alpha})\}$ is given by

$$\boldsymbol{\alpha}^* = \boldsymbol{\Sigma}_{CC}{}^{-1}\boldsymbol{\sigma}_{XC}, \tag{1.3.15}$$

where $\boldsymbol{\Sigma}_{CC}$ denotes the $m \times m$ covariance matrix of $\boldsymbol{C}$ and $\boldsymbol{\sigma}_{XC}$ denotes the $m \times 1$ vector whose components are the covariances between X and C_i, $i = 1, \ldots, m$. The corresponding minimal variance equals

$$\text{Var}\{X(\alpha^*)\} = (1 - R_{XC}^2)\text{Var}\{X\}, \tag{1.3.16}$$

where

$$R^2_{XC} = \boldsymbol{\sigma}_{XC}' \boldsymbol{\Sigma}_{CC}^{-1} \boldsymbol{\sigma}_{XC} / \mathrm{Var}\{X\}$$

is the square of the multiple correlation coefficient between X and $\mathbf{C}$. Again the larger is the absolute value of the multiple correlation coefficient between $\boldsymbol{C}$ and X, the greater is the reduction in variance.

Consider now the multidimensional version of the control random variables where we seek to estimate a p-dimensional mean vector $\boldsymbol{\mu} = \mathbb{E}(\mathbf{X})$ of the $p \times 1$ random vector $\mathbf{X}$. As before, let $\boldsymbol{C} = (C_1, \ldots, C_m)'$ be a vector of control variates with known mean vector $\mathbf{r} = \mathbb{E}(\boldsymbol{C})$. Then for a $p \times m$ parameter matrix $\mathbf{A}$ the multidimensional analogue of (1.3.11) can be defined as

$$\boldsymbol{X}(\mathbf{A}) = \boldsymbol{X} - \mathbf{A}(\boldsymbol{C} - \mathbf{r}). \tag{1.3.17}$$

Let $\boldsymbol{\Sigma}_{XX}$, $\boldsymbol{\Sigma}_{\mathbf{CC}}$ and $\boldsymbol{\Sigma}_{X(A)}$ denote the respective covariance matrices of $\boldsymbol{X}$, $\boldsymbol{C}$ and $\boldsymbol{X}(\mathbf{A})$, and let $\boldsymbol{\Sigma}_{XC}$ denote the $p \times m$ covariance matrix between $\boldsymbol{X}$ and $\mathbf{C}$. Then it is not difficult to show (see exercise 4(b)) that

$$\boldsymbol{\Sigma}_{X(A)} = \boldsymbol{\Sigma}_{XX} + \mathbf{A}\boldsymbol{\Sigma}_{CC}\mathbf{A}' - \mathbf{A}\boldsymbol{\Sigma}_{XC}' - \boldsymbol{\Sigma}_{XC}\mathbf{A}'. \tag{1.3.18}$$

Suppose that the matrix $\boldsymbol{\Sigma}_{CC}$ is nonsingular and consider the matrix

$$\mathbf{A}^* = \boldsymbol{\Sigma}_{XC}\boldsymbol{\Sigma}_{CC}^{-1}. \tag{1.3.19}$$

Then the following result holds.

Proposition 1.3.1 *For any $p \times 1$ vector $\mathbf{b}$ and any $p \times m$ matrix $\mathbf{A}$,*

$$\mathbf{b}'\boldsymbol{\Sigma}_{X(A)}\mathbf{b} \geq \mathbf{b}'\boldsymbol{\Sigma}_{X(A^*)}\mathbf{b}. \tag{1.3.20}$$

Proof Substituting $\mathbf{A}^*$ from (1.3.19) into the right-hand side of (1.3.18) we obtain

$$\boldsymbol{\Sigma}_{X(A^*)} = \boldsymbol{\Sigma}_{XX} - \boldsymbol{\Sigma}_{XC}\boldsymbol{\Sigma}_{CC}^{-1}\boldsymbol{\Sigma}_{XC}'. \tag{1.3.21}$$

Consequently (1.3.20) is equivalent to

$$\mathbf{b}'\boldsymbol{\Sigma}_{CC}\mathbf{A}'\mathbf{b} + \mathbf{b}'\boldsymbol{\Sigma}_{XC}\boldsymbol{\Sigma}_{CC}^{-1}\boldsymbol{\Sigma}_{XC}'\mathbf{b} \geq 2\mathbf{b}'\mathbf{A}\boldsymbol{\Sigma}_{XC}'\mathbf{b}. \tag{1.3.22}$$

Consider next the vectors

$$\mathbf{f} = \boldsymbol{\Sigma}_{CC}^{1/2}\mathbf{A}'\mathbf{b} \quad \text{and} \quad \mathbf{g} = \boldsymbol{\Sigma}_{CC}^{-1/2}\boldsymbol{\Sigma}_{XC}'\mathbf{b}.$$

Taking into account that

$$f_i^2 + g_i^2 \geq 2f_i g_i, \quad i = 1, \ldots, m,$$

we obtain

$$\mathbf{f}'\mathbf{f} + \mathbf{g}'\mathbf{g} \geq 2\mathbf{f}'\mathbf{g}. \tag{1.3.23}$$

The inequality (1.3.20) follows now from the equivalence of (1.3.23) to (1.3.22) and of (1.3.22) to (1.3.20). □

Definition 1.1 We say that a symmetric matrix $\mathbf{S}$ is greater than or equal to a symmetric matrix $\mathbf{T}$ in the Loewner sense of inequality (denoted by $\mathbf{S} \geq \mathbf{T}$ (see, for example, Marshall and Olkin (1979)), if the matrix $\mathbf{S} - \mathbf{T}$ is positive semidefinite (non-negative definite).

We need the following two well-known inequalities associated with definition 1.1.

(i) If $\mathbf{S} \geq \mathbf{T}$, then tr $\mathbf{S} \geq$ tr $\mathbf{T}$.

(ii) If $\mathbf{S} \geq \mathbf{T}$ and $\mathbf{T} \geq \mathbf{0}$, then $\det(\mathbf{S}) \geq \det(\mathbf{T})$.

Returning to proposition 1.3.1 we have now that $\mathbf{A}^*$ minimizes the covariance matrix of $\boldsymbol{X}(\mathbf{A})$ in the Loewner sense of inequality. By properties (i) and (ii) it follows then that $\mathbf{A}^*$ also minimizes the trace and the determinant of $\boldsymbol{\Sigma}_{X(A)}$ as well. The determinant of $\boldsymbol{\Sigma}_{X(A)}$ is called *the generalized variance* of $\boldsymbol{X}(\mathbf{A})$. Therefore we have shownthat $\mathbf{A}^*$ minimizes the generalized variance of $\boldsymbol{X}(\mathbf{A})$ (cf. Rubinstein and Marcus (1985)).

The optimal matrix $\mathbf{A}^*$ is typically unknown and must be estimated from simulation. A consistent estimator of $\mathbf{A}^*$ is given by its sample equivalent

$$\hat{\mathbf{A}}^* = \mathbf{S}_{XC}\,\mathbf{S}_{CC}^{-1}, \tag{1.3.24}$$

where $\mathbf{S}_{XC}$ and $\mathbf{S}_{CC}$ are the respective sample covariance matrices. That is, if $\boldsymbol{X}_1, \ldots, \boldsymbol{X}_N$ and $\boldsymbol{C}_1, \ldots, \boldsymbol{C}_N$ are random samples of $\mathbf{X}$ and $\mathbf{C}$, respectively, then

$$\mathbf{S}_{XC} = (N-1)^{-1}\sum_{j=1}^{N}(\boldsymbol{X}_j - \bar{\boldsymbol{X}})(\mathbf{C}_j - \bar{\boldsymbol{C}})',$$

$$\mathbf{S}_{CC} = (N-1)^{-1}\sum_{j=1}^{N}(\boldsymbol{C}_j - \bar{\boldsymbol{C}})(\mathbf{C}_j - \bar{\boldsymbol{C}})',$$

where

$$\bar{\boldsymbol{X}} = N^{-1}\sum_{j=1}^{N}\boldsymbol{X}_j \quad \text{and} \quad \bar{\mathbf{C}} = N^{-1}\sum_{j=1}^{N}\boldsymbol{C}_j.$$

With $\hat{\mathbf{A}}^*$ in hand, the linear control random variables $\boldsymbol{X}_j(\hat{\mathbf{A}}^*)$ can be written as

$$\boldsymbol{X}_j(\hat{\mathbf{A}}^*) = \boldsymbol{X}_j - \hat{\mathbf{A}}^*(\mathbf{C}_j - \mathbf{r}) \tag{1.3.25}$$

and thus $\boldsymbol{\mu} = \mathbb{E}(\boldsymbol{X})$ can be estimated by

$$\bar{\boldsymbol{X}}(\hat{\mathbf{A}}^*) = N^{-1} \sum_{j=1}^{N} \boldsymbol{X}_j(\hat{\mathbf{A}}^*) = \bar{\boldsymbol{X}} - \hat{\mathbf{A}}^*(\bar{\boldsymbol{C}} - \mathbf{r}). \tag{1.3.26}$$

Consider now confidence intervals and regions for μ. Let X_N be a sequence of random variables such that $N^{1/2}(X_N - \mu)$ converges in distribution to the normal distribution with mean zero and variance σ^2. We have that

$$\lim_{N\to\infty} \mathrm{P}\left\{N^{1/2}|X_N - \mu| \leq z_{1-\alpha/2}\sigma\right\} = 1 - \alpha.$$

Typically σ is unknown and must be replaced by a consistent estimator $\hat{\sigma}_N$. Then, for large N, an approximate $(1 - \alpha)$ 100% confidence interval for μ is given by $\left\{x_N \pm N^{-1/2} z_{1-\alpha/2}\hat{\sigma}_N\right\}$, where x_N is a realization of X_N. We also write

$$\mathrm{Var}\left\{N^{1/2}(X_N - \mu)\right\} \approx \sigma^2,$$

where the notation "$\approx$" emphasizes that σ^2 is the variance of the *limiting* distribution of $N^{1/2}(X_N - \mu)$.

In the multivariate case an analogue of the confidence interval is the *confidence region*. More specifically, suppose that $\mathbf{X}$ is a p-dimensional random vector having a normal distribution with mean vector $\boldsymbol{\mu}$ and covariance matrix $\boldsymbol{\Sigma}$, i.e. $\boldsymbol{X} \sim N(\boldsymbol{\mu}, \boldsymbol{\Sigma})$. Then the random variable $(\boldsymbol{X} - \boldsymbol{\mu})'\boldsymbol{\Sigma}^{-1}(\boldsymbol{X} - \boldsymbol{\mu})$ has a (central) chi-square distribution with p degrees of freedom. Let $\chi_p^2(\alpha)$ be a number such that

$$\mathrm{P}\left\{Y \leq \chi_p^2(\alpha)\right\} = 1 - \alpha,$$

where Y is a chi-squared random variable with p degrees of freedom. We have then that

$$P\{(\boldsymbol{X} - \boldsymbol{\mu})'\boldsymbol{\Sigma}^{-1}(\boldsymbol{X} - \boldsymbol{\mu}) \leq \chi_p^2(\alpha)\} = 1 - \alpha.$$

This suggests taking the ellipsoid

$$\left\{\mathbf{m} : (\boldsymbol{X} - \mathbf{m})'\boldsymbol{\Sigma}^{-1}(\boldsymbol{X} - \mathbf{m}) \leq \chi_p^2(\alpha)\right\}$$

as an $(1 - \alpha)$ 100% confidence region for the mean vector $\boldsymbol{\mu}$. Here again $\mathbf{x}$ is a realization of $\mathbf{X}$.

Let now $\boldsymbol{X}_N$ be a sequence of random vectors such that $N^{1/2}(\boldsymbol{X}_N - \boldsymbol{\mu})$ converges in distribution to $N(\mathbf{0}, \boldsymbol{\Sigma})$. The corresponding asymptotic confidence region (ellipsoid) for $\boldsymbol{\mu}$ is given by

$$\{\mathbf{m} : N(\boldsymbol{X}_N - \mathbf{m})'\boldsymbol{\Sigma}^{-1}(\boldsymbol{X}_N - \mathbf{m}) \leq \chi^2(\alpha)\}.$$

Note finally that if $\boldsymbol{\Sigma}$ is unknown it should be replaced by a consistent estimate $\hat{\boldsymbol{\Sigma}}_N$.

1.4 The framework for sensitivity analysis

Consider the model (1.1.1), that is $\ell(\boldsymbol{v}) = \mathbb{E}_{\boldsymbol{v}} L$. Suppose that we want to estimate the expected performance $\ell(\boldsymbol{v})$ and the associated sensitivities, $\nabla \ell(\boldsymbol{v})$, for different values of the parameter vector $\boldsymbol{v}$. For typical DEDS, like queueing networks, the sample performance L is calculated from the underlying stochastic process Z_t, say the waiting time process, which is driven by an input sequence of random vectors $Y_1, Y_2, \ldots$. The process Z_t, which changes the states at discrete times, might also depend on the parameter vector $\boldsymbol{v}$.

We can view the dependence of the stochastic process Z_t and hence of the sample performance L_t on $\boldsymbol{v}$ in two somewhat different ways. We can assume that for every $\boldsymbol{v}$, the process $Z_t(\boldsymbol{v})$ is defined on a measurable space $(\Xi, \mathcal{A})$ with the corresponding probability measure $P_{\boldsymbol{v}}$. In this case, *only the probability measure* depends on $\boldsymbol{v}$ and

$$\ell(\boldsymbol{v}) = \int_{\Xi} L(\xi) \mathrm{d} P_{\boldsymbol{v}}(\xi) . \tag{1.4.1}$$

On the other hand, in some situations we can construct $Z_t = Z_t(\boldsymbol{v})$ on a *single probability space* $(\Omega, \mathcal{F}, Q)$ and hence to express $\ell(\boldsymbol{v})$ in the form

$$\ell(\boldsymbol{v}) = \int_{\Omega} H(\omega, \boldsymbol{v}) \mathrm{d} Q(\omega) , \tag{1.4.2}$$

where $H(\boldsymbol{v}) = H(\omega, \boldsymbol{v})$ is a random function depending on the parameter vector $\boldsymbol{v}$.

In order to derive estimators of the sensitivities $\nabla^k \ell(\boldsymbol{v})$, $k \geq 1$, we shall need to verify the interchangeability of integration and differentiation in (1.4.2), that is to verify the validity of

$$\nabla^k \ell(\boldsymbol{v}) = \int_{\Omega} \nabla^k H(\omega, \boldsymbol{v}) \mathrm{d} Q(\omega) . \tag{1.4.3}$$

Such conditions are given, for example, in Chapter 1 of Glasserman (1991c); L'Ecuyer (1990a); and theorem 6.1.1 of chapter 6. A basic requirement of these conditions is that for almost every ω the function $H(\omega, \boldsymbol{v})$ is continuous in $\boldsymbol{v}$. Although this continuity condition is not absolutely necessary (see Proposition 4 of L'Ecuyer and Perron (1992)), it is straightforward to construct a discontinuous function $H(\omega, \boldsymbol{v})$ in $\boldsymbol{v}$ for which the interchangeability conditions are violated, that is (1.4.3) does not hold. Think, for instance, of a function $H(\omega, \boldsymbol{v})$ which for every ω is piecewise constant in some (all) components of $\boldsymbol{v}$.

There are two somewhat different ways of constructing representations of the form (1.4.2) from (1.4.1). The first way is based on a transformation of the probability space $(\Xi, \mathcal{A}, P_{\boldsymbol{v}})$ (each transformation depends on the parameter vector $\boldsymbol{v}$). For example, let $F(y, \boldsymbol{v})$ be the *common* cdf of the input sequence $Y_1, Y_2, \ldots,$ of iid random vectors. Assume that this input sequence can be generated by the inverse transformation, that is $Y_i = F^{-1}(U_i, \boldsymbol{v})$, where $U_1, U_2, \ldots$ is an iid sequence of uniformly distributed on the unit interval random variables. The required single probability space is then associated with the sequence $U_1, U_2, \ldots$ and similarly for the multidimensional case of $\boldsymbol{Y}$. In this case the derivative of $H(\omega, \boldsymbol{v})$ is expressed in terms of the derivative of the inverse transformation $F^{-1}(\omega, \boldsymbol{v})$. This approach is at the heart of *infinitesimal perturbation analysis* (IPA) and has a serious drawback (see Glasserman (1991c); Heidelberger et al. (1988); L'Ecuyer (1990a); and chapter 5). This is so, since performing a transformation depending on $\boldsymbol{v}$, we make the sample performance $L_t(\boldsymbol{v})$ dependent on the parameter vector $\boldsymbol{v}$, regardless of the fact that in the original form (without the transformation $Y = F^{-1}(U, \boldsymbol{v})$) it is not so. As a result, in many interesting situations the such obtained function $H(\omega, \boldsymbol{v})$ becomes nondifferential and even discontinuous and consequently formula (1.4.3) does not hold.

In this monograph we discuss an alternative to the IPA approach which is based on the Radon-Nikodym derivative, or change of the probability measure $P_{\boldsymbol{v}}$ (see section 2.7 of chapter 2). In this case (1.4.2) is derived by constructing the so-called *score function* (likelihood ratio) process (see chapter 3 below) with the corresponding function $H(\omega, \boldsymbol{v})$ in (1.4.2) depending on the parameter $\boldsymbol{v}$ *only* through the generated likelihood ratio process and being typically smooth (differentiable) in $\boldsymbol{v}$.

1.5 Exercises

1. Let L be the steady-state sojourn time of a customer in the $M/M/1$ queue with traffic intensity $\rho < 1$. Let $\rho = 0.6$. Find

   (a) The correlation factor and the squared coefficient of variation.

   (b) The 95% confidence interval for $\ell = \mathbb{E}L$, the absolute and the relative widths of the confidence interval, assuming that the sample size (the number of simulated customers) $T = 10^4$.

   (c) Find the estimated length of the simulation run $T_a(\Delta, \alpha)$ and $T_r(\Delta, \alpha)$ assuming that the width Δ of the confidence interval for ℓ equals 0.1 and $100(1-\alpha) = 95$.

   (d) Repeat (a), (b) and (c) for $\rho = 0.8$ and discuss (a), (b) and (c) for $\rho \to 1$.

2. Prove formula (1.2.6)

3. Table 1.5.1 presents a possible realization (sample path) of the steady-state waiting time process $\{L_t\}$ in the $GI/G/1$ queue during six regeneretive cycles with the total of $t = 15$ customers.

Table 1.5.1 A possible realization of the steady-state waiting time L_t in the $GI/G/1$ queue as a function of t.

t	1	2	3	4	5	6	7	8	9	10	11	12	13	14	15
L_t	1	2	0	1	2	1	0	1	0	1	0	1	0	1	0

Find the point estimators $\bar{\ell}_N$ and the 95% confidence intervals for $\ell = \mathbb{E}L$ while using (i) the regenerative simulation and (ii) the batch mean method, assuming that the length of each batch equals 5 customers.

4. Table 1.5.2 presents three possible sample paths (realizations) of the process $\{L_{ti}\}$, $i = 1, 2, 3$, presenting the number of customers in the $GI/G/1$ system (including the customer in the service facility) during the first 23 units of time starting at the origin (each time starting with an empty system).

Table 1.5.2 A possible realization of the steady-state waiting time L_t in the $GI/G/1$ queue as a function of t.

t	0-2	2-3	3-8	8-10	10-11	11-13	13-16	16- 18	18-21	21-23
L_{t1}	1	2	1	0	1	0	1	0	1	2
t	0-3	3-5	5-8	8-10	10-11	11-12	12-16	16- 20	20-21	21-25
L_{t2}	1	2	3	2	1	0	1	2	1	0
t	0-1	1-5	5-6	6-7	7-11	11-12	12-13	13- 18	18-20	20-21
L_{t3}	1	0	1	2	1	0	1	0	1	2

(a) Find the point estimator $\bar{\ell}_N$ and the 95% confidence interval for the expected number of customers in the $GI/G/1$ system during (i) the first 10 units of time and (ii) for $15 < t \leq 23$.

(b Find the point estimator $\bar{\ell}_N$ and the 95% confidence interval for the expected sojourn time of a customer in the $GI/G/1$ queue (i) for the first 5 customers and (ii) for customers 10–14.

(c) Repeat (b) for the expected waiting time of a customer in the $GI/G/1$ queue.

5. Prove formula (1.3.5)

6. (a) Consider the estimator $X(\boldsymbol{\alpha})$ given in (1.3.14) Show that

$$\mathrm{Var}\{X(\boldsymbol{\alpha})\} = \mathrm{Var}\{X\} - 2\boldsymbol{\alpha}'\boldsymbol{\sigma}_{XC} + \boldsymbol{\alpha}'\boldsymbol{\Sigma}_{CC}\boldsymbol{\alpha}$$

and that the value $\boldsymbol{\alpha}^*$ given in (1.3.15) minimizes the variance of $X(\boldsymbol{\alpha})$.

(b) Consider the estimator $\boldsymbol{X}(\mathbf{A})$ given in (1.3.17). Show that its covariance matrix $\boldsymbol{\Sigma}_{X(A)}$ is given by the right-hand side of (1.3.18).

(c) Let $\mathbf{S}$ and $\mathbf{T}$ be symmetric matrices. Show that $\mathbf{S} \geq \mathbf{T}$ in the Loewner sense of inequality if and only if $\mathbf{b}'\mathbf{S}\mathbf{b} \geq \mathbf{b}'\mathbf{T}\mathbf{b}$ for any vector $\mathbf{b}$.

(d) Show that if $\mathbf{S} \geq \mathbf{T}$, then $\mathrm{tr}\,\mathbf{S} \geq \mathrm{tr}\,\mathbf{T}$.

(e) Show that if $\mathbf{S} \geq \mathbf{T}$ and $\mathbf{T} \geq \mathbf{0}$, then $\det(\mathbf{S}) \geq \det(\mathbf{T})$.

7. Suppose that the random vectors $\mathbf{X}$ and $\mathbf{C}$ in (1.3.17) have the same dimensionality, i.e. $p = m$, and for a scalar parameter α define

$$\boldsymbol{X}(\alpha) = \boldsymbol{X} - \alpha(\boldsymbol{C} - \mathbf{r}) .$$

(a) Show that the value α^* of the parameter α that minimizes the trace of $\boldsymbol{\Sigma}_{X(\alpha)}$ is

$$\alpha^* = \frac{\mathrm{tr}\ \boldsymbol{\Sigma}_{XC}}{\mathrm{tr}\ \boldsymbol{\Sigma}_{CC}} .$$

(b) Show that α^* which minimizes the determinant of $\boldsymbol{\Sigma}_{X(\alpha)}$ is

$$\alpha^* = \frac{\mathrm{tr}\ \boldsymbol{\Sigma}_{XC}\ \boldsymbol{\Sigma}_{XX}^{-1}}{\mathrm{tr}\ \boldsymbol{\Sigma}_{CC}\ \boldsymbol{\Sigma}_{XX}^{-1}} .$$

Chapter 2

Sensitivity Analysis and Stochastic Optimization of Discrete Event Static Systems (DESS)

2.1 Introduction

In this chapter we deal with performance evaluation, sensitivity analysis and optimization of DESS with the expected performance given as $\ell(\boldsymbol{v}) = \mathbb{E}_{\boldsymbol{v}} L(\boldsymbol{Y}) = \int L(\boldsymbol{y}) \, \mathrm{d} F(\boldsymbol{y}, \boldsymbol{v})$, $\boldsymbol{v} \in V \subset \mathbb{R}^n$. As examples of such DESS consider

(i) **Reliability system**
The mean lifetime of a coherent reliability system can be written as

$$\ell(\boldsymbol{v}) = \mathbb{E}_{\boldsymbol{v}} \{ \max_{j=1,\ldots,p} \min_{i \in \mathcal{L}_j} Y_i \} , \qquad (2.1.1)$$

where $\mathcal{L}_j$ is the j-th complete path from a source to a sink in the system; Y_i, $i = 1, \ldots, m$, are the durations (lifetimes) of the components having a cdf $F_i(y_i, v_i)$; and p is the number of complete paths in the system.

(ii) **Stochastic PERT network**
The mean shortest path (minimal project duration) in a stochastic PERT network can be written as

$$\ell(\boldsymbol{v}) = \mathbb{E}_{\boldsymbol{v}} \{ \min_{j=1,\ldots,p} \sum_{i \in \mathcal{L}_j} Y_i \} , \qquad (2.1.2)$$

where $\mathcal{L}_j$, Y_i and p are similar to their counterparts in (2.1.1).

As far as sensitivity analysis is concerned, in the case of model (2.1.1) we might be interested in estimating the sensitivities (gradients, Hessians, etc.) of the mean lifetime of the coherent system with respect to the expected life of components and similarly for (2.1.2). As far as optimization is concerned, we might be interested in maximizing the mean lifetime and minimizing the mean shortest path in the two models, subject to some constraints. As for a motivating problem associated with optimization of the reliability model consider

Example 2.1.1

$$
(\mathrm{P}_0) \qquad \begin{array}{lll} \text{maximize} & \ell_0(\boldsymbol{v}), & \boldsymbol{v} \in V, \\ \text{subject to} & \ell_1(\boldsymbol{v}) \leq 0, & \\ & \ell_2(\boldsymbol{v}) \leq 0, & \end{array} \tag{2.1.3}
$$

where $V = \mathbb{R}^n_+ = \{\boldsymbol{v} = (v_1, \ldots, v_n) : \ v_j \geq 0, \ j = 1, \ldots, n\}$, $\ell_0(\boldsymbol{v})$ coincides with $\ell(\boldsymbol{v})$ in (2.1.1),

$$
\ell_1(\boldsymbol{v}) = \sum_{j=1}^{n} c_j v_j - b_1
$$

is the "deterministic" constraint function, c_j being the cost per unit increase in v_j, $v_j = \mathbb{E}_{\boldsymbol{v}}(Y_j)$, and $\ell_2(\boldsymbol{v}) = \mathbb{E}_{\boldsymbol{v}}\{L^2(\boldsymbol{Y}) - b_2\}$, where

$$
L(\boldsymbol{Y}) = \max_{j=1,\ldots,p} \min_{i \in \mathcal{L}_j} Y_i, \quad Y_i \sim F_i(y_i, v_i).
$$

A similar optimization problem can be formulated for a stochastic PERT network as well.

The remainder of this chapter is organized as follows. Section 2.2 deals with sensitivity analysis (estimation of the gradients $\nabla \ell(\boldsymbol{v})$, Hessians $\nabla^2 \ell(\boldsymbol{v})$, etc.). We show that the SF method allows us to estimate *all* the sensitivities *simultaneously* from a *single* simulation experiment. This becomes possible since the sensitivities $\nabla^k \ell(\boldsymbol{v})$, $k = 1, 2, \ldots$, are expressed as expectations with respect to the same probability measure (distribution) as the expected performance.

Section 2.3 deals with the "what if" problem, which can be formulated as follows: "*What* will be the expected performance of the system $\ell(\boldsymbol{v})$ and the associated sensitivities $\nabla^k \ell(\boldsymbol{v})$, $k = 1, 2, \ldots$, *if* we perturb some (all) of the parameters of the vector $\boldsymbol{v}$?" Here again we show how the SF method resolves this problem in the sense that it permits estimation of the performance $\ell(\boldsymbol{v})$ at several values (scenarios) of $\boldsymbol{v}$, say $\boldsymbol{v}_1, \ldots, \boldsymbol{v}_r$, on the basis of a *single* simulation experiment, and discuss conditions under which it performs well.

In section 2.4 we show how the SF method optimizes a DESS and in particular how it evaluates the optimal solution of the problem (P_0) from a *single* simulation run. The method is based on a probability measure transformation and construction of statistical counterparts for the conventional deterministic optimization procedures. We also discuss the statistical properties of the derived estimators and present numerical results.

In section 2.5 we show how the control random variables techniques can be applied to the SF estimators in order to improve their accuracy. Finally, in sections 2.6 and 2.7 (appendix A and appendix B) we discuss convergence of the the optimal solution of the program (P_0) and the basic concepts of the change-of-measure techniques based on the Radon-Nikodym derivative, respectively.

2.2 Sensitivity Analysis of the System Performance

This section deals with sensitivity analysis (evaluation of the gradients $\nabla \ell(\boldsymbol{v})$, Hessians $\nabla^2 \ell(\boldsymbol{v})$, etc.) of DESS by the SF method. We assume that the expected performance $\ell(\boldsymbol{v})$ can be represented in the form

$$\ell(\boldsymbol{v}) = \mathbb{E}_{\boldsymbol{v}} L(\boldsymbol{Y}) = \int L(\boldsymbol{y})\, \mathrm{d}F(\boldsymbol{y}, \boldsymbol{v}), \tag{2.2.1}$$

where $F(\boldsymbol{y}, \boldsymbol{v})$ belongs to a family of distributions that are absolutely continuous with respect to the Lebesgue measure. That is, $\mathrm{d}F(\boldsymbol{y}, \boldsymbol{v}) = f(\boldsymbol{y}, \boldsymbol{v})\, \mathrm{d}\boldsymbol{y}$, where $f(\boldsymbol{y}, \boldsymbol{v})$ is a pdf (probability density function). Treatment of the case where $F(\boldsymbol{y}, \boldsymbol{v})$ belongs to a family of discrete or mixture distributions is similar.

Let us start by considering the situation where the parameter v is a scalar and the parameter set V is an open interval of the real line. Suppose that for all $\boldsymbol{y}$ the function $f(\boldsymbol{y}, v)$ is continuously differentiable in v and that there exists an integrable (with respect to the Lebesgue measure) function $h(\boldsymbol{y})$ such that

$$|L(\boldsymbol{y}) \partial f(\boldsymbol{y}, v)/\partial v| \leq h(\boldsymbol{y}) \tag{2.2.2}$$

for all $v \in V$. Then by the Lebesgue dominated convergence theorem the operators of differentiation and expectation (integration) are interchangeable:

$$\frac{\mathrm{d}\ell(v)}{\mathrm{d}v} = \frac{\mathrm{d}}{\mathrm{d}v} \int L(\boldsymbol{y}) f(\boldsymbol{y}, v) \mathrm{d}\boldsymbol{y} = \int L(\boldsymbol{y}) \frac{\partial f(\boldsymbol{y}, v)}{\partial v} \mathrm{d}\boldsymbol{y}.$$

Furthermore, we can write

$$\int L(y)\frac{\partial f(y,v)}{\partial v}\mathrm{d}y = \int L(y)\frac{\partial f(y,v)}{\partial v}\frac{f(y,v)}{f(y,v)}\mathrm{d}y$$

$$= \int L(y)\,\frac{\partial \log f(y,v)}{\partial v}\,f(y,v)dy = \mathbb{E}_v\left\{L(Y)\frac{\partial \log f(Y,v)}{\partial v}\right\}.$$

Consequently we obtain

$$\frac{\mathrm{d}\ell(v)}{\mathrm{d}v} = \mathbb{E}_v\left\{L(Y)\frac{\partial \log f(y,v)}{\partial v}\right\}. \tag{2.2.3}$$

The extension to the multidimensional case where $\boldsymbol{v} \in \mathbb{R}^n$ is straightforward. Indeed, by similar arguments we can write the gradient of $\ell(\boldsymbol{v})$ in the form

$$\nabla\ell(\boldsymbol{v}) = \mathbb{E}_{\boldsymbol{v}}\{L(\boldsymbol{Y})\nabla \log f(\boldsymbol{Y},\boldsymbol{v})\} = \mathbb{E}_{\boldsymbol{v}}\{L(\boldsymbol{Y})\boldsymbol{S}^{(1)}(\boldsymbol{Y},\boldsymbol{v})\}, \tag{2.2.4}$$

where

$$\boldsymbol{S}^{(1)}(\boldsymbol{y},\boldsymbol{v}) = \frac{\nabla f(\boldsymbol{y},\boldsymbol{v})}{f(\boldsymbol{y},\boldsymbol{v})} = \nabla \log f(\boldsymbol{y},\boldsymbol{v}) \tag{2.2.5}$$

is called the *efficient score function.*

Let us find now the Hessian matrix $\nabla^2\ell(\boldsymbol{v})$ of the second-order partial derivatives $\partial^2\ell(\boldsymbol{v})/\partial v_i \partial v_j$, $i,j = 1,\ldots,n$. Assuming that the second-order partial derivatives can be taken inside the expected value (integral) we obtain

$$\begin{aligned}
\frac{\partial^2\ell(\boldsymbol{v})}{\partial v_i \partial v_j} &= \int L(\boldsymbol{y})\,\frac{\partial^2 f(\boldsymbol{y},\boldsymbol{v})}{\partial v_i \partial v_j}\mathrm{d}\boldsymbol{y} = \int L(\boldsymbol{y})\frac{\partial}{\partial v_i}\left[\frac{\partial \log f(\boldsymbol{y},\boldsymbol{v})}{\partial v_j} f(\boldsymbol{y},\boldsymbol{v})\right]\mathrm{d}\boldsymbol{y} \\
&= \int L(\boldsymbol{y})\left[\frac{\partial^2 \log f(\boldsymbol{y},\boldsymbol{v})}{\partial v_i \partial v_j} + \frac{\partial \log f(\boldsymbol{y},\boldsymbol{v})}{\partial v_i}\,\frac{\partial \log f(\boldsymbol{y},\boldsymbol{v})}{\partial v_j}\right] f(\boldsymbol{y},\boldsymbol{v})\mathrm{d}\boldsymbol{y} \\
&= \mathbb{E}_{\boldsymbol{v}}\left\{L(\boldsymbol{Y})\left[\frac{\partial^2 \log f(\boldsymbol{Y},\boldsymbol{v})}{\partial v_i \partial v_j} + \frac{\partial \log f(\boldsymbol{Y},\boldsymbol{v})}{\partial v_i}\,\frac{\partial \log f(\boldsymbol{Y},\boldsymbol{v})}{\partial v_j}\right]\right\}.
\end{aligned}$$

Denoting

$$\begin{aligned}
\boldsymbol{S}^{(2)}(\boldsymbol{y},\boldsymbol{v}) &= \nabla \boldsymbol{S}^{(1)}(\boldsymbol{y},\boldsymbol{v}) + \boldsymbol{S}^{(1)}(\boldsymbol{y},\boldsymbol{v})\boldsymbol{S}^{(1)}(\boldsymbol{y},\boldsymbol{v})' \\
&= \nabla^2 \log f(\boldsymbol{y},\boldsymbol{v}) + \nabla \log f(\boldsymbol{y},\boldsymbol{v})\nabla \log f(\boldsymbol{y},\boldsymbol{v})',
\end{aligned} \tag{2.2.6}$$

the above equality can be written in a more compact form as

$$\nabla^2\ell(\boldsymbol{v}) = \mathbb{E}_{\boldsymbol{v}}\{L(\boldsymbol{Y})\boldsymbol{S}^{(2)}(\boldsymbol{Y},\boldsymbol{v})\}. \tag{2.2.7}$$

Here "a prime" stands for the transpose operator applied to a vector or a matrix and all partial derivatives are taken with respect to the components of the parameter vector $\boldsymbol{v}$. Employing similar arguments it is possible to derive higher order partial derivatives of $\ell(\boldsymbol{v})$ as well (see exercise 2.2.1).

Typically the sensitivities $\nabla^k \ell(\boldsymbol{v})$, $k = 1, 2, \ldots$, are not available analytically since the performance $\ell(\boldsymbol{v})$ is not either. They can be evaluated, however, either by conventional deterministic numerical methods (e.g. Szidarovszky and Yakowitz (1978)) or via simulation. Both methods are applicable here since formulas (2.2.4)–(2.2.7) present *closed form expressions* for $\nabla^k \ell(\boldsymbol{v})$, $k = 1, 2, \ldots$. Simulation is particularly convenient since the performance $\ell(\boldsymbol{v})$ and *all* the sensitivities $\nabla^k \ell(\boldsymbol{v})$ are expressed as expectations with respect to the same underlying distribution $f(\boldsymbol{y}, \boldsymbol{v})$.

To proceed with simulation let $\boldsymbol{Y}_1, \ldots, \boldsymbol{Y}_N$ be a sample from $f(\boldsymbol{y}, \boldsymbol{v})$. Then $\ell(\boldsymbol{v})$, $\nabla \ell(\boldsymbol{v})$ and $\nabla^2 \ell(\boldsymbol{v})$ can be estimated *simultaneously* from a *single simulation* by

$$\bar{\ell}_N(\boldsymbol{v}) = N^{-1} \sum_{i=1}^{N} L(\boldsymbol{Y}_i) \,, \tag{2.2.8}$$

$$\bar{\nabla} \ell_N(\boldsymbol{v}) = N^{-1} \sum_{i=1}^{N} L(\boldsymbol{Y}_i) \boldsymbol{S}^{(1)}(\boldsymbol{Y}_i, \boldsymbol{v}) \tag{2.2.9}$$

and

$$\bar{\nabla}^2 \ell_N(\boldsymbol{v}) = N^{-1} \sum_{i=1}^{N} L(\boldsymbol{Y}_i) \boldsymbol{S}^{(2)}(\boldsymbol{Y}_i, \boldsymbol{v}) \,, \tag{2.2.10}$$

with $\boldsymbol{S}^{(1)}(\boldsymbol{y}, \boldsymbol{v})$ and $\boldsymbol{S}^{(2)}(\boldsymbol{y}, \boldsymbol{v})$ being defined in (2.2.5) and (2.2.6), respectively.

Since the sensitivity estimators $\bar{\nabla} \ell_N(\boldsymbol{v})$, $\bar{\nabla}^2 \ell_N(\boldsymbol{v})$, etc. are based on the *efficient score* $\boldsymbol{S}^{(1)}(\boldsymbol{Y}, \boldsymbol{v}) = \nabla \log f(\boldsymbol{Y}, \boldsymbol{v})$ we shall call our method the *score function* (SF) method and $\boldsymbol{S}^{(k)}(\boldsymbol{Y}, \boldsymbol{v})$ the k-th order *score function*, $k = 1, 2, \ldots$, or just the *score*. The algorithm for the simultaneous estimation of both the performance $\ell(\boldsymbol{v})$ and all the sensitivities (gradient $\nabla \ell(\boldsymbol{v})$, Hessian $\nabla^2 \ell(\boldsymbol{v})$, etc.) from a single simulation is straightforward and can be written as follows.

Algorithm 2.2.1 :

1. *Generate a sample* $\boldsymbol{Y}_1, \ldots, \boldsymbol{Y}_N$ *from* $f(\boldsymbol{y}, \boldsymbol{v})$.

2. *Calculate the sample performances* $L(\boldsymbol{Y}_i)$ *and the sample scores* $\boldsymbol{S}^{(k)}(\boldsymbol{Y}_i, \boldsymbol{v})$.

3. *Calculate* $\bar{\ell}_N(\boldsymbol{v})$, $\bar{\nabla} \ell_N(\boldsymbol{v})$ *and* $\bar{\nabla}^2 \ell_N(\boldsymbol{v})$ *according to (2.2.8)–(2.2.10), respectively.*

It is important to note that in order to estimate the sensitivities $\nabla^k \ell(\boldsymbol{v})$, $k = 1, 2, \ldots$, we *do not need* to differentiate $L(\boldsymbol{Y})$, which in most practical applications is not a smooth (differentiable) function. What we *only need* to have at our disposal is a sample $\boldsymbol{Y}_i$, $i = 1, \ldots, N$, from the underlying pdf $F(\boldsymbol{y}, \boldsymbol{v})$, the realizations of the sample performance $L(\boldsymbol{Y}_i)$ and the scores $\boldsymbol{S}^{(k)}(\boldsymbol{Y}_i, \boldsymbol{v})$.

Remark 2.2.1 If we take $L(\boldsymbol{y}) = 1$, then $\ell(\boldsymbol{v}) = 1$ for all $\boldsymbol{v}$ and hence all partial derivatives of $\ell(\boldsymbol{v})$ are zero. It follows that $\mathbb{E}_{\boldsymbol{v}}\{\boldsymbol{S}^{(k)}(Y, \boldsymbol{v})\} = \boldsymbol{0}$, $k = 1, 2, \ldots$. Consequently we can write

$$\nabla^k \ell(\boldsymbol{v}) = \mathbb{E}_{\boldsymbol{v}}\{L\boldsymbol{S}^{(k)}\} - \mathbb{E}_{\boldsymbol{v}}\{L\}\mathbb{E}_{\boldsymbol{v}}\{\boldsymbol{S}^{(k)}\} = \text{Cov}_{\boldsymbol{v}}\ \{L, \boldsymbol{S}^{(k)}\}. \quad (2.2.11)$$

Thus, the sensitivities of $\ell(\boldsymbol{v})$ can be represented as the *covariances* between the sample performance L and the associated scores $\boldsymbol{S}^{(k)}$.

Example 2.2.1 Let Y_k, $k = 1, \ldots, m$, be independent random variables each distributed gamma, i.e.

$$f_k(y_k, \lambda_k, \beta_k) = \frac{\lambda_k \mathrm{e}^{-\lambda_k y_k}(\lambda_k y_k)^{\beta_k - 1}}{\Gamma(\beta_k)}, \quad y_k > 0,\ \lambda_k > 0,\ \beta_k > 0, \quad (2.2.12)$$

and

$$f(\boldsymbol{y}, \boldsymbol{\lambda}, \boldsymbol{\beta}) = \prod_{k=1}^{m} f_k(y_k, \lambda_k, \beta_k).$$

Assume that we are interested in the sensitivities with respect to $\boldsymbol{\lambda} = (\lambda_1, \ldots, \lambda_m)'$ only. Then

$$\boldsymbol{S}^{(1)}(\boldsymbol{y}, \boldsymbol{\lambda}) = \boldsymbol{\beta}\boldsymbol{\lambda}^{-1} - \boldsymbol{y}, \quad (2.2.13)$$

$$\boldsymbol{S}^{(2)}(\boldsymbol{y}, \boldsymbol{\lambda}) = (\boldsymbol{\beta}\boldsymbol{\lambda}^{-1} - \boldsymbol{y})(\boldsymbol{\beta}\boldsymbol{\lambda}^{-1} - \boldsymbol{y})' - \text{diag}(\boldsymbol{\beta}\boldsymbol{\lambda}^{-2}), \quad (2.2.14)$$

where $\boldsymbol{\beta}\boldsymbol{\lambda}^{-1}$ is the m-dimensional vector with components $\beta_1\lambda_1^{-1}, \ldots, \beta_m\lambda_m^{-1}$, and $\text{diag}(\boldsymbol{\beta}\boldsymbol{\lambda}^{-2})$ is the $m \times m$ diagonal matrix with diagonal elements $\beta_1\lambda_1^{-2}, \ldots, \beta_m\lambda_m^{-2}$.

Example 2.2.2 Let $\boldsymbol{Y} \sim N(\boldsymbol{\mu}, \boldsymbol{\Sigma})$, i.e.

$$f(\boldsymbol{y}, \boldsymbol{\mu}, \boldsymbol{\Sigma}) = \frac{1}{(2\pi)^{m/2}\det(\boldsymbol{\Sigma})^{1/2}} \exp\left\{-\frac{1}{2}\left[(\boldsymbol{y} - \boldsymbol{\mu})'\boldsymbol{\Sigma}^{-1}(\boldsymbol{y} - \boldsymbol{\mu})\right]\right\},$$

where $\boldsymbol{\Sigma}$ is a positive definite matrix. Assume that we are interested in the sensitivities with respect to $\boldsymbol{\mu}$ only. We have

$$\boldsymbol{S}^{(1)}(\boldsymbol{y}, \boldsymbol{\mu}) = \boldsymbol{\Sigma}^{-1}(\boldsymbol{y} - \boldsymbol{\mu}) \quad (2.2.15)$$

and

$$\boldsymbol{S}^{(2)}(\boldsymbol{y}, \boldsymbol{\mu}) = \boldsymbol{\Sigma}^{-1}(\boldsymbol{y} - \boldsymbol{\mu})(\boldsymbol{y} - \boldsymbol{\mu})'\boldsymbol{\Sigma}^{-1} - \boldsymbol{\Sigma}^{-1}. \quad (2.2.16)$$

Example 2.2.3 Let Y_k, $k = 1, \ldots, m$, be independent random variables each distributed Bernoulli with parameter p_k, i.e.

$$P_k(y_k, p_k) = p_k^{y_k}(1-p_k)^{1-y_k}, \quad y_k = 0, 1.$$

Notice that here we are dealing with a discrete rather than a continuous distribution. It is straightforward to see that replacing the integral in (2.2.1) by an appropriate sum we obtain analogues of (2.2.3)–(2.2.7) for the Bernoulli distribution with the parameter vector $\boldsymbol{p} = (p_1, \ldots, p_m)'$. We have then

$$\left[\boldsymbol{S}^{(1)}(\boldsymbol{y}, \boldsymbol{p})\right]_k = \frac{y_k - p_k}{p_k(1-p_k)}, \tag{2.2.17}$$

$k = 1, \ldots, m$, and

$$\left[\boldsymbol{S}^{(2)}(\boldsymbol{y}, \boldsymbol{p})\right]_{jk} = \frac{y_j - p_j}{p_j(1-p_j)} \frac{y_k - p_k}{p_k(1-p_k)} - \delta_{jk} \frac{y_k - 2p_k y_k + p_k^2}{p_k^2(1-p_k)^2}, \tag{2.2.18}$$

$j, k = 1, \ldots, m$, with $\delta_{jk} = 1$ if $j = k$, and $\delta_{jk} = 0$ if $j \neq k$.

The above examples are particular cases of the following exponential families.

Example 2.2.4 (Exponential family) Let $\boldsymbol{Y}$ be a random vector distributed according to an exponential family. That is

$$f(\boldsymbol{y}, \boldsymbol{v}) = a(\boldsymbol{v}) \exp\left\{\sum_{k=1}^{s} b_k(\boldsymbol{v}) t_k(\boldsymbol{y})\right\} h(\boldsymbol{y}), \tag{2.2.19}$$

where $a(\boldsymbol{v}) > 0$ and $b_k(\boldsymbol{v})$ are real valued functions of the parameter vector $\boldsymbol{v}$ and $t_k(\boldsymbol{y})$, and $h(\boldsymbol{y})$ are real valued functions of $\boldsymbol{y}$. Then

$$\boldsymbol{S}^{(1)}(\boldsymbol{y}, \boldsymbol{v}) = a(\boldsymbol{v})^{-1} \nabla a(\boldsymbol{v}) + \sum_{k=1}^{s} t_k(\boldsymbol{y}) \nabla b_k(\boldsymbol{v}) \tag{2.2.20}$$

and

$$\begin{aligned} \boldsymbol{S}^{(2)}(\boldsymbol{y}, \boldsymbol{v}) &= a(\boldsymbol{v})^{-1} \nabla^2 a(\boldsymbol{v}) - a(\boldsymbol{v})^{-2} \left[\nabla a(\boldsymbol{v})\right] \left[\nabla a(\boldsymbol{v})\right]' \\ &+ \sum_{k=1}^{s} t_k(\boldsymbol{y}) \nabla^2 b_k(\boldsymbol{v}) + \boldsymbol{S}^{(1)}(\boldsymbol{y}, \boldsymbol{v}) \boldsymbol{S}^{(1)}(\boldsymbol{y}, \boldsymbol{v})'. \end{aligned} \tag{2.2.21}$$

Notice that in the examples above the function $\ell(\boldsymbol{v})$ is differentiable and its derivatives can be taken inside the expected value provided that the corresponding expectations do exist.

Table 2.2.1 presents score functions for several commonly used pdf's.

No	pdf	$f(y,v)$	$S^{(1)}$	$S^{(2)}$
1	gamma	$\frac{\lambda^\beta y^{\beta-1}}{\Gamma(\beta)}\exp(-\lambda y),\ y \geq 0$	$\left(\beta\lambda^{-1} - y,\ \log\lambda + \log y - \frac{\Gamma'(\beta)}{\Gamma(\beta)}\right)$	$\left[\begin{array}{c\|c} -\beta\lambda^{-2} & \lambda^{-1} \\ * & \Gamma(\beta)^{-2}\{\Gamma''(\beta)\Gamma(\beta) - \Gamma'(\beta)^2\} \end{array}\right] + \left[\begin{array}{c\|c} (\beta\lambda^{-1} - y)^2 & (\beta\lambda^{-1} - y)\left(\log\lambda + \log y - \frac{\Gamma'(\beta)}{\Gamma(\beta)}\right) \\ * & \left(\log\lambda + \log y - \frac{\Gamma'(\beta)}{\Gamma(\beta)}\right)^2 \end{array}\right]$
2	normal	$\frac{1}{(2\pi)^{1/2}\sigma}\exp\left\{\frac{-(y-\mu)^2}{2\sigma^2}\right\}$	$(\sigma^{-2}(y-\mu),\ \ -\sigma^{-1} + \sigma^{-3}(y-\mu)^2)$	$\left[\begin{array}{c\|c} -\sigma^{-2} & -2\sigma^{-3}(y-\mu) \\ * & \sigma^{-2} - 3\sigma^{-4}(y-\mu)^2 \end{array}\right] + \left[\begin{array}{c\|c} \sigma^{-4}(y-\mu)^2 & \sigma^{-2}(\sigma^{-3} - \sigma^{-1})(y-\mu)^3 \\ * & (\sigma^{-3}(y-\mu)^2 - \sigma^{-1})^2 \end{array}\right]$
3	Weibull	$\alpha\beta^{-1}y^{\alpha-1}\exp\left(-\beta^{-1}y^\alpha\right),\ y \geq 0$	$(\alpha^{-1} + \log y - \beta^{-1}y^\alpha \log y,\ \ -\beta^{-1} + \beta^{-2}y^\alpha)$	$\left[\begin{array}{c\|c} -\alpha^{-2} - \beta^{-1}y^\alpha(\log y)^2 & \beta^{-2}y^\alpha \log y \\ * & \beta^{-2} - 2\beta^{-3}y^\alpha \end{array}\right] + \left[\begin{array}{c\|c} (\alpha^{-1} + \log y - \beta^{-1}y^\alpha \log y)^2 & (\alpha^{-1} + \log y - \beta^{-1}y^\alpha \log y)(\beta^{-2}y^\alpha - \beta^{-1}) \\ * & (\beta^{-2}y^\alpha - \beta^{-1})^2 \end{array}\right]$
4	binomial	$\binom{n}{y}p^y(1-p)^{n-y},\ y = 0,1,\ldots,n$	$\frac{y-np}{p(1-p)}$	$\frac{np^2-(1-2p)y-(y-np)^2}{p^2(1-p)^2}$
5	Poisson	$\frac{\lambda^y e^{-\lambda}}{y!}\ ,\ y = 0,1,2,\ldots$	$\frac{y}{\lambda} - 1$	$-\frac{y}{\lambda^2} + \left(\frac{y}{\lambda} - 1\right)^2$
6	geometric	$p(1-p)^{y-1},\ y = 1,2,\ldots,$	$\frac{1-py}{p(1-p)}$	$\frac{2p-p^2y-2py+p^2y^2}{p^2(1-p)^2}$

Table 2.2.1 Score Functions $\boldsymbol{S}^{(1)}(y,\boldsymbol{v}) = \nabla\log f(y,\boldsymbol{v})$ and $\boldsymbol{S}^{(2)}(y,\boldsymbol{v}) = \nabla^2\log f(y,\boldsymbol{v}) + \boldsymbol{S}^{(1)}(y,\boldsymbol{v})\boldsymbol{S}^{(1)}(y,\boldsymbol{v})'$ for commonly used exponential families

Example 2.2.5 (Reliability system) We shall derive the sensitivities $\nabla \ell(\boldsymbol{v})$ and $\nabla^2 \ell(\boldsymbol{v})$, and their corresponding estimates $\bar{\nabla} \ell_N(\boldsymbol{v})$ and $\bar{\nabla}^2 \ell_N(\boldsymbol{v})$, for the reliability system (2.1.1). We have

$$\nabla \ell(\boldsymbol{v}) = \mathbb{E}_{\boldsymbol{v}}\{(\max_{j=1,\ldots,p} \min_{i \in \mathcal{L}_j} Y_i) \boldsymbol{S}^{(1)}(\boldsymbol{Y}, \boldsymbol{v})\}$$

and

$$\nabla^2 \ell(\boldsymbol{v}) = \mathbb{E}_{\boldsymbol{v}}\{(\max_{j=1,\ldots,p} \min_{i \in \mathcal{L}_j} Y_i) \boldsymbol{S}^{(2)}(\boldsymbol{Y}, \boldsymbol{v})\},$$

respectively.

Assume that Y_k are independent, each distributed gamma. Taking (2.2.13) and (2.2.14) into account we can estimate $\nabla \ell(\boldsymbol{\lambda})$ and $\nabla^2 \ell(\boldsymbol{\lambda})$ by

$$\bar{\nabla} \ell_N(\boldsymbol{\lambda}) = N^{-1} \sum_{t=1}^{N} (\max_{j=1,\ldots,p} \min_{i \in \mathcal{L}_j} Y_{it})(\boldsymbol{\beta} \boldsymbol{\lambda}^{-1} - \boldsymbol{Y}_t) \quad (2.2.22)$$

and

$$\begin{aligned} \bar{\nabla}^2 \ell_N(\boldsymbol{\lambda}) &= N^{-1} \sum_{t=1}^{N} (\max_{j=1,\ldots,p} \min_{i \in \mathcal{L}j} Y_{it}) \\ &\quad \times \left[(\boldsymbol{\beta} \boldsymbol{\lambda}^{-1} - \boldsymbol{Y}_t)(\boldsymbol{\beta} \boldsymbol{\lambda}^{-1} - \boldsymbol{Y}_t)' - \operatorname{diag}(\boldsymbol{\beta} \boldsymbol{\lambda}^{-2}) \right], \quad (2.2.23) \end{aligned}$$

respectively.

Note also that in order to obtain the sensitivities for the stochastic PERT network we need to replace only the expression "$\max_{j=1,\ldots,p} \min_{i \in \mathcal{L}_j}$" by its counterpart "$\min_{j=1,\ldots,p} \sum_{i \in \mathcal{L}_j}$" (see (2.1.2)), while all other data remain the same.

Remark 2.2.2 The estimates of $\nabla \ell(\boldsymbol{v})$ and $\nabla^2 \ell(\boldsymbol{v})$ can be used in a Taylor series expansion to approximate the performance measure $\ell(\boldsymbol{v} + \boldsymbol{h})$ for small perturbations $\boldsymbol{h}$, as

$$\ell(\boldsymbol{v} + \boldsymbol{h}) \cong \bar{\ell}_N(\boldsymbol{v}) + \boldsymbol{h}' \bar{\nabla} \ell_N(\boldsymbol{v}) + \tfrac{1}{2} \boldsymbol{h}' \bar{\nabla}^2 \ell_N(\boldsymbol{v}) \boldsymbol{h}.$$

The accuracy of such an approximation needs further investigation, however.

Confidence regions for $\nabla^k \ell(\boldsymbol{v})$ can be obtained by standard techniques based on the Central Limit Theorem for $\bar{\nabla}^k \ell_N(\boldsymbol{v})$ (see section 1.3 of chapter 1). In particular, $N^{1/2} [\bar{\nabla} \ell_N(\boldsymbol{v}) - \nabla \ell(\boldsymbol{v})]$ converges in distribution to a multivariate normal with mean zero and the covariance matrix

$$\begin{aligned} \operatorname{Cov}\{L \boldsymbol{S}^{(1)}\} &= \mathbb{E}_{\boldsymbol{v}}\{L(\boldsymbol{Y})^2 \boldsymbol{S}^{(1)}(\boldsymbol{Y}, \boldsymbol{v}) \boldsymbol{S}^{(1)}(\boldsymbol{Y}, \boldsymbol{v})'\} \\ &\quad - [\nabla \ell(\boldsymbol{v})][\nabla \ell(\boldsymbol{v})]'. \quad (2.2.24) \end{aligned}$$

Consider now the particular case where the components $Y_1, \ldots, Y_m$ of the random vector $\boldsymbol{Y}$ are independent random variables, $Y_k \sim f_k(y_k, v_k)$, with the corresponding scalar parameters v_k, $k = 1, \ldots, m$. In this case $\boldsymbol{Y} \sim f(\boldsymbol{y}, \boldsymbol{v})$, where $f(\boldsymbol{y}, \boldsymbol{v}) = \prod_{k=1}^{m} f_k(y_k, v_k)$ and $\boldsymbol{v} = (v_1, \ldots, v_m)'$. We have then that

$$\boldsymbol{S}^{(1)}(\boldsymbol{Y}, \boldsymbol{v}) = \nabla \log f(\boldsymbol{Y}, \boldsymbol{v}) = (S^{(1)}(Y_1, v_1), \ldots, S^{(1)}(Y_m, v_m))' ,$$

where

$$S^{(1)}(Y_k, v_k) = \partial \log f_k(Y_k, v_k) / \partial v_k ,$$

$k = 1, \ldots, m$. It follows that in the considered situation the variance of the k-th component of $\boldsymbol{S}^{(1)}(\boldsymbol{Y}, \boldsymbol{v})$ is determined by the distribution of Y_k alone and does not increase with m.

On the other hand, consider the situation where the random variables $Y_1, \ldots, Y_m$ are independent and $Y_k \sim f(y_k, \boldsymbol{v})$ with the common pdf and the same parameter vector $\boldsymbol{v}$. We have then that, for a random sample $\boldsymbol{Y}_i = (Y_{1i}, \ldots, Y_{mi})'$, $i = 1, \ldots, N$,

$$\bar{\nabla} \ell_N(\boldsymbol{v}) = N^{-1} \sum_{i=1}^{N} L_i \boldsymbol{S}_i^{(1)},$$

where $L_i = L(\boldsymbol{Y}_i)$ and

$$\boldsymbol{S}_i^{(1)} = \sum_{k=1}^{m} \boldsymbol{S}_{ki}^{(1)} = \sum_{k=1}^{m} \nabla \log f(Y_{ki}, \boldsymbol{v}).$$

It follows that the variances of the components of $\boldsymbol{S}_i^{(1)}$ are m times the variances of the corresponding components of $\boldsymbol{S}_{ki}^{(1)}$ and hence are of order $O(m)$, provided the variance of $\boldsymbol{S}_{ki}^{(1)}$ is of order $0(1)$. This indicates that in this case the variances of the components of $\bar{\nabla} \ell_N(\boldsymbol{v})$ can be very substantial provided m is large.

It is important to note that the estimators $\bar{\nabla} \ell_N(\boldsymbol{v})$ and $\bar{\nabla}^2 \ell_N(\boldsymbol{v})$, given in (2.2.9) and (2.2.10), allow us to evaluate the performance $\ell(\boldsymbol{v})$ and its sensitivities $\nabla \ell(\boldsymbol{v})$ and $\nabla^2 \ell(\boldsymbol{v})$ only at a *fixed* point $\boldsymbol{v}$. We present now an extended version of the above estimators which allow us to evaluate $\ell(\boldsymbol{v})$ and $\nabla^k \ell(\boldsymbol{v}), k = 1, 2, \ldots$, essentially at any point $\boldsymbol{v}$, provided some regularity conditions are met.

Let G be a probability measure (distribution) on $\mathbb{R}^m$ having a density function $g(\boldsymbol{y})$, i.e. $\mathrm{d}G(\boldsymbol{y}) = g(\boldsymbol{y})\,\mathrm{d}\boldsymbol{y}$. Suppose that for every permissible value of the parameter vector, the support of $f(\boldsymbol{y}, \boldsymbol{v})$ lies in the support of $g(\boldsymbol{y})$, that is

$$\operatorname{supp}\{f(\boldsymbol{y}, \boldsymbol{v})\} \subset \operatorname{supp}\{g(\boldsymbol{y})\} , \quad \boldsymbol{v} \in V. \tag{2.2.25}$$

(Recall that $\text{supp}\{g(\boldsymbol{y})\}$ is the set of those values of $\boldsymbol{y}$ for which $g(\boldsymbol{y})$ is strictly greater than zero.) In this case we can write $\ell(\boldsymbol{v})$ as follows:

$$\begin{aligned}\ell(\boldsymbol{v}) &= \int L(\boldsymbol{y})f(\boldsymbol{y},\boldsymbol{v})\,\mathrm{d}\boldsymbol{y} = \int L(\boldsymbol{y})W(\boldsymbol{y},\boldsymbol{v})\,\mathrm{d}G(\boldsymbol{y}) \\ &= \mathbb{E}_g\{L(\mathbf{Z})W(\mathbf{Z},\boldsymbol{v})\},\end{aligned} \tag{2.2.26}$$

where $W(\boldsymbol{y},\boldsymbol{v}) = f(\boldsymbol{y},\boldsymbol{v})/g(\boldsymbol{y})$ and $\mathbf{Z} \sim g(\boldsymbol{z})$. (By definition, zero divided by zero is zero.) Notice that the function $W(\boldsymbol{y},\boldsymbol{v})$ is well defined for all $\boldsymbol{v} \in V$ because of the assumption (2.2.25) . In the statistical literature $W(\mathbf{Z},\boldsymbol{v})$ is called the *likelihood ratio* or the Radon-Nikodym derivative. For more details see appendix B.

It is important to note that the original expectation of $L(\boldsymbol{Y})$ in (2.2.26) is taken with respect to the underlying pdf $f(\boldsymbol{y},\boldsymbol{v})$, while that given in the last expression of (2.2.26) is taken with respect to the pdf $g(\boldsymbol{y})$. It follows that changing the probability density from $f(\boldsymbol{y},\boldsymbol{v})$ to $g(\boldsymbol{y})$ we can express the performance measure $\ell(\boldsymbol{v})$ for all $\boldsymbol{v} \in V$ as an expectation with respect to $g(\boldsymbol{y})$ and then estimate it accordingly. We shall call $g(\boldsymbol{y})$, satisfying condition (2.2.25), the *dominating* pdf.

Let $f(\boldsymbol{y},\boldsymbol{v})$ be differentiable in $\boldsymbol{v}$. Then

$$\begin{aligned}\nabla\ell(\boldsymbol{v}) &= \int L(\boldsymbol{y})\nabla f(\boldsymbol{y},\boldsymbol{v})\,\mathrm{d}\boldsymbol{y} \\ &= \int L(\boldsymbol{y})\nabla W(\boldsymbol{y},\boldsymbol{v})\,\mathrm{d}G(\boldsymbol{y}) = \mathbb{E}_g\{L(\mathbf{Z})\nabla W(\mathbf{Z},\boldsymbol{v})\},\end{aligned}$$

provided the operators of the integration and differentiation are interchangeable. In a similar way we can write

$$\nabla^k\ell(\boldsymbol{v}) = \mathbb{E}_g\{L(\mathbf{Z})\nabla^k W(\mathbf{Z},\boldsymbol{v})\}\ , \tag{2.2.27}$$

$k = 1, 2, \ldots$. It can be readily seen that

$$\nabla^k W(\boldsymbol{y},\boldsymbol{v}) = W(\boldsymbol{y},\boldsymbol{v})\boldsymbol{S}^{(k)}(\boldsymbol{y},\boldsymbol{v})\,. \tag{2.2.28}$$

Note that our previously derived sensitivities $\nabla^k\ell(\boldsymbol{v}) = \mathbb{E}_{\boldsymbol{v}}\{L\boldsymbol{S}^{(k)}\}$ represent a particular case of (2.2.27) with $g(\boldsymbol{y}) = f(\boldsymbol{y},\boldsymbol{v})$ and $W(\boldsymbol{y},\boldsymbol{v}) = 1$. Formula (2.2.27) is also valid for $k = 0$ if we take $\nabla^0\ell(\boldsymbol{v}) = \ell(\boldsymbol{v})$ and $\nabla^0 W(\boldsymbol{y},\boldsymbol{v}) = W(\boldsymbol{y},\boldsymbol{v})$.

Formula (2.2.27) provides a closed form expression for both the *performance* $\ell(\boldsymbol{v})$ and *all* the *sensitivities* $\nabla^k\ell(\boldsymbol{v})$, $k = 1, 2, \ldots$. It can be employed for approximation purposes either by standard numerical deterministic techniques or via simulation. In the last case we can estimate $\nabla^k\ell(\boldsymbol{v})$ from a *single* simulation run for *different* values of $\boldsymbol{v} \in V$ as follows:

$$\bar{\nabla}^k\ell_N(\boldsymbol{v}) = N^{-1}\sum_{i=1}^{N} L(\mathbf{Z}_i)\nabla^k W(\mathbf{Z}_i,\boldsymbol{v})\,, \tag{2.2.29}$$

$k = 0, 1, \ldots$, where $\mathbf{Z}_1, \ldots, \mathbf{Z}_N$ is a sample from $g(\mathbf{z})$.

It is interesting to note that the above estimator $\bar{\nabla}\ell_N(\boldsymbol{v})$ of the gradient $\nabla\ell(\boldsymbol{v})$ coincides with the gradient $\nabla\bar{\ell}_N(\boldsymbol{v})$ of the estimator $\bar{\ell}_N(\boldsymbol{v})$, where

$$\bar{\ell}_N(\boldsymbol{v}) = N^{-1}\sum_{i=1}^{N} L(\mathbf{Z}_i)W(\mathbf{Z}_i, \boldsymbol{v}) , \tag{2.2.30}$$

and similarly for the higher order derivatives. This important property can be written symbolically in the form

$$\bar{\nabla}^k\ell_N(\boldsymbol{v}) = \nabla^k\bar{\ell}_N(\boldsymbol{v}) , \quad k = 1, 2, \ldots. \tag{2.2.31}$$

We shall call $\nabla^k W(\mathbf{Z}, \boldsymbol{v})$, $k = 1, 2, \ldots$, the *generalized scores* and, as mentioned before, $W(\mathbf{Z}, \boldsymbol{v}) = \nabla^0 W(\mathbf{Z}, \boldsymbol{v})$ is called the *likelihood ratio.* Note that $\bar{\nabla}^k\ell_N(\boldsymbol{v})$, $k = 0, 1, \ldots$, allow us to estimate $\nabla^k\ell(\boldsymbol{v})$ at virtually any point $\boldsymbol{v} \in V$ provided the interchangeability conditions hold. This makes the above estimators particularly suitable for solving the optimization problem (P_0) introduced in chapter 1. Note also that when $g(\boldsymbol{y}) = f(\boldsymbol{y}, \boldsymbol{v})$ we obtain $W(\boldsymbol{y}, \boldsymbol{v}) = 1$ and thus the generalized scores $\nabla^k W$ reduce to the standard scores $\boldsymbol{S}^{(k)}$, $k = 1, 2, \ldots$.

Since $\mathbb{E}_g\{W(\mathbf{Z}, \boldsymbol{v})\} = 1$ for all $\boldsymbol{v}$, we have that $\mathbb{E}_g\{\nabla^k W(\mathbf{Z}, \boldsymbol{v})\} = \mathbf{0}$, $k = 1, 2, \ldots$. Therefore, similar to (2.2.11),

$$\nabla^k\ell(\boldsymbol{v}) = \operatorname{Cov}_g\{L(\mathbf{Z}), \nabla^k W(\mathbf{Z}, \boldsymbol{v})\}, \quad k = 1, 2, \ldots. \tag{2.2.32}$$

That is, the sensitivities $\nabla^k\ell(\boldsymbol{v})$ can be represented as the covariances between the sample performance L and the generalized scores $\nabla^k W$.

For a given dominating pdf $g(\mathbf{z})$ the algorithm for estimating $\nabla^k\ell(\boldsymbol{v})$ from a single simulation run can be written as follows.

Algorithm 2.2.2 :

1. *Generate a sample* $\mathbf{Z}_1, \ldots, \mathbf{Z}_N$, *from the dominanting pdf* $g(\mathbf{z})$.
2. *Calculate the sample performance* $L(\mathbf{Z}_i)$ *and the generalized scores* $\nabla^k W(\mathbf{Z}_i, \boldsymbol{v})$, $i = 1, \ldots, N$.
3. *Calculate* $\bar{\nabla}^k\ell_N(\boldsymbol{v})$ *according to (2.2.29).*

We shall call the problem of evaluating the performance measure $\ell(\boldsymbol{v})$ and the associated sensitivities $\nabla^k\ell(\boldsymbol{v})$ $k = 1, 2 \ldots$, for different values of $\boldsymbol{v}$ while using the estimators $\nabla^k\bar{\ell}_N(\boldsymbol{v})$, $k = 0, 1, \ldots$, the "what if" problem (see, for example, Rubinstein (1986b)). It can be formulated as follows: "What will be the values of the performance measures $\nabla^k\ell(\boldsymbol{v})$, $k = 0, 1, \ldots$, if we perturb some (all) of the components of the parameters vector $\boldsymbol{v}$?"

Clearly, the accuracy (variance) of the estimators $\bar{\nabla}^k \ell_N(\boldsymbol{v})$, $k = 0, 1, \ldots,$ depends on the particular choice of the dominating density $g(\mathbf{z})$. Consider for example $\bar{\nabla}\ell_N(\boldsymbol{v})$. Its covariance matrix is given by N^{-1} times the covariance matrix of $L(\mathbf{Z})\nabla W(\mathbf{Z}, \boldsymbol{v})$, where

$$\begin{aligned} &\mathrm{Cov}_g \left\{L(\mathbf{Z})\nabla W(\mathbf{Z}, \boldsymbol{v})\right\} \\ &= \mathbb{E}_g\{L(\mathbf{Z})^2 \nabla W(\mathbf{Z}, \boldsymbol{v}) \nabla W(\mathbf{Z}, \boldsymbol{v})'\} - [\nabla \ell(\boldsymbol{v})][\nabla \ell(\boldsymbol{v})]' . \end{aligned} \tag{2.2.33}$$

Notice that in the particular case where $g(\boldsymbol{y}) = f(\boldsymbol{y}, \boldsymbol{v})$, (2.2.33) reduces to (2.2.24). We can try to choose the density $g(\mathbf{z})$ in order to minimize, for a given $\boldsymbol{v}$, some criterion function associated with the above covariance matrix. For example, we can minimize the trace, determinant or the maximal diagonal element of this covariance matrix. In the one-dimensional case the optimal pdf $g^*(\mathbf{z})$ reduces to the *importance sampling density* (see also section 1.3.1), i.e. to

$$g^*(\mathbf{z}) = c^{-1} |L(\mathbf{z}) \partial f(\mathbf{z}, v) / \partial v| , \tag{2.2.34}$$

where

$$c = \int |L(\mathbf{z}) \partial f(\mathbf{z}, v) / \partial v| \, \mathrm{d}\, \mathbf{z}.$$

Notice that the implementation of importance sampling for the SF estimators $\bar{\nabla}^k \ell_N(\boldsymbol{v})$ is not an easy task. In section 2.6 we discuss instead some variance reduction techniques in order to improve the accuracy of the estimators $\bar{\nabla}^k \ell_N(\boldsymbol{v})$.

2.3 The "What if" Problem

Throughout this section we assume that the dominating density function $g(\mathbf{z})$ is of the form $g(\mathbf{z}) = f(\mathbf{z}, \boldsymbol{v}_0)$ for some fixed value $\boldsymbol{v}_0$ of the parameter vector which is called the *reference parameter*. Then

$$\nabla^k \bar{\ell}_N(\boldsymbol{v}) = N^{-1} \sum_{i=1}^{N} L(\mathbf{Z}_i) \nabla^k W(\mathbf{Z}_i, \boldsymbol{v}), \tag{2.3.1}$$

$k = 0, 1, \ldots,$ where

$$W(\boldsymbol{Z}_i, \boldsymbol{v}) = \frac{f(\boldsymbol{Z}_i, \boldsymbol{v})}{f(\boldsymbol{Z}_i, \boldsymbol{v}_0)}, \tag{2.3.2}$$

and $\boldsymbol{Z}_i \sim f(\mathbf{z}, \boldsymbol{v}_0)$.

The accuracy of the estimator $\bar{\ell}_N(\boldsymbol{v})$ is determined by its variance. We have that

$$\mathrm{Var}\{\bar{\ell}_N(\boldsymbol{v})\} = N^{-1} \mathrm{Var}_{\boldsymbol{v}_0}\{L(\boldsymbol{Z}) W(\boldsymbol{Z}, \boldsymbol{v})\} , \tag{2.3.3}$$

where the subscript $\boldsymbol{v}_0$ on the right-hand side of (2.3.3) indicates that all expectations involved in the calculation of this variance are taken with respect to the density $f(\boldsymbol{y}, \boldsymbol{v}_0)$.

To get a better insight into the problem, let us consider the following case. Suppose that the components $Y_1, \ldots, Y_m$, of the random vector $\boldsymbol{Y}$ are independent, each distributed gamma, i.e.

$$f(\boldsymbol{y}, \boldsymbol{\lambda}, \boldsymbol{\beta}) = \prod_{k=1}^{m} f_k(y_k, \lambda_k, \beta_k)$$

with densities $f_k(y_k, \lambda_k, \beta_k)$ given in (2.2.12). Suppose further that we are interested in sensitivities with respect to the parameters λ_k, $k = 1, \ldots, m$, only. That is, our parameter vector $\boldsymbol{v} = \boldsymbol{\lambda} = (\lambda_1, \ldots, \lambda_m)'$, while parameters β_k, $k = 1, \ldots, m$, are supposed to be fixed and known. Then

$$\mathrm{Var}_{\boldsymbol{\lambda}_0}\{LW\} = \prod_{k=1}^{m} \left(1 + \frac{\alpha_k^2}{1 + 2\alpha_k}\right)^{\beta_k} \mathbb{E}_{\boldsymbol{\lambda}_0 + 2\Delta\boldsymbol{\lambda}}\{L(\boldsymbol{Z})^2\} - \ell(\boldsymbol{v})^2, \quad (2.3.4)$$

where $\boldsymbol{\lambda}_0 = (\lambda_{01}, \ldots, \lambda_{0m})'$, $\Delta\boldsymbol{\lambda} = \boldsymbol{\lambda} - \boldsymbol{\lambda}_0$ and $\alpha_k = (\Delta\lambda_k)/\lambda_k$. Notice that the subscripts $\boldsymbol{\lambda}_0$ and $\boldsymbol{\lambda}_0 + 2\Delta\boldsymbol{\lambda}$ in the left- and right-hand sides of (2.3.4), respectively, stand for values of the parameter vector $\boldsymbol{\lambda}$ with respect to which the expectations are taken. We shall call $\Delta\lambda_k$ and α_k the *absolute* and *relative* perturbations.

Let us outline the derivation of (2.3.4) for $m = 1$. An extension to $m > 1$ is straightforward and is left as an exercise. We have

$$W(z, \lambda, \beta) = \frac{f(z, \lambda, \beta)}{f(z, \lambda_0, \beta)} = \left(\frac{\lambda}{\lambda_0}\right)^{\beta} \exp\{-(\lambda - \lambda_0)z\}$$

and

$$\begin{aligned} \mathbb{E}_{\lambda_0}\{(LW)^2\} &= \int L(y)^2 \lambda^{2\beta} \lambda_0^{-\beta} \Gamma(\beta)^{-1} y^{\beta-1} \exp\{-(\lambda_0 + 2\Delta\lambda)y\}\, dy \\ &= \left[\frac{\lambda^2}{\lambda_0(\lambda_0 + 2\Delta\lambda)}\right]^{\beta} \mathbb{E}_{\lambda_0 + 2\Delta\lambda}\{L(Z)^2\}. \end{aligned}$$

It remains to note that

$$\frac{\lambda^2}{\lambda_0(\lambda_0 + 2\Delta\lambda)} = 1 + \frac{\alpha^2}{1 + 2\alpha}$$

and that

$$\mathbb{E}_{\lambda_0}\{LW\} = \mathbb{E}_{\lambda}\{L\} = \ell(\lambda) = \ell(v).$$

By taking $L = 1$ we obtain from (2.3.4) that

$$\mathrm{Var}_{\boldsymbol{\lambda}_0}\{W\} = \prod_{k=1}^{m}\left(1+\frac{\alpha_k^2}{1+2\alpha_k}\right)^{\beta_k} - 1 .$$

Therefore formula (2.3.4) can be written in the form

$$\mathrm{Var}_{\boldsymbol{\lambda}_0}\{LW\} = \mathbb{E}_{\boldsymbol{\lambda}_0}\{W^2\}\mathbb{E}_{\boldsymbol{\lambda}_0+2\Delta\boldsymbol{\lambda}}\{L^2\} - \ell(\boldsymbol{v})^2, \qquad (2.3.5)$$

where

$$\mathbb{E}_{\boldsymbol{\lambda}_0}\{W^2\} = \prod_{k=1}^{m}\left(1+\frac{\alpha_k^2}{1+2\alpha_k}\right)^{\beta_k} . \qquad (2.3.6)$$

Note that formula (2.3.4) is derived under the assumption that $1+2\alpha_k > 0$ or, equivalently,

$$\lambda_k > \tfrac{1}{2}\lambda_{0k}, \quad k = 1,\dots,m . \qquad (2.3.7)$$

If the expectation $\mathbb{E}_{\boldsymbol{\lambda}_0+2\Delta\boldsymbol{\lambda}}\{L^2\}$ is finite, then condition (2.3.7) is sufficient for the corresponding "what if" estimator to have a finite variance. This condition is also necessary if $\mathbb{E}_{\boldsymbol{\lambda}_0+2\Delta\boldsymbol{\lambda}}\{L^2\}$ does not tend to zero as $\boldsymbol{\lambda}_0$ approaches $2\boldsymbol{\lambda}$. (Note that this condition is not necessary in general. Take, for example, $L(y) = \exp(-ay)$, $a > 0$, $k = 1$ and $\beta = 1$. Then it is not difficult to see that $\mathbb{E}_{\lambda_0}\{(LW)^2\}$ is finite if and only if $2(\lambda + a) > \lambda_0$.) This shows how sensitive the "what if" estimators can be to the choice of the reference parameter. This will be discussed further in chapter 3.

Suppose now that all α_k and all β_k are equal, say $\alpha_k = \alpha$ and $\beta_k = \beta$, $k = 1,\dots,m$. Then (2.3.4) reduces to

$$\mathrm{Var}_{\boldsymbol{\lambda}_0}\{LW\} = \left(1+\frac{\alpha^2}{1+2\alpha}\right)^{m\beta} \mathbb{E}_{\boldsymbol{\lambda}_0+2\Delta\boldsymbol{\lambda}}\{L^2\} - \ell(\boldsymbol{\lambda})^2 . \qquad (2.3.8)$$

We see that for fixed α and β, even with $1+2\alpha > 0$, the variance of LW increases exponentially in m. For small values of α the first term in the right-hand side of (2.3.8) can be approximated as follows:

$$\left(1+\frac{\alpha^2}{1+2\alpha}\right)^{m\beta} = \exp\left\{m\beta\log\left(1+\frac{\alpha^2}{1+2\alpha}\right)\right\} \cong \exp\left\{\frac{m\beta\alpha^2}{1+2\alpha}\right\} .$$

(We use here the fact that for small x, $\log(1+x) \cong x$.) This shows that in order for the variance of LW to be meaningful, the value $m\beta\alpha^2/(1+2\alpha)$ should be not "too big". That is, as m increases, $\alpha^2\beta$ should be approximately of order $O(m^{-1})$.

Table 2.3.1 presents the exact value of $\mathrm{Var}_{\boldsymbol{\lambda}_0}\{LW\}$ as a function of α and m for the particular cases of $\beta = 2$ and $L = 1$ ($\mathcal{L}$ denotes a large number, greater than 10^3, say). It can be seen that the results of table 2.3.1 match well with the requirement that $\alpha^2\beta = O(m^{-1})$

Table 2.3.1 $\mathrm{Var}_{\boldsymbol{\lambda}_0}\{LW\}$ as a function of m and α for $\beta = 2$ and $L = 1$.

α/m	1	10	10^2	10^3	10^4	10^5
-0.50	$\mathcal{L}$	$\mathcal{L}$	$\mathcal{L}$	$\mathcal{L}$	$\mathcal{L}$	$\mathcal{L}$
-0.40	2.2400	$\mathcal{L}$	$\mathcal{L}$	$\mathcal{L}$	$\mathcal{L}$	$\mathcal{L}$
-0.20	0.1378	2.636	$\mathcal{L}$	$\mathcal{L}$	$\mathcal{L}$	$\mathcal{L}$
-0.10	0.0252	0.282	10.99	$\mathcal{L}$	$\mathcal{L}$	$\mathcal{L}$
0.00	0.0000	0.000	0.000	0.00	0.0	0
0.01	0.0002	0.002	0.020	0.21	7.1	$\mathcal{L}$
0.02	0.0008	0.008	0.080	1.16	$\mathcal{L}$	$\mathcal{L}$
0.03	0.0017	0.017	0.185	4.46	$\mathcal{L}$	$\mathcal{L}$
0.04	0.0030	0.030	0.340	18.31	$\mathcal{L}$	$\mathcal{L}$
0.05	0.0046	0.046	0.570	92.72	$\mathcal{L}$	$\mathcal{L}$
0.10	0.0167	0.181	4.260	$\mathcal{L}$	$\mathcal{L}$	$\mathcal{L}$
0.20	0.0580	0.750	278.8	$\mathcal{L}$	$\mathcal{L}$	$\mathcal{L}$
0.50	0.2656	49.500	$\mathcal{L}$	$\mathcal{L}$	$\mathcal{L}$	$\mathcal{L}$
1.00	0.7770	314.34	$\mathcal{L}$	$\mathcal{L}$	$\mathcal{L}$	$\mathcal{L}$

Take, for example, $m = 10$, $\alpha = 20\%$; $m = 100$, $\alpha = 10\%$; and $m = 1000$, $\alpha = 3\%$. We have then from table 2.3.1 that $\mathrm{Var}_{\boldsymbol{\lambda}_0} W$ is given by 0.75, 4.26 and 4.46, respectively. If we take now larger values of α, say $\alpha = 50\%$, 20% and 5%, for the same values of m we obtain from table 2.3.1, $\mathrm{Var}_{\boldsymbol{\lambda}_0} W = 49.5$, 278.8 and 92.72, respectively, and thus a very substantial increase in the variance of W.

Now let us consider the covariance matrix (2.2.33) for the case of gamma distributions. We have

$$\mathrm{Cov}_{\boldsymbol{\lambda}_0}\{L\nabla W\} = \mathbb{E}_{\boldsymbol{\lambda}_0}\{L^2(\nabla W)(\nabla W)'\} - (\nabla \ell)(\nabla \ell)' .$$

Arguing similarly to (2.3.4)–(2.3.6) and taking into account that

$$W(\mathbf{z}, \boldsymbol{\lambda}, \boldsymbol{\beta}) = \prod_{k=1}^{m} \left(\frac{\lambda_k}{\lambda_{0k}}\right)^{\beta_k} \exp\{-(\lambda_k - \lambda_{0k}) z_k\}$$

and hence

$$\frac{\partial W(\mathbf{z}, \boldsymbol{\lambda}, \boldsymbol{\beta})}{\partial \lambda_k} = (\beta_k \lambda_k^{-1} - z_k) \frac{f(\mathbf{z}, \boldsymbol{\lambda}, \boldsymbol{\beta})}{f(\mathbf{z}, \boldsymbol{\lambda}_0, \boldsymbol{\beta})} ,$$

$k = 1, \ldots, m$, we obtain that

$$\mathbb{E}_{\boldsymbol{\lambda}_0}\{L^2(\nabla W)(\nabla W)'\} = \mathbb{E}_{\boldsymbol{\lambda}_0}\{W^2\} \mathbb{E}_{\boldsymbol{\lambda}_0 + 2\Delta\boldsymbol{\lambda}}\{L(\mathbf{Z})^2(\boldsymbol{\beta}\boldsymbol{\lambda}^{-1} - \mathbf{Z})(\boldsymbol{\beta}\boldsymbol{\lambda}^{-1} - \mathbf{Z})'\} .$$

It follows that

$$\begin{aligned} &\mathrm{Cov}_{\boldsymbol{\lambda}_0}\{L\nabla W\} && (2.3.9) \\ &= \mathbb{E}_{\boldsymbol{\lambda}_0}\{W^2\}\mathbb{E}_{\boldsymbol{\lambda}+2\Delta\boldsymbol{\lambda}}\{L(\boldsymbol{Z})^2(\beta\boldsymbol{\lambda}^{-1} - \boldsymbol{Z})(\beta\boldsymbol{\lambda}^{-1} - \boldsymbol{Z})'\} - (\nabla\ell)(\nabla\ell)', \end{aligned}$$

where $\mathbb{E}_{\boldsymbol{\lambda}_0}\{W^2\}$ is given in (2.3.6).

Consider the following two cases with $\beta = 2$.

(i) Let $\alpha = 0.1$ and $m = 100$. Then we have from (2.3.5), (2.3.6) and table 2.3.1 that

$$\mathrm{Var}_{\boldsymbol{\lambda}_0}\{LW\} = 5.26\mathbb{E}_{\boldsymbol{\lambda}_0+2\Delta\boldsymbol{\lambda}}\{L^2\} - \ell(\boldsymbol{v})^2 .$$

It is reasonable to expect that in this case the "what if" estimator $\bar{\ell}_N(\boldsymbol{v})$ will perform satisfactorily provided the expectation $\mathbb{E}_{\boldsymbol{\lambda}_0+2\Delta\boldsymbol{\lambda}}\{L^2\}$ is not too big.

(ii) Let again $\alpha = 0.1$, but $m = 1000$. Then

$$\mathrm{Var}_{\boldsymbol{\lambda}_0}\{LW\} = \left(1 + \frac{0.1^2}{1+0.2}\right)^{2000} \mathbb{E}_{\boldsymbol{\lambda}_0+2\Delta\boldsymbol{\lambda}}\{L^2\} - \ell(\boldsymbol{v})^2,$$

where

$$\left(1 + \frac{0.1^2}{1.02}\right)^{2000} = 5.26^{10} \cong 1.6 \times 10^7 .$$

In this case the variance of LW and the elements of $\mathrm{Cov}_{\boldsymbol{\lambda}_0}\{L\nabla W\}$ are very large.

We now extend (2.3.5) and (2.3.9) to the case of an m-parameter exponential family given in the following *canonical* form:

$$f(\boldsymbol{y}, \boldsymbol{v}) = c(\boldsymbol{v}) \exp\left\{\sum_{k=1}^{m} v_k t_k(\boldsymbol{y})\right\} h(\boldsymbol{y}) . \qquad (2.3.10)$$

The density $f(\boldsymbol{y}, \boldsymbol{v})$ is parameterized by the vector $\boldsymbol{v} = (v_1, \ldots, v_m)' \in V$. If the parameters v_k were functions of a parameter vector $\boldsymbol{\mu}$, say $v_k = b_k(\boldsymbol{\mu})$, we would obtain an exponential family parameterized in the form (2.2.19) introduced in section 2.2. Notice that the function $c(\boldsymbol{v})$ in (2.3.10) is determined by the functions $t_1(\boldsymbol{y}), \ldots, t_m(\boldsymbol{y})$ and $h(\boldsymbol{y})$ through the identity

$$c(\boldsymbol{v}) \int \exp\left\{\sum_{k=1}^{m} v_k t_k(\boldsymbol{y})\right\} h(\boldsymbol{y})\, \mathrm{d}\boldsymbol{y} = 1. \qquad (2.3.11)$$

(For a detailed discussion of exponential families see, for example, Lehmann (1983)).

We have then that

$$W(\boldsymbol{y},\boldsymbol{v})=\frac{f(\boldsymbol{y},\boldsymbol{v})}{f(\boldsymbol{y},\boldsymbol{v}_0)}=\frac{c(\boldsymbol{v})}{c(\boldsymbol{v}_0)}\exp\left\{\sum_{k=1}^{m}(v_k-v_{0k})t_k(\boldsymbol{y})\right\} \tag{2.3.12}$$

and

$$\begin{aligned}\mathrm{Var}_{\boldsymbol{v}_0}\{LW\} &= \frac{c(\boldsymbol{v})^2}{c(\boldsymbol{v}_0)c(\boldsymbol{v}_0+2\Delta\boldsymbol{v})}\mathbb{E}_{\boldsymbol{v}_0+2\Delta\boldsymbol{v}}\{L^2\}-\ell(\boldsymbol{v})^2\\ &= \mathbb{E}_{\boldsymbol{v}_0}\{W^2\}\mathbb{E}_{\boldsymbol{v}_0+2\Delta\boldsymbol{v}}\{L^2\}-\ell(\boldsymbol{v})^2, \end{aligned}\tag{2.3.13}$$

where $\Delta\boldsymbol{v}=\boldsymbol{v}-\boldsymbol{v}_0$. The derivation of (2.3.13) is similar to that of (2.3.4)–(2.3.6) and is left as an exercise. Of course, formula (2.3.13) holds under the condition that the vector $\boldsymbol{v}_0+2\Delta\boldsymbol{v}$ lies in the permissible parameter space V. In the case of gamma distributions this reduces to the condition (2.3.7).

The gradient of the function $W(\boldsymbol{y},\boldsymbol{v})$ given by the right-hand side of (2.3.12) can be written in the form

$$\nabla W(\boldsymbol{y},\boldsymbol{v})=W(\boldsymbol{y},\boldsymbol{v})\boldsymbol{S}^{(1)}(\boldsymbol{y},\boldsymbol{v}), \tag{2.3.14}$$

where

$$\boldsymbol{S}^{(1)}(\boldsymbol{y},\boldsymbol{v})=c(\boldsymbol{v})^{-1}\nabla c(\boldsymbol{v})+\mathbf{t}(\boldsymbol{y}) \tag{2.3.15}$$

and $\mathbf{t}(\boldsymbol{y})=(t_1(\boldsymbol{y}),\ldots,t_m(\boldsymbol{y}))'$. It follows that

$$\begin{aligned}&\mathrm{Cov}_{\boldsymbol{v}_0}\{L\nabla W\}\\ &\quad=\mathbb{E}_{\boldsymbol{v}_0}\{W^2\}\mathbb{E}_{\boldsymbol{v}_0+2\Delta\boldsymbol{v}}\{L(\mathbf{Z})^2\boldsymbol{S}^{(1)}(\mathbf{Z},\boldsymbol{v})\boldsymbol{S}^{(1)}(\mathbf{Z},\boldsymbol{v})'\}-(\nabla\ell)(\nabla\ell)'.\end{aligned}\tag{2.3.16}$$

It is interesting to note that the second moments of LW and $L\nabla W$ (see (2.3.13) and (2.3.16)) can be represented as the products of $\mathbb{E}_{\boldsymbol{v}_0}\{W^2\}$ with $\mathbb{E}_{\boldsymbol{v}_0+2\Delta\boldsymbol{v}}\{L^2\}$ and $\mathbb{E}_{\boldsymbol{v}_0+2\Delta\boldsymbol{v}}\{L^2\boldsymbol{S}^{(1)}\boldsymbol{S}^{(1)\prime}\}$, respectively. For large m, the term $\mathbb{E}_{\boldsymbol{v}_0}\{W^2\}$ in (2.3.13) and (2.3.16) is typically dominating.

With these remarks at hand let us derive now an approximation for $\mathbb{E}_{\boldsymbol{v}_0}\{W^2\}$ in the case of the general pdf $f(\boldsymbol{y},\boldsymbol{v})$. To do so consider the function

$$\Psi(\boldsymbol{v})=\mathbb{E}_{\boldsymbol{v}_0}\{W(\mathbf{Z},\boldsymbol{v})^2\}.$$

By the definition of W we have $\Psi(\boldsymbol{v}_0)=1$. Assuming now that the derivatives of $\Psi(\boldsymbol{v})$ can be taken inside the expected value we obtain

$$\begin{aligned}\nabla\Psi(\boldsymbol{v}) &= \mathbb{E}_{\boldsymbol{v}_0}\{\nabla(W(\mathbf{Z},\boldsymbol{v})^2)\}=2\mathbb{E}_{\boldsymbol{v}_0}\{W(\mathbf{Z},\ \boldsymbol{v})\nabla W(\boldsymbol{Z},\boldsymbol{v})\},\\ \nabla^2\Psi(\boldsymbol{v}) &= 2\mathbb{E}_{\boldsymbol{v}_0}\{\nabla W(\boldsymbol{Z},\boldsymbol{v})\nabla W(\boldsymbol{Z},\boldsymbol{v})'+W(\boldsymbol{Z},\boldsymbol{v})\nabla^2 W(\boldsymbol{Z},\boldsymbol{v})\}.\end{aligned}$$

Since for all $\boldsymbol{v}$ the integral of $f(\boldsymbol{y}, \boldsymbol{v})$ is equal to one, the integrals of $\nabla f(\boldsymbol{y}, \boldsymbol{v}_0)$ and $\nabla^2 f(\boldsymbol{y}, \boldsymbol{v}_0)$ are zero and hence

$$\begin{aligned}
\nabla \Psi(\boldsymbol{v}_0) &= 2\int \nabla f(\boldsymbol{y}, \boldsymbol{v}_0)\,\mathrm{d}\boldsymbol{y} = 0\,, \\
\nabla^2 \Psi(\boldsymbol{v}_0) &= 2\mathbb{E}_{\boldsymbol{v}_0}\{\nabla W(\boldsymbol{Z}, \boldsymbol{v}_0)\nabla W(\boldsymbol{Z}, \boldsymbol{v}_0)'\} + 2\int \nabla^2 f(\boldsymbol{y}, \boldsymbol{v}_0)\,\mathrm{d}\boldsymbol{y} \\
&= 2\mathbb{E}_{\boldsymbol{v}_0}\{\nabla \log f(\boldsymbol{Z}, \boldsymbol{v}_0)\nabla \log f(\boldsymbol{Z}, \boldsymbol{v}_0)'\} = 2\mathbf{I}(\boldsymbol{v}_0)\,,
\end{aligned}$$

where

$$\mathbf{I}(\boldsymbol{v}_0) = \mathbb{E}_{\boldsymbol{v}_0}\{\nabla \log f(\boldsymbol{Z}, \boldsymbol{v}_0)\nabla \log f(\mathbf{Z}, \boldsymbol{v}_0)'\}. \tag{2.3.17}$$

From a Taylor series expansion we have then that

$$\Psi(\boldsymbol{v}) = 1 + \tfrac{1}{2}(\boldsymbol{v} - \boldsymbol{v}_0)'\nabla^2\Psi(\boldsymbol{v}_0)(\boldsymbol{v} - \boldsymbol{v}_0) + r(\boldsymbol{v}),$$

where the remainder term $r(\boldsymbol{v})$ is of order $o(\|\boldsymbol{v}-\boldsymbol{v}_0\|^2)$, i.e. $\|\boldsymbol{v}-\boldsymbol{v}_0\|^{-2}r(\boldsymbol{v}) \to 0$ as $\boldsymbol{v} \to \boldsymbol{v}_0$. Combining all this we obtain the following result:

$$\mathbb{E}_{\boldsymbol{v}_0}\{W^2\} = 1 + (\boldsymbol{v} - \boldsymbol{v}_0)'\mathbf{I}(\boldsymbol{v}_0)(\boldsymbol{v} - \boldsymbol{v}_0) + o(\|\boldsymbol{v} - \boldsymbol{v}_0\|^2)\,. \tag{2.3.18}$$

Note that in the statistical literature the matrix $\mathbf{I}(\boldsymbol{v}_0)$, given in (2.3.17), is called *Fisher's information matrix.* In the particular case where v is a scalar, (2.3.17) reduces to

$$I(v_0) = \mathbb{E}_{v_0}\left\{[\mathrm{d}\log f(\mathbf{Z}, v_0)/\mathrm{d}v]^2\right\}\,,$$

and is called *Fisher's information.*

Consider now the situation where the components Y_k, $k = 1,\ldots,m$, of the random vector $\boldsymbol{Y}$ are independent, each distributed according to the same parametric family, $Y_k \sim f(y_k, v_k)$, $v_k \in \mathbb{R}$. Then the density function of $\boldsymbol{Y}$ is given by the product of $f(y_k, v_k)$, $k = 1,\ldots,m$, $\boldsymbol{v} = (v_1,\ldots,v_m)'$ and

$$\mathbb{E}_{\boldsymbol{v}_0}\{W(\mathbf{Z}, \boldsymbol{v})^2\} = \prod_{k=1}^{m} \mathbb{E}_{v_{0k}}\{W_k(Z_k, v_k)^2\}\,,$$

where

$$W_k(z_k, v_k) = f(z_k, v_k)/f(z_k, v_{0k}).$$

It follows that

$$\begin{aligned}
\mathbb{E}_{\boldsymbol{v}_0}\{W^2\} &= \prod_{k=1}^{m}\{1 + (v_k - v_{0k})^2 I(v_{0k}) + o(|v_k - v_{0k}|^2)\} \\
&= \exp\left\{\sum_{k=1}^{m}[(v_k - v_{0k})^2 I(v_{0k}) + o(|v_k - v_{0k}|^2)]\right\}\,,
\end{aligned} \tag{2.3.19}$$

where $I(v_{0k})$ is Fisher's information corresponding to the random variable Y_k, $k = 1, \ldots, m$. (We again use here the fact that $\log(1 + x)$ equals to $x + o(x)$.) Therefore we can write

$$\mathbb{E}_{\boldsymbol{v}_0}\{W^2\} = \exp\left\{\sum_{k=1}^{m}[\alpha_k^2 J(v_{0k}) + o(\alpha_k^2)]\right\} , \tag{2.3.20}$$

where $\alpha_k = (\Delta v_k)/v_{0k}$, $\Delta v_k = v_k - v_{0k}$, and $J(v_{0k}) = v_{0k}^2 I(v_{0k})$, $k = 1, \ldots, m$. We consider several examples. Let

(i) $Y_k \sim \exp(\lambda_k)$. Then $I(\lambda_{0k}) = \lambda_{0k}^{-2}$, and therefore

$$\mathbb{E}_{\boldsymbol{\lambda}_0}\{W^2\} \cong \exp\left\{\sum_{k=1}^{m} \alpha_k^2\right\} .$$

(ii) $Y_k \sim \text{Poisson}(\lambda_k)$. Then $I(\lambda_{0k}) = \lambda_{0k}^{-1}$ and

$$\mathbb{E}_{\boldsymbol{\lambda}_0}\{W^2\} \cong \exp\left\{\sum_{k=1}^{m} \alpha_k^2 \lambda_{0k}\right\}.$$

(iii) $Y_k \sim \text{Ber}(p_k)$. Then $I(p_{0k}) = [p_{0k}(1 - p_{0k})]^{-1}$, and

$$\mathbb{E}_{\boldsymbol{p}_0}\{W^2\} \cong \exp\left\{\sum_{k=1}^{m} \alpha_k^2 p_{0k}(1 - p_{0k})^{-1}\right\} .$$

In the particular case where all α_k and all v_{0k} are equal, say $\alpha_k = \alpha$ and $v_{0k} = u$, $k = 1, \ldots, m$, we obtain from (2.3.20) that

$$\mathbb{E}_{\boldsymbol{v}_0}\{W^2\} \cong \exp\{m\alpha^2 J(u)\} . \tag{2.3.21}$$

Note that in order for the variance of LW to be meaningful the parameter α should be of order $O(m^{-1/2})$ for large values of m.

Finally, consider the following *separable* case. Suppose that the sample performance function $L(\boldsymbol{y})$ can be written in the form

$$L(\boldsymbol{y}) = \sum_{k=1}^{m} L_k(y_k). \tag{2.3.22}$$

We have then that

$$\ell(\boldsymbol{v}) = \sum_{k=1}^{m} \ell_k(v_k) ,$$

where $\ell_k(v_k) = \mathbb{E}_{v_k}\{L_k(Y_k)\}$ and $Y_k \sim f_k(y_k, v_k)$, $k = 1, \ldots, m$. Assume furthermore that the components $Y_1, \ldots, Y_m$ of the random vector $\boldsymbol{Y}$ are independent random variables. We can then represent each $\ell_k(v_k)$, $k = 1,\dots,m$ as

$$\ell_k(v_k) = \mathbb{E}_{v_{0k}}\{L_k(Z_k)W_k(Z_k,v_k)\}\ , \tag{2.3.23}$$

where $Z_k \sim f_k(z_k, v_{0k})$ and

$$W_k(z_k,v_k) = f(z_k,v_k)/f(z_k,v_{0k})\ ,$$

and then estimate $\ell_k(v_k)$ by the corresponding "what if" estimator $\bar{\ell}_{kN}(v_k)$. Consequently, $\ell(\boldsymbol{v})$ can be estimated by $\sum_{k=1}^{m}\bar{\ell}_{kN}(v_k)$. Notice that the variance

$$\mathrm{Var}_{\boldsymbol{v}_0}\left\{\sum_{k=1}^{m}\bar{\ell}_{kN}(v_k)\right\} = \sum_{k=1}^{m}\mathrm{Var}_{v_{0k}}\left\{\bar{\ell}_{kN}(v_k)\right\}$$

of the so obtained estimator is typically much smaller than the corresponding variance of the "what if" estimator $\bar{\ell}_N(\boldsymbol{v})$. Also note that the covariance matrix of the gradient of $\sum_{k=1}^{m}\bar{\ell}_{kN}(v_k)$ is diagonal.

2.4 Optimization of DESS

In this section we consider the mathematical programming problem (P_0) introduced in section 1.1 of chapter 1. That is,

$$(\mathrm{P}_0) \qquad \begin{array}{lll} \text{minimize} & \ell_0(\boldsymbol{v}), & \boldsymbol{v}\in V, \\ \text{subject to} & \ell_j(\boldsymbol{v}) \le 0, & j=1,\dots,k, \\ & \ell_j(\boldsymbol{v}) = 0, & j=k+1,\dots,M, \end{array} \tag{2.4.1}$$

where

$$\ell_j(\boldsymbol{v}) = \mathbb{E}_{\boldsymbol{v}}\{L_j(\boldsymbol{Y})\} = \int L_j(\boldsymbol{y})f(\boldsymbol{y},\boldsymbol{v})\,\mathrm{d}\boldsymbol{y}, \tag{2.4.2}$$

$j = 0,1,\dots,M$. We assume that the objective function $\ell_0(\boldsymbol{v})$ and some of the constraints $\ell_j(\boldsymbol{v})$ are not available analytically and in order to solve (P_0) we need to resort either to deterministic numerical techniques or to the Monte Carlo method. We show that the SF method allows us to estimate an optimal solution of the program (P_0) from a single simulation experiment. The underlying idea is to replace the deterministic program (P_0) by what is called its *stochastic counterpart*, which then can be solved by standard techniques of mathematical programming. The obtained optimal solution will provide an estimator of the corresponding optimal solution of the "true" (original) program (P_0).

In order to construct the stochastic counterpart of (P_0) we argue as follows. Consider first the "what if" estimators of $\ell_j(\boldsymbol{v})$, that is

$$\bar{\ell}_{jN}(\boldsymbol{v}) = N^{-1}\sum_{i=1}^{N} L_j(\mathbf{Z}_i)W(\mathbf{Z}_i,\boldsymbol{v}) \ , \tag{2.4.3}$$

$j = 0,1,\ldots,M$, where $\mathbf{Z}_1,\ldots,\mathbf{Z}_N$ is a random sample from the dominating pdf $g(\mathbf{z})$ and

$$W(\mathbf{z},\boldsymbol{v}) = f(\mathbf{z},\boldsymbol{v})/g(\mathbf{z}).$$

Viewing next $\bar{\ell}_{jN}(\boldsymbol{v})$ as functions of $\boldsymbol{v}$ rather than estimators for fixed $\boldsymbol{v}$, we can define the stochastic counterpart of (P_0) as follows,

$$(\bar{P}_N) \qquad \begin{array}{lll} \text{minimize} & \bar{\ell}_{0N}(\boldsymbol{v}), & \boldsymbol{v} \in V, \\ \text{subject to} & \bar{\ell}_{jN}(\boldsymbol{v}) \le 0, & j = 1,\ldots,k, \\ & \ell_{jN}(\boldsymbol{v}) = 0, & j = k+1,\ldots,M. \end{array} \tag{2.4.4}$$

Notice that as soon as the sample $\mathbf{Z}_1,\ldots,\mathbf{Z}_N$, is generated, the functions $\bar{\ell}_{jN}(\boldsymbol{v})$, $j = 0,\ldots,M$, are given *explicitly* through the known density functions $f(\mathbf{Z}_i,\boldsymbol{v})$. The corresponding gradients $\nabla\bar{\ell}_{jN}(\boldsymbol{v})$ and Hessian matrices $\nabla^2\bar{\ell}_{jN}(\boldsymbol{v})$ can be calculated according to (2.2.29) from a single simulation by the SF method. Consequently the optimization problem $(\bar{P}_N)$ can be solved in principle by standard methods of mathematical programming. The obtained optimal value $\bar{\varphi}_N$ and the optimal solution $\bar{\boldsymbol{v}}_N$ of the program $(\bar{P}_N)$ provide estimators of the optimal value φ^* and the optimal solution $\boldsymbol{v}^*$ of the program (P_0), respectively. Note that all this is feasible since we assumed that the sample functions $L_j(\boldsymbol{y})$ *do not depend* on $\boldsymbol{v}$.

Example 2.4.1 Consider the optimization problem (2.1.3) of example 2.1.1. Its stochastic counterpart can be written as

$$(\bar{P}_N) \qquad \begin{array}{lll} \text{maximize} & \bar{\ell}_{0N}(\boldsymbol{v}), & \boldsymbol{v} \in V, \\ \text{subject to} & \ell_1(\boldsymbol{v}) \le 0, \ \bar{\ell}_{2N}(\boldsymbol{v}) \le 0 \ , & \end{array} \tag{2.4.5}$$

where

$$\begin{aligned} \bar{\ell}_{0N}(\boldsymbol{v}) &= N^{-1}\sum_{i=1}^{N} L(\mathbf{Z}_i)W(\mathbf{Z}_i,\boldsymbol{v}) \ , \\ \bar{\ell}_{2N}(\boldsymbol{v}) &= N^{-1}\sum_{i=1}^{N} L(\mathbf{Z}_i)^2 W(\mathbf{Z}_i,\boldsymbol{v}) - b_2 \ , \\ L(\mathbf{Z}_i) &= \max_{j=1,\ldots,p}\ \min_{k\in\mathcal{L}_j} Z_{ki} \end{aligned}$$

and the function $\ell_1(\boldsymbol{v})$ is the same as in example 2.1.1. Note that the only difference between (2.4.5) and (2.1.3) is that $\ell_0(\boldsymbol{v})$ and $\ell_2(\boldsymbol{v})$ in (2.1.3) are replaced by $\bar{\ell}_{0N}(\boldsymbol{v})$ and $\bar{\ell}_{2N}(\boldsymbol{v})$, while the deterministic constraint function $\ell_1(\boldsymbol{v})$ remains the same.

The algorithm for estimating the optimal solution $\boldsymbol{v}^*$ of the program (P_0) while using the stochastic counterpart $(\bar{\mathrm{P}}_N)$ can be written as follows.

Algorithm 2.4.1 :

1. *Generate a random sample* $\mathbf{Z}_1, \ldots, \mathbf{Z}_N$ *from* $g(\mathbf{z})$.
2. *Calculate* $L_j(\mathbf{Z}_i)$, $j = 0, \ldots, M;\ i = 1, \ldots, N$.
3. *Solve* $(\bar{\mathrm{P}}_N)$ *by methods of mathematical programming.*
4. *Deliver the obtained optimal solution* $\bar{\boldsymbol{v}}_N$ *of* $(\bar{\mathrm{P}}_N)$ *as an estimator of* $\boldsymbol{v}^*$.

Notice that the third step of the above algorithm (the solution of the optimization problem $(\bar{\mathrm{P}}_N)$) typically requires iterative numerical procedures, which in turn assumes calculation of the functions $\bar{\ell}_{jN}(\boldsymbol{v})$, $j = 0, \ldots, M$, and possibly their gradients and Hessians for different values of the parameter vector $\boldsymbol{v}$. Notice also that the CPU time of the program $(\bar{\mathrm{P}}_N)$ increases in the sample size N. Our extensive simulation studies with typical DES show that $\bar{\boldsymbol{v}}_N$ of $(\bar{\mathrm{P}}_N)$ presents a reliable estimator of the optimal solution $\boldsymbol{v}^*$, provided the sample size N is of the order of 1000 or more, and the program $(\bar{\mathrm{P}}_N)$ is convex. For nonconvex programs we can use methods of global Monte Carlo optimization (e.g. Katkovnik (1976), Katkovnik and Antonov (1972) and Rubinstein (1986)).

To provide a better insight into the estimators $\bar{\boldsymbol{v}}_N$ and their statistical properties let us consider the unconstrained case. In this case the program (P_0) reduces to

$$(\mathrm{P}_0) \qquad \text{minimize } \ell(\boldsymbol{v}), \quad \boldsymbol{v} \in \mathrm{V}, \tag{2.4.6}$$

with $\ell(\boldsymbol{v}) = \mathbb{E}_{\boldsymbol{v}}\{L(\boldsymbol{Y})\}$. Suppose for a moment that the set V is open and that $\ell(\boldsymbol{v})$ is continuously differentiable on V. Then, by the first-order necessary conditions, the gradient of $\ell(\boldsymbol{v})$ at the optimal solution point $\boldsymbol{v}^*$ must be zero. Therefore in this case one may try to solve the system of equations $\nabla \ell(\boldsymbol{v}) = \mathbf{0}$ and to search for the minimizer $\boldsymbol{v}^*$ among the solutions of these equations. The stochastic counterpart of (2.4.6) is the program

$$(\bar{\mathrm{P}}_N) \qquad \text{minimize } \bar{\ell}_N(\boldsymbol{v}), \quad \boldsymbol{v} \in V, \tag{2.4.7}$$

where for any fixed $\boldsymbol{v}$, $\bar{\ell}_N(\boldsymbol{v})$ is the "what if" estimator of $\ell(\boldsymbol{v})$. Similarly to the case of program (2.4.6), the optimal solution $\bar{\boldsymbol{v}}_N$ of (2.4.7) can be found among the solutions of the corresponding system of equations

$$\nabla \bar{\ell}_N(\boldsymbol{v}) = \mathbf{0}, \quad \boldsymbol{v} \in V, \tag{2.4.8}$$

which represents the stochastic counterpart of the deterministic system $\nabla \ell(\boldsymbol{v}) = \mathbf{0}$.

In the statistical literature optimal solutions of the program (2.4.7) correspond to the so called M-estimators (Huber (1964), (1981)), which are extensions of the conventional maximum likelihood estimators. The statistical properties of the M-estimators have been studied in Huber (1967). Note that the traditional M-estimators used in statistics differ in their main goals from the estimators $\bar{\boldsymbol{v}}_N$ considered here. In the first case the M-estimators are employed to estimate the unknown parameters while the estimators $\bar{\boldsymbol{v}}_N$ are used to evaluate the unknown optimal solution $\boldsymbol{v}^*$ of the program (P_0).

In the following theorem we summarize the basic statistical properties of $\bar{\boldsymbol{v}}_N$ (consistency, asymptotic normality). Proofs and additional technical details are given in appendix A.

Theorem 2.4.1 Let $\boldsymbol{v}^*$ be a unique minimizer of $\ell(\boldsymbol{v})$ over V.
(a) Suppose that

(i) *The set V is compact.*

(ii) *For almost every $\boldsymbol{y}$, the function $f(\boldsymbol{y}, \cdot)$ is continuous on V*

(iii) *The family $\{|L(\boldsymbol{y})f(\boldsymbol{y}, \boldsymbol{v})|, \boldsymbol{v} \in V\}$ is dominated by an integrable (with respect to the Lebesgue measure)* function $h(\boldsymbol{y})$, *i.e.*

$$|L(\boldsymbol{y})f(\boldsymbol{y}, \boldsymbol{v})| \leq h(\boldsymbol{y}) \quad \textit{for all } \boldsymbol{v} \in V.$$

Then an optimal solution $\bar{\boldsymbol{v}}_N$ of (2.4.7) *converges with probability one to $\boldsymbol{v}^*$ as $N \to \infty$.*
(b) Furthermore, suppose that

(i) *$\boldsymbol{v}^*$ is an interior point of V.*

(ii) *For almost every $\boldsymbol{y}, f(\boldsymbol{y}, \cdot)$ is twice continuously differentiable in a neighborhood U of $\boldsymbol{v}^*$ and the families $\{\|L(\boldsymbol{y})\nabla^s f(\boldsymbol{y}, \boldsymbol{v})\|, \boldsymbol{v} \in U\}$, $s = 1, 2$, are dominated by integrable functions.*

(iii) *The matrix*

$$\boldsymbol{B} = \mathbb{E}_g\{L(\mathbf{Z})\nabla^2 W(\mathbf{Z}, \boldsymbol{v}^*)\} \tag{2.4.9}$$

is nonsingular.

(iv) *The covariance matrix*

$$\mathrm{Cov}\{L(\mathbf{Z})\nabla W(\mathbf{Z},\boldsymbol{v}^*)\} = \mathbb{E}_g\{L(\mathbf{Z})^2\nabla W(\mathbf{Z},\boldsymbol{v}^*)\nabla W(\mathbf{Z},\boldsymbol{v}^*)'\} \quad (2.4.10)$$

exists.

Then the random variable $N^{1/2}(\bar{\boldsymbol{v}}_N - \boldsymbol{v}^*)$ converges in distribution to a normal random variable with zero mean and the covariance matrix

$$\mathrm{Cov}\{N^{1/2}(\bar{\boldsymbol{v}}_N - \boldsymbol{v}^*)\} \approx \boldsymbol{B}^{-1}\mathrm{Cov}\{L(\mathbf{Z})\nabla W(\mathbf{Z},\boldsymbol{v}^*)\}\boldsymbol{B}^{-1}. \quad (2.4.11)$$

Some remarks about theorem 2.4.1 are now in order. In statistics an estimator is called weakly and strongly consistent if it converges to the "true" unknown parameter vector in probability and with probability one, respectively. In part (a) of theorem 2.4.1 it is stated that, under assumptions (i)–(iii), $\bar{\boldsymbol{v}}_N$ is a strongly consistent estimator of $\boldsymbol{v}^*$. Assumption (i) of compactness of V is required in order to prevent the possible escaping of $\bar{\boldsymbol{v}}_N$ to infinity. (Recall that a subset of a finite-dimensional vector space is compact if and only if it is bounded and closed.) Often this can be verified by ad hoc methods (cf. Huber (1967)). Assumption (iii) is needed to ensure the *uniform* convergence w.p.1 of $\bar{\ell}_N(\boldsymbol{v})$ to $\ell(\boldsymbol{v})$ (see also lemma A.1 of appendix A). In assumption (i) of part (b) we suppose that $\boldsymbol{v}^*$ is an interior point of V. This means that there is a neighborhood U of $\boldsymbol{v}^*$ such that $U \subset V$. Then, by the consistency of $\bar{\boldsymbol{v}}_N$, we have that w.p.1, for N large enough, $\bar{\boldsymbol{v}}_N$ lies in U and hence the local behavior of $\bar{\boldsymbol{v}}_N$ is not effected by the boundary of the parameter set V. We also have that $\nabla\ell(\boldsymbol{v}^*) = \mathbf{0}$ and that $\nabla\bar{\ell}_N(\bar{\boldsymbol{v}}_N) = \mathbf{0}$ if $\bar{\boldsymbol{v}}_N$ lies in the interior of V. Notice that $\mathbb{E}_g\{L(\mathbf{Z})\nabla W(\mathbf{Z},\boldsymbol{v}^*)\}$ is equal to $\nabla\ell(\boldsymbol{v}^*)$ and hence is zero. Therefore it does not appear in the expression for the covariance matrix given by the right-hand side of (2.4.10). Because of the assumption (ii) of part (b) the second-order derivatives can be taken inside the expected value and therefore the matrix $\boldsymbol{B}$, given in (2.4.9), is equal to the Hessian matrix $\nabla^2\ell(\boldsymbol{v}^*)$. Finally notice that the covariance matrix given by the right-hand side of (2.4.11) represents the covariance matrix of the *limiting* distribution of $N^{1/2}(\bar{\boldsymbol{v}}_N - \boldsymbol{v}^*)$. Therefore the symbol "$\approx$" appears.

It can be shown (see appendix A) that under the assumptions of theorem 2.4.1, the estimator $\bar{\boldsymbol{v}}_N$, satisfying the first-order necessary conditions $\nabla\bar{\ell}_N(\bar{\boldsymbol{v}}_N) = \mathbf{0}$, is *unique* w.p.1 for sufficiently large N. Note, however, that if the sample size N is small, the stochastic counterpart (2.4.8) may have multiple roots. This means that for small N a difficulty of identifying an appropriate solution $\bar{\boldsymbol{v}}_N$ may arise. The situation is simplified considerably if the program (2.4.7) is *convex* regardless the sample size N, i.e. the set V and the function $\bar{\ell}_N(\boldsymbol{v})$ are convex. In this case the first-order necessary conditions $\nabla\bar{\ell}_N(\bar{\boldsymbol{v}}_N) = \mathbf{0}$ are also sufficient for $\bar{\boldsymbol{v}}_N$ to be an optimal solution of

(2.4.7) provided $\bar{\boldsymbol{v}}_N$ is an interior point of V. Also the optimal solution $\bar{\boldsymbol{v}}_N$ is unique if, in addition, $\bar{\ell}_N(\boldsymbol{v})$ is *strictly* convex. We leave this as an exercise to show that $\ell(\boldsymbol{v})$ and $\bar{\ell}_N(\boldsymbol{v})$ are convex if for every $\boldsymbol{y}$ the function $L(\boldsymbol{y})f(\boldsymbol{y},\cdot)$ is convex (see exercise 2.8). Notice that if $L(\boldsymbol{y}) > 0$, then $L(\boldsymbol{y})f(\boldsymbol{y},\cdot)$ is convex if and only if $f(\boldsymbol{y},\cdot)$ is convex. Not all pdf's of exponential families satisfy this condition. For example, the exponential pdf $f(y,v) = v\exp(-vy)$ is not convex in $v \in \mathbb{R}$. Consequently $\ell(v)$ can be nonconvex. As an example consider $L(y) = y\exp(-y)$, $y \geq 0$. Then $\ell(v) = v/(v+1)^2$, which is not convex.

The asymptotic efficiency of the estimator is determined by the covariance matrix given in the right-hand side of (2.4.11). Under the assumptions of theorem 4.1, this covariance matrix can be consistently estimated by $\mathbf{B}_N^{-1}\boldsymbol{\Sigma}_N\boldsymbol{B}_N^{-1}$, where

$$\boldsymbol{B}_N = N^{-1}\sum_{i=1}^{N} L(\mathbf{Z}_i)\nabla^2 W(\mathbf{Z}_i, \bar{\boldsymbol{v}}_N) \tag{2.4.12}$$

and

$$\boldsymbol{\Sigma}_N = N^{-1}\sum_{i=1}^{N} L(\mathbf{Z}_i)^2\nabla W(\mathbf{Z}_i, \bar{\boldsymbol{v}}_N)\nabla W(\mathbf{Z}_i, \bar{\boldsymbol{v}}_N)', \tag{2.4.13}$$

provide consistent estimators of the matrices $\boldsymbol{B}$ and $\mathrm{Cov}\{L(\mathbf{Z})\nabla W(\mathbf{Z}, \boldsymbol{v}^*)\}$, respectively. Notice that these matrices can be estimated from the same sample $\mathbf{Z}_1, \ldots, \mathbf{Z}_N$ simultaneously with the estimator $\bar{\boldsymbol{v}}_N$.

Note also that the matrix $\boldsymbol{B}$ is equal to the Hessian matrix $\nabla^2\ell(\boldsymbol{v}^*)$ and therefore is independent of a particular choice of the pdf $g(\mathbf{z})$. On the other hand, the covariance matrix given in (2.4.10) can be very sensitive to the choice of $g(\mathbf{z})$. Note finally that all corresponding results of the previous section are directly applicable to this covariance matrix as well. In particular for the exponential family (2.3.10) and $g(\boldsymbol{y}) = f(\boldsymbol{y}, \boldsymbol{v}_0)$, $\boldsymbol{v}_0 \in V$, we have (see (2.3.16))

$$\begin{aligned} &\mathrm{Cov}_{\boldsymbol{v}_0}\{L(\mathbf{Z})\nabla W(\mathbf{Z}, \boldsymbol{v}^*)\} \\ &= \mathbb{E}_{\boldsymbol{v}_0}\{W(\mathbf{Z}, \boldsymbol{v}^*)^2\}\mathbb{E}_{\boldsymbol{v}_0+2\Delta\boldsymbol{v}}\{L(\mathbf{Z})^2\boldsymbol{S}^{(1)}(\mathbf{Z}, \boldsymbol{v}^*)\boldsymbol{S}^{(1)}(\mathbf{Z}, \boldsymbol{v}^*)'\}, \end{aligned} \tag{2.4.14}$$

where $\Delta\boldsymbol{v} = \boldsymbol{v}^* - \boldsymbol{v}_0$ and $\boldsymbol{S}^{(1)}(\mathbf{z}, \boldsymbol{v})$ is given in (2.3.15). In the case of gamma distributions with $\boldsymbol{v} = (\lambda_1, \ldots, \lambda_m)'$, formula (2.4.14) reduces to (see (2.3.4) and (2.3.9))

$$\begin{aligned} &\mathrm{Cov}_{\boldsymbol{\lambda}_0}\{L(\mathbf{Z})\nabla W(\mathbf{Z}, \boldsymbol{\lambda}^*)\} \\ &= \mathbb{E}_{\boldsymbol{\lambda}_0}\{W(\mathbf{Z}, \boldsymbol{\lambda}^*)^2\}\mathbb{E}_{\boldsymbol{\lambda}_0+2\Delta\boldsymbol{\lambda}}\{L(\mathbf{Z})^2(\beta\boldsymbol{\lambda}^{*^{-1}} - \mathbf{Z})(\beta\boldsymbol{\lambda}^{*^{-1}} - \mathbf{Z})'\}, \end{aligned} \tag{2.4.15}$$

where

$$\mathbb{E}_{\boldsymbol{\lambda}_0}\{W(\mathbf{Z},\boldsymbol{\lambda}^*)^2\} = \prod_{k=1}^{m}\left(1+\frac{\alpha_k^{*2}}{1+2\alpha_k^*}\right)^{\beta_k} \qquad (2.4.16)$$

and $\alpha_k^* = (\lambda_k^* - \lambda_{0k})/\lambda_{0k}$.

Note that formula (2.4.14) is derived under the assumption that $\boldsymbol{v}_0 + 2(\boldsymbol{v}^* - \boldsymbol{v}_0)$ lies in the permissible set V. In the case of gamma distributions this leads to the condition (see (2.3.7))

$$\lambda_k^* > \frac{1}{2}\lambda_{0k}\,, \quad k = 1,\ldots,m.$$

We present now numerical results for the unconstrained minimization problem (2.4.6) with

$$\ell(\boldsymbol{v}) = \mathbb{E}_{\boldsymbol{v}}\{L(\boldsymbol{Y})\} + \psi(\boldsymbol{v})\,, \qquad (2.4.17)$$

where the components $Y_1,\ldots,Y_n$ of $\boldsymbol{Y}$ are independent, $Y_k \sim \exp(v_k)$, $k = 1,\ldots,n$,

$$L(\boldsymbol{Y}) = n\min\{Y_1,\ldots,Y_n\} \qquad (2.4.18)$$

and

$$\psi(\boldsymbol{v}) = n^{-1}b\sum_{k=1}^{n} v_k^3\,. \qquad (2.4.19)$$

Notice that here the parameter set V is given by $V = \mathbb{R}_+^n = \{\boldsymbol{v} = (v_1,\ldots,v_n) : v_k \geq 0,\, k = 1,\ldots,n\}$. For the considered problem the stochastic counterpart of $\nabla\ell(\boldsymbol{v}) = \mathbf{0}$ can be written as

$$\nabla\bar{\ell}_N(\boldsymbol{v}) = N^{-1}\sum_{i=1}^{N} L(\boldsymbol{Z}_i)\nabla W(\boldsymbol{Z}_i,\boldsymbol{v}) + 3n^{-1}b\boldsymbol{v}^2 = \mathbf{0}\,, \qquad (2.4.20)$$

where

$$\nabla W(\mathbf{z},\boldsymbol{v}) = \frac{\nabla f(\mathbf{z},\boldsymbol{v})}{f(\mathbf{z},\boldsymbol{v}_0)} = (\boldsymbol{v}^{-1} - \mathbf{z})\prod_{k=1}^{n}\frac{v_k}{v_{0k}}\exp\left\{-\sum_{k=1}^{n}(v_k - v_{0k})z_k\right\}\,,$$

$\boldsymbol{v}^2 = (v_1^2,\ldots,v_n^2)$ and $\boldsymbol{v}^{-1} = (v_1^{-1},\ldots,v_n^{-1})$.

Consider the minimizer $\boldsymbol{v}^*$ of $\ell(\boldsymbol{v})$ over V and the corresponding SF estimator $\bar{\boldsymbol{v}}_N$ obtained as a solution of the system of equations (2.4.20). Notice that in this case all components v_k^* of $\boldsymbol{v}^*$ are equal to each other and are denoted v^*. Table 2.4.1 presents the theoretical values of v^*, the point estimators, $\bar{v}_{1N}$ of the first component of $\boldsymbol{v}^*$, the corresponding sample variances $\hat{\sigma}^2$ and the 95% confidence intervals for $\boldsymbol{v}^*$ as functions of b and the relative perturbations $\alpha_k = (v_{0k} - v_k^*)/v_k^*)$ for $n = 5$. We chose the reference parameter vector $\boldsymbol{v}_0$ with all components equal to one, i.e. $v_{0k} = 1$,

$k = 1, \ldots, 5$; assumed $\alpha_k = \alpha$ and took the sample size $N = 1000$. Note that *all* estimators were obtained *simultaneously from a single simulation run* by solving the system of equations (2.4.20) for different values of b, provided the reference parameters of the vector $\boldsymbol{v}_0$ are fixed and equal to one.

Table 2.4.1 Performance of the SF method for the unconstrained minimization problem with $\ell(\boldsymbol{v})$ defined in (2.4.17)–(2.4.19).

b	α	v^*	$\bar{v}_{1N}$	$\hat{\sigma}^2$	Confidence intervals
2.441	-0.250	0.800	0.852	0.34	(0.816, 0.887)
1.000	0.000	1.000	1.004	0.11	(0.984, 1.024)
0.482	0.167	1.200	1.190	0.10	(1.170, 1.210)
0.153	0.375	1.600	1.584	0.21	(1.557, 1.609)
0.063	0.500	2.000	1.980	0.29	(1.948, 2.012)

We found that the SF method performs better for positive perturbations ($\alpha > 0$) rather than for negative perturbations ($\alpha < 0$), (see table 2.4.1 and also chapter 4 below). We also found that for $\alpha \le -0.5$ the SF method fails to work. This is in agreement with the theoretical results of section 2.3 and in particular with (2.3.7).

2.5 Variance Reduction Techniques

In this section we apply the (a) *linear* and (b) *nonlinear* control random variables techniques (see subsection 1.3.2) to improve the accuracy (reduce the variance) of the SF "what if" estimators $\nabla^k \bar{\ell}_N(\boldsymbol{v}),\ k = 0, 1, \ldots$. We start with

(a) **Linear control random variables (LCRV).**

For a given $\boldsymbol{v} \in V$ consider the random variables $L(\mathbf{Z})W(\mathbf{Z}, \boldsymbol{v})$ and $W(\mathbf{Z}, \boldsymbol{v})$, where $\mathbf{Z} \sim g(\mathbf{z})$. Since the expectation of $W(\mathbf{Z}, \boldsymbol{v})$ is equal to one, we can take $W(\mathbf{Z}, \boldsymbol{v})$ as a control variate for $L(\boldsymbol{Z})W(\boldsymbol{Z}, \boldsymbol{v})$. This leads to the following linear control random variable:

$$X(\alpha) = L(\boldsymbol{Z})W(\boldsymbol{Z}, \boldsymbol{v}) - \alpha[W(\boldsymbol{Z}, \boldsymbol{v}) - 1]\,. \tag{2.5.1}$$

Taking into account that for any $\alpha \in \mathbb{R}$ the expectation of $X(\alpha)$ is $\ell(\boldsymbol{v})$, we obtain that

$$\hat{\ell}_N(\boldsymbol{v}, \alpha) \quad = \quad N^{-1} \sum_{i=1}^{N} \{L(\boldsymbol{Z}_i)W(\boldsymbol{Z}_i, \boldsymbol{v}) - \alpha[W(\mathbf{Z}_i, \boldsymbol{v}) - 1]\}$$

$$= \bar{\ell}_N(\boldsymbol{v}) - \alpha(\bar{W}_N - 1) \tag{2.5.2}$$

is an unbiased estimator of $\ell(\boldsymbol{v})$. Here $\boldsymbol{Z}_1, \ldots, \boldsymbol{Z}_N$ is a sample from the pdf $g(\mathbf{z})$ and

$$\bar{W}_N = \bar{W}_N(\boldsymbol{v}) = N^{-1} \sum_{i=1}^{N} W(\boldsymbol{Z}_i, \boldsymbol{v}) .$$

The value α^* of the parameter α which minimizes the variance of $X(\alpha)$ and hence the variance of $\hat{\ell}_N(\boldsymbol{v}, \alpha)$, is then

$$\alpha^* = \frac{\text{Cov}\{LW, W\}}{\text{Var}\{W\}} \tag{2.5.3}$$

(see equation (1.3.12) of chapter 1).

Notice that since neither Var$\{W\}$ nor Cov$\{LW,\ W\}$ are typically available we can estimate them from the same sample $\boldsymbol{Z}_1, \ldots, \boldsymbol{Z}_N$. The resulting consistent estimator of α^* can be written as

$$\hat{\alpha}_N^* = \frac{N^{-1} \sum_{i=1}^{N} L(\boldsymbol{Z}_i) W(\boldsymbol{Z}_i, \boldsymbol{v})^2 - \bar{\ell}_N(\boldsymbol{v})}{N^{-1} \sum_{i=1}^{N} W(\boldsymbol{Z}_i, \boldsymbol{v})^2 - 1} \tag{2.5.4}$$

and thus the resulting LCRV estimator of $\ell(\boldsymbol{v})$ is given by $\hat{\ell}_N(\boldsymbol{v}, \hat{\alpha}_N^*)$.

We now show that the LCRV estimator is typically more accurate than the "what if" estimator $\bar{\ell}_N(\boldsymbol{v})$. To see this, note first that since $\hat{\alpha}_N^*$ is a consistent estimator of α^* we can write $\hat{\alpha}_N^* = \alpha^* + o_p(1)$, where $o_p(1)$ denotes a random variable converging in probability to zero as $N \to \infty$ (see appendix A for a discussion of the notation o_p and O_p). We have then that

$$\hat{\alpha}_N^*(\bar{W}_N - 1) = \alpha^*(\bar{W}_N - 1) + o_p(1)(\bar{W}_N - 1) . \tag{2.5.5}$$

Note further that by the Central Limit Theorem, $N^{1/2}(\bar{W}_N - 1)$ converges in distribution (to a normal) and hence is bounded in probability. It follows that the last term in the right-hand side of (2.5.5) is of stochastic order $o_p(N^{-1/2})$ and hence

$$\hat{\ell}_N(\boldsymbol{v}, \hat{\alpha}_N^*) = \hat{\ell}_N(\boldsymbol{v}, \alpha^*) + o_p(N^{-1/2}) . \tag{2.5.6}$$

Applying finally the Central Limit Theorem to $\hat{\ell}_N(\boldsymbol{v}, \alpha^*)$ and using (2.5.6) we obtain that $N^{1/2}[\hat{\ell}_N(\boldsymbol{v}, \hat{\alpha}_N^*) - \ell(\boldsymbol{v})]$ converges in distribution to a normal with the mean zero and the variance Var$\{X(\alpha^*)\}$, where (see equation (1.3.13) of chapter 1)

$$\text{Var}\{X(\alpha^*)\} = (1 - \text{Corr}\{LW, W\}^2)\ \text{Var}\{LW\} . \tag{2.5.7}$$

Here Corr$\{LW, W\}$ denotes the correlation coefficient between LW and W. This shows that the asymptotic variance of the LCRV estimator

$\hat{\ell}_N(\boldsymbol{v}, \hat{\alpha}_N^*)$ will be considerably smaller than the asymptotic variance of the corresponding "what if" estimator $\bar{\ell}_N(\boldsymbol{v})$ provided the two random variables $L(\boldsymbol{Z})W(\boldsymbol{Z}, \boldsymbol{v})$ and $W(\boldsymbol{Z}, \boldsymbol{v})$ are highly correlated.

Let us extend now the above analysis to the SF estimators $\nabla\bar{\ell}_N(\boldsymbol{v})$. An extension to higher order derivatives is similar. Taking into account that $\mathbb{E}\{\nabla W(\boldsymbol{Z}, \boldsymbol{v})\} = 0$ we can define the following LCRV for $L(\boldsymbol{Z})\nabla W(\boldsymbol{Z}, \boldsymbol{v})$:

$$\boldsymbol{X}(\boldsymbol{A}) = L(\boldsymbol{Z})\nabla W(\boldsymbol{Z}, \boldsymbol{v}) - \boldsymbol{A}\nabla W(\boldsymbol{Z}, \boldsymbol{v}) \,, \tag{2.5.8}$$

where $\boldsymbol{A}$ is an $n \times n$ matrix. The corresponding unbiased estimator of $\nabla\ell(\boldsymbol{v})$ is

$$\begin{aligned} \hat{\nabla}\ell_N(\boldsymbol{v}, \boldsymbol{A}) &= N^{-1}\sum_{i=1}^{N}\{L(\boldsymbol{Z}_i)\nabla W(\boldsymbol{Z}_i, \boldsymbol{v}) - \boldsymbol{A}\nabla W(\boldsymbol{Z}_i, \boldsymbol{v})\} \\ &= \nabla\bar{\ell}_N(\boldsymbol{v}) - \boldsymbol{A}\nabla\bar{W}_N(\boldsymbol{v}). \end{aligned} \tag{2.5.9}$$

The matrix $\boldsymbol{A}^*$ which minimizes the covariance matrices of $\boldsymbol{X}(\boldsymbol{A})$ and $\hat{\nabla}\ell_N(\boldsymbol{v}, \boldsymbol{A})$ (in the Loewner sense of inequality), is given (see proposition 1.1 of chapter 1) by

$$\boldsymbol{A}^* = \text{Cov}\{L\nabla W,\ \nabla W\}[\text{Cov}\{\nabla W\}]^{-1} \,. \tag{2.5.10}$$

Notice that since L is a scalar the matrix

$$\text{Cov}\{L\nabla W,\ \nabla W\} = \mathbb{E}_g\{L(\nabla W)\,(\nabla W)'\}$$

of covariances between $L(\mathbf{Z})\nabla W(\boldsymbol{Z}, \boldsymbol{v})$ and $\nabla W(\boldsymbol{Z}, \boldsymbol{v})$ is symmetric. The matrix $\boldsymbol{A}^*$ can be consistently estimated by

$$\hat{\boldsymbol{A}}_N^* = \left[\sum_{i=1}^{N} L_i(\nabla W_i)(\nabla W_i)'\right]\left[\sum_{i=1}^{N}(\nabla W_i)(\nabla W_i)'\right]^{-1} \,, \tag{2.5.11}$$

where $L_i = L(\boldsymbol{Z}_i)$ and $\nabla W_i = \nabla W(\boldsymbol{Z}_i, \boldsymbol{v})$. Thus the LCRV estimator of $\nabla\ell(\boldsymbol{v})$ is given by $\hat{\nabla}\ell_N(\boldsymbol{v}, \hat{\boldsymbol{A}}_N^*)$. Similarly to (2.5.6) we have that $\hat{\nabla}\ell_N(\boldsymbol{v}, \hat{\boldsymbol{A}}_N^*)$ is asymptotically equivalent to $\hat{\nabla}\ell_N(\boldsymbol{v}, \boldsymbol{A}^*)$ and, by the Central Limit Theorem, $N^{1/2}[\hat{\nabla}\ell_N(\boldsymbol{v}, \hat{\boldsymbol{A}}_N^*) - \nabla\ell(\boldsymbol{v})]$ converges in distribution to a normal with zero mean and the covariance matrix $\text{Cov}\{\boldsymbol{X}(\boldsymbol{A})^*)\}$ given by

$$\begin{aligned} &\text{Cov}\{\boldsymbol{X}(\boldsymbol{A}^*)\} \\ &= \text{Cov}\{L\nabla W\} - \text{Cov}\{L\nabla W, \nabla W\}[\text{Cov}\{\nabla W\}]^{-1}\,\text{Cov}\{L\nabla W, \nabla W\}. \end{aligned} \tag{2.5.12}$$

Let us apply now the LCRV techniques to the unconstrained optimization problem (P_0) given in (2.4.6). It is readily seen that in this case the stochastic counterpart of the deterministic system $\nabla\ell(\boldsymbol{v}) = \mathbf{0}$ can be written as follows:

$$\hat{\nabla}\ell_N(\boldsymbol{v}, \boldsymbol{A}) = \nabla\bar{\ell}_N(\boldsymbol{v}) - \boldsymbol{A}\nabla\bar{W}_N(\boldsymbol{v}) = \mathbf{0}, \quad \boldsymbol{v} \in V \,. \tag{2.5.13}$$

Let $\hat{\boldsymbol{v}}_N$ be a solution of the system (2.5.13). Notice that for $\boldsymbol{A} = \boldsymbol{0}$, the system (2.5.13) coincides with (2.4.8) and $\hat{\boldsymbol{v}}_N = \bar{\boldsymbol{v}}_N$.

The statistical properties of $\hat{\boldsymbol{v}}_N$ are similar to those of $\bar{\boldsymbol{v}}_N$. In particular, under regularity conditions similar to those specified in theorem 2.4.1, $\hat{\boldsymbol{v}}_N$ is a consistent estimator of $\boldsymbol{v}^*$ and $N^{1/2}(\hat{\boldsymbol{v}}_N - \boldsymbol{v}^*)$ converges in distribution to a normal with zero mean and the covariance matrix

$$\text{Cov}\{N^{1/2}(\hat{\boldsymbol{v}}_N - \boldsymbol{v}^*)\} \approx \boldsymbol{B}^{-1}\,\text{Cov}\{\boldsymbol{X}(\boldsymbol{A})\}\boldsymbol{B}^{-1}\,, \tag{2.5.14}$$

where

$$\boldsymbol{X}(\boldsymbol{A}) = L(\boldsymbol{Z})\nabla W(\boldsymbol{Z},\boldsymbol{v}^*) - \boldsymbol{A}\nabla W(\boldsymbol{Z},\boldsymbol{v}^*)$$

and $\boldsymbol{B} = \nabla^2 \ell(\boldsymbol{v}^*)$. Notice that the asymptotic covariance matrix of $\hat{\boldsymbol{v}}_N$ in (2.5.14) differs from the corresponding asymptotic covariance matrix of $\bar{\boldsymbol{v}}_N$ in (2.4.11) in that $\text{Cov}\{L\nabla W\}$ is replaced by $\text{Cov}\{\boldsymbol{X}(\boldsymbol{A})\}$. If we take now $\boldsymbol{A} = \boldsymbol{A}^*$, where $\boldsymbol{A}^*$ is given in (2.5.10) with $\boldsymbol{v} = \boldsymbol{v}^*$, we shall minimize the covariance matrix $\text{Cov}\{\boldsymbol{X}(\boldsymbol{A})\}$, in the Loewner sense of inequality, from which it follows that $\boldsymbol{A}^*$ minimizes the matrix $\boldsymbol{B}^{-1}\,\text{Cov}\{\boldsymbol{X}(\boldsymbol{A})\}\boldsymbol{B}^{-1}$ as well. That is, the inequality

$$\boldsymbol{a}'\boldsymbol{B}^{-1}\,\text{Cov}\{\boldsymbol{X}(\boldsymbol{A})\}\boldsymbol{B}^{-1}\boldsymbol{a} \geq \boldsymbol{a}'\boldsymbol{B}^{-1}\,\text{Cov}\{\boldsymbol{X}(\boldsymbol{A}^*)\}\boldsymbol{B}^{-1}\boldsymbol{a} \tag{2.5.15}$$

holds for any vector $\boldsymbol{a}$ and matrix $\boldsymbol{A}$. In particular (2.5.15) holds for $\boldsymbol{A} = \boldsymbol{0}$. This means that for any vector $\boldsymbol{a}$ the asymptotic variance of $\boldsymbol{a}'\hat{\boldsymbol{v}}_N$ is less than or equal to the asymptotic variance of $\boldsymbol{a}'\bar{\boldsymbol{v}}_N$. Notice that both $\boldsymbol{a}'\hat{\boldsymbol{v}}_N$ and $\boldsymbol{a}'\bar{\boldsymbol{v}}_N$ are consistent estimators of $\boldsymbol{a}'\boldsymbol{v}^*$.

The above aymptotic results still hold if the, generally unknown, matrix $\boldsymbol{A}^*$ is replaced by its consistent estimator. For example, we can use the estimator $\hat{\boldsymbol{A}}^*_N$ defined in (2.5.11) for $\boldsymbol{v} = \tilde{\boldsymbol{v}}_N$ with $\tilde{\boldsymbol{v}}_N$ being a consistent estimator of $\boldsymbol{v}^*$. In this case the stochastic counterpart of $\nabla \ell(\boldsymbol{v}) = \boldsymbol{0}$ can be written as

$$\hat{\nabla}\ell_N(\boldsymbol{v},\hat{\boldsymbol{A}}^*_N) = \nabla\bar{\ell}_N(\boldsymbol{v}) - \hat{\boldsymbol{A}}^*_N\nabla\bar{W}_N(\boldsymbol{v}) = \boldsymbol{0}\,,\quad \boldsymbol{v}\in V\,. \tag{2.5.16}$$

We have then that $N^{1/2}(\hat{\boldsymbol{v}}_N - \boldsymbol{v}^*)$ is asymptotically normal with

$$\text{Cov}\{N^{1/2}(\hat{\boldsymbol{v}}_N - \boldsymbol{v}^*)\} \approx \boldsymbol{B}^{-1}\,\text{Cov}\{\boldsymbol{X}(\boldsymbol{A}^*)\}\boldsymbol{B}^{-1}\,, \tag{2.5.17}$$

where the covariance matrix of $\boldsymbol{X}(\boldsymbol{A}^*)$ is given in (2.5.12). It is important to note that the root $\hat{\boldsymbol{v}}_N = \hat{\boldsymbol{v}}_N(\hat{\boldsymbol{A}}^*_N)$ of the system of equations (2.5.16) is asymptotically the *best LCRV estimator* of $\boldsymbol{v}^*$ among the LCRV estimators obtained as solutions of the equations (2.5.13).

(b) **Nonlinear control random variables (NCRV)**

Taking into account again that the expectation of $W(\boldsymbol{Z},\boldsymbol{v})$ is equal to one, we can write $\ell(\boldsymbol{v})$ as

$$\ell(\boldsymbol{v}) = \frac{\mathbb{E}_g\{L(\boldsymbol{Z})W(\boldsymbol{Z},\boldsymbol{v})\}}{\mathbb{E}_g\{W(\boldsymbol{Z},\boldsymbol{v})\}} . \tag{2.5.18}$$

This suggests that we consider

$$\tilde{\ell}_N(\boldsymbol{v}) = \frac{\sum_{i=1}^N L(Z_i)W(Z_i,\boldsymbol{v})}{\sum_{i=1}^N W(Z_i,\boldsymbol{v})} = \frac{\bar{\ell}_N(\boldsymbol{v})}{\bar{W}_N(\boldsymbol{v})} \tag{2.5.19}$$

as a consistent estimator of $\ell(\boldsymbol{v})$. We refer to $\tilde{\ell}_N(\boldsymbol{v})$ as the NCRV estimator of $\ell(\boldsymbol{v})$ (see for example, Glynn and Whitt (1989)).

Let us consider the accuracy (asymptotic variance) of $\tilde{\ell}_N(\boldsymbol{v})$. Since $\mathbb{E}_g\{W\} = 1$ and $\mathbb{E}_g\{LW\} = \ell(\boldsymbol{v})$, we have by the Strong Law of Large Numbers that $\bar{W}_N(\boldsymbol{v})$ and $\bar{\ell}_N(\boldsymbol{v})$ converge w.p.1 to 1 and $\ell(\boldsymbol{v})$, respectively. Consequently $\tilde{\ell}_N(\boldsymbol{v})$ converges with probability one to $\ell(\boldsymbol{v})$. Moreover, by the delta method (see, for example, chapter 6 for a discussion of the delta method), $N^{1/2}[\tilde{\ell}_N(\boldsymbol{v}) - \ell(\boldsymbol{v})]$ converges in distribution to a normal with zero mean and the variance

$$\sigma^2 = \text{Var}\{LW\} - 2\ell(\boldsymbol{v}) \text{ Cov}\{LW, W\} + \ell(\boldsymbol{v})^2 \text{Var}\{W\} . \tag{2.5.20}$$

Therefore the NCRV estimator $\tilde{\ell}_N(\boldsymbol{v})$ will be asymptotically more efficient than $\bar{\ell}_N(\boldsymbol{v})$ if and only if

$$2\ell(\boldsymbol{v}) \text{ Cov}\{LW,\ W\} - \ell(\boldsymbol{v})^2 \text{Var}\{W\} > 0 . \tag{2.5.21}$$

Suppose that $\ell(\boldsymbol{v}) > 0$. Then (2.5.21) is equivalent to

$$2 \text{ Corr}\{LW, W\} \geq \ell(\boldsymbol{v}) \left[\frac{\text{Var}\{W\}}{\text{Var}\{LW\}}\right]^{1/2} . \tag{2.5.22}$$

Note that condition (2.5.22) holds if $\ell(\boldsymbol{v})$ and $\text{Var}\{W\}/\ \text{Var}\{LW\}$ are small and LW and W are highly positively correlated.

2.6 Appendix A. Convergence of the Estimators of the Optimal Solutions

In this section we give some preliminary results concerning convergence properties of the estimators $\bar{\varphi}_N$ and $\bar{\boldsymbol{v}}_N$. For the sake of simplicity we discuss here only the unconstrained case. A more general and detailed discussion will be postponed until chapter 6. Consider a real-valued function $h(\boldsymbol{z},\boldsymbol{v})$ of two vector variables $\boldsymbol{z}$ and $\boldsymbol{v}$ and let $\boldsymbol{Z}$ be a random vector with a probability measure (distribution) $G(\boldsymbol{z})$. Suppose that the expected value function

$$\ell(\boldsymbol{v}) = \mathbb{E}\{h(\boldsymbol{Z}, \boldsymbol{v})\} = \int h(\boldsymbol{z}, \boldsymbol{v}) \, \mathrm{d}G(\boldsymbol{z})$$

exists for all $\boldsymbol{v}$ in a region $V \subset \mathbb{R}^n$ and consider the optimization problem

$$(\mathrm{P}_0) \qquad \text{minimize } \ell(\boldsymbol{v}), \quad \boldsymbol{v} \in \mathrm{V} .$$

Of course, problem (P_0) corresponds to the program defined in (2.4.6) if we take $h(\boldsymbol{z}, \boldsymbol{v}) = L(\boldsymbol{z})W(\boldsymbol{z}, \boldsymbol{v})$, where $W(\boldsymbol{z}, \boldsymbol{v}) = f(\boldsymbol{z}, \boldsymbol{v})/g(\boldsymbol{z})$ and $g(\boldsymbol{z})$ is the density function of the distribution $G(\boldsymbol{z})$. Now let $\boldsymbol{Z}_1, \ldots, \boldsymbol{Z}_N$ be a random sample of independent identically distributed (iid) random vectors with the common distribution $G(\boldsymbol{z})$. Then a stochastic counterpart of the program (P_0) is given by

$$(\bar{\mathrm{P}}_N) \qquad \text{minimize } \bar{\ell}_N(\boldsymbol{v}) \ , \quad \boldsymbol{v} \in V,$$

where

$$\bar{\ell}_N(\boldsymbol{v}) = N^{-1} \sum_{i=1}^{N} h(\boldsymbol{Z}_i, \boldsymbol{v}) .$$

Denote by $\bar{\varphi}_N$ the optimal value

$$\bar{\varphi}_N = \inf \{\bar{\ell}_N(\boldsymbol{v}) \ : \ \boldsymbol{v} \in V\}$$

and by $\bar{\boldsymbol{v}}_N$ the optimal solution of the program $(\bar{\mathrm{P}}_N)$. Notice that $\bar{\varphi}_N$ and $\bar{\boldsymbol{v}}_N$ depend on the random sample $\boldsymbol{Z}_1, \ldots, \boldsymbol{Z}_N$, and therefore are stochastic. We consider $\bar{\varphi}_N$ and $\bar{\boldsymbol{v}}_N$ to be statistical estimators of the optimal value φ^* and an optimal solution $\boldsymbol{v}^*$ of the program (P_0).

For any fixed $\boldsymbol{v} \in V$, we have by the Strong Law of Large Numbers that $\bar{\ell}_N(\boldsymbol{v})$ converges to $\ell(\boldsymbol{v})$ with probability one (w.p.1). It is possible to show that in fact the above convergence is uniform if the set V is compact and the following two conditions hold.

Assumption C1 For almost every $\boldsymbol{z}$, with respect to the probability measure $G(\boldsymbol{z})$, the function $h(\boldsymbol{z}, \cdot)$ is continuous on V.

Assumption C2 The family $\{|h(\boldsymbol{z}, \boldsymbol{v})|, \ \boldsymbol{v} \in V\}$ is dominated by an integrable (with respect to $G(\boldsymbol{z})$) function.

Lemma A1 *Suppose that assumptions C1 and C2 hold. Then the expected value function $\ell(\boldsymbol{v})$ is continuous on V. If, in addition, the set V is compact, then w.p.1 $\bar{\ell}_N(\boldsymbol{v})$ converges to $\ell(\boldsymbol{v})$ uniformly on V.*

Proof Consider a point $\boldsymbol{v} \in V$ and a sequence of points $\boldsymbol{v}_k$ in V converging to $\boldsymbol{v}$. By the Lebesgue dominated convergence theorem it follows from assumption C2 that

$$\lim_{k\to\infty} \mathbb{E}\{h(\boldsymbol{Z}, \boldsymbol{v}_k)\} = \mathbb{E}\{\lim_{k\to\infty} h(\boldsymbol{Z}, \boldsymbol{v})\}. \tag{2.6.1}$$

Because of assumption C1,

$$\lim_{k\to\infty} h(\boldsymbol{Z}, \boldsymbol{v}_k) = h(\boldsymbol{Z}, \boldsymbol{v}) \quad \text{w.p.1}$$

and hence it follows from (2.6.1) that $\ell(\boldsymbol{v}_k)$ tends to $\ell(\boldsymbol{v})$. This shows that $\ell(\boldsymbol{v})$ is continuous on V.

Now consider a sequence W_k of neighborhoods of $\boldsymbol{v}$, in V, shrinking to $\{\boldsymbol{v}\}$ and the function

$$b_k(\boldsymbol{z}) = \sup\{|h(\boldsymbol{z}, \boldsymbol{w}) - h(\boldsymbol{z}, \boldsymbol{v})| \;:\; \boldsymbol{w} \in W_k\}.$$

It follows from assumption C1 that for almost every $\boldsymbol{z}$, $b_k(\boldsymbol{z})$ tends to zero as $k \to \infty$. Furthermore, assumption C2 implies that the family $\{b_k(\boldsymbol{z}),\ k = 1, 2 \ldots\}$ is dominated by an integrable function and therefore by the Lebesgue dominated convergence theorem

$$\lim_{k\to\infty} \mathbb{E}\{b_k(\boldsymbol{Z})\} = \mathbb{E}\{\lim_{k\to\infty} b_k(\boldsymbol{Z})\} = 0. \tag{2.6.2}$$

Now we have that

$$|\bar{\ell}_N(\boldsymbol{w}) - \bar{\ell}_N(\boldsymbol{v})| \le N^{-1} \sum_{i=1}^{N} |h(\mathbf{Z}_i, \boldsymbol{w}) - h(\mathbf{Z}_i, \boldsymbol{v})|$$

and hence

$$\sup_{\boldsymbol{w}\in W_k} |\bar{\ell}_N(\boldsymbol{w}) - \bar{\ell}_N(\boldsymbol{v})| \le N^{-1} \sum_{i=1}^{N} b_k(\boldsymbol{Z}_i). \tag{2.6.3}$$

By the Strong Law of Large Numbers the right-hand side of (2.6.3) converges w.p.1 to $\mathbb{E}\{b_k(\boldsymbol{Z})\}$. Together with (2.6.2) this implies that for any given $\epsilon > 0$ there exists a neighborhood W of $\boldsymbol{v}$ such that w.p.1 for sufficiently large N,

$$\sup\{|\bar{\ell}_N(\boldsymbol{w}) - \bar{\ell}_N(\boldsymbol{v})| \;:\; \boldsymbol{w} \in W\} < \epsilon.$$

Since V is compact, there exists a finite number of points $\boldsymbol{v}_1, \ldots, \boldsymbol{v}_m \in V$ and corresponding neighborhoods $W_1, \ldots, W_m$ covering V such that w.p.1 for sufficiently large N,

$$\sup\{|\bar{\ell}_N(\boldsymbol{w}) - \bar{\ell}_N(\boldsymbol{v}_j)| \;:\; \boldsymbol{w} \in W_j\} < \epsilon,$$

$j = 1, \ldots, m$. Moreover, since $\ell(\boldsymbol{v})$ is continuous on V, these neighborhoods can be chosen in such a way that

$$\sup\{|\ell(\boldsymbol{w}) - \ell(\boldsymbol{v}_j)| \; : \; \boldsymbol{w} \in W_j\} < \epsilon,$$

$j = 1, \ldots, m$. Again by the Strong Law of Large Numbers

$$|\bar{\ell}_N(\boldsymbol{v}_j) - \ell(\boldsymbol{v}_j)| < \epsilon, \quad j = 1, \ldots, m,$$

w.p.1 for sufficiently large N. It follows that w.p.1 for N large enough

$$|\bar{\ell}_N(\boldsymbol{v}) - \ell(\boldsymbol{v})| < 3\epsilon$$

for all $\boldsymbol{v} \in V$ and hence the proof is complete. □

With the help of lemma A1 it is not difficult to prove strong consistency of the estimators $\bar{\varphi}_N$ and $\bar{\boldsymbol{v}}_N$.

Theorem A1 *Suppose that assumptions C1 and C2 hold and that V is compact. Then $\bar{\varphi}_N$ converges to φ^* w.p.1. Moreover, if $\boldsymbol{v}^*$ is a unique minimizer of $\ell(\boldsymbol{v})$ over V, then $\bar{\boldsymbol{v}}_N$ converges to $\boldsymbol{v}^*$ w.p.1.*

Proof From lemma A1 we know that for any $\epsilon > 0$ and all sufficiently large N and all $\boldsymbol{v} \in V$,

$$|\bar{\ell}_N(\boldsymbol{v}) - \ell(\boldsymbol{v})| < \epsilon \tag{2.6.4}$$

w.p.1. It follows from (2.6.4) that $|\bar{\varphi}_N - \varphi^*|$ is less than ϵ and hence $\bar{\varphi}_N$ converges to φ^* w.p.1.

Now suppose that the minimizer $\boldsymbol{v}^*$ is unique and consider a neighborhood W of $\boldsymbol{v}^*$ in V. Since $\ell(\boldsymbol{v})$ is continuous and V is compact we have that there exists $\epsilon > 0$ such that

$$\ell(\boldsymbol{v}) > \ell(\boldsymbol{v}^*) + 2\epsilon$$

for all $\boldsymbol{v} \in V$ and $\boldsymbol{v} \notin W$. Together with (2.6.4) this implies that

$$\bar{\ell}_N(\boldsymbol{v}) > \ell(\boldsymbol{v}^*) + \epsilon \tag{2.6.5}$$

for all $\boldsymbol{v} \in V$ and $\boldsymbol{v} \notin W$. On the other hand, we have by (2.6.4) that

$$\bar{\ell}_N(\bar{\boldsymbol{v}}_N) = \bar{\varphi}_N < \ell(\boldsymbol{v}^*) + \epsilon. \tag{2.6.6}$$

It follows from (2.6.5) and (2.6.6) that $\bar{\boldsymbol{v}}_N \in W$ and since the neighborhood W was arbitrary we obtain that $\bar{\boldsymbol{v}}_N$ converges to $\boldsymbol{v}^*$ w.p.1. □

If the set V^* of optimal solutions of the program (P_0) contains more than one point, then the second result of theorem A1 can be modified by showing that the distance from $\bar{\boldsymbol{v}}_N$ to V^* tends to zero w.p.1. This means that for N large enough, $\bar{\boldsymbol{v}}_N$ will be arbitrarily close to the solution set V^*. But we cannot guarantee then that $\bar{\boldsymbol{v}}_N$ will converge to a particular point in V^*.

We established that under mild regularity conditions, the estimator $\bar{\boldsymbol{v}}_N$ converges to the optimal solution of the program (P_0). However, this result does not give us any indication of the rate of the convergence. For that purpose, we evaluate now the asymptotic distribution of $\bar{\boldsymbol{v}}_N$. In order to investigate the asymptotic behavior of $\bar{\boldsymbol{v}}_N$ it will be necessary to utilize the differentiability properties of the function $h(\boldsymbol{z}, \boldsymbol{v})$. First we establish the following useful result about interchangeability of the expectation and differentiation operators. Let W be an open set in $\mathbb{R}^n$.

Assumption C3 There exists an integrable function $k(\boldsymbol{z})$, with respect to $G(\boldsymbol{z})$, such that for all $\boldsymbol{v}, \boldsymbol{w} \in W$,

$$|h(\boldsymbol{z}, \boldsymbol{w}) - h(\boldsymbol{z}, \boldsymbol{v})| \leq k(\boldsymbol{z}) \|\boldsymbol{w} - \boldsymbol{v}\|. \tag{2.6.7}$$

Assumption C4 For a given $\boldsymbol{v} \in W$ and almost every $\boldsymbol{z}$, with respect to $G(\boldsymbol{z})$, the function $h(\boldsymbol{z}, \cdot)$ is differentiable at $\boldsymbol{v}$.

Notice that the inequality (2.6.7) of assumption C3 implies that for almost every (fixed) $\boldsymbol{z}$, the function $h(\boldsymbol{z}, \cdot)$ is Lipschitz continuous on W.

Lemma A2 *Suppose that assumptions C3 and C4 hold. Then the expected value function $\ell(\boldsymbol{v})$ is Lipschitz continuous and differentiable on W and*

$$\nabla \ell(\boldsymbol{v}) = \mathbb{E}\{\nabla h(\boldsymbol{Z}, \boldsymbol{v})\}. \tag{2.6.8}$$

Proof We have that for any $\boldsymbol{w}, \boldsymbol{v} \in W$,

$$|\ell(\boldsymbol{w}) - \ell(\boldsymbol{v})| \leq \mathbb{E}\{|h(\boldsymbol{Z}, \boldsymbol{w}) - h(\boldsymbol{Z}, \boldsymbol{v})|\} ,$$

and because of assumption C3,

$$\mathbb{E}\{|h(\boldsymbol{Z}, \boldsymbol{w}) - h(\boldsymbol{Z}, \boldsymbol{v})|\} \leq \|\boldsymbol{w} - \boldsymbol{v}\| \mathbb{E}\{k(\boldsymbol{Z})\}.$$

It follows that

$$|\ell(\boldsymbol{w}) - \ell(\boldsymbol{v})| \leq K \|\boldsymbol{w} - \boldsymbol{v}\|,$$

where $K = \mathbb{E}\{k(\boldsymbol{Z})\}$, and hence $\ell(\boldsymbol{v})$ is Lipschitz continuous on W.

Now consider a point $\boldsymbol{v} \in W$ and a vector $\boldsymbol{d}$. It follows from assumption C3 that for all numbers t sufficiently close to zero,

$$t^{-1} |h(\boldsymbol{z}, \boldsymbol{v} + t\boldsymbol{d}) - h(\boldsymbol{z}, \boldsymbol{v})| \leq k(\boldsymbol{z}) \|\boldsymbol{d}\|.$$

Then by the Lebesgue dominated convergence theorem we have that

$$\lim_{t \to 0} \frac{\ell(\boldsymbol{v}+t\boldsymbol{d}) - \ell(\boldsymbol{v})}{t} = \mathbb{E}\left\{\lim_{t \to 0} \frac{h(\boldsymbol{Z}, \boldsymbol{v}+t\boldsymbol{d}) - h(\boldsymbol{Z}, \boldsymbol{v})}{t}\right\}. \tag{2.6.9}$$

Furthermore, assumption C4 implies that the limit in the right-hand side of (2.6.9) is equal to $\boldsymbol{d}'\nabla h(\boldsymbol{Z}, \boldsymbol{v})$ w.p.1 . It follows that the limit in the left-hand side of (2.6.9) is equal to $\boldsymbol{d}'E\{\nabla h(\boldsymbol{Z}, \boldsymbol{v})\}$. Together with Lipschitz continuity of $\ell(\boldsymbol{v})$ this implies that $\ell(\boldsymbol{v})$ is differentiable at $\boldsymbol{v}$ and (2.6.8) follows. □

In particular, it follows from lemma A2 that if for every $\boldsymbol{z}$ the function $h(\boldsymbol{z}, \cdot)$ is continuously differentiable on W and the family $\{\|\nabla h(\boldsymbol{z}, \boldsymbol{v})\|, \boldsymbol{v} \in W\}$ is dominated by an integrable function, with respect to $G(\boldsymbol{z})$, then $\ell(\boldsymbol{v})$ is continuously differentiable on W and (2.6.8) holds.

Now in order to evaluate the asymptotic behavior of $\bar{\boldsymbol{v}}_N$ we proceed as follows. Consider the empirical measure (distribution) $G_N(\boldsymbol{z})$ corresponding to the sample $\boldsymbol{Z}_1, \ldots, \boldsymbol{Z}_N$. That is

$$G_N = N^{-1} \sum_{i=1}^{N} \delta(\boldsymbol{Z}_i),$$

where $\delta(\boldsymbol{z})$ denotes the probability measure of mass one at the point $\boldsymbol{z}$. In the one-dimensional case the empirical distribution function associated with the empirical measure presents a stepwise function with jumps of N^{-1} at each Z_i, $i = 1, \ldots, N$. We have that the function $\bar{\ell}_N(\boldsymbol{v})$ can be expressed as the expected value with respect to G_N,

$$\bar{\ell}_N(\boldsymbol{v}) = \int h(\boldsymbol{z}, \boldsymbol{v})\, \mathrm{d}G_N(\boldsymbol{z}) = \mathbb{E}_{G_N}\{h(\boldsymbol{Z}, \boldsymbol{v})\}. \tag{2.6.10}$$

Let us consider a distribution P and a minimizer of the corresponding expected value function $\mathbb{E}_P\{h(\boldsymbol{Z}, \boldsymbol{v})\}$ over V. We denote this minimizer by $\bar{\boldsymbol{v}}(P)$ to emphasize that it is considered as a function of P. In particular, $\boldsymbol{v}^* = \bar{\boldsymbol{v}}(G)$ and it follows from (2.6.10) that $\bar{\boldsymbol{v}}_N = \bar{\boldsymbol{v}}(G_N)$. The idea of considering an estimator as a function of the empirical measure goes back to von Mises (1947) and such functions are called (von Mises) statistical functionals.

Clearly, the empirical distribution G_N presents an approximation of the "true" distribution G. If we could approximate $\bar{\boldsymbol{v}}(G_N)$ by, say, a linear function of $G_N - G$, then could derive the corresponding asymptotics for the estimator $\bar{\boldsymbol{v}}_N$. This idea is to calculate the directional derivatives of $\bar{\boldsymbol{v}}(P)$. In particular, consider a family of the distributions

$$P_t = (1-t)G + tG_N = G + t(G_N - G)$$

depending on the parameter t, $0 \leq t \leq 1$. Clearly for $t = 0$ and $t = 1$ we obtain the distributions G and G_N, respectively. It also follows that $\bar{\boldsymbol{v}}(P_t)$ is the minimizer of the function

$$q(\boldsymbol{v}, t) = \mathbb{E}_{P_t}\{h(\boldsymbol{Z}, \boldsymbol{v})\} = (1-t)\ell(\boldsymbol{v}) + t\bar{\ell}_N(\boldsymbol{v})$$

over V. Consequently if $\bar{\boldsymbol{v}}(P_t)$ is an interior point of V, then it must satisfy the system of, generally nonlinear, equations

$$\nabla_{\boldsymbol{v}} q(\boldsymbol{v}, t) = \mathbf{0}, \tag{2.6.11}$$

representing the first-order necessary optimality conditions. Now in order to calculate the derivative $\mathrm{d}\bar{\boldsymbol{v}}(P_t)/\mathrm{d}t$, at $t = 0$, we can apply the Implicit Function Theorem to equations (2.8.11). Notice that

$$\nabla^2_{\boldsymbol{v}\boldsymbol{v}} q(\boldsymbol{v}, 0) = \nabla^2 \ell(\boldsymbol{v})$$

and

$$\nabla^2_{\boldsymbol{v}t} q(\boldsymbol{v}, 0) = \nabla \bar{\ell}_N(\boldsymbol{v}) - \nabla \ell(\boldsymbol{v})\,.$$

Suppose that $\bar{\boldsymbol{v}}(P_t)$ tends to $\boldsymbol{v}^* = \bar{\boldsymbol{v}}(P_0)$ as $t \to 0$, such that $\boldsymbol{v}^*$ is an interior point of V and that the Hessian matrix $\boldsymbol{B} = \nabla^2 \ell(\boldsymbol{v}^*)$ is nonsingular. Then we obtain

$$\left.\frac{\mathrm{d}\bar{\boldsymbol{v}}(P_t)}{\mathrm{d}t}\right|_{t=0} = -\boldsymbol{B}^{-1} \nabla \bar{\ell}_N(\boldsymbol{v}^*)\,. \tag{2.6.12}$$

The term $\nabla \ell(\boldsymbol{v}^*)$ drops out from the right-hand side of (2.6.12) since it equals zero by the first-order optimality conditions. The derivative in the left-hand side of (2.6.12) represents the directional derivative of $\bar{\boldsymbol{v}}(P)$ at G in the direction $G_N - G$. It gives an approximation of the difference

$$\bar{\boldsymbol{v}}(P_1) - \bar{\boldsymbol{v}}(P_0) = \bar{\boldsymbol{v}}_N - \boldsymbol{v}^*$$

and therefore we can write

$$\bar{\boldsymbol{v}}_N - \boldsymbol{v}^* = -\boldsymbol{B}^{-1} \nabla \bar{\ell}_N(\boldsymbol{v}^*) + \mathbf{r}_N\,, \tag{2.6.13}$$

where $\mathbf{r}_N$ denotes the error of approximation. The term $\mathbf{B}^{-1} = [\nabla^2 \ell(\boldsymbol{v}^*)]^{-1}$ is independent of the sample and therefore the asymptotic behavior of the expression in the right-hand side of (2.6.13) is determined by the term $\nabla \bar{\ell}_N(\boldsymbol{v})^*$. We have that

$$\nabla \bar{\ell}_N(\boldsymbol{v}^*) = N^{-1} \sum_{i=1}^{N} \nabla h(\boldsymbol{Z}_i, \boldsymbol{v}^*)\,.$$

Suppose that the operators of expectation and differentiation are interchangeable (see lemma A2). Then

$$\mathbb{E}\{\nabla \bar{\ell}_N(\boldsymbol{v}^*)\} = \nabla \ell(\boldsymbol{v}^*) = \mathbf{0}\,.$$

Moreover, by the Central Limit Theorem, $N^{1/2}\nabla\bar{\ell}_N(\boldsymbol{v}^*)$ converges in distribution to multivariate normal with zero mean vector and the covariance matrix

$$\boldsymbol{\Sigma} = \mathbb{E}_G\{\nabla h(\boldsymbol{Z}, \boldsymbol{v}^*)\nabla h(\boldsymbol{Z}, \boldsymbol{v}^*)'\} . \tag{2.6.14}$$

Now suppose that the error term $\mathbf{r}_N$ in (2.6.13) is asymptotically negligible. Then it follows that $N^{1/2}(\bar{\boldsymbol{v}}_N - \boldsymbol{v}^*)$ converges in distribution to multivariate normal with zero mean and the covariance matrix $\boldsymbol{B}^{-1}\boldsymbol{\Sigma}\boldsymbol{B}^{-1}$.

The above arguments give us correct formulas for the asymptotics of $\bar{\boldsymbol{v}}_N$ but are insufficient for a rigorous proof of it. The main difficulty is to show that the error $\mathbf{r}_N$ is indeed "asymptotically negligible". Although these arguments can be made precise, it will require rather sophisticated mathematical tools. Therefore, we prefer at this stage to give a direct proof of the obtained result. Let us first state some regularity conditions.

Assumption C5 The minimizer $\boldsymbol{v}^*$ is an interior point of V.

Assumption C6 For almost every $\boldsymbol{z}$, $h(\boldsymbol{z}, \cdot)$ is twice continuously differentiable in a neighborhood of $\boldsymbol{v}^*$.

Assumption C7 The families $\{\|\nabla h(\boldsymbol{z}, \boldsymbol{v})\|,\ \boldsymbol{v} \in V\}$ and $\{\|\nabla^2 h(\boldsymbol{z}, \boldsymbol{v})\|,\ \boldsymbol{v} \in V\}$ are dominated by integrable functions.

Assumption C8 The matrix $\boldsymbol{B} = \mathbb{E}\{\nabla^2 h(\boldsymbol{Z}, \boldsymbol{v}^*)\}$ is nonsingular.

Assumption C9 The covariance matrix $\boldsymbol{\Sigma}$, given in (2.8.14), does exist.

Notice that the probability distribution in assumptions C6–C9 is $G(\boldsymbol{z})$. By lemma A2 assumptions C6 and C7 imply that the first- and second-order derivatives can be taken inside the expected value and hence the matrix $\boldsymbol{B}$ in assumption C8 can be also written in the form $\boldsymbol{B} = \nabla^2\ell(\boldsymbol{v}^*)$.

Consistency and asymptotic normality of $\bar{\boldsymbol{v}}_N$ are proved by different methods under different assumptions. Consistency is a global property while asymptotic normality holds locally for the estimates $\bar{\boldsymbol{v}}_N$ sufficiently close to $\boldsymbol{v}^*$. Therefore we *assume* in the following theorem that $\bar{\boldsymbol{v}}_N$ is a consistent estimator of $\boldsymbol{v}^*$, referring to theorem A1 for required regularity conditions. It will be convenient to use the symbols $o_p(\cdot)$ and $O_p(\cdot)$ as the stochastic analogues of $O(\cdot)$ and $o(\cdot)$. Let $\{\boldsymbol{X}_n\}$ and $\{\mathbf{Y}_n\}$ be sequences of random vectors. The notation $\boldsymbol{X}_n = o_p(\mathbf{Y}_n)$ means that $\mathrm{P}\{\|\boldsymbol{X}_n\| \geq \epsilon\|\mathbf{Y}_n\|\} \to 0$ for each $\epsilon > 0$. The notation $\boldsymbol{X}_n = O_p(\mathbf{Y}_n)$ means that for each $\epsilon > 0$ there exists a real number k such that $\mathrm{P}\{\|\boldsymbol{X}_n\| \geq k\|\mathbf{Y}_n\|\} < \epsilon$ if n is large enough. It is possible to deal with $o_p(\cdot)$ and $O_p(\cdot)$ in a way similar to their deterministic counterparts $o(\cdot)$ and $O(\cdot)$ (Mann and Wald (1943)). Typically $\{\mathbf{Y}_n\}$ will be nonrandom. For example, $\boldsymbol{X}_n = o_p(1)$ is another way of writing that $\boldsymbol{X}_n$ converges to zero in probability, and $\boldsymbol{X}_n = O_p(1)$ means that $\boldsymbol{X}_n$ is bounded in probability.

Theorem A2 *Suppose that assumptions C5–C9 hold and that $\bar{\boldsymbol{v}}_N$ converges in probability (w.p.1) to $\boldsymbol{v}^*$. Then $N^{1/2}(\bar{\boldsymbol{v}}_N - \boldsymbol{v}^*)$ converges in distribution to multivariate normal with zero mean and the covariance matrix $\boldsymbol{B}^{-1}\boldsymbol{\Sigma}\boldsymbol{B}^{-1}$.*

Proof First we show that $N^{1/2}(\bar{\boldsymbol{v}}_N - \boldsymbol{v}^*)$ is bounded in probability. Consider a convex, compact neighborhood W of $\boldsymbol{v}^*$ in V. Since $\bar{\boldsymbol{v}}_N$ converges in probability to $\boldsymbol{v}^*$, we have that with probability tending to one, $\bar{\boldsymbol{v}}_N \in W$. Then employing Taylor expansion we obtain

$$\bar{\ell}_N(\bar{\boldsymbol{v}}_N) = \bar{\ell}_N(\boldsymbol{v}^*) + (\bar{\boldsymbol{v}}_N - \boldsymbol{v}^*)'\nabla\bar{\ell}_N(\boldsymbol{v}^*) + \tfrac{1}{2}(\bar{\boldsymbol{v}}_N - \boldsymbol{v}^*)'\boldsymbol{B}_N(\bar{\boldsymbol{v}}_N - \boldsymbol{v}^*)\,,$$

where $\boldsymbol{B}_N$ is the Hessian matrix of $\bar{\ell}_N(\cdot)$ calculated at a point on the segment joining $\bar{\boldsymbol{v}}_N$ and $\boldsymbol{v}^*$. Since $\bar{\boldsymbol{v}}_N$ is a minimizer of $\bar{\ell}_N(\cdot)$ over V, we have that

$$\bar{\ell}_N(\boldsymbol{v}^*) \geq \bar{\ell}_N(\bar{\boldsymbol{v}}_N)$$

and hence

$$0 \geq (\bar{\boldsymbol{v}}_N - \boldsymbol{v}^*)'\nabla\bar{\ell}_N(\boldsymbol{v}^*) + \tfrac{1}{2}(\bar{\boldsymbol{v}}_N - \boldsymbol{v}^*)'\boldsymbol{B}_N(\bar{\boldsymbol{v}}_N - \boldsymbol{v}^*)\,.$$

It follows that

$$\|\bar{\boldsymbol{v}}_N - \boldsymbol{v}^*\|\;\|\nabla\bar{\ell}_N(\boldsymbol{v}^*)\| \;\geq\; \tfrac{1}{2}(\bar{\boldsymbol{v}}_N - \boldsymbol{v}^*)'\boldsymbol{B}_N(\bar{\boldsymbol{v}}_N - \boldsymbol{v}^*)\,. \tag{2.6.15}$$

We know from lemma A1 that w.p.1, $\nabla^2\bar{\ell}_N(\boldsymbol{v})$ converges to $\nabla^2\ell(\boldsymbol{v})$ uniformly on W. It follows that for sufficiently small W, all $\boldsymbol{v} \in W$ and N large enough the Hessian matrix $\nabla^2\bar{\ell}_N(\boldsymbol{v})$ will be arbitrarily close to the matrix $\boldsymbol{B}$ w.p.1. Consequently we obtain that

$$\boldsymbol{B}_N = \boldsymbol{B} + o_p(1)\,.$$

Since $\boldsymbol{v}^*$ is the minimizer of $\ell(\boldsymbol{v})$ we have that the matrix $\boldsymbol{B} = \nabla^2\ell(\boldsymbol{v}^*)$ is non-negative definite. By assumption C8 the matrix $\boldsymbol{B}$ is nonsingular and hence $\boldsymbol{B}$ is positive definite. It follows that there exists a positive number ϵ such that $\boldsymbol{u}'\boldsymbol{B}\boldsymbol{u} \geq \epsilon\|\boldsymbol{u}\|^2$ for any vector $\boldsymbol{u}$. We obtain then from (2.6.15) that

$$2\|\bar{\boldsymbol{v}}_N - \boldsymbol{v}^*\|\;\|\nabla\bar{\ell}_N(\boldsymbol{v}^*)\| \;\geq \epsilon\|\bar{\boldsymbol{v}}_N - \boldsymbol{v}^*\|^2 + o_p(\|\bar{\boldsymbol{v}}_N - \boldsymbol{v}^*\|^2)$$

and hence

$$2\|\nabla\bar{\ell}_N(\boldsymbol{v}^*)\| \geq [\epsilon + o_p(1)]\|\bar{\boldsymbol{v}}_N - \boldsymbol{v}^*\|.$$

It remains to point out that by the Central Limit Theorem, $N^{1/2}\nabla\bar{\ell}_N(\boldsymbol{v}^*)$ converges in distribution and therefore is bounded in probability. This proves that

$$\bar{\boldsymbol{v}}_N - \boldsymbol{v}^* = O_p(N^{-1/2}). \tag{2.6.16}$$

Now by employing the Taylor formula and using the uniform convergence result of lemma A1 we obtain

$$\nabla\bar{\ell}_N(\bar{\boldsymbol{v}}_N) - \nabla\bar{\ell}_N(\boldsymbol{v}^*) = \mathbf{B}(\bar{\boldsymbol{v}}_N - \boldsymbol{v}^*) + o_p(\bar{\boldsymbol{v}}_N - \boldsymbol{v}^*).$$

Since $\nabla\bar{\ell}_N(\bar{\boldsymbol{v}}_N)$ is zero and because of (2.6.16), it follows that

$$\bar{\boldsymbol{v}}_N - \boldsymbol{v}^* = -\boldsymbol{B}^{-1}\nabla\bar{\ell}_N(\boldsymbol{v}^*) + o_p(N^{-1/2}). \tag{2.6.17}$$

Consequently $N^{1/2}(\bar{\boldsymbol{v}}_N - \boldsymbol{v}^*)$ converges in distribution to the same limit as $-\boldsymbol{B}^{-1}[N^{1/2}\nabla\bar{\ell}_N(\boldsymbol{v}^*)]$, which together with the Central Limit Theorem, applied to $N^{1/2}\nabla\bar{\ell}_N(\boldsymbol{v}^*)$, completes the proof. □

Formula (2.6.17) shows that the error term $\mathbf{r}_N$ in (2.6.13) is of order $o_p(N^{-1/2})$ and therefore is indeed "asymptotically negligible". In situations where the optimization problem involves (stochastic) constraints the analysis becomes more complicated. In this respect we will exploit both approaches outlined in this appendix.

Finally, let us show that under the assumptions of theorems A1 and A2 the estimator $\bar{\boldsymbol{v}}_N$, satisfying the first-order necessary conditions $\nabla\bar{\ell}_N(\bar{\boldsymbol{v}}_N) = \mathbf{0}$, is *unique* w.p.1 for N large enough. Let $\tilde{\boldsymbol{v}}_N$ be another (strongly) consistent estimator of $\boldsymbol{v}^*$ satisfying the first-order necessary conditions. Then we have that

$$\nabla\bar{\ell}_N(\bar{\boldsymbol{v}}_N) - \nabla\bar{\ell}_N(\tilde{\boldsymbol{v}}_N) = \mathbf{0}. \tag{2.6.18}$$

We know that the left-hand side of equation (2.6.18) can be written as $\mathbf{B}_N(\bar{\boldsymbol{v}}_N - \tilde{\boldsymbol{v}}_N)$, where the matrix $\boldsymbol{B}_N$ will be arbitrarily close to $\boldsymbol{B}$ and hence nonsingular w.p.1 for N large enough. It remains to note that

$$\boldsymbol{B}_N(\bar{\boldsymbol{v}}_N - \tilde{\boldsymbol{v}}_N) = \mathbf{0}$$

and nonsingularity of $\boldsymbol{B}_N$ imply that $\bar{\boldsymbol{v}}_N - \tilde{\boldsymbol{v}}_N$ is zero and hence $\bar{\boldsymbol{v}}_N = \tilde{\boldsymbol{v}}_N$.

2.7 Appendix B. The Radon-Nikodym Derivative

In this appendix we briefly discuss the basic concepts of the *change-of-measure* techniques which are crucial for the SF method. Let $(\Omega, \mathcal{F})$ be a measurable space, i.e. $\mathcal{F}$ is a σ-algebra of subsets of the set Ω. Let P and Q be two probability (finite) measures on $(\Omega, \mathcal{F})$. It is said that the measure P is *absolutely continuous* with respect to Q if for every A in $\mathcal{F}$, $Q(A) = 0$ implies that $P(A) = 0$. In this case it is also said that the measure P is *dominated* by the measure Q.

If there exists a Q-measurable function $p(\omega)$, $\omega \in \Omega$, such that $P(A) = \int_A p\mathrm{d}Q$ for all $A \in \mathcal{F}$, then it follows that P is dominated by Q. The converse of this is known as the Radon-Nikodym theorem. That is, if P is absolutely continuous with respect to Q, then there exists a non-negative, Q-integrable function $p(\omega)$, a density, such that $P(A) = \int_A p\mathrm{d}Q$ for all $A \in \mathcal{F}$. The density $p(\omega)$ is called the *Radon-Nikodym derivative* of P with respect to Q and is denoted $\mathrm{d}P/\mathrm{d}Q$. Notice also that if $h(\omega)$ is a P-integrable function and $p(\omega) = \mathrm{d}P/\mathrm{d}Q$, then

$$\int_\Omega h(\omega)P(\mathrm{d}\omega) = \int_\Omega h(\omega)p(\omega)Q(d\omega). \tag{2.7.1}$$

Now, let $P_{\boldsymbol{v}}$ be a family of probability measures, on $(\Omega, \mathcal{F})$, depending on the parameter vector $\boldsymbol{v} \in V$ and let $L(\omega)$ be a real-valued function, $P_{\boldsymbol{v}}$-integrable for all $\boldsymbol{v} \in V$. Let

$$\ell(\boldsymbol{v}) = \int_\Omega L(\omega)P_{\boldsymbol{v}}(\mathrm{d}\omega) = \mathbb{E}_{P_{\boldsymbol{v}}}\{L\}$$

be the corresponding expected value function (expected performance). Suppose that there exists a probability measure Q, on $(\Omega, \mathcal{F})$, such that $P_{\boldsymbol{v}}$ is dominated by Q for all $\boldsymbol{v} \in V$. Let $p(\omega, \boldsymbol{v}) = \mathrm{d}P_{\boldsymbol{v}}/\mathrm{d}Q$ be the associated densities depending on $\boldsymbol{v}$. Then we can write

$$\ell(\boldsymbol{v}) = \int_\Omega L(\omega)p(\omega, \boldsymbol{v})\mathrm{d}Q = \mathbb{E}_Q\{L(\omega)p(\omega, \boldsymbol{v})\}. \tag{2.7.2}$$

and thus express the function $\ell(\boldsymbol{v})$ as the expected value of $L(\omega)p(\omega, \boldsymbol{v})$ with respect to the dominating measure Q.

Of course, in order to apply the change-of-measure formula (2.7.2) we have to find a dominating measure Q and to calculate the associated Radon-Nikodym derivative $\mathrm{d}P_{\boldsymbol{v}}/\mathrm{d}Q$. In this respect we would like to note that chapter 2 deals with a rather simple application of the Radon-Nikodym derivative. Indeed, let $\Omega = \mathbb{R}^m$ and let $\mathcal{F}$ be the Borel σ-algebra of $\mathbb{R}^m$. Let further $F_{\boldsymbol{v}}(\boldsymbol{y}) = F(\boldsymbol{y}, \boldsymbol{v})$, $\boldsymbol{v} \in V$, and $G(\boldsymbol{y})$ be probability measures on $(\mathbb{R}^m, \mathcal{F})$ having densities $f(\boldsymbol{y}, \boldsymbol{v}) = \mathrm{d}F(\boldsymbol{y}, \boldsymbol{v})/\mathrm{d}\boldsymbol{y}$ and $g(\boldsymbol{y}) = \mathrm{d}G(\boldsymbol{y})/\mathrm{d}\boldsymbol{y}$, respectively. Then the condition (2.2.25) stating that the support of $f(\boldsymbol{y}, \boldsymbol{v})$ lies in the support of $g(\boldsymbol{y})$, merely implies that $F(\boldsymbol{y}, \boldsymbol{v})$ is dominated by $G(\boldsymbol{y})$. The corresponding Radon-Nikodym derivative is given by (see (2.2.26))

$$\frac{\mathrm{d}F(\boldsymbol{y}, \boldsymbol{v})}{\mathrm{d}G(\boldsymbol{y})} = \frac{f(\boldsymbol{y}, \boldsymbol{v})}{g(\boldsymbol{y})} = W(\boldsymbol{y}, \boldsymbol{v})\,.$$

Note finally that more sophisticated applications of the Radon-Nikodym derivative are presented in chapter 3.

2.8 Exercises

1. Consider the higher order partial derivatives $\partial^k \ell(\boldsymbol{v})/\partial v_{i_1} \ldots \partial v_{i_k}$, $1 \leq i_1, \ldots, i_k \leq n$. Assume that these derivatives can be taken inside the expected value and show that

$$\partial^k \ell(\boldsymbol{v})/\partial v_{i_1} \cdots \partial v_{i_k} = \mathbb{E}_{\boldsymbol{v}}\{L(\boldsymbol{Y})[\boldsymbol{S}^{(k)}(\boldsymbol{Y}, \boldsymbol{v})]_{i_1 \ldots i_k}\},$$

where $\boldsymbol{S}^{(k)}(\boldsymbol{y}, \boldsymbol{v})$ are calculated by the following recursive formulas:

$$[\boldsymbol{S}^{(1)}(\boldsymbol{y}, \boldsymbol{v})]_i = \partial \log f(\boldsymbol{y}, \boldsymbol{v})/\partial v_i \,, \quad i = 1, \ldots, n,$$

$$\left[\boldsymbol{S}^{(k+1)}(\boldsymbol{y}, \boldsymbol{v})\right]_{i_1 \ldots i_{k+1}} = \frac{\partial [\boldsymbol{S}^{(k)}(\boldsymbol{y}, \boldsymbol{v})]_{i_1 \ldots i_k}}{\partial v_{i_{k+1}}} + [\boldsymbol{S}^{(k)}(\boldsymbol{y}, \boldsymbol{v})]_{i_1 \ldots i_k} [\boldsymbol{S}^{(1)}(\boldsymbol{y}, \boldsymbol{v})]_{i_{k+1}}.$$

2. Let $Y_k, \ k = 1, \ldots, m$, be independent random variables and let $\boldsymbol{Y} = (Y_1, \ldots, Y_m)'$ be the corresponding random vector.

   (i) Show that if each Y_k is distributed Weibull, i.e.

$$f_k(y_k, \alpha_k) = \alpha_k y_k^{\alpha_k - 1} \exp(-y_k^{\alpha_k}) \,,$$

   then, for $\boldsymbol{\alpha} = (\alpha_1, \ldots, \alpha_m)'$, the efficient scores are given by

$$[\boldsymbol{S}^{(1)}(\boldsymbol{y}, \boldsymbol{\alpha})]_k = \alpha_k^{-1} + \log y_k - y_k^{\alpha_k} \log y_k \,,$$

   $k = 1, \ldots, m$, and

$$\left[\boldsymbol{S}^{(2)}(\boldsymbol{y}, \boldsymbol{\alpha})\right]_{jk} = \delta_{jk} \left[-\alpha_k^{-2} - y_k^{\alpha_k} (\log y_k)^2\right]$$

$$+(\alpha_j^{-1} + \log y_j - y_j^{\alpha_j} \log y_j)(\alpha_k^{-1} + \log y_k - y_k^{\alpha_k} \log y_k) \,,$$

   $j, \ k = 1, \ldots, m$, with $\delta_{jk} = 1$ if $j = k$ and $\delta_{jk} = 0$ if $j \neq k$.

   (ii) Show that if each Y_k is distributed Poisson, i.e.

$$P_k(y_k, \lambda_k) = \frac{\lambda_k^{y_k}}{y_k!} \, \mathrm{e}^{-\lambda_k} \,, \quad y_k = 0, 1, \ldots,$$

   then

$$[\boldsymbol{S}^{(1)}(\boldsymbol{y}, \boldsymbol{\lambda})]_k = \lambda_k^{-1} \, y_k - 1 \,,$$

   $k = 1, \ldots, m$, and

$$[\boldsymbol{S}^{(2)}(\boldsymbol{y}, \boldsymbol{\lambda})]_{jk} = -\delta_{jk} \lambda_k^{-2} y_k + (\lambda_j^{-1} y_j - 1)(\lambda_k^{-1} y_k - 1) \,,$$

   $j, \ k = 1, \ldots, m$.

3. Consider the function $W(\boldsymbol{y}, \boldsymbol{v})$ in the form (2.3.12). Show that

$$\nabla^2 W(\boldsymbol{y}, \boldsymbol{v}) = W(\boldsymbol{y}, \boldsymbol{v}) \boldsymbol{S}^{(2)}(\boldsymbol{y}, \boldsymbol{v}) ,$$

where $\mathbf{S}^{(2)}(\mathbf{y}, \boldsymbol{v})$ is an $m \times m$ matrix given by

$$\begin{aligned} \boldsymbol{S}^{(2)}(\boldsymbol{y}, \boldsymbol{v}) &= c(\boldsymbol{v})^{-1} \nabla^2 c(\boldsymbol{v}) - c(\boldsymbol{v})^{-2} \nabla c(\boldsymbol{v}) \nabla c(\boldsymbol{v})' \\ &+ [c(\boldsymbol{v})^{-1} \nabla c(\boldsymbol{v}) + \mathbf{t}(\boldsymbol{y})][c(\boldsymbol{v})^{-1} \nabla c(\boldsymbol{v}) + \mathbf{t}(\boldsymbol{y})]'. \end{aligned}$$

4. Let Y_k, $k = 1 \ldots, m$, be independent random variables each distributed normally with the mean μ_k and variance σ_k^2 , i.e.

$$f_k(y_k, \mu_k, \sigma_k) = \frac{1}{(2\pi)^{1/2} \sigma_k} \exp \left\{ \frac{-(y_k - \mu_k)^2}{2\sigma_k^2} \right\} .$$

Suppose that we are interested in the sensitivities with respect to $\boldsymbol{\mu} = (\mu_1, \ldots, \mu_m)'$ only. Show that formulas (2.3.13) and (2.3.16) hold with

$$\mathbb{E}_{\boldsymbol{\mu}_0}\{W^2\} = \exp \left\{ \sum_{k=1}^{m} \frac{(\Delta \mu_k)^2}{\sigma_k^2} \right\}$$

and

$$[\boldsymbol{S}^{(1)}(\boldsymbol{y}, \boldsymbol{\mu})]_k = \sigma_k^{-2} (y_k - \mu_k) ,$$

$k = 1, \ldots, m$.

5. Let the components Y_k, $k = 1, \ldots, m$, of the random vector $\boldsymbol{Y}$ be independent, each distributed according to the pdf

$$f_k(y_k, \boldsymbol{v}_k) = a_k(\boldsymbol{v}_k) \exp\{b_k(\boldsymbol{v}_k) t_k(y_k)\} h_k(y_k)$$

where $a_k(\boldsymbol{v}_k)$ is a positive valued function of $\boldsymbol{v}_k$ and $b_k(\boldsymbol{v}_k)$ and $t_k(y_k)$, $h_k(y_k)$ are real-valued functions of $\boldsymbol{v}_k$ and y_k, respectively. Consider the corresponding pdf of $\boldsymbol{Y}$ given by

$$f(\boldsymbol{y}, \boldsymbol{v}) = a(\boldsymbol{v}) \exp \left\{ \sum_{k=1}^{m} b_k(\boldsymbol{v}_k) t_k(y_k) \right\} h(\boldsymbol{y}) ,$$

where $\boldsymbol{v} = (\boldsymbol{v}_1', \ldots, \boldsymbol{v}_m')'$, $a(\boldsymbol{v}) = \prod_{k=1}^m a_k(\boldsymbol{v}_k)$ and $h(\boldsymbol{y}) = \prod_{k=1}^m h_k(y_k)$.

(i) Show that

$$\mathrm{Var}_{\boldsymbol{v}_0}\{LW\} = \mathbb{E}_{\boldsymbol{v}_0}\{W^2\} \mathbb{E}_{(a^*, b^*)}\{L^2\} - \ell(\boldsymbol{v})^2 ,$$

where the subscript (a^*, b^*) means that the expectation is taken with respect to the exponential family determined by the functions

$$b_k^*(\boldsymbol{v}_k, \boldsymbol{v}_{0k}) = 2b_k(\boldsymbol{v}_k) - b_k(\boldsymbol{v}_{0k}) ,$$

$k = 1, \ldots, m$, and the function $a^*(\boldsymbol{v}, \boldsymbol{v}_o)$ satisfying the condition

$$a^*(\boldsymbol{v}, \boldsymbol{v}_0) \int \exp\left\{\sum_{k=1}^{m} b_k^*(\boldsymbol{v}_k, \boldsymbol{v}_{ok}) t_k(y_k)\right\} h(\boldsymbol{y}) \mathrm{d}\boldsymbol{y} = 1 .$$

(ii) Show that

$$\mathbb{E}_{\boldsymbol{v}_0}\{W^2\} = \frac{a(\boldsymbol{v})^2}{a(\boldsymbol{v}_0) a^*(\boldsymbol{v}, \boldsymbol{v}_0)} .$$

(iii) Show that

$$a^*(\boldsymbol{v}, \boldsymbol{v}_0) = \prod_{k=1}^{m} a_k^*(\boldsymbol{v}_k, \boldsymbol{v}_{0k}) ,$$

where $a_k^*(\boldsymbol{v}_k, \boldsymbol{v}_{0k})$ are defined by the equations

$$a_k^*(\boldsymbol{v}_k, \boldsymbol{v}_{0k}) \int \exp\{b_k^*(\boldsymbol{v}_k, \boldsymbol{v}_{0k}) t_k(y_k)\} h_k(y_k) dy_k = 1 .$$

(iv) Show that

$$\mathrm{Cov}_{\boldsymbol{v}_0}\{L \nabla W\} = \mathbb{E}_{\boldsymbol{v}_0}\{W^2\} \mathbb{E}_{(a^*, b^*)}\{L^2 \boldsymbol{S}^{(1)} \boldsymbol{S}^{(1)'}\} - (\nabla \ell)(\nabla \ell)' ,$$

where $\boldsymbol{S}^{(1)} = (\boldsymbol{S}_1^{(1)}(Z_1, \boldsymbol{v}_1)', \ldots, \boldsymbol{S}_m^{(1)}(Z_m, \boldsymbol{v}_m)')'$ with

$$\boldsymbol{S}_k^{(1)}(z_k, \boldsymbol{v}_k) = a_k(\boldsymbol{v}_k)^{-1} \nabla a_k(\boldsymbol{v}_k) + t_k(z_k) \nabla b_k(\boldsymbol{v}_k) ,$$

$k = 1, \ldots, m$ (the gradients are taken with respect to $\boldsymbol{v}_k$).

(v) Show that $\boldsymbol{S}^{(2)}(\boldsymbol{y}, \boldsymbol{v})$ is a block diagonal matrix with the corresponding blocks

$$\boldsymbol{S}_k^{(2)}(y_k, \boldsymbol{v}_k) = a_k(\boldsymbol{v}_k)^{-1} \nabla^2 a_k(\boldsymbol{v}_k) - a_k(\boldsymbol{v}_k)^{-2} \nabla a_k(\boldsymbol{v}_k) \nabla a_k(\boldsymbol{v}_k)'$$

$$+ \boldsymbol{S}_k^{(1)}(y_k, \boldsymbol{v}_k) \boldsymbol{S}_k^{(1)}(y_k, \boldsymbol{v}_k)',$$

$k = 1, \ldots, m$.

6. Based on the results of Problem 5 present a table similar to that of table 2.2.1 while replacing the score functions $\boldsymbol{S}^{(1)}(\boldsymbol{y}, \boldsymbol{v})$ and $\boldsymbol{S}^{(2)}(\boldsymbol{y}, \boldsymbol{v})$ by $\mathrm{Var}_{\boldsymbol{v}_0}\{LW\}$ and $\mathrm{Cov}_{\boldsymbol{v}_0}\{L \nabla W\}$, respectively.

7. Consider the information matrix $\mathbf{I}(\boldsymbol{v})$ defined in (2.3.17).

   (i) Show that for every permissible value of the parameter vector $\boldsymbol{v}$, the matrix $\mathbf{I}(\boldsymbol{v})$ is nonnegative definite.

   (ii) Show that $\mathbf{I}(\boldsymbol{v})$ can be represented in the alternative form

   $$\mathbf{I}(\boldsymbol{v}) = -\mathbb{E}_{\boldsymbol{v}}\{\nabla^2 \log f(\mathbf{Z}, \boldsymbol{v})\}$$

   provided the second-order derivatives can be taken inside the expected value.

8. Suppose that assumptions (i)–(iii) of theorem 2.4.1 hold and let V^* be the set of optimal solutions of the program (P_0) given in (2.4.6). (We do not assume here that the optimal set V^* contains only one element.) Show that w.p.1 the distance $\mathrm{dist}(\bar{\boldsymbol{v}}_N, V^*)$, from $\bar{\boldsymbol{v}}_N$ to V^*, tends to zero (see theorem A1 in appendix A).

9. Let $f(\boldsymbol{y}, \boldsymbol{v})$, $\boldsymbol{v} \in V$, be a family of probability density functions. Suppose that the parameter set $V \subset \mathbb{R}^n$ is convex.

   (i) Show that if for every $\boldsymbol{y}$ the function $f(\boldsymbol{y}, \cdot)$ is convex on V, then $\ell(\boldsymbol{v}) = \mathbb{E}_{\boldsymbol{v}} L(\boldsymbol{Y})$ is convex on V for any non-negative valued, integrable function $L(\boldsymbol{y})$.

   (ii) Conversely, suppose in addition that V is open, $f(\boldsymbol{y}, \boldsymbol{v})$ is differentiable in $\boldsymbol{v}$ and that the gradient $\nabla f(\boldsymbol{y}, \boldsymbol{v})$ is continuous jointly in $\boldsymbol{y}$ and $\boldsymbol{v}$. Show that if for every non-negative valued, integrable function $L(\boldsymbol{y})$ the corresponding expected value function $\ell(\boldsymbol{v})$ is convex on V, then $f(\boldsymbol{y}, \cdot)$ is convex on V for every $\boldsymbol{y}$.

10. Consider the exponential pdf $f(y, v) = v \exp(-vy)$. Show that if $L(y)$ is a monotonically increasing function, then the expected performance $\ell(v) = \mathbb{E}_v L(Y)$ is a monotonically decreasing, convex function of $v \in (0, \infty)$.

Chapter 3

Sensitivity Analysis and Stochastic Optimization of Discrete Event Dynamic Systems (DEDS)

3.1 Introduction

This chapter deals with performance evaluation, sensitivity analysis and optimization of DEDS and in particular with nonproduct queueing networks. As we shall see below the results of this chapter extend those of chapter 2 to DEDS in the sense that DESS can be viewed as a particular case of DEDS when the length of the busy period equals one.

Assume that the DEDS is driven by an iid (independent identically distributed) sequence of random vectors $\boldsymbol{Y}_t$ generated from the pdf $f(\boldsymbol{y}, \boldsymbol{v})$ with $\boldsymbol{v}$ being a vector of parameters, say the vector of the interarrival rates, service rates and routing probabilities. Assume also that the output process $\{L_t : t > 0\}$, say the sojourn time process, the number of customers in the system, the throughput and utilization processes, which is called the *sample performance process*, settles into the steady-state and becomes a stationary and ergodic process. In typical situations the operation of DEDS starts at some initial state and it runs until stationarity is approximately obtained. Thereafter T consecutive observations $L_1, L_2, \ldots, L_T$ are taken to evaluate (estimate) the expected performance, the associated sensitivities and to optimize an entire queueing network, i.e. to solve an optimization problem of the type (P_0) (see (1.1.3)). In particular we might be interested in evaluating the expected steady-state sojourn time in the sytem and the associated sensitivities, say with respect to the parameters of the service rates and routing probabilities in an open queueing network. In addition we might be interested in minimizing the expected throughput with respect to the service rate vector in a closed queueing network subject to some cost constraints.

It is important to note that for Markovian (product form) queueing networks the program (P_0) can be solved by the conventional methods of mathematical programming since in this case both the objective function and the constraints are available analytically. Examples of such networks are given in Bertsekas and Gallager (1987) and Kleinrock (1975). We, however, abandon below the product form, that is, we assume arbitrary distributions of the interarrival and service time random variables, limited buffer sizes, priorities, etc. In this case, $\ell_j(\boldsymbol{v}) = \mathbb{E}_{\boldsymbol{v}} L_j, j = 0, 1, \ldots, M$, are no longer available analytically and we have to evaluate them either using deterministic numerical techniques or via simulation. We shall deal with simulation only.

Before proceeding further, we shall distinguish as in subsection 1.1.2 between finite-horizon (terminating) and steady-state simulations. We shall first show that sensitivity analysis and optimization of DEDS for finite-horizon is similar to that of static models discussed in chapter 2. To see this consider, for example, the conditional expectation $\ell = \ell(t_1, m)$ given in (1.2.3), which can be written as

$$\ell(\boldsymbol{v}) = \mathbb{E}_{\boldsymbol{v}} X, \tag{3.1.1}$$

and assume that we want to estimate $\nabla^k \ell(\boldsymbol{v}), k = 0, 1, \ldots$ for different values of $\boldsymbol{v}$. Here

$$X = t_1^{-1} \sum_{t=1}^{t_1} L_t(\underline{\boldsymbol{Y}}_t), \tag{3.1.2}$$

given the initial state of the system $\mathcal{L}(0) = m$; $\underline{\boldsymbol{Y}}_t = (\boldsymbol{Y}_1, \ldots, \boldsymbol{Y}_t)$; $t = 1, \ldots, t_1$, is a sequence of iid random vectors generated from the pdf $f(\boldsymbol{y}, \boldsymbol{v})$; t_1 and m are fixed.

It is straightforward to see that in this case the "what if" estimator of $\nabla^k \ell(\boldsymbol{v})$ can be written in analogy to (2.2.29) as

$$\overline{\nabla}^k \ell_N(\boldsymbol{v}) = N^{-1} \sum_{i=1}^{N} t_1^{-1} \sum_{t=1}^{t_1} L_{ti}(\underline{\boldsymbol{Z}}_{ti}) \nabla^k \widetilde{W}_{ti}(\underline{\boldsymbol{Z}}_{ti}, \boldsymbol{v}), \tag{3.1.3}$$

where $k = 0, 1, \ldots$,

$$\widetilde{W}_{ti}(\underline{\boldsymbol{Z}}_{ti}, \boldsymbol{v}) = \frac{f_t(\underline{\boldsymbol{Z}}_{ti}, \boldsymbol{v})}{g_t(\underline{\boldsymbol{Z}}_{ti})},$$

$$f_t(\underline{\boldsymbol{z}}_t, \boldsymbol{v}) = \prod_{j=1}^{t} f(\boldsymbol{z}_j, \boldsymbol{v}), \quad g_t(\underline{\boldsymbol{z}}_t) = \prod_{j=1}^{t} g(\boldsymbol{z}_j),$$

$t = 1, \ldots, t_1$, and $Z_{ti} \sim g(z)$.

With (3.1.3) at hand the algorithm for estimating the terminating parameters $\nabla^k \ell(v)$ everywhere in v can be written in analogy to algorithm 2.2.2 as follows.

Algorithm 3.1.1 : Finite-horizon simulation

1. *Perform N independent replications (simulation runs), each of length t_1, starting the simulations at the initial state $\mathcal{L}(0) = m$, and then generate at each replication i, $i = 1, \ldots, N$, a sample $\mathbf{Z}_{1i}, \ldots, \mathbf{Z}_{t_1 i}$, from the dominating pdf $g(z)$.*

2. *At each replication i, $i = 1, \ldots, N$, generate the output processes L_{ti} and $\nabla^k \widetilde{W}_{ti}$.*

3. *Estimate $\nabla^k \ell(v)$ according to (3.1.3).*

Consider now the program (P_0) (see (2.4.1)) for DEDS in the finite-horizon regime. It is readily seen that in order to construct a stochastic counterpart ($\bar{\mathrm{P}}_N$) similar to that given in (2.4.4) for static models, we have to replace $\bar{\ell}_{jN}(v)$ in (2.4.3) by the corresponding estimators $\ell_j(v)$ obtained via (3.1.3) with $k = 0$ and $L = L_j$ with all other data remaining the same. The reader is asked to modify algorithm 2.4.1 and theorem 2.4.1 for DESS by making them suitable for DEDS in the terminating state.

The rest of this chapter deals with sensitivity analysis and stochastic optimization of queueing models in the steady-state; its organization is similar to that of chapter 2. In particular, section 3.2 deals with evaluation of the performance measures $\nabla^k \ell(v), k = 0, 1, \ldots$, that is with evaluation of the system performance $\ell(v)$ and the associated sensitivities $\nabla^k \ell(v), k \geq 1$, provided that L_t is a regenerative process. It is shown that $\nabla^k \ell(v)$ can be represented as the covariance of two processes: the standard output process L_t and what we call the *k-th order score process*. The last is based on the score function. Also a convenient recursive formula for computing higher order sensitivities via the lower ones (for different values of the parameter vector v) is presented and it is shown how to calculate all of them from a single simulation experiment (sample path).

Section 3.3 deals with the so-called *decomposable and truncated estimators* introduced by Rubinstein (1992b). We show that although in general these estimators introduce some bias, the resulting variance reduction compared with the regenerative score function (SF) estimators of section 3.2 might be dramatic. We finally show how to use decomposable and truncated estimators in order to solve (from a single sample path) the so-called network topological design problem or the so-called network "what if" design problem which can be stated as follows: What will be the expected performance and the associated sensitivities of the network if we perturb the topology of the system, that is if we add (eliminate) a node (several nodes) to (from) the network.

In section 3.4 we present the *conditional score function* (CSF) method due to McLeish and Rollans (1992). We show that it typically leads to substantial variance reduction compared with the standard SF method, provided the input sequence is generated from an exponential family of distributions.

In section 3.5, which is based on Rubinstein (1991), we show how to optimize a non-Markovian queueing model from a *single* simulation experiment, or in other words how to estimate the associated optimal solution of the program (P_0). Again as in section 2.4 of chapter 2 we show that in order to do so we have to replace the deterministic program (P_0) by its stochastic counterpart ($\bar{\mathrm{P}}_N$) (see, for example, (3.4.4) below) generated from simulation. Then solving the obtained stochastic counterpart by the standard deterministic procedures we can estimate the optimal solution, say $\boldsymbol{v}^*$, of the program (P_0).

Finally, in section 3.6 we discuss the sensitivity analysis of the optimal solutions of the programs (P_0) and ($\bar{\mathrm{P}}_N$) with respect to system parameters, and robustness of the optimal solution $\boldsymbol{v}^*$ with respect to contamination of the underlying pdf $f(\boldsymbol{y}, \boldsymbol{v})$.

3.2 Sensitivity Analysis of the System Performance

3.2.1 The Score Function Process

Let $\boldsymbol{Y}_1, \boldsymbol{Y}_2, \ldots$ be an input sequence of independent identically distributed (iid) m-dimensional random vectors generated from a density function $f(\boldsymbol{y}, \boldsymbol{v})$ depending on the n-dimensional parameter vector $\boldsymbol{v}$. Consider an output process $\{L_t : t > 0\}$ driven by the input sequence $\{\boldsymbol{Y}_t\}$. That is, $L_t = L_t(\underline{\boldsymbol{Y}}_t)$, where the vector $\underline{\boldsymbol{Y}}_t = (\boldsymbol{Y}_1, \boldsymbol{Y}_2, \ldots, \boldsymbol{Y}_t)$ represents a history of the input process up to time t and $L_t(\cdot)$ is a sequence of real-valued functions. Assume that $\{L_t\}$ is a discrete time regenerative process with a regenerative cycle of length τ. It is well known in the renewal theory of regenerative processes (see, for example, Asmussen (1987)) that the expected steady-state performance can be written as

$$\ell(\boldsymbol{v}) = \frac{\mathbb{E}_{\boldsymbol{v}} X}{\mathbb{E}_{\boldsymbol{v}} \tau}, \tag{3.2.1}$$

where $X = \sum_1^{\tau} L_t$. Similar results hold when $\{L_t\}$ is a continuous time process where the sum in the definition of X is replaced by the corresponding integral.

Notice that $\ell(\boldsymbol{v})$ can be viewed as the expected reward $\mathbb{E}_{\boldsymbol{v}} X$ obtained during a cycle divided by the expected length $\mathbb{E}_{\boldsymbol{v}}\tau$ of the cycle. Unless stated otherwise we assume below that $\{L_t : t > 0\}$ is a discrete time regenerative process, for example the sojourn time process in a $GI/G/1$ queue with the FIFO discipline. In the last case, τ is the number of customers served during the busy period (the cycle). An extension of our results to different queueing disciplines and more complex queueing models will be given later on.

Before proceeding with the calculation of $\nabla^k \ell(\boldsymbol{v})$ we first consider $\ell_1(\boldsymbol{v}) = \mathbb{E}_{\boldsymbol{v}} X$. Let $\psi(x)$ be a real-valued function and let $\psi(X)$ be the associated random variable. We shall show that the gradient of $\mathbb{E}_{\boldsymbol{v}}\{\psi(X)\}$ can be represented as

$$\nabla \mathbb{E}_{\boldsymbol{v}}\{\psi(X)\} = \mathbb{E}_{\boldsymbol{v}}\{\psi(X)\widetilde{\boldsymbol{S}}_{\tau}^{(1)}(\underline{\boldsymbol{Y}}_{\tau}, \boldsymbol{v})\}, \tag{3.2.2}$$

where

$$\widetilde{\boldsymbol{S}}_t^{(1)}(\underline{\boldsymbol{y}}_t, \boldsymbol{v}) = \frac{\nabla f_t(\underline{\boldsymbol{y}}_t, \boldsymbol{v})}{f_t(\underline{\boldsymbol{y}}_t, \boldsymbol{v})} = \nabla \log f_t(\underline{\boldsymbol{y}}_t, \boldsymbol{v}) \tag{3.2.3}$$

and $f_t(\underline{\boldsymbol{y}}_t, \boldsymbol{v}) = \prod_{i=1}^t f(\boldsymbol{y}_i, \boldsymbol{v})$, $\underline{\boldsymbol{y}}_t = (\boldsymbol{y}_1, \ldots, \boldsymbol{y}_t)$.

Indeed, we can represent $\psi(X)$ as

$$\psi(X) = \sum_{r=1}^{\infty} \psi\left(\sum_{t=1}^{r} L_t\right) I_{[\tau=r]}, \tag{3.2.4}$$

where I_A denotes the indicator function of an event A. Since τ is a stopping time for $\boldsymbol{Y}_1, \boldsymbol{Y}_2, \ldots$, we have that $I_{[\tau=r]}$ depends on $\underline{\boldsymbol{Y}}_r$ only and hence

$$\mathbb{E}_{\boldsymbol{v}}\left\{\psi\left(\sum_{t=1}^{r} L_t\right) I_{[\tau=r]}\right\} = \int \psi\left[\sum_{t=1}^{r} L_t(\underline{\boldsymbol{y}}_t)\right] I_{[\tau=r]}(\underline{\boldsymbol{y}}_r) f_r(\underline{\boldsymbol{y}}_r, \boldsymbol{v}) \mathrm{d}\underline{\boldsymbol{y}}_r. \tag{3.2.5}$$

Suppose that the differentiation and integration operators in the right-hand side of (3.2.5) are interchangeable. It follows then that

$$\begin{aligned}
\nabla \mathbb{E}_{\boldsymbol{v}}\left\{\psi\left(\sum_{t=1}^{r} L_t\right) I_{[\tau=r]}\right\} &= \int \psi\left[\sum_{t=1}^{r} L_t(\underline{\boldsymbol{y}}_t)\right] I_{[\tau=r]}(\underline{\boldsymbol{y}}_r) \nabla f_r(\underline{\boldsymbol{y}}_r, \boldsymbol{v}) \mathrm{d}\underline{\boldsymbol{y}}_r \\
&= \int \psi\left[\sum_{t=1}^{r} L_t(\underline{\boldsymbol{y}}_t)\right] I_{[\tau=r]}(\underline{\boldsymbol{y}}_r) \widetilde{\boldsymbol{S}}_r^{(1)}(\underline{\boldsymbol{y}}_r, \boldsymbol{v}) f_r(\underline{\boldsymbol{y}}_r, \boldsymbol{v}) \mathrm{d}\underline{\boldsymbol{y}}_r \\
&= \mathbb{E}_{\boldsymbol{v}}\left\{\psi\left(\sum_{t=1}^{r} L_t\right) \widetilde{\boldsymbol{S}}_r^{(1)} I_{[\tau=r]}\right\}
\end{aligned}$$

which together with (3.2.4) implies (3.2.2).

Notice that for any $t = 1, 2, \ldots$, we can write

$$\widetilde{\boldsymbol{S}}_t^{(1)}(\underline{\boldsymbol{y}}_t, \boldsymbol{v}) = \sum_{j=1}^{t} \boldsymbol{S}_j^{(1)}(\boldsymbol{y}_j, \boldsymbol{v}),$$

where $\boldsymbol{S}_j^{(1)}(\boldsymbol{y}_j, \boldsymbol{v}) = \nabla \log f(\boldsymbol{y}_j, \boldsymbol{v})$.

In a similar way the higher order derivatives can be written as

$$\nabla^k \mathbb{E}_{\boldsymbol{v}}\{\psi(X)\} = \mathbb{E}_{\boldsymbol{v}}\{\psi(X)\widetilde{\boldsymbol{S}}_\tau^{(k)}\}, \tag{3.2.6}$$

where

$$\widetilde{\boldsymbol{S}}_t^{(k)} = \widetilde{\boldsymbol{S}}_t^{(k)}(\underline{\boldsymbol{Y}}_t, \boldsymbol{v}) = \frac{\nabla^k f_t(\underline{\boldsymbol{Y}}_t, \boldsymbol{v})}{f_t(\underline{\boldsymbol{Y}}_t, \boldsymbol{v})}, \quad k = 1, 2, \ldots$$

In particular for $\psi(\boldsymbol{x}) = \boldsymbol{x}$ and $\mathbb{E}_{\boldsymbol{v}} X = \ell_1(\boldsymbol{v})$ we have from (3.2.6) that

$$\nabla^k \mathbb{E}_{\boldsymbol{v}} X = \nabla^k \ell_1(\boldsymbol{v}) = \mathbb{E}_{\boldsymbol{v}} \left\{ X \widetilde{\boldsymbol{S}}_\tau^{(k)} \right\}. \tag{3.2.7}$$

We shall show now (see also Asmussen and Rubinstein (1992a)) that $\nabla^k \ell_1(\boldsymbol{v})$ can be written in the following alternative form:

$$\nabla^k \ell_1(\boldsymbol{v}) = \mathbb{E}_{\boldsymbol{v}} \left\{ \sum_{t=1}^{\tau} L_t \widetilde{\boldsymbol{S}}_t^{(k)} \right\}. \tag{3.2.8}$$

where $\widetilde{\boldsymbol{S}}_t^{(k)} = \nabla^k f_t(\underline{\boldsymbol{Y}}_t, \boldsymbol{v}) / f_t(\underline{\boldsymbol{Y}}_t, \boldsymbol{v})$. In particular,

$$\begin{aligned} \widetilde{\boldsymbol{S}}_t^{(2)} &= \nabla \widetilde{\boldsymbol{S}}_t^{(1)} + \left(\widetilde{\boldsymbol{S}}_t^{(1)}\right)\left(\widetilde{\boldsymbol{S}}_t^{(1)}\right)' = \sum_{j=1}^{t} \nabla^2 \log f(\boldsymbol{Y}_j, \boldsymbol{v}) \\ &+ \left(\sum_{j=1}^{t} \nabla \log f(\boldsymbol{Y}_j, \boldsymbol{v})\right)\left(\sum_{j=1}^{t} \nabla \log f(\boldsymbol{Y}_j, \boldsymbol{v})\right)'. \end{aligned} \tag{3.2.9}$$

To prove (3.2.8) we argue as follows. We write first $\sum_{t=1}^{\tau} L_t$ as

$$\sum_{t=1}^{\tau} L_t = \sum_{t=1}^{\infty} L_t I_{[\tau \geq t]}. \tag{3.2.10}$$

Taking into account next that $I_{[\tau \geq t]}$ depends on $\underline{\boldsymbol{Y}}_t$ only we have

$$\mathbb{E}_{\boldsymbol{v}}\{L_t I_{[\tau \geq t]}\} = \int L_t(\underline{\boldsymbol{y}}_t) I_{[\tau \geq t]}(\underline{\boldsymbol{y}}_t) f_t(\underline{\boldsymbol{y}}_t, \boldsymbol{v}) \mathrm{d}\underline{\boldsymbol{y}}_t. \tag{3.2.11}$$

Differentiating (3.2.11) with respect to $\boldsymbol{v}$ we obtain

$$\begin{aligned}
\nabla^k \mathbb{E}_{\boldsymbol{v}}\{L_t I_{[\tau \geq t]}\} &= \int L_t(\underline{\boldsymbol{y}}_t) I_{[\tau \geq t]}(\underline{\boldsymbol{y}}_t) \nabla^k f_t(\underline{\boldsymbol{y}}_t, \boldsymbol{v}) \mathrm{d}\underline{\boldsymbol{y}}_t \\
&= \int L_t(\underline{\boldsymbol{y}}_t) I_{[\tau \geq t]}(\underline{\boldsymbol{y}}_t) \widetilde{\boldsymbol{S}}_t^{(k)}(\underline{\boldsymbol{y}}_t, \boldsymbol{v}) f_t(\underline{\boldsymbol{y}}_t, \boldsymbol{v}) \mathrm{d}\underline{\boldsymbol{y}}_t \\
&= \mathbb{E}_{\boldsymbol{v}}\{L_t \widetilde{\boldsymbol{S}}_t^{(k)} I_{[\tau \geq t]}\},
\end{aligned}$$

which together with (3.2.10) implies (3.2.8).

Note that each component of $\nabla^k \ell_1(\boldsymbol{v}) = \mathbb{E}_{\boldsymbol{v}}\{\sum_1^\tau L_t \widetilde{\boldsymbol{S}}_t^{(k)}\}$ can be viewed as the expected reward associated with the process $\{L_t \widetilde{\boldsymbol{S}}_t^{(k)}, t > 0\}$ obtained during a cycle of length τ.

Example 3.2.1 Let Y be distributed gamma, that is,

$$f(y, \lambda, \beta) = \frac{\lambda^\beta y^{\beta-1} \mathrm{e}^{-\lambda y}}{\Gamma(\beta)}, \quad y > 0.$$

Assume that we are interested in the sensitivities with respect to λ. We have then that

$$\tilde{S}_t^{(1)}(\underline{Y}_t, \lambda) = \frac{\partial}{\lambda} \log f_t(\underline{Y}_t, \lambda, \beta) = t\beta\lambda^{-1} - \sum_{i=1}^t Y_i$$

and

$$\tilde{S}_t^{(2)}(\underline{Y}_t, \lambda) = \left(t\beta\lambda^{-1} - \sum_{i=1}^t Y_i\right)^2 - t\beta\lambda^{-2}.$$

Note that for $\tau = 1$, (3.2.8) corresponds to the k-th derivative of the expected system performance in *static queueing models*. As examples of such static models consider the $GI/G/1/0$ queue (without waiting room), the $GI/G/\infty$ queue (the infinite server queue) or a closed queueing network with a single customer travelling in it.

With (3.2.7) and (3.2.8) at hand we can estimate $\nabla^k \ell_1(\boldsymbol{v})$, $k = 1, \ldots,$ as

$$\widetilde{\nabla}^k \ell_{1N}(\boldsymbol{v}) = N^{-1} \sum_{i=1}^N \sum_{t=1}^{\tau_i} L_{ti} \widetilde{\boldsymbol{S}}_{\tau_i}^{(k)}, \tag{3.2.12}$$

and as

$$\overline{\nabla}^k \ell_{1N}(\boldsymbol{v}) = N^{-1} \sum_{i=1}^N \sum_{t=1}^{\tau_i} L_{ti} \widetilde{\boldsymbol{S}}_{ti}^{(k)}, \tag{3.2.13}$$

respectively. Here τ_i and N are the length of the i-th cycle and the number of generated cycles, respectively, and $\boldsymbol{Y}_{11}, \ldots, \boldsymbol{Y}_{\tau_1 1}, \ldots, \boldsymbol{Y}_{1N}, \ldots, \boldsymbol{Y}_{\tau_N N}$ is the corresponding sample. In particular

$$\widetilde{\boldsymbol{S}}_{ti}^{(1)} = \sum_{j=1}^{t} \nabla \log f(\boldsymbol{Y}_{ji}, \boldsymbol{v}), \;\; t = 1, \ldots, \tau_i, \quad i = 1, \ldots, N, \tag{3.2.14}$$

and

$$\widetilde{\boldsymbol{S}}_{ti}^{(2)} = \nabla \widetilde{\boldsymbol{S}}_{ti}^{(1)} + \left(\widetilde{\boldsymbol{S}}_{ti}^{(1)}\right)\left(\widetilde{\boldsymbol{S}}_{ti}^{(1)}\right)'. \tag{3.2.15}$$

We shall call $\widetilde{\boldsymbol{S}}_t^{(1)}$ the *generalized score function process* and $\widetilde{\nabla}^k \ell_{1N}(\boldsymbol{v})$ and $\overline{\nabla}^k \ell_{1N}(\boldsymbol{v})$ the *straightforward* and the *modified* SF estimators of $\nabla^k \ell_1(\boldsymbol{v})$, respectively. The accuracy of these estimators is discussed in Asmussen and Rubinstein (1992a). In particular they show that in heavy traffic ($\rho \to 1$) the score process $\widetilde{\boldsymbol{S}}_t^{(1)}$, when properly normalized, has a heavy traffic limit involving a certain variant of two-dimensional Brownian motion (see also below). Based on this heavy traffic (diffusion) theory they find approximations for the variance of both the modified $\overline{\nabla}^k \ell_{1N}(\boldsymbol{v})$ and the staightforward $\widetilde{\nabla}^k \ell_{1N}(\boldsymbol{v})$ estimators and show that the first is superior (has a smaller variance) to the second. Because of this we shall deal below with the modified estimators only.

With (3.2.8) and (3.2.9) at hand let us turn now to $\ell(\boldsymbol{v}) = \mathbb{E}_{\boldsymbol{v}} X / \mathbb{E}_{\boldsymbol{v}} \tau$. Differentiating $\ell(\boldsymbol{v})$ and taking into account (3.2.8) we obtain

$$\begin{aligned} \nabla \ell(\boldsymbol{v}) &= \frac{\nabla \mathbb{E}_{\boldsymbol{v}} X}{\mathbb{E}_{\boldsymbol{v}} \tau} - \frac{\mathbb{E}_{\boldsymbol{v}} X}{\mathbb{E}_{\boldsymbol{v}} \tau} \cdot \frac{\nabla \mathbb{E}_{\boldsymbol{v}} \tau}{\mathbb{E}_{\boldsymbol{v}} \tau} \\ &= \frac{\mathbb{E}_{\boldsymbol{v}} \{\sum_1^\tau L_t \widetilde{\boldsymbol{S}}_t^{(1)}\}}{\mathbb{E}_{\boldsymbol{v}} \tau} - \frac{\mathbb{E}_{\boldsymbol{v}} \{\sum_1^\tau L_t\}}{\mathbb{E}_{\boldsymbol{v}} \tau} \cdot \frac{\mathbb{E}_{\boldsymbol{v}} \{\sum_1^\tau \widetilde{\boldsymbol{S}}_t^{(1)}\}}{\mathbb{E}_{\boldsymbol{v}} \tau}. \end{aligned} \tag{3.2.16}$$

Note that $\mathbb{E}_{\boldsymbol{v}} \tau$ and $\nabla \mathbb{E}_{\boldsymbol{v}} \tau$ represent particular cases of $\mathbb{E}_{\boldsymbol{v}} X$ and $\nabla \mathbb{E}_{\boldsymbol{v}} X$, respectively, with $L_t = 1$.

It is important to note that the process $\{\widetilde{\boldsymbol{S}}_t^{(k)}\}$ can be modified to a regenerative process without altering formulas (3.2.7)–(3.2.9). To do so, let $\tau_{(1)}$ be the length of the second cycle and let for $\tau + 1 \le t \le \tau + \tau_{(1)}$ the modified process $\{\hat{\boldsymbol{S}}_t^{(k)}\}$ be defined from $\boldsymbol{Y}_{\tau+1}, \ldots, \boldsymbol{Y}_t$ in the same way as $\{\widetilde{\boldsymbol{S}}_t^{(k)}\}$ is defined from $\boldsymbol{Y}_1, \ldots, \boldsymbol{Y}_t$ when $1 \le t \le \tau$. That is,

$$\hat{\boldsymbol{S}}_t^{(k)} = \frac{\nabla^k \left\{\prod_{j=1}^{t-\tau} f(\boldsymbol{Y}_{\tau+j}, \boldsymbol{v})\right\}}{\prod_{j=1}^{t-\tau} f(\boldsymbol{Y}_{\tau+j}, \boldsymbol{v})}, \quad \tau + 1 \le t \le \tau + \tau_{(1)},$$

and so on. Obviously, this definition makes the process $\{\hat{\boldsymbol{S}}_t^{(k)}\}$ regenerative. Moreover, in all the above formulas, like (3.2.7)–(3.2.9) and (3.2.16), we may always replace $\widetilde{\boldsymbol{S}}_t^{(k)}$ by $\hat{\boldsymbol{S}}_t^{(k)}$ and vice versa. We assume subsequently that the score function process is always modified to be regenerative and shall not distinguish between $\widetilde{\boldsymbol{S}}_t^{(k)}$ and its regenerative modification $\hat{\boldsymbol{S}}_t^{(k)}$.

By $\tilde{\boldsymbol{s}}^{(k)} = \tilde{\boldsymbol{s}}^{(k)}(\boldsymbol{v})$ we denote the expected long-run average (the expected steady-state performance) of the regenerative process $\tilde{\boldsymbol{S}}_t^{(k)}$. Note again that by the renewal theory

$$\tilde{\boldsymbol{s}}^{(k)} = \frac{\mathbb{E}_{\boldsymbol{v}}\{\sum_1^{\tau} \tilde{\boldsymbol{S}}_t^{(k)}\}}{\mathbb{E}_{\boldsymbol{v}}\tau}.$$

Consider now the regenerative process $L_t\boldsymbol{S}_t^{(1)}$. It follows from (3.2.16) that $\nabla\ell(\boldsymbol{v})$ can be represented as the expected steady-state performance of the process

$$\boldsymbol{Q}_t^{(1)} = (L_t - \ell(\boldsymbol{v}))\left(\widetilde{\boldsymbol{S}}_t^{(1)} - \tilde{\boldsymbol{s}}^{(1)}\right). \tag{3.2.17}$$

That is, we can rewrite (3.2.16) as

$$\nabla\ell(\boldsymbol{v}) = \mathbb{E}_{\boldsymbol{v}}\{\boldsymbol{Q}^{(1)}\} = \mathrm{Cov}_{\boldsymbol{v}}\{L, \widetilde{\boldsymbol{S}}^{(1)}\} = \frac{\mathbb{E}_{\boldsymbol{v}}\{\sum_1^{\tau} \boldsymbol{Q}_t^{(1)}\}}{\mathbb{E}_{\boldsymbol{v}}\tau}. \tag{3.2.18}$$

Thus, the gradient $\nabla\ell(\boldsymbol{v})$ can be expressed as the *covariance* between the steady-state sample performance process $\{L_t\}$ and the score function processes $\{\widetilde{\boldsymbol{S}}_t^{(1)}\}$, respectively. The last is based on the score function $\nabla \log f(\boldsymbol{Y}, \boldsymbol{v})$. It is also important to note that the only difference between (3.2.1) and (3.2.18) is that L_t is replaced by

$$\boldsymbol{Q}_t^{(1)} = (L_t - \ell(\boldsymbol{v}))(\widetilde{\boldsymbol{S}}_t^{(1)} - \tilde{\boldsymbol{s}}^{(1)}).$$

We say that in this case the gradient $\nabla\ell(\boldsymbol{v})$ is *embedded* in the regenerative framework in the sense that it is expressed as the ratio between the expected reward $\mathbb{E}_{\boldsymbol{v}}\{\sum_1^{\tau} \boldsymbol{Q}_t^{(1)}\}$ obtained during the cycle of length τ and the expected length of the cycle $\mathbb{E}_{\boldsymbol{v}}\tau$. Notice that the process $\boldsymbol{Q}_t^{(1)}$ is regenerative as the product of two regenerative processes. Let $\tau = 1$ (e.g. the $GI/G/1/0$ queue). In this case $\widetilde{\boldsymbol{S}}_t^{(1)} = \boldsymbol{S}_t^{(1)}$ and (3.2.16) reduces to

$$\nabla\ell(\boldsymbol{v}) = \mathbb{E}_{\boldsymbol{v}}\{L\boldsymbol{S}^{(1)}\} = \mathrm{Cov}_{\boldsymbol{v}}\{L, \boldsymbol{S}^{(1)}\},$$

which coincides with the representation of $\nabla\ell(\boldsymbol{v})$ for *static models*.

Proceeding recursively with (3.2.18) and taking into account (3.2.8) we obtain

$$\nabla^k\ell(\boldsymbol{v}) = \mathbb{E}_{\boldsymbol{v}}\boldsymbol{Q}^{(k)} = \frac{\mathbb{E}_{\boldsymbol{v}}\{\sum_1^{\tau} \boldsymbol{Q}_t^{(k)}\}}{\mathbb{E}_{\boldsymbol{v}}\tau}. \tag{3.2.19}$$

In particular for $k = 2$ we have

$$\boldsymbol{Q}_t^{(2)} = (L_t - \ell)\left[\left(\widetilde{\boldsymbol{S}}_t^{(1)} - \widetilde{\boldsymbol{s}}^{(1)}\right)\left(\widetilde{\boldsymbol{S}}_t^{(1)} - \widetilde{\boldsymbol{s}}^{(1)}\right)' + \left(\nabla\widetilde{\boldsymbol{S}}_t^{(1)} - \widetilde{\boldsymbol{s}}^{(2)}\right)\right],$$

where $\widetilde{\boldsymbol{s}}^{(2)}$ can also be written in the form $\tilde{\boldsymbol{s}}^{(2)} = \mathbb{E}_{\boldsymbol{v}}\{\nabla\tilde{\boldsymbol{S}}_t^{(1)}\}$.

Example 3.2.2 Let L_t be the steady-state sojourn time of a customer in the $GI/G/1$ system. In this case $\ell(\boldsymbol{v}) = \mathbb{E}_{\boldsymbol{v}} L_t$ can be written as

$$\ell(\boldsymbol{v}) = \frac{\mathbb{E}_{\boldsymbol{v}}\{\sum_1^\tau L_t\}}{\mathbb{E}_{\boldsymbol{v}}\tau} = \frac{\mathbb{E}_{\boldsymbol{v}}\left\{\sum_{t=1}^\tau \sum_{j=1}^t Y_{1j} - \sum_{t=2}^\tau \sum_{j=2}^t Y_{2j}\right\}}{\mathbb{E}_{\boldsymbol{v}}\tau},$$

where Y_{1j} is the service time of the j-th customer, $Y_{2j} = 0$ for $j = 1$, $Y_{2j} = A_j - A_{j-1}$ for $j \geq 2$, A_j is the time of the arrival of the j-th customer and $\tau = \min\{t : \sum_1^t (Y_{1j} - Y_{2j+1}) \leq 0\}$ is the number of customers served during the busy period. Denoting $U_j = Y_{1j} - Y_{2j}$ we can rewrite $\ell(\boldsymbol{v})$ as

$$\ell(\boldsymbol{v}) = \frac{\mathbb{E}_{\boldsymbol{v}}\{\sum_{t=1}^\tau \sum_{j=1}^t U_j\}}{\mathbb{E}_{\boldsymbol{v}}\tau}.$$

In this case

$$\nabla\ell(\boldsymbol{v}) = \frac{\mathbb{E}_{\boldsymbol{v}}\{\sum_1^\tau \boldsymbol{Q}_t^{(1)}\}}{\mathbb{E}_{\boldsymbol{v}}\tau},$$

where

$$\boldsymbol{Q}_t^{(1)} = \left(\sum_{j=1}^t U_j - \ell\right)\left(\widetilde{\boldsymbol{S}}_t^{(1)} - \widetilde{\boldsymbol{s}}^{(1)}\right)$$

and

$$\widetilde{\boldsymbol{S}}_t^{(1)} = \sum_{j=1}^t \nabla \log f(\boldsymbol{Y}_j, \boldsymbol{v}), \;\; f(\boldsymbol{y}, \boldsymbol{v}) = f_1(y_1, v_1) f_2(y_2, v_2),$$

$$\boldsymbol{Y} = (Y_1, Y_2), \;\; Y_1 \sim f_1(y_1, v_1), \;\; Y_2 \sim f_2(y_2, v_2).$$

Denoting $\ell(\boldsymbol{v}) \equiv \nabla^0\ell(\boldsymbol{v})$ and $L_t \equiv \boldsymbol{Q}_t^{(0)}$ we can extend $\nabla^k\ell(\boldsymbol{v})$ in (3.2.19) to the case k =0 as well. Since $\nabla^k\ell(\boldsymbol{v})$ is given in the *closed form* , both the *performance* $\ell(\boldsymbol{v})$ and *all* the *sensitivities* $\nabla^k\ell(\boldsymbol{v})$, $k = 1, 2, \ldots$, can be evaluated either numerically, say by using the approach taken by Whitt (1983) in his "Queueing Network Analyzer", or through simulation. Simulation is particularly convenient here since all the quantities $\nabla^k\ell(\boldsymbol{v})$, $k = 0, 1, \ldots$, (see (3.2.19)) are expressed as the expectations taken with respect to the same underlying pdf $f(\boldsymbol{y}, \boldsymbol{v})$. We shall deal with simulation only.

Let $\boldsymbol{Y}_{11}, \ldots, \boldsymbol{Y}_{\tau_1 1}, \ldots, \boldsymbol{Y}_{1N}, \ldots, \boldsymbol{Y}_{\tau_N N}$ be a sample of N regenerative cycles generated from the pdf $f(\boldsymbol{y}, \boldsymbol{v})$. Then taking into account (3.2.19) we can estimate all the quantities $\nabla^k \ell(\boldsymbol{v}),\ k = 0, 1, \ldots$, from a *single* simulation run as

$$\overline{\nabla}^k \ell_N(\boldsymbol{v}) = \frac{\sum_{i=1}^N \sum_{t=1}^{\tau_i} \widetilde{\boldsymbol{Q}}_{ti}^{(k)}}{\sum_{i=1}^N \sum_{t=1}^{\tau_i} 1}, \quad k = 0, 1, \ldots, \tag{3.2.20}$$

where

$$\widetilde{\boldsymbol{Q}}_{ti}^{(0)} = \boldsymbol{Q}_{ti}^{(0)} = L_{ti},\ \widetilde{\boldsymbol{Q}}_{ti}^{(1)} = \left(L_{ti} - \overline{\ell}_N\right)\left(\widetilde{\boldsymbol{S}}_{ti}^{(1)} - \overline{\boldsymbol{s}}_N^{(1)}\right), \tag{3.2.21}$$

and so on for $k > 1$. Here $\overline{\ell}_N$ and $\overline{\boldsymbol{s}}_N^{(1)}$ are the sample estimators of $\ell = \mathbb{E}_{\boldsymbol{v}}\{\sum_1^\tau L_t\}/\mathbb{E}_{\boldsymbol{v}}\tau$ and $\widetilde{\boldsymbol{s}}^{(1)} = \mathbb{E}_{\boldsymbol{v}}\{\sum_1^\tau \widetilde{\boldsymbol{S}}_t^{(1)}\}/\mathbb{E}_{\boldsymbol{v}}\tau$, respectively.

The algorithm for estimating $\nabla^k \ell(\boldsymbol{v}),\ k = 0, 1, \ldots$, from a single simulation can be written as follows:

Algorithm 3.2.1 :

1. *Generate a random sample* $\boldsymbol{Y}_1, \ldots, \boldsymbol{Y}_T$, $T = \sum_1^N \tau_i$, *from* $f(\boldsymbol{y}, \boldsymbol{v})$.

2. *Generate the output process* L_t *and the score function process* $\widetilde{\boldsymbol{S}}_t^{(k)}$, $t = 1, \ldots, T$.

3. *Calculate the estimator* $\overline{\nabla}^k \ell_N(\boldsymbol{v})$ *according to (3.2.20).*

Note again that in order to estimate the performance $\ell(\boldsymbol{v})$ and *all* the sensitivities $\nabla^k \ell(\boldsymbol{v}), k = 0, 1, \ldots$, we *neither* have to differentiate the output process L_t (in most practical applications $L_t(\underline{\boldsymbol{y}}_t)$ are *not smooth* functions), *nor* have to know the analytic expression for L_t. What we *only* need to have at out disposal are the realizations (samples) of the processes L_t and $\widetilde{\boldsymbol{S}}_t^{(k)}$. It is important to note that the SF estimators $\overline{\nabla}^k \ell_N(\boldsymbol{v}),\ k = 0, 1, \ldots$ (see (3.2.20) and (3.2.21)) allow us to evaluate the performance $\ell(\boldsymbol{v})$ and all the sensitivities $\nabla^k \ell(\boldsymbol{v}),\ k = 1, \ldots$, at a *fixed* point $\boldsymbol{v}$ *only*.

We present now the so-called extended or "what-if" version of the above estimators which allow us to evaluate $\ell(\boldsymbol{v})$ and $\nabla^k \ell(\boldsymbol{v}),\ k = 1, \ldots$, at *any* point $\boldsymbol{v}$, provided some regularity conditions hold. To do so let us choose a probability distribution G on $\mathbb{R}^m$ having a density function $g(\boldsymbol{z})$, that is $\mathrm{d}G(\boldsymbol{z}) = g(\boldsymbol{z})\mathrm{d}\boldsymbol{z}$. Assume that $g(\boldsymbol{z})$ dominates the densities $f(\boldsymbol{z}, \boldsymbol{v})$ in the sense that

$$\operatorname{supp} f(\boldsymbol{z}, \boldsymbol{v}) \subset \operatorname{supp} g(\boldsymbol{z}) \ \text{ for all } \boldsymbol{v} \in V.$$

Before proceeding with $\nabla^k \ell(\boldsymbol{v})$ we consider again $\ell_1(\boldsymbol{v}) = \mathbb{E}_{\boldsymbol{v}} X$. We have

$$\sum_{t=1}^{\tau} L_t = \sum_{t=1}^{\infty} L_t I_{[\tau \geq t]} \tag{3.2.22}$$

and

$$\begin{aligned}\mathbb{E}_{\boldsymbol{v}}\{L_t I_{[\tau\geq t]}\} &= \int L_t(\underline{\boldsymbol{z}}_t) I_{[\tau\geq t]}(\underline{\boldsymbol{z}}_t) f_t(\underline{\boldsymbol{z}}_t, \boldsymbol{v}) \mathrm{d}\underline{\boldsymbol{z}}_t \\ &= \int L_t(\underline{\boldsymbol{z}}_t) I_{[\tau\geq t]}(\underline{\boldsymbol{z}}_t) \widetilde{W}_t(\underline{\boldsymbol{z}}_t, \boldsymbol{v}) g_t(\underline{\boldsymbol{z}}_t) \mathrm{d}\underline{\boldsymbol{z}}_t, \quad (3.2.23)\end{aligned}$$

where

$$\widetilde{W}_t(\underline{\boldsymbol{z}}_t, \boldsymbol{v}) = \frac{f_t(\underline{\boldsymbol{z}}_t, \boldsymbol{v})}{g_t(\underline{\boldsymbol{z}}_t)}$$

and

$$f_t(\underline{\boldsymbol{z}}_t, \boldsymbol{v}) = \prod_{i=1}^{t} f(\boldsymbol{z}_i, \boldsymbol{v}), \quad g_t(\underline{\boldsymbol{z}}_t) = \prod_{i=1}^{t} g(\boldsymbol{z}_i).$$

It follows from (3.2.22) and (3.2.23) that

$$\ell_1(\boldsymbol{v}) = \mathbb{E}_g \left\{ \sum_{t=1}^{\tau} L_t(\underline{\boldsymbol{Z}}_t) \widetilde{W}_t(\underline{\boldsymbol{Z}}_t, \boldsymbol{v}) \right\}, \quad (3.2.24)$$

where

$$\underline{\boldsymbol{Z}}_t = (\boldsymbol{Z}_1, \ldots, \boldsymbol{Z}_t) \sim g_t(\underline{\boldsymbol{z}}_t), \ \widetilde{W}_t = \prod_1^t W_j \text{ and } W_j = f(\boldsymbol{Z}_j, \boldsymbol{v})/g(\boldsymbol{Z}_j).$$

Similarly

$$\ell_2(\boldsymbol{v}) = \mathbb{E}_{\boldsymbol{v}} \tau = \mathbb{E}_g \left\{ \sum_{t=1}^{\tau} \widetilde{W}_t(\underline{\boldsymbol{Z}}_t, \boldsymbol{v}) \right\}.$$

Under standard regularity conditions ensuring the interchangeability of the differentiation and the expected value operators (see theorem 6.1.1 and discussion in section 6.1 of chapter 6), we can now write

$$\nabla^k \ell_1(\boldsymbol{v}) = \mathbb{E}_g \left\{ \sum_{t=1}^{\tau} L_t(\underline{\boldsymbol{Z}}_t) \nabla^k \widetilde{W}_t(\underline{\boldsymbol{Z}}_t, \boldsymbol{v}) \right\}. \quad (3.2.25)$$

For $k = 1, 2, \ldots$, we call $\boldsymbol{G}_t^k = L_t \nabla^k \widetilde{W}_t$ and $\nabla^k \widetilde{W}_t$ the *sensitivity* and the *score function processes of order* k, and for $k = 0$ we call $\boldsymbol{G}_t^0 = L_t \nabla^0 \widetilde{W}_t = L_t \widetilde{W}_t$ and $\nabla^0 \widetilde{W}_t = \widetilde{W}_t$ the *sample performance* and the *likelihood ratio processes*, respectively. Note that for $k = 1, 2, \ldots$ we have $\nabla^k \widetilde{W}_t = \widetilde{W}_t \widetilde{\boldsymbol{S}}_t^{(k)}$ (assuming that the process $\nabla^k \widetilde{W}_t$ is made regenerative in a way similar to the process $\tilde{\boldsymbol{S}}_t^{(k)}$). For the particular case where $g(\boldsymbol{z}) = f(\boldsymbol{z}, \boldsymbol{v})$ the process $\nabla^k \widetilde{W}_t$ reduces to $\widetilde{\boldsymbol{S}}_t^{(k)}$, that is to the score function process.

Let now $\boldsymbol{Z}_{11},\ldots,\boldsymbol{Z}_{\tau_1 1},\ldots,\boldsymbol{Z}_{1N},\ldots,\boldsymbol{Z}_{\tau_N N}$ be a sample of N regenerative cycles generated from the pdf $g(\boldsymbol{z})$. Then taking into account (3.2.25), we can estimate $\nabla^k \ell_1(\boldsymbol{v}),\quad k=0,1,\ldots$, from a *single simulation* as follows

$$\overline{\nabla}^k \ell_{1N}(\boldsymbol{v}) = N^{-1}\sum_{i=1}^{N}\sum_{t=1}^{\tau_i} L_t(\underline{\boldsymbol{Z}}_{ti})\nabla^k \widetilde{W}_t(\underline{\boldsymbol{Z}}_{ti},\boldsymbol{v}). \tag{3.2.26}$$

Notice that here $\nabla^k \bar{\ell}_{1N}(\boldsymbol{v}) = \overline{\nabla}^k \ell_{1N}(\boldsymbol{v})$. Notice also that the estimator (3.1.3) for finite-horizon simulation can be obtained from the estimator (3.2.26) while assuming $\tau_i = t_1$ and $L = X$.

Let us turn now to $\ell(\boldsymbol{v}) = \mathbb{E}_{\boldsymbol{v}} X/\mathbb{E}_{\boldsymbol{v}}\tau$. Taking into account (3.2.25) and noting that $\tau_i = \sum_1^\tau 1$ is a particular case of $X = \sum_1^\tau L_t$ with $L_t = 1$ we can write $\ell(\boldsymbol{v})$ and $\nabla \ell(\boldsymbol{v})$ as

$$\ell(\boldsymbol{v}) = \frac{\mathbb{E}_g\{\sum_1^\tau L_t \widetilde{W}_t\}}{\mathbb{E}_g\{\sum_1^\tau \widetilde{W}_t\}} \tag{3.2.27}$$

and

$$\nabla \ell(\boldsymbol{v}) = \frac{\mathbb{E}_g\{\sum_1^\tau L_t \nabla \widetilde{W}_t\}}{\mathbb{E}_g\{\sum_1^\tau \widetilde{W}_t\}} - \frac{\mathbb{E}_g\{\sum_1^\tau L_t \widetilde{W}_t\}}{\mathbb{E}_g\{\sum_1^\tau \widetilde{W}_t\}}\cdot\frac{\mathbb{E}_g\{\sum_1^\tau \nabla \widetilde{W}_t\}}{\mathbb{E}_g\{\sum_1^\tau \widetilde{W}_t\}}, \tag{3.2.28}$$

respectively.

Proceeding with (3.2.28) we can obtain higher order partial derivatives. Notice that when $g(\boldsymbol{z}) = f(\boldsymbol{z},\boldsymbol{v})$, (3.2.27) and (3.2.28) reduce to (3.2.19) with $k = 0$ and $k = 1$, respectively. Note, however, that for $g(\boldsymbol{z}) \neq f(\boldsymbol{z},\boldsymbol{v})$ the sample sensitivities $\nabla^k \ell(\boldsymbol{v})$ (see, for example, (3.2.28) for $k = 1$) cannot be presented similarly to (3.2.19), that is, as the covariance between the sample performance process L_t and the score function process $\nabla^k \widetilde{W}_t$.

Taking into account (3.2.27) and (3.2.28) we can estimate $\ell(\boldsymbol{v})$ and $\nabla \ell(\boldsymbol{v})$ as

$$\bar{\ell}_N(\boldsymbol{v}) = \frac{\sum_1^N \sum_1^{\tau_i} L_{ti}\widetilde{W}_{ti}}{\sum_1^N \sum_1^{\tau_i} \widetilde{W}_{ti}} \tag{3.2.29}$$

and

$$\overline{\nabla} \ell_N(\boldsymbol{v}) = \frac{\sum_1^N \sum_1^{\tau_i} L_{ti}\nabla \widetilde{W}_{ti}}{\sum_1^N \sum_1^{\tau_i} \widetilde{W}_{ti}} - \frac{\sum_1^N \sum_1^{\tau_i} L_{ti}\widetilde{W}_{ti}}{\sum_1^N \sum_1^{\tau_i} \widetilde{W}_{ti}}\cdot\frac{\sum_1^N \sum_1^{\tau_i} \nabla \widetilde{W}_{ti}}{\sum_1^N \sum_1^{\tau_i} \widetilde{W}_{ti}}, \tag{3.2.30}$$

respectively. In this case again $\nabla \bar{\ell}_N(\boldsymbol{v}) = \overline{\nabla} \ell_N(\boldsymbol{v})$, where $\nabla \bar{\ell}_N(\boldsymbol{v})$ represents the gradient of the estimator $\bar{\ell}_N(\boldsymbol{v})$. Note also that similar estimators can be derived for $\nabla^k \ell(\boldsymbol{v}),\ k = 2,3\ldots$, by differentiating $\bar{\ell}_N(\boldsymbol{v})$ k times.

The algorithm for estimating the gradient $\nabla \ell(\boldsymbol{v})$ with the sensitivity estimator (3.2.30) can be written as follows.

Algorithm 3.2.2 :

1. *Generate a random sample* $\boldsymbol{Z}_1, \ldots, \boldsymbol{Z}_T$, $T = \sum_1^N \tau_i$, *from* $g(\boldsymbol{z})$.
2. *Generate the output processes* L_t *and* $\nabla \widetilde{W}_t = \widetilde{W}_t \widetilde{\boldsymbol{S}}_t^{(1)}$.
3. *Calculate* $\nabla \bar{\ell}_N(\boldsymbol{v})$ *according to (3.2.30).*

Remark 3.2.1 With any dominating pdf $g(\boldsymbol{z})$ at hand there are two alternative ways of estimating $\nabla \ell(\boldsymbol{v})$ with the sensitivity estimator $\nabla \bar{\ell}_N(\boldsymbol{v})$:

(a) estimate $\ell(\boldsymbol{v})$ as in (3.2.29) and then differentiate $\bar{\ell}_N(\boldsymbol{v})$ to obtain $\nabla \bar{\ell}_N(\boldsymbol{v})$;

(b) derive first (3.2.28) by applying the Radon-Nikodym derivative to each term of $\nabla \ell(\boldsymbol{v})$ in (3.2.16). Then estimate $\nabla \ell(\boldsymbol{v})$ in (3.2.28) by using $\overline{\nabla \ell}_N(\boldsymbol{v})$ which is equal to $\nabla \bar{\ell}_N(\boldsymbol{v})$.

Let us define now SF estimators for more general queueing models assuming as before that $\{L_t : t > 0\}$ is the sojourn time process. It is not difficult to see that the basic formulas (3.2.27)–(3.2.30) developed for the $GI/G/1$ queue with the FIFO discipline can be extended to rather general queueing models in the sense that (3.2.27)–(3.2.30) still hold provided the *indexing* in the likelihood ratio process $\widetilde{W}_t$ and the associated quantities like $\sum_1^\tau L_t \widetilde{W}_t$ and $\sum_1^\tau L_t \nabla \widetilde{W}_t$ are defined in a more sophisticated way. More specifically consider

(i) a $GI/G/1/b$ queue, where b is the buffer size,

(ii) a $GI/G/1$ queue in various queueing disciplines,

(iii) a general r-node queueing networks.

In case (i) we assume that we vary (perturb) the parameter vector $\boldsymbol{v}$ of the interarrival time pdf, while in cases (ii) and (iii) we assume that we vary the parameter vector $\boldsymbol{v} = (v_1, \ldots, v_r)$ of the service time pdf's.

(i) In this case the index $t = 1, \ldots, \tau$ in L_t and $\widetilde{W}_t$ must correspond to the order in which the customers *arrive* at the system, and in which they *enter service*, that is the interarrival random variables associated with the rejected customers must also be taken into account in both L_t and $\widetilde{W}_t$. As an example, consider the $GI/G/1/2$ queue. Assume that 6 customers attempted to enter the system during a given regenerative cycle. Assume that the customers numbered 4 and 5 were prevented from entering the system (the buffer was full). In this case, clearly, the cycle has length $\tau = 4$, and we have

$$\widetilde{W}_t = \prod_1^t W_i, \quad t = 1,\ 2\ ,3, \quad \widetilde{W}_4 = \prod_1^6 W_i.$$

(ii) Consider, for instance, the nonpreemptive LIFO (Last In First Out) $GI/G/1$ queue. In this case the index $t = 1, \ldots, r$ in L_t and $\widetilde{W}_t$ must correspond to the order in which the customers *enter the service*, and not to the order in which they *arrive* at the system. That is, the likelihood ratio process $\widetilde{W}_t$ associated with the t-th customer is the product of the likelihood ratios of customers entering service before him. Note that exactly the same conclusion holds for the PS (Processor Sharing) queueing discipline.

(iii) Consider, for example, the sojourn time in a general open queueing network with r nodes, and suppose that only the service time densities $f_i(\cdot, \boldsymbol{v}_i)$, $i = 1, \ldots, r$, are varied. We here let the index t correspond to all events of the event clock of the discrete events simulation. These events should include at least all instances of the time when a customer enters service and all departure epochs. If t corresponds to the time a customer enters the service at node i, a service time is drawn according to $f_i(\cdot, \boldsymbol{v})$. Note that it may not always be necessary to draw a random vector $\underline{\boldsymbol{Y}}_t$, but for formal reasons we can always assume that this is the case (certain $\underline{\boldsymbol{Y}}_t$ may be dummy). If t corresponds to a departure epoch, we let L_t be the sojourn time of the departing customer, in all other cases we let $L_t = 0$.

More formally, let

$$\underline{\boldsymbol{Y}}_t = \left(\underline{\boldsymbol{Y}}_t^{\,1}, \underline{\boldsymbol{Y}}_t^{\,2}, \ldots, \underline{\boldsymbol{Y}}_t^{\,r}\right)$$

be an iid sequence, where each $\underline{\boldsymbol{Y}}_t^{\,i}$ is by itself a sequence of independent random vectors generated from the same distribution function, say $f_i(\boldsymbol{y}, \boldsymbol{v}_i)$, and $\boldsymbol{v}_i$, $i = 1, \ldots, r$, is a vector of parameters. Let also C_t^i be the set of indices of the random variables Y in $\underline{\boldsymbol{Y}}_t^{\,i}$ such that $\underline{\boldsymbol{Y}}_t^{\,i} = \{Y_j, j \in C_t^i\}$. (Note that each Y_j presents a random variable rather than a random vector.) With these definitions at hand we can now present $\widetilde{W}_t$ as

$$\widetilde{W}_t = \prod_{i=1}^{r} \prod_{j \in C_t^i} W_j \tag{3.2.31}$$

and similarly $\nabla^k \widetilde{W}_t$. Note that for r $GI/G/1$ queues in tandem with the FIFO discipline (3.2.31) reduces to $\widetilde{W}_t = \prod_{i=1}^{r} \prod_{j=1}^{t} W_{ji}$. More details on the issue of indexing the likelihood ratio process $\widetilde{W}_t$ for open queueing networks with various topologies and their implementation in the QNSO simulation package are given in Perez-Luna (1990).

Remark 3.2.2 Similar methods can also be applied for closed queueing networks, provided some conditions on the service time distributions are imposed (see Borovkov (1986), Daduna (1988), Kaspi and Mandelbaum (1991), and Sigman (1990)). Although these conditions vary in the above references, all of them require that the regeneration scheme is positive recurrent with the regeneration points occuring when all the customers queue up in front of one specific server and he finishes serving the customer at that particular queue.

At this end we whish point out that Pflug (1988), (1989), (1990), (1991), (1992) introduces an alternative method for sensitivity analysis and stochastic optimization of DEDS, called the *weak derivatives*. He proves convergence of his method and presents examples (see Pflug (1992)) where, according to certain criteria, weak derivatives outperform the SF method. Although Pflug's method certainly has potential, his main disadvantage is that *two* simulations are required for estimating a derivative.

3.2.2 The Case of Unknown Parameters

Consider the estimator (3.2.13) which can be rewritten as

$$\bar{\nabla}^k \ell_{1N}(\boldsymbol{v}) = N^{-1} \sum_{i=1}^{N} \sum_{t=1}^{\tau_i} L_t(\underline{\boldsymbol{Y}}_{ti}) \tilde{\boldsymbol{S}}_t^{(k)}(\underline{\boldsymbol{Y}}_{ti}, \boldsymbol{v}). \tag{3.2.32}$$

Assume that the true value of the parameter vector $\boldsymbol{v}$ is unknown and must be estimated. Let $\hat{\boldsymbol{v}} = \hat{\boldsymbol{v}}(\underline{\boldsymbol{Y}}_{\tau_1 1}, \ldots, \underline{\boldsymbol{Y}}_{\tau_N N})$ be a consistent estimator of $\boldsymbol{v}$. Then it is natural to use

$$\bar{\nabla}^k \ell_{1N}(\hat{\boldsymbol{v}}) = N^{-1} \sum_{i=1}^{N} \sum_{t=1}^{\tau_i} L_t(\underline{\boldsymbol{Y}}_{ti}) \tilde{\boldsymbol{S}}_t^{(k)}(\underline{\boldsymbol{Y}}_{ti}, \hat{\boldsymbol{v}}) \tag{3.2.33}$$

as an estimator of $\nabla^k \ell(\boldsymbol{v})$. Note that $\bar{\nabla}^k \ell_{1N}(\hat{\boldsymbol{v}})$ differs from the standard SF estimator $\bar{\nabla}^k \ell_{1N}(\boldsymbol{v})$ only in that the unknown value $\boldsymbol{v}$ of the parameter vector is replaced by its estimator $\hat{\boldsymbol{v}}$.

We discuss below the statistical properties of the estimator $\bar{\nabla}^k \ell_{1N}(\hat{\boldsymbol{v}})$ (for similar results see also Heidelberger and Towsley (1989)). The statistical properties of the "what if" estimator $\bar{\nabla}^k \ell_N(\boldsymbol{v})$ (see (3.2.26) and (3.2.30)) with $\boldsymbol{v}$ replaced by $\hat{\boldsymbol{v}}$ can be treated similarly and are left as an exercise.

We can write $\bar{\nabla}^k \ell_{1N}(\hat{\boldsymbol{v}})$ as

$$\bar{\nabla}^k \ell_{1N}(\hat{\boldsymbol{v}}) = \bar{\nabla}^k \ell_{1N}(\boldsymbol{v}) + \boldsymbol{R}_N, \tag{3.2.34}$$

where

$$\boldsymbol{R}_N = \bar{\nabla}^k \ell_{1N}(\hat{\boldsymbol{v}}) - \bar{\nabla}^k \ell_{1N}(\boldsymbol{v}) = N^{-1} \sum_{i=1}^{N} \sum_{t=1}^{\tau_i} L_t(\underline{\boldsymbol{Y}}_{ti}) [\tilde{\boldsymbol{S}}_t^{(k)}(\underline{\boldsymbol{Y}}_{ti}, \hat{\boldsymbol{v}}) - \tilde{\boldsymbol{S}}_{ti}^{(k)}(\underline{\boldsymbol{Y}}_{ti}, \boldsymbol{v})] \tag{3.2.35}$$

We have (by the Law of Large Numbers) that $\bar{\nabla}^k \ell_{1N}(\boldsymbol{v})$ converges w.p.1 to $\nabla^k \ell_1(\boldsymbol{v})$. Therefore, in order to establish strong consistency of $\bar{\nabla}^k \ell_{1N}(\hat{\boldsymbol{v}})$ we need to verify that $\boldsymbol{R}_N$ converges w.p.1 to $\boldsymbol{0}$. To do so, suppose that:

(i) $\hat{\boldsymbol{v}}$ is a strongly consistent estimator of $\boldsymbol{v}$.

(ii) There exist functions $K_t(\underline{\boldsymbol{y}}_t)$, $t = 1, 2, \ldots$, such that

$$\|\tilde{\boldsymbol{S}}_t^{(k)}(\underline{\boldsymbol{y}}_t, \boldsymbol{u}) - \tilde{\boldsymbol{S}}_t^{(k)}(\underline{\boldsymbol{y}}_t, \boldsymbol{v})\| \le K_t(\underline{\boldsymbol{y}}_t) \|\boldsymbol{u} - \boldsymbol{v}\| \tag{3.2.36}$$

for all $\boldsymbol{u}$ in a neighborhood of $\boldsymbol{v}$.

(iii) The expectation

$$\mathbb{E}_{\boldsymbol{v}} \left\{ \sum_{t=1}^{\tau} |L_t(\underline{\boldsymbol{Y}}_t) K_t(\underline{\boldsymbol{Y}}_t)| \right\}$$

is finite.

We have then that for all $\hat{\boldsymbol{v}}$ sufficiently close to $\boldsymbol{v}$

$$\begin{aligned} \|\boldsymbol{R}_N\| &\le N^{-1} \sum_{i=1}^{N} \sum_{t=1}^{\tau_i} |L_t(\underline{\boldsymbol{Y}}_{ti})| \, \|\tilde{\boldsymbol{S}}_t^{(k)}(\underline{\boldsymbol{Y}}_{ti}, \hat{\boldsymbol{v}}) - \tilde{\boldsymbol{S}}_t^{(k)}(\underline{\boldsymbol{Y}}_{ti}, \boldsymbol{v})\| \\ &\le \|\hat{\boldsymbol{v}} - \boldsymbol{v}\| N^{-1} \sum_{i=1}^{N} \sum_{t=1}^{\tau_i} |L_t(\underline{\boldsymbol{Y}}_{ti}) K_t(\underline{\boldsymbol{Y}}_{ti})| \,. \end{aligned} \tag{3.2.37}$$

Since $\hat{\boldsymbol{v}}$ converges w.p.1 to $\boldsymbol{v}$ and, by the Law of Large Numbers, $N^{-1} \sum_{i=1}^{N} \sum_{t=1}^{\tau_i} |L_t(\underline{\boldsymbol{Y}}_{ti}) K_t(\underline{\boldsymbol{Y}}_{ti})|$ converges w.p.1 to the corresponding expected value, we obtain from (3.2.35) that $\|\boldsymbol{R}_N\| \to 0$ w.p.1. It follows then that $\bar{\nabla}^k \ell_{1N}(\hat{\boldsymbol{v}})$ is a strongly consistent estimator of $\nabla^k \ell_1(\boldsymbol{v})$. Similar analysis can be applied to the estimator $\bar{\nabla}^k \ell_{2N}(\hat{\boldsymbol{v}})$ of $\nabla^k \ell_2(\boldsymbol{v})$, as well, and hence to the estimator $\bar{\nabla}^k \ell_N(\hat{\boldsymbol{v}})$ of $\nabla^k \ell(\boldsymbol{v})$.

We establish now the asymptotic normality of the vector $\bar{\nabla} \ell_{1N}(\hat{\boldsymbol{v}})$ for $k = 1$ and $\tau = 1$ (static model), while considering the exponential family in the canonical form (2.3.10). Note that here $\bar{\nabla} \ell_{1N} = \bar{\nabla} \ell_N(\boldsymbol{v})$.

In order to proceed further note first (see (2.3.15)) that for any $\boldsymbol{y}$ we have

$$\boldsymbol{S}^{(1)}(\boldsymbol{y}, \boldsymbol{u}) - \boldsymbol{S}^{(1)}(\boldsymbol{y}, \boldsymbol{v}) = c(\boldsymbol{u})^{-1} \nabla c(\boldsymbol{u}) - c(\boldsymbol{v})^{-1} \nabla c(\boldsymbol{v})$$

and hence

$$\boldsymbol{R}_N = \bar{\nabla}\ell_N(\hat{\boldsymbol{v}}) - \bar{\nabla}\ell_N(\boldsymbol{v}) = [c(\hat{\boldsymbol{v}})^{-1}\nabla c(\hat{\boldsymbol{v}}) - c(\boldsymbol{v})^{-1}\nabla c(\boldsymbol{v})]\bar{\ell}_N(\boldsymbol{v}). \quad (3.2.38)$$

(In this case, condition (3.2.36) holds with the corresponding constant $K = K(\boldsymbol{y})$ independent of $\boldsymbol{y}$.)

Suppose next that the estimator $\hat{\boldsymbol{v}}$ is asymptotically normal, i.e. $\{N^{1/2}[\bar{\nabla}\ell_N(\boldsymbol{v}) - \nabla\ell(\boldsymbol{v})],\ N^{1/2}(\hat{\boldsymbol{v}} - \boldsymbol{v})\}$ converges in distribution to a multivariate normal random variable. Since $\bar{\ell}_N(\boldsymbol{v}) = \ell(\boldsymbol{v}) + o_p(1)$ we can write $\boldsymbol{R}_N$ as

$$\boldsymbol{R}_N = \boldsymbol{A}(\boldsymbol{v})(\hat{\boldsymbol{v}} - \boldsymbol{v}) + o_p(N^{-1/2}) \ , \quad (3.2.39)$$

where

$$\begin{aligned} \boldsymbol{A}(\boldsymbol{v}) &= \ell(\boldsymbol{v})\nabla[c(\boldsymbol{v})^{-1}\nabla c(\boldsymbol{v})] = \\ &= \ell(\boldsymbol{v})[-c(\boldsymbol{v})^{-2}\nabla c(\boldsymbol{v})\nabla c(\boldsymbol{v})' + c(\boldsymbol{v})^{-1}\nabla^2 c(\boldsymbol{v})] \ . \end{aligned}$$

It follows therefore that

$$N^{1/2}[\bar{\nabla}\ell_N(\hat{\boldsymbol{v}}) - \nabla\ell(\boldsymbol{v})] = N^{1/2}[\bar{\nabla}\ell_N(\boldsymbol{v}) - \nabla\ell(\boldsymbol{v})] + N^{1/2}\boldsymbol{A}(\boldsymbol{v})(\hat{\boldsymbol{v}} - \boldsymbol{v}) + o_p(1)$$

and hence $\bar{\nabla}\ell_N(\hat{\boldsymbol{v}})$ also is asymptotically normal.

Suppose, for example, that the estimator $\hat{\boldsymbol{v}}$ is derived using the maximum likelihood method. That is, $\hat{\boldsymbol{v}}$ is calculated as a solution of the equations

$$\sum_{i=1}^{N} \boldsymbol{S}^{(1)}(\boldsymbol{Y}_i, \boldsymbol{v}) = \boldsymbol{0}$$

and thus

$$c(\hat{\boldsymbol{v}})^{-1}\nabla c(\hat{\boldsymbol{v}}) = -N^{-1}\sum_{i=1}^{N} \boldsymbol{t}(\boldsymbol{Y}_i) \ . \quad (3.2.40)$$

(Note that the right-hand side of (3.2.40) represents an unbiased estimator of $c(\boldsymbol{v})^{-1}\nabla c(\boldsymbol{v})$.) Consequently, we obtain

$$\bar{\nabla}\ell_N(\hat{\boldsymbol{v}}) = N^{-1}\sum_{i=1}^{N} L(\boldsymbol{Y}_i)\left[\boldsymbol{t}(\boldsymbol{Y}_i) - N^{-1}\sum_{j=1}^{N}\boldsymbol{t}(\boldsymbol{Y}_j)\right] . \quad (3.2.41)$$

It readily follows now that $N^{1/2}[\bar{\nabla}\ell_N(\hat{\boldsymbol{v}}) - \nabla\ell(\boldsymbol{v})]$ converges in distribution to a multivariate normal random variable with zero mean vector and covariance matrix

$$\mathrm{Cov}\{N^{1/2}(\bar{\nabla}\ell_N(\hat{\boldsymbol{v}}) - \nabla\ell(\boldsymbol{v}))\} \approx \mathrm{Cov}_{\boldsymbol{v}}\{L(\boldsymbol{Y})\boldsymbol{S}^{(1)}(\boldsymbol{Y}, \boldsymbol{v}) - \ell(\boldsymbol{v})\boldsymbol{t}(\boldsymbol{Y})\} \ .$$

Example 3.2.3 Let $f(y, v) = v\,\mathrm{e}^{-vy}$. Then the maximum likelihood estimator of v^{-1} is given by

$$\hat{v}^{-1} = N^{-1} \sum_{i=1}^{N} Y_i = \bar{Y}_N$$

and thus

$$\bar{\nabla} \ell_N(\hat{v}) = N^{-1} \sum_{i=1}^{N} L(Y_i)(\hat{v}^{-1} - Y_i) = N^{-1} \sum_{i=1}^{N} L(Y_i)(\bar{Y}_N - Y_i) \, .$$

The asymptotic variance of $\bar{\nabla} \ell_N(\hat{v})$ is then

$$N^{-1} \mathrm{Var}_v \{ L(Y)(v^{-1} - Y) + \ell(v) Y \} \, .$$

Consider now dynamic models. Let $\boldsymbol{Y}_{11}, \ldots, \boldsymbol{Y}_{\tau_1 1}, \ldots, \boldsymbol{Y}_{1N}, \ldots, \boldsymbol{Y}_{\tau_N N}$ be a sample of N regenerative cycles generated from an exponential family pdf $f(\boldsymbol{y}, \boldsymbol{v})$ in the canonical form (2.3.10). Note that in this case (for $k = 1$) condition (3.2.36) holds with $K_t(\underline{\boldsymbol{y}}_t) = Kt$, where the constant K is independent of $\underline{\boldsymbol{y}}_t$. Consequently condition (iii) reduces here to the existence of the expectation $\mathbb{E}_{\boldsymbol{v}}\{\sum_1^{\tau} t |L_t(\underline{\boldsymbol{Y}}_t)|\}$. The corresponding maximum likelihood estimator $\hat{\boldsymbol{v}}$ of $\boldsymbol{v}$ is given by the root of the equations

$$\sum_{i=1}^{N} \tilde{\boldsymbol{S}}_{\tau_i}^{(1)} (\underline{\boldsymbol{Y}}_{\tau_i i}, \boldsymbol{v}) = \boldsymbol{0},$$

where $\underline{\boldsymbol{Y}}_{\tau_i i} = (\boldsymbol{Y}_{1i}, \ldots, \boldsymbol{Y}_{\tau_i i})$. It follows that

$$c(\hat{\boldsymbol{v}})^{-1} \nabla c(\hat{\boldsymbol{v}}) = \frac{- \sum_{i=1}^{N} \sum_{j=1}^{\tau_i} \boldsymbol{t}(\boldsymbol{Y}_{ji})}{\sum_{i=1}^{N} \tau_i}. \tag{3.2.42}$$

Consequently

$$\bar{\nabla} \ell_{1N}(\hat{\boldsymbol{v}}) = N^{-1} \sum_{i=1}^{N} \sum_{t=1}^{\tau_i} L_t(\underline{\boldsymbol{Y}}_{ti}) \tilde{\boldsymbol{S}}_t^{(1)} (\underline{\boldsymbol{Y}}_{ti}, \hat{\boldsymbol{v}}),$$

with

$$\tilde{\boldsymbol{S}}_t^{(1)} (\underline{\boldsymbol{Y}}_{ti}, \hat{\boldsymbol{v}}) = \sum_{k=1}^{t} \boldsymbol{t}(\boldsymbol{Y}_{ki}) + t\, c(\hat{\boldsymbol{v}})^{-1} \nabla c(\hat{\boldsymbol{v}})$$

and $c(\hat{\boldsymbol{v}})^{-1} \nabla c(\boldsymbol{v})$ given in (3.2.42). It is not difficult then to show that $\bar{\nabla} \ell_{1N}(\hat{\boldsymbol{v}})$ is asymptotically normal. The expression for the corresponding covariance matrix will, hovewer, be messy.

We close this subsection by mentioning the nonparametric case, namely where *only a sample* $\underline{\boldsymbol{Y}}_{\tau_1 1}, \ldots, \underline{\boldsymbol{Y}}_{\tau_N}$ but not the pdf $f(\boldsymbol{y}, \boldsymbol{v})$ is available. In this case one can still perform sensitivity analysis and stochastic optimization of DEDS by adopting the methods of nonparametric regression, and in particular those presented in Katkovnik (1985), (1992).

3.2.3 Asymptotic Properties of the SF Estimators

We derive first confidence intervals (regions) for (i) $\ell(\boldsymbol{v})$ and (ii) $\nabla\ell(\boldsymbol{v})$.

(i) Denote

$$\begin{aligned}
\chi_1 &= \sum_{t=1}^{\tau} L_t \widetilde{W}_t, \\
\chi_2 &= \sum_{t=1}^{\tau} \widetilde{W}_t, \\
\widehat{\chi}_1 &= N^{-1}\sum_{i=1}^{N}\sum_{t=1}^{\tau_i} L_{ti}\widetilde{W}_{ti}, \\
\widehat{\chi}_2 &= N^{-1}\sum_{i=1}^{N}\sum_{t=1}^{\tau_i} \widetilde{W}_{ti}.
\end{aligned}$$

Then $(\widehat{\chi}_1, \widehat{\chi}_2)$ is asymptotically distributed as a two-dimensional normal random vector with mean

$$(\mathbb{E}_g\chi_1, \mathbb{E}_g\chi_2)$$

and covariance matrix $N^{-1}\boldsymbol{\Sigma}$, where

$$\boldsymbol{\Sigma} = \begin{pmatrix} \mathrm{Var}_g\chi_1 & \mathrm{Cov}_g\{\chi_1, \chi_2\} \\ \mathrm{Cov}_g\{\chi_2, \chi_1\} & \mathrm{Var}_g\chi_2 \end{pmatrix}. \tag{3.2.43}$$

Denoting $\varphi(x_1, x_2) = x_1/x_2$, it follows from the *delta method* (see chapter 6) that $\bar{\ell}_N(\boldsymbol{v})$ is asymptotically distributed as a normal random variable with mean $\ell(\boldsymbol{v})$ and variance σ^2/N, where

$$\sigma^2 = (\nabla\varphi)'\boldsymbol{\Sigma}\,(\nabla\varphi),$$

$$\nabla\varphi = \begin{pmatrix} \partial\varphi/\partial x_1 \\ \partial\varphi/\partial x_2 \end{pmatrix} = \begin{pmatrix} 1/x_2 \\ -x_1/x_2^2 \end{pmatrix},$$

and $(x_1, x_2) = (\mathbb{E}_g\chi_1, \mathbb{E}_g\chi_2)$.

Denoting now $s^2 = (\nabla\widehat{\varphi})'\boldsymbol{S}\,(\nabla\widehat{\varphi})$ the sample equivalent of σ^2, where

$$\nabla\widehat{\varphi} = \begin{pmatrix} 1/\widehat{\chi}_2 \\ -\widehat{\chi}_1/\widehat{\chi}_2^2, \end{pmatrix}$$

and $\boldsymbol{S}$ is the sample covariance matrix of $\boldsymbol{\Sigma}$, that is

$$\boldsymbol{S} = \begin{pmatrix} S_{11} & S_{12} \\ S_{21} & S_{22} \end{pmatrix}, \tag{3.2.44}$$

$$S_{kj} = \frac{1}{N-1}\sum_{i=1}^{N}(\chi_{ki} - \widehat{\chi}_k)(\chi_{ji} - \widehat{\chi}_j), \quad k, j = 1, 2,$$

$$\chi_{1i} = \sum_{t=1}^{\tau_i} L_{ti}\widetilde{W}_{ti},$$

$$\chi_{2i} = \sum_{t=1}^{\tau_i} \widetilde{W}_{ti},$$

we obtain by straightforward calculation that

$$s = \frac{1}{\widehat{\chi}_2^2}S_{11} - 2\frac{\widehat{\chi}_1}{\widehat{\chi}_2^3}S_{12} + \frac{\widehat{\chi}_1^2}{\widehat{\chi}_2^4}S_{22}. \tag{3.2.45}$$

The asymptotic $(1-\alpha)$ 100% confidence interval for $\overline{\ell}_N(\boldsymbol{v})$ can be written finally as

$$\{\overline{\ell}_N(\boldsymbol{v}) \pm z_{1-\alpha/2}s/\sqrt{N}\}, \tag{3.2.46}$$

where $\Phi(z_{1-\alpha}) = 1-\alpha$, Φ denotes the standard normal cumulative distribution function.

(ii) To obtain the asymptotic confidence region for $\nabla\overline{\ell}_N(\boldsymbol{v})$ we proceed as follows. Let

$$\begin{aligned}
\chi_1 &= \sum_{t=1}^{\tau} L_t\widetilde{W}_t, \\
\chi_2 &= \sum_{t=1}^{\tau} \widetilde{W}_t, \\
\boldsymbol{\chi}_3 &= \sum_{t=1}^{\tau} L_t\nabla\widetilde{W}_t, \\
\boldsymbol{\chi}_4 &= \sum_{t=1}^{\tau} \nabla\widetilde{W}_t, \\
\widehat{\chi}_1 &= \frac{1}{N}\sum_{i=1}^{N}\sum_{t=1}^{\tau_i} L_{ti}\widetilde{W}_{ti}, \\
\widehat{\chi}_2 &= \frac{1}{N}\sum_{i=1}^{N}\sum_{t=1}^{\tau_i} \widetilde{W}_{ti}, \\
\widehat{\boldsymbol{\chi}}_3 &= \frac{1}{N}\sum_{i=1}^{N}\sum_{t=1}^{\tau_i} L_{ti}\nabla\widetilde{W}_{ti}, \\
\widehat{\boldsymbol{\chi}}_4 &= \frac{1}{N}\sum_{i=1}^{N}\sum_{t=1}^{\tau_i} \nabla\widetilde{W}_{ti}.
\end{aligned}$$

Then $(\widehat{\chi}_1, \widehat{\chi}_2, \widehat{\chi}_3, \widehat{\chi}_4)$ is asymptotically normally distributed with mean

$$\boldsymbol{\mu} = (\mu_1, \mu_2, \boldsymbol{\mu}_3, \boldsymbol{\mu}_4), \quad \mu_i = \mathbb{E}_g \chi_i,$$

and covariance matrix $N^{-1}\boldsymbol{\Sigma}$, where $\boldsymbol{\Sigma} = (\sigma_{ij})$, and

$$\sigma_{ij} = \mathrm{Cov}_g(\chi_i, \chi_j), \quad i,j = 1, \ldots, 4. \tag{3.2.47}$$

We call the matrix $\boldsymbol{\Sigma}$ defined in (3.2.43) and (3.2.47) the *performance* and the *sensitivity* matrix, respectively.

Assume now that v is a scalar. Taking into account that in this case $\widehat{\chi}_3$ and $\widehat{\chi}_4$ are scalars (and thus $(\widehat{\chi}_1, \widehat{\chi}_2, \widehat{\chi}_3, \widehat{\chi}_4)$ is a four-dimensional vector) and denoting $\varphi(x_1, x_2, x_3, x_4) = x_3/x_2 - x_1 x_4/x_2^2$, it follows from the delta method (see chapter 6) that $\nabla \bar{\ell}_N$ is asymptotically normally distributed with mean $\nabla \ell(v)$ and variance $N^{-1}\sigma^2$, where

$$\sigma^2 = (\nabla\varphi)' \boldsymbol{\Sigma} (\nabla\varphi),$$

$$\nabla\varphi = \begin{pmatrix} \partial\varphi/\partial x_1 \\ \partial\varphi/\partial x_2 \\ \partial\varphi/\partial x_3 \\ \partial\varphi/\partial x_4 \end{pmatrix} = \begin{pmatrix} x_4/x_2^2 \\ -x_3/x_2^2 + 2x_1x_4/x_2^3 \\ 1/x_2 \\ -x_1/x_2^2 \end{pmatrix},$$

and $x_i = \mu_i = \mathbb{E}_g(\chi_i)$.

Denoting now $s^2 = (\nabla\hat{\varphi})' \boldsymbol{S} (\nabla\hat{\varphi})$, where $\boldsymbol{S}$ is the sample covariance matrix of $\boldsymbol{\Sigma}$ and

$$\nabla\widehat{\varphi} = \begin{pmatrix} \widehat{\chi}_4/\widehat{\chi}_2^2 \\ -\widehat{\chi}_3/\widehat{\chi}_2^2 + 2\widehat{\chi}_1\widehat{\chi}_4/\widehat{\chi}_2^3 \\ 1/\widehat{\chi}_2 \\ -\widehat{\chi}_1/\widehat{\chi}_2^2, \end{pmatrix}$$

we can obtain in analogy to (3.2.46) an asymptotic $(1-\alpha)$ 100% confidence interval for $\nabla \bar{\ell}_N(\boldsymbol{v})$. For alternative confidence interval estimators, with lower bias, see Glynn, L'Ecuyer and Adés (1991).

As in chapter 2, there is no easy way to choose the dominating density $g(\boldsymbol{z})$ in order to minimize the variance of the performance and the sensitivity matrix $\boldsymbol{\Sigma}$ (see (3.2.43) and (3.2.47), respectively). Efficient variance reduction techniques like control random variables and the conditional Monte Carlo method can be successfully used here (see, for example, Asmussen and Rubinstein (1992a), Glynn and Iglehart (1989) and McLeish and Rollans (1992)). In chapter 4 we discuss how to choose a "good" (optimal) parameter $\boldsymbol{v}_0$ to improve the performance of the "what if" estimators $\nabla^k \bar{\ell}_N(\boldsymbol{v})$, provided $g(\boldsymbol{y}) = f(\boldsymbol{y}, \boldsymbol{v}_0)$.

We close this subsection by citing several recent results related to the accuracy (variance) of our estimators. We first cite some results related to the case $g(\boldsymbol{y}) = f(\boldsymbol{y}, \boldsymbol{v})$, that is without changing the probability measure, and then we cite a result related to the case $g(\boldsymbol{y}) = f(\boldsymbol{y}, \boldsymbol{v}_0)$.

Whitt (1989b) gives analytic formulas for the elements of the performance matrix $\boldsymbol{\Sigma}$ in (3.2.43) for the $M/M/1$ queue and the heuristic for the $GI/G/1$ queue in heavy traffic. Asmussen (1992) presents a rigorous treatment of the heavy traffic approximations of the elements of the performance matrix $\boldsymbol{\Sigma}$ in (3.2.43) for the $GI/G/1$ queue. The standard heavy traffic setup is summarized in Asmussen (1992) as follows:

(a) There exist constants c_1, c_2 such that

$$\left\{((1-\rho)/c_1)L_{[tc_2/(1-\rho)^2]} : t = 0, 1, \ldots\right\} \xrightarrow{\mathcal{D}} \left\{\bar{\xi}_t : t \geq 0\right\}$$

in $D[0, \infty)$ as $\rho \uparrow 1$. Here $\left\{\bar{\xi}_t\right\}$ presents a reflected Brownian motion (RBM) in its standard form, that is with drift -1 and variance 1.

(b) The stationary distributions and the moments converge as well in the sense that

$$\frac{1-\rho}{c_1} L \xrightarrow{\mathcal{D}} \bar{\xi}$$

and

$$\mathbb{E}L^p \approx \frac{c_1^p}{(1-\rho)^p}\gamma, \tag{3.2.48}$$

respectively. In particular, $\mathbb{E}L \approx c_1/2(1-\rho)$, $\mathbb{E}L^2 \approx c_1^4/2(1-\rho)^2$. Here $\bar{\xi}$ is the limiting stationary exponential random variable for RBM such that

$$P(\bar{\xi} > u) = \mathrm{e}^{-2u} \text{ and } \gamma = \mathbb{E}\bar{\xi}^p = p!/2^p.$$

Taking into account that the score $\nabla \log f(\boldsymbol{y}, \boldsymbol{v})$ has mean zero, Asmussen and Rubinstein (1992a) extended the results of (a) and (b) as follows. Let $\left\{\tilde{\xi}_t\right\}$ and $\{\tilde{\eta}_t\}$ denote Brownian motions defined on the same probability space such that $\left\{\tilde{\xi}_t\right\}$ has unit variance and drift -1, while $\{\tilde{\eta}_t\}$ has drift zero and the covariance structure given by $\mathrm{Var}(\tilde{\eta}_t) = t(c_3^2 + c_4^2)$ and $\mathrm{Cov}(\bar{\xi}_t, \tilde{\eta}_t) = tc_3$. Here c_3 and c_4 are some constants given in Asmussen and Rubinstein (1992a), depending on the queueing process under the study. Define next the following two-dimensional process $\left\{(\bar{\xi}_t, \eta_t) : t \geq 0\right\}$, where

$$\left\{\bar{\xi}_t = \tilde{\xi}_t - \min_{0 \leq u \leq t} \tilde{\xi}_u = \tilde{\xi}_t - \tilde{\xi}_{\nu(t)}, \; \eta_t = \tilde{\eta}_t - \tilde{\eta}_{\nu(t)}\right\}.$$

Here $\nu(t) = \max\{u \in [0, t] : \tilde{\xi}_u \leq \tilde{\xi}_v, 0 \leq v \leq u\}$ is the time of the last minimum before t. Note that $\{\eta_t\}$ evolves as a standard Brownian motion

$\{\tilde{\eta}_t\}$ but restarted at zero whenever $\left\{\bar{\xi}_t\right\}$ hits zero. Note also that since the $\{L_t\}$ and $\{\widetilde{\boldsymbol{S}}_t^{(1)}\}$ are typically dependent, $\left\{\bar{\xi}_t\right\}$ and $\{\eta_t\}$ will be so as well.

With these definitions at hand Asmussen and Rubinstein (1992a) show that under appropriate regularity conditions, there exist constants c_1, c_2, c_3 and c_4 such that

$$\left\{\frac{1-\rho}{c_1} L_{tc_2/(1-\rho)^2}, (1-\rho)\widetilde{\boldsymbol{S}}^{(1)}_{tc_2/(1-\rho)^2}) : t = 0, 1, \ldots\right\} \xrightarrow{\mathcal{D}} \left\{(\bar{\xi}_t, \eta_t) : t \geq 0\right\}$$

on $D[0, \infty)$. Furthermore,

$$\left(\frac{1-\rho}{c_1} L, (1-\rho)\widetilde{\boldsymbol{S}}^{(1)}\right) \xrightarrow{\mathcal{D}} (\bar{\xi}, \eta),$$

$$\mathbb{E}\{L^p \widetilde{\boldsymbol{S}}^{(1)^m}\} \approx \frac{c_1}{(1-\rho)^{p+m}} \mathbb{E}\{\bar{\xi}^p \eta^m\}$$

and

$$\nabla(\mathbb{E}_{\boldsymbol{v}} L^p) \approx \frac{c_3 c_1^p p! p}{2^p (1-\rho)^{p+1}}. \tag{3.2.49}$$

Shalmon and Rubinstein (1990) give analytic expressions for the elements of the sensitivity matrix $\boldsymbol{\Sigma}$ in (3.2.47) for any $\rho \leq 1$ in a FIFO $M/M/1$ queue, provided v is the service rate, and $\{L_t : t > 0\}$ is the waiting time process, while Shalmon (1991) derives similar results for the $M/G/1$ queue.

3.3 Decomposable and Truncated Estimators

In the previous section we presented a framework for performance evaluation and sensitivity analysis of queueing models using the SF method, provided the underlying output process $\{L_t\}$ is a regenerative one. It follows from these results that the SF estimators $\nabla^k \bar{\ell}_N(\boldsymbol{v})$, $k = 0, 1, \ldots$, ($\nabla^0 \bar{\ell}_N(\boldsymbol{v}) \equiv \bar{\ell}_N(\boldsymbol{v})$), perform well (have a reasonably small variance) if the regenerative cycles τ_i, $i = 1, \ldots, N$, are not large.

Consider, for example, r $GI/G/1$ queues in tandem. As before, let L_t be the steady-state sojourn time process. In this case the SF estimators $\nabla^k \bar{\ell}_N(\boldsymbol{v})$, $k = 0, 1, \ldots$, become useless (have a large variance), if r is large. This is because for large r the probability of regeneration (the event that all queues are simultaneously empty) is very small. This typically causes long regenerative cycles τ_i and thus large variance of random variables like

$$\sum_{t=1}^{\tau} L_t \widetilde{W}_t, \quad \sum_{t=1}^{\tau} \widetilde{W}_t, \quad \sum_{t=1}^{\tau} L_t \nabla \widetilde{W}_t, \quad \text{and} \quad \sum_{t=1}^{\tau} \nabla \widetilde{W}_t,$$

and of the associated estimators $\nabla^k \bar{\ell}_N(\boldsymbol{v})$, $k = 1, 2, \ldots$

To give details on this issue, assume that $g(\boldsymbol{y}) = f(\boldsymbol{y}, \boldsymbol{v}_0)$. In this case the likelihood ratio $\widetilde{W}_t \to 0$ w.p.1 as the sample size $t \to \infty$ for all $\boldsymbol{v} \neq \boldsymbol{v}_0$. On the other hand, for each t, the expectation $\mathbb{E}_{\boldsymbol{v}} \widetilde{W}_t = 1$ for all $\boldsymbol{v}_0$. The apparent contradiction is resolved upon recalling that the likelihood ratios are not uniformly integrable, and they do not behave (asymptotically) in probability in the same way as they do in expectation, since they have decreasingly small probability mass at larger and larger values. Also, the moment-type estimators based on $\widetilde{W}_t$ will tend to be very unstable in mean and variance, particularly when $\boldsymbol{v}$ is far from $\boldsymbol{v}_0$. This actually implies that the SF method can be useful only for queueing networks of *small or moderate size*, provided that all queues are not congested queues. In all other cases the SF method, as presented in the previous section, will perform poorly.

To overcome this difficulty we present below the so-called *decomposable score function* (DSF) estimators (see Lirov and Melamed (1992), Perez-Luna (1990), and Rubinstein (1991), (1992b)) which are based on *local* regenerative cycles at each separate queue, rather than on the *global* ones τ of the entire system, and the so-called *truncated score function* (TSF) estimators, which are based on truncation of the SF process (see Rubinstein (1992b)). We show that

1. Although for general networks the DSF and TSF estimators contain some *bias*, they lead to *dramatic variance reduction* compared with the SF estimators of the previous section, which we call below the *regenerative score function* (RSF) estimators. Here we also give some insight into why the bias is typically not too large.

2. The efficiencies of the DSF and TSF estimators compared to their RSF counterparts increase with the size of the network.

3. Unlike the RSF estimators, the DSF and TSF estimators can be used not only for regenerative processes $\{L_t\}$ but for *stationary and ergodic processes* as well. This actually means that the DSF and TSF estimators are suitable for both the regenerative method and the batch means method.

4. DSF and TSF estimators can be used to solve the "what if" problem for the topological design of queueing networks.

The RSF, DSF and TSF estimators have been implemented into a simulation package, called the *queueing network stabilizer and optimizer* (QNSO) discussed in Perez-Luna (1990) and Rubinstein (1992b). This package is suitable for performance evaluation, sensitivity analysis and optimization of general open non-Markovian queueing networks with respect to the parameter vector $\boldsymbol{v}$ of an exponential family of distributions.

The rest of this section is organized as follows. In subsections 3.3.1 and 3.3.2 we define the DSF and TSF estimators and discuss their efficiency. Subsection 3.3.3 deals with the "what if" problem for the topological design of queueing networks and subsection 3.3.4 presents numerical results with both DSF and RSF estimators for several Markovian and non-Markovian queueing models.

3.3.1 Decomposable Score Function Estimators

The idea behind the DSF estimators is to decompose the queueing network into smaller units, called *modules*, such that each module contains several connected queues, and then to approximate (estimate) the unknown quantities $\nabla^k \ell(\boldsymbol{v})$, $k = 1, 2, \ldots$, by treating these modules as if they were completely independent. In other words we want to use frequently occurring *local* regenerative cycles at each *individual module* instead of *true* but seldom occurring *global* ones of the *entire system.* Although the local cycles at each module interact with their neighbors, our numerical studies below show that, if properly chosen, the contribution from the neighbors might be quite small and thus DSF estimators will approximate the unknown quantities $\nabla^k \ell(\boldsymbol{v})$ rather well, in the sense that their variance will be manageable, the bias will not be too large and the variance reduction obtained by using DSF estimators instead of their RSF counterparts typically is dramatic.

To proceed further first consider the case where each module represents a single queue. Let $\{L_t : t > 0\}$ be a discrete-time regenerative process, say the sojourn time process of a target customer, and let τ be the length of a regenerative cycle of the queueing system. Assume that the interarrival, service, and routing decision random variables are independent, then L_t can be written (see also Lirov and Melamed (1992)) as

$$L_t = \sum_{q=1}^{r} L_{qt} = \sum_{q=1}^{r} \sum_{k=1}^{p_q} L_{qkt}, \tag{3.3.1}$$

where r is the number of nodes in the system, $\{L_{qkt} : t > 0\}$ is the partial sojourn time of the target customer on its k-th visit at node q, and p_q is the number of visits to node q.

Taking into account (3.3.1) and assuming that L_t is a regenerative process, we first rewrite $\ell(\boldsymbol{v})$ in (3.2.27) as

$$\ell(\boldsymbol{v}) = \sum_{q=1}^{r} \frac{\mathbb{E}_g\{\sum_{t=1}^{\tau} L_{qt} \widetilde{W}_t\}}{\mathbb{E}_g\{\sum_{t=1}^{\tau} \widetilde{W}_t\}} \tag{3.3.2}$$

and similarly for $\nabla^k \ell(\boldsymbol{v})$, $k = 1, 2, \ldots$

Here

$$L_{qt} = L_{qt}(\underline{Z}_t), \quad \widetilde{W}_t = \prod_{q=1}^{r} \prod_{j \in C_t^q} W_j, \quad \text{and} \quad W_j = W(Z_j, v) = \frac{f(Z_j, v)}{g(Z_j)},$$

where C_t^q, $q = 1, \ldots, r$, is the set of indeces defined in (3.2.31), and r is the number of nodes in the network. Note that for r $GI/G/1$ queues in tandem with the FIFO discipline, $\widetilde{W}_t$ reduces to $\widetilde{W}_t = \prod_{k=1}^{r} \prod_{j=1}^{t} W_{jk}$, where

$$W_{jk} = W(Z_{jk}, v) = \frac{f(Z_{jk}, v)}{g(Z_{jk})}.$$

Clearly $\ell(v)$ in (3.3.2) can now be *approximated* as follows:

$$\ell^d(v) = \sum_{q=1}^{r} \frac{\mathbb{E}_g\{\sum_{t=1}^{\tau_q} L_{qt} \widetilde{W}_{qt}^d\}}{\mathbb{E}_g\{\sum_{t=1}^{\tau_q} \widetilde{W}_{qt}^d\}}. \tag{3.3.3}$$

Here τ_q, $q = 1, \ldots, r$, is the length of the regenerative (*local*) cycle at the q-th queue:

$$\widetilde{W}_{qt}^d = \prod_{j \in \tilde{C}_t^q} W_{qj} \quad \text{and} \quad W_{qj} = W(Z_{qj}, v) = \frac{f(Z_{qj}, v)}{g(Z_{qj})},$$

where $\tilde{C}_t^q$, $q = 1, \ldots, r$, is the set of indexes defined in analogy to (3.2.31). Note that for r tandem $GI/G/1$ queues with the FIFO discipline, $\widetilde{W}_{qt}^d$ reduces to $\widetilde{W}_{qt}^d = \prod_{j=1}^{t} W_{qj}$, which can also be written as

$$\widetilde{W}_{qt}^d = \widetilde{W}_{qt}^d(\underline{Y}_{qt}, v_q) = f_{qt}(\underline{Y}_{qt}, v_q)/g_{qt}(\underline{Y}_{qt}), \; f_{qt}(\underline{Y}_{qt}, v_q) = \prod_{i=1}^{t} f_q(Y_{qi}, v_q),$$

$$g_{qt}(\underline{Y}_{qt}) = \prod_{i=1}^{t} g_q(Y_{qi}), \; \underline{Y}_{qt} = (Y_{q1}, \ldots, Y_{qt}), \quad t = 1, \ldots, \tau_q.$$

Note also that (3.3.3) is obtained from (3.3.2) by replacing the *global* cycle τ with the local ones τ_q, $q = 1, \ldots, r$.

Differentiating $\ell^d(v)$ with respect to v, we obtain

$$\begin{aligned} \nabla \ell^d(v) \;=\; & \sum_{q=1}^{r} \left(\frac{\mathbb{E}_g\{\sum_{t=1}^{\tau_q} L_{qt} \nabla \widetilde{W}_{qt}^d\}}{\mathbb{E}_g\{\sum_{t=1}^{\tau_q} \widetilde{W}_{qt}^d\}} \right. \\ & \left. - \frac{\mathbb{E}_g\{\sum_{t=1}^{\tau_q} L_{qt} \widetilde{W}_{qt}^d\}}{\mathbb{E}_g\{\sum_{t=1}^{\tau_q} \widetilde{W}_{qt}^d\}} \cdot \frac{\mathbb{E}_g\{\sum_{t=1}^{\tau_q} \nabla \widetilde{W}_{qt}^d\}}{\mathbb{E}_g\{\sum_{t=1}^{\tau_q} \widetilde{W}_{qt}^d\}} \right), \end{aligned} \tag{3.3.4}$$

and similarly for higher order partial derivatives.

Based on (3.3.3) and (3.3.4), we can estimate $\ell^d(\boldsymbol{v})$ and $\nabla \ell^d(\boldsymbol{v})$ as

$$\bar{\ell}_N^d(\boldsymbol{v}) = \sum_{q=1}^{r} \frac{\sum_{i=1}^{N} \sum_{t=1}^{\tau_{qi}} L_{qti} \widetilde{W}_{qti}^d}{\sum_{i=1}^{N} \sum_{t=1}^{\tau_{qi}} \widetilde{W}_{qti}^d} \tag{3.3.5}$$

and

$$\begin{aligned} \nabla \bar{\ell}_N^d(\boldsymbol{v}) &= \sum_{q=1}^{r} \left(\frac{\sum_{i=1}^{N} \sum_{t=1}^{\tau_{qi}} L_{qti} \nabla \widetilde{W}_{qti}^d}{\sum_{i=1}^{N} \sum_{t=1}^{\tau_{qi}} \widetilde{W}_{qti}^d} \right. \\ &\quad - \left. \frac{\sum_{i=1}^{N} \sum_{t=1}^{\tau_{qi}} L_{qti} \widetilde{W}_{qti}^d}{\sum_{i=1}^{N} \sum_{t=1}^{\tau_{qi}} \widetilde{W}_{qti}^d} \cdot \frac{\sum_{i=1}^{N} \sum_{t=1}^{\tau_{qi}} \nabla \widetilde{W}_{qti}^d}{\sum_{i=1}^{N} \sum_{t=1}^{\tau_{qi}} \widetilde{W}_{qti}^d} \right), \end{aligned} \tag{3.3.6}$$

and similarly for $\nabla^k \bar{\ell}_N^d(\boldsymbol{v}),\ k > 1$.

Since $\tau_q \leq \tau$ the estimators $\nabla^k \bar{\ell}_N^d(\boldsymbol{v}),\ k = 0, 1, \ldots$, represent *truncated* versions of the RSF estimators $\nabla^k \bar{\ell}_N(\boldsymbol{v}), k = 0, 1, \ldots$ (see (3.2.29) and (3.2.30)) and thus will typically be biased. It is important to note, however, that $\nabla^k \bar{\ell}_N^d(\boldsymbol{v}),\ k = 0, 1, \ldots$, give consistent estimators for $\nabla^k \ell(\boldsymbol{v})$ for tandem $M/M/1$ queues and for static queueing networks containing $G/G/\infty$ and $GI/G/1/0$ queues. In the first case this follows directly from Burke's theorem, which states that the interdeparture times at each stable tandem $M/M/1$ queue are exponentially distributed with the same parameter as the interarrival times, or in other words, the steady-state outputs at each stable tandem $M/M/1$ queue can be analyzed as independent Poisson processes with the same interarrival rate λ (e.g. Kleinrock (1976, p. 149)); while in the second case it follows directly from (2.3.24) for decomposable static models. Note also that for $r = 1,\ \nabla^k \bar{\ell}_N^d(\boldsymbol{v}),\ k = 0, 1, \ldots$, reduces to $\nabla^k \bar{\ell}_N(\boldsymbol{v}),\ k = 0, 1, \ldots$, and thus to the RSF estimators.

In order to reduce the bias of the DSF estimators, (3.3.5) and (3.3.6), we can use larger units, namely modules, rather than single queues and then treat the data at different modules as if they were completely independent. In other words, we can first decompose the underlying network into modules (pairs, triplets, etc.), and then produce DSF estimators while combining several local cycles of the individual queues in a given module. To do so, let us represent $\widetilde{W}_{qt}^d$ as a function of γ and Δ_γ, that is, $\widetilde{W}_{qt}^d = \widetilde{W}_{qt}^d(\gamma, \Delta_\gamma)$, where γ and $\Delta_\gamma = (\delta_1, \ldots, \delta_\gamma),\ \gamma = 1, \ldots, r$, denote the number of nodes in each module including the current node, and $\delta_m,\ m = 1, \ldots, r$, denote the number of local regenerative cycles. With such a representation of $\widetilde{W}_{qt}^d(\gamma, \Delta_\gamma)$, we can define for $k = 0$ and $k = 1$ (in analogy to (3.3.5) and (3.3.6)) the following decomposable estimators:

$$\bar{\ell}_N^d(\boldsymbol{v}, \gamma, \Delta_\gamma) = \sum_{q=1}^{r} \frac{\sum_{i=1}^N \sum_{t=1}^{\tau_i} L_{qti} \widetilde{W}_{qti}^d(\gamma, \Delta_\gamma)}{\sum_{i=1}^N \sum_{t=1}^{\tau_i} \widetilde{W}_{qti}^d(\gamma, \Delta_\gamma)} \tag{3.3.7}$$

and

$$\begin{aligned} \nabla \bar{\ell}_N^d(\boldsymbol{v}, \gamma, \Delta_\gamma) &= \sum_{q=1}^{r} \left(\frac{\sum_{i=1}^N \sum_{t=1}^{\tau_i} L_{qti} \nabla \widetilde{W}_{qti}^d(\gamma, \Delta_\gamma)}{\sum_{i=1}^N \sum_{t=1}^{\tau_i} \widetilde{W}_{qti}^d(\gamma, \Delta_\gamma)} \right. \\ &- \frac{\sum_{i=1}^N \sum_{t=1}^{\tau_i} L_{qti} \widetilde{W}_{qti}^d(\gamma, \Delta_\gamma)}{\sum_{i=1}^N \sum_{t=1}^{\tau_i} \widetilde{W}_{qti}^d(\gamma, \Delta_\gamma)} \\ &\cdot \left. \frac{\sum_{i=1}^N \sum_{t=1}^{\tau_i} \nabla \widetilde{W}_{qti}^d(\gamma, \Delta_\gamma)}{\sum_{i=1}^N \sum_{t=1}^{\tau_i} \widetilde{W}_{qti}^d(\gamma, \Delta_\gamma)} \right), \end{aligned} \tag{3.3.8}$$

and similarly for $\nabla^k \bar{\ell}_N^d(\boldsymbol{v}, \gamma, \Delta_\gamma)$, $k > 1$.

From (3.3.7) and (3.3.8) we can identify the following two extreme cases. At the one extreme when $\gamma = 1$ and $\delta_1 = 1$ we have the DSF estimators (3.3.5) and (3.3.6), which are based on the local cycles of the current node only. At the other extreme, when $\gamma = r$ and all δ_m correspond to $\tau = \sum_{k=1}^{\gamma_m} \tau_{mk}$ (γ_m is the total number of local cycles at the node m and τ_{mk} is the length of the k-th local cycle at node m, $m = 1, \ldots, r$), we obtain the RSF estimators (3.2.29) and (3.2.30). Between these two extremes of γ and $\Delta_\gamma = (\delta_1, \ldots, \delta_\gamma)$, $\gamma = 1, \ldots, r$, we have an infinite set of estimators of $\nabla^k \ell(\boldsymbol{v})$. Note again that for the above two extremes, the estimators $\nabla^k \bar{\ell}_N^d(\boldsymbol{v}, \gamma, \Delta_\gamma)$ reduce to $\nabla^k \bar{\ell}_N^d(\boldsymbol{v}) = \nabla^k \bar{\ell}_N^d(\boldsymbol{v}, 1, 1)$ and $\nabla^k \bar{\ell}_N(\boldsymbol{v})$, and for $r = 1$ we have

$$\nabla^k \bar{\ell}_N^d(\boldsymbol{v}, \gamma, \Delta_\gamma) = \nabla^k \bar{\ell}_N^d(\boldsymbol{v}) = \nabla^k \bar{\ell}_N(\boldsymbol{v}).$$

Since the estimators $\nabla^k \bar{\ell}_N^d(\boldsymbol{v}, \gamma, \Delta_\gamma)$, $k = 0, 1, \ldots$, are based on network decomposition into modules using the *local* cycles, we shall call them DSF estimators of order (γ, Δ_γ) or simply DSF estimators. Note that $\nabla^k \bar{\ell}_N^d(\boldsymbol{v})$ represents a DSF estimator of order $(1, 1)$.

We now give some guidance on how to choose a "good" set of parameters (γ, Δ_γ) in the sense that the variance of the DSF estimator $\bar{\ell}_N^d(\boldsymbol{v}, \gamma, \Delta_\gamma)$ is still manageable and the relative bias of the system performance, defined as

$$b = \left| \frac{\mathbb{E} \bar{\ell}_N^d(\boldsymbol{v}, \gamma, \Delta_\gamma) - \ell(\boldsymbol{v})}{\ell(\boldsymbol{v})} \right|,$$

is relatively small, say no greater than 0.1. Formally speaking, we would like to minimize the mean squared error of $\nabla^k \bar{\ell}_N^d(\boldsymbol{v}, \gamma, \Delta_\gamma)$ with respect to (γ, Δ_γ). Since the analytic solution of such a minimization problem is not feasible (even for simple queueing networks) we resorted to simulation studies. Our extensive simulation studies show that typically the bias of the estimators $\nabla^k \bar{\ell}_N^d(\boldsymbol{v}, \gamma, \Delta_\gamma)$ is a nonincreasing function in each component of the vector (γ, Δ_γ), and in order for b to be less than 0.05 one can use in light, moderate and heavy traffic the following choices for (γ, Δ_γ):

$$\gamma = 2, \delta = 2; \quad \gamma = 2, \delta = 5; \quad \gamma = 3, \delta = 10, \tag{3.3.9}$$

respectively. Here $\delta = \delta_\gamma$, $\gamma = 1, \ldots, r$. It follows, for example, that if the queueing network works in moderate traffic, then in order to keep the variance of DSF estimators reasonably small and the relative bias b less than 0.1, one can decompose the network into modules (in quite arbitrary fashion) such that the size of each module (the number of queues in each module) is $\gamma = 2$, and the number of local regenerative cycles at each queue in a fixed module is $\delta = 4$.

Since the DSF estimators use local rather than global regenerative cycles they can be used for both the regenerative and the standard batch means methods (see subection 1.2.2) and that variance reduction obtained while using DSF estimators versus their RSF counterparts will typically be very substantial. Note finally that confidence intervals (regions) for $\bar{\ell}_N^d(\boldsymbol{v}, \gamma, \Delta_\gamma)$ and $\nabla \bar{\ell}_N^d(\boldsymbol{v}, \gamma, \Delta_\gamma)$ can be derived by using standard statistical methods similar to those given in subection 3.3.2. This is left as an exercise for the reader.

3.3.2 Truncated Score Function Estimators

We now introduce the so-called *truncated score function* (TSF) estimators for sensitivity analysis of DEDS, by citing some material from Rubinstein (1992b).

Assume first that $L_t = L_t(\underline{\boldsymbol{Y}}_t) = L_t(\boldsymbol{Y}_t, \ldots, \boldsymbol{Y}_{t-M+1})$ is an M-dependent process with autocovariance function (see, for example, Fishman (1978))

$$R_L = \begin{cases} \text{Cov}\{L_t, L_{t-h}\} \neq 0, & \text{if } |h| \leq M, \\ 0, & \text{otherwise.} \end{cases}$$

Assuming further for simplicity that the random vectors $\boldsymbol{Y}_j$, $j = 1, \ldots,$ are iid with a common pdf $f(\boldsymbol{y}, \boldsymbol{v})$ we can represent the expected performance measures $\nabla^k \ell(\boldsymbol{v})$ as

$$\nabla^k \ell(\boldsymbol{v}, M) = \mathbb{E}_g\{L_t \nabla^k \widehat{W}_t(\boldsymbol{v}, M)\}, \quad k = 0, 1, \ldots, \tag{3.3.10}$$

and then estimate $\nabla^k \ell(\boldsymbol{v})$ by using the following batch means estimator (see subsection 1.2.2):

$$\nabla^k \bar{\ell}_N(\boldsymbol{v}, m) = N^{-1}T^{-1} \sum_{i=1}^{N} \sum_{t=1}^{T} L_{ti} \nabla^k \widehat{W}_{ti}(\boldsymbol{v}, M), \tag{3.3.11}$$

where T is the batch size and N is the number of batches. Here

$$\widehat{W}_t(\boldsymbol{v}, M) = \prod_{j=t-M+1}^{t} W_j, \quad \text{and} \quad W_j = W(\boldsymbol{Z}_j, \boldsymbol{v}) = \frac{f(\boldsymbol{Z}_j, \boldsymbol{v})}{g(\boldsymbol{Z}_j)}.$$

Suppose now that M is large. In this case the variance of $L_t \nabla^k \widehat{W}_t(\boldsymbol{v}, M)$ becomes large and thus the corresponding estimator of $\nabla^k \ell(\boldsymbol{v}, M)$ is useless.

To overcome this difficulty we define the so-called *truncated version* (approximation) of $\nabla^k \ell(\boldsymbol{v}, M)$, which can be written as

$$\nabla^k \ell^{tr}(\boldsymbol{v}, m) = \mathbb{E}_g\{L_t \nabla^k \widetilde{W}_t^{tr}(\boldsymbol{v}, m)\}, \quad k = 0, 1, \ldots \tag{3.3.12}$$

Here

$$\widetilde{W}_t^{tr}(\boldsymbol{v}, m) = \prod_{j=t-m+1}^{t} W_j, \quad m < M,$$

where m is called the *truncation parameter*. The estimator of $\nabla^k \ell^{tr}(\boldsymbol{v}, m)$, namely

$$\nabla^k \bar{\ell}_N^{tr}(\boldsymbol{v}, m) = N^{-1}T^{-1} \sum_{i=1}^{N} \sum_{t=1}^{T} L_{ti} \nabla^k \widetilde{W}_{ti}^{tr}(\boldsymbol{v}, m), \tag{3.3.13}$$

will be biased. This is because instead of the "true" likelihood ratio $\widehat{W}_t(\boldsymbol{v}, M) = \prod_{j=t-M+1}^{t} W_j$, we use the "truncated" one $\widetilde{W}_t^{tr}(\boldsymbol{v}, m) = \prod_{j=t-m+1}^{t} W_j$, which contains only the last m $(m < M)$ terms of the product $\prod_{j=t-M+1}^{t} W_j$. Note that $\widetilde{W}_t^{tr}(\boldsymbol{v}, m) = \prod_{j=t-m+1}^{t} W_j$ in (3.3.13) assumes that before starting to collect data one needs to delete initially at least m customers. In order to reduce the bias of the estimator $\nabla^k \bar{\ell}_N^{tr}(\boldsymbol{v}, m)$, we have to increase m. By doing so, however, we increase the variance of $\nabla^k \bar{\ell}_N^{tr}(\boldsymbol{v}, m)$. Therefore, there exists a trade-off between the bias and the variance of $\nabla^k \bar{\ell}_N^{tr}(\boldsymbol{v}, m)$.

We briefly discuss both (i) the bias and (ii) the variance. For more details, see Gadrich (1989) and Rubinstein (1991).

(i) Let $b^k(M, m) = |\nabla^k \ell^{tr}(\boldsymbol{v}, m) - \nabla^k \ell(\boldsymbol{v}, M)|$ be the bias of the estimator $\nabla^k \bar{\ell}_N^{tr}(\boldsymbol{v}, m)$, where v is a single dimensional parameter. Consider for simplicity $k = 1$. It is shown in Rubinstein (1991) that

$$b^1(M, m) = \begin{cases} R_{\{L \nabla \widetilde{W}^{tr}\}}(m), & \text{if } m \le M, \\ 0, & \text{otherwise}, \end{cases} \tag{3.3.14}$$

where

$$R_{\{L\widetilde{W}^{tr}\}}(h) = \mathbb{E}_g\{L_t, \nabla \widetilde{W}_{t+h}^{tr}(v, M-m)\} = \mathrm{Cov}\{L_t, \nabla \widetilde{W}_{t+h}^{tr}(v, M-m)\}$$

. That is, the bias introduced by using the truncated estimator $\nabla \bar{\ell}_N^{tr}(\boldsymbol{v}, m)$ equals the value of the *cross-covariance* function $R_{\{L\widetilde{W}^{tr}\}}(h)$ between the processes L_t and $\nabla \widetilde{W}_{t+h}^{tr}(v, M-m)\}$ at the point $h = m$.

Let $v \in V \subset R^1$. Define

$$\gamma^k(M, m) = \left| \frac{b^k(M, m)}{\nabla^k \ell(v, M)} \right|$$

as the relative bias of $\nabla^k \bar{\ell}_N^{tr}(\boldsymbol{v}, m)$, and similarly for $\boldsymbol{v} \in V \subset R^n$.

Example 3.3.1 Moving average process. Let $L_t = \sum_{j=0}^{M-1} a_j Y_{t-j}$, where $Y_k,\ k = 1, \ldots, M$, are iid rv's. Then it is readily seen that for $k = 1$ we have

$$\gamma^1(M, m) = \frac{\sum_{j=m}^{M-1} a_j}{\sum_{j=0}^{M-1} a_j}.$$

In the particular case where $a_j = a^j$ we have

$$\gamma^1(M, m) = \frac{a^m - a^M}{1 - a^M}.$$

Solving $\gamma^1(M, m)$ with respect to m we obtain

$$m = \frac{\ln[(1 - a^M)\gamma^1 + a^M]}{\ln a}.$$

Table 3.3.1 represents m as a function of a and γ^1 for $M = \infty$.

Table 3.3.1 m as a function of a and γ^1 for $M = \infty$.

γ^1	a						
	0.1	0.3	0.5	0.7	0.9	0.95	0.99
0.01	2	4	7	13	44	90	458
0.05	1	2	4	8	28	58	298
0.1	1	2	3	6	21	45	229

It follows from the results of table 3.3.1 that in order to keep the relative bias $\gamma^1(M, m)$ small we can choose the truncated parameter m small relative to M, provided M is large and $0 < a < 1$. Also m increases in M, and therefore the data in table 3.3.1 with $M = \infty$ provide the upper values of m for $M < \infty$.

(ii) Consider now the variance of the estimator $\overline{\ell}_N^{tr}(\boldsymbol{v}, m)$ (see (3.3.13)). Assume as before that the M-dependent process $L_t = L_t(Y_t, \ldots, Y_{t-M+1})$ is driven by an iid sequence of random variables $Y_t, \ldots, Y_{t-M+1}$ and consider for simplicity the case where each $Y_j, j = 1, \ldots, M$, is distributed as a gamma random variable with the parameters λ and β. Assume also that λ is the sensitivity parameter, while β is fixed and known. In this case (see section 2.3 of chapter 2) it is possible to show (see section 2.3) that $\mathrm{Var}_{\lambda_0}\{L_t \widehat{W}_t(\boldsymbol{v}, M)\}$ and $\mathrm{Var}_{\lambda_0}\{L_t \widetilde{W}_t^{tr}(\boldsymbol{v}, m)\}$ are equal to

$$\mathrm{Var}_{\lambda_0}\{L_t \widehat{W}_t(\boldsymbol{v}, M)\} = \left(1 + \frac{\alpha^2}{\cdot 1 + 2\alpha}\right)^{M\beta} \mathbb{E}_{\lambda_0 + 2\Delta\lambda}\{L^2\} - \ell(\lambda)^2 \quad (3.3.15)$$

and

$$\mathrm{Var}_{\lambda_0}\{L_t \widetilde{W}_t^{tr}(\boldsymbol{v}, m)\} = \left(1 + \frac{\alpha^2}{1 + 2\alpha}\right)^{m\beta} \mathbb{E}_{\lambda_0 + 2\Delta\lambda}\{L^2\} - \ell(\lambda)^2, \quad (3.3.16)$$

provided the dominating pdf is of the form $g(\boldsymbol{y}) = f(\boldsymbol{y}, \boldsymbol{v}_0)$, $\boldsymbol{v}_0 = (\lambda_0, \beta)$. Here $\Delta\lambda = \lambda - \lambda_0$ and $\alpha = \Delta\lambda/\lambda_0$. Similar expressions can be obtained for $\mathrm{Var}_{\lambda_0}\{L_t \nabla \widetilde{W}_t^{tr}(\boldsymbol{v}, m)\}$.

It follows directly from (3.3.15) and (3.3.16) that the variance reduction obtained while using the truncated estimators instead of the untruncated ones will be enormous, especially when $M >> m$.

Let us now suppress the M-dependence assumption and assume that L_t is a stationary and ergodic process. In this case we can directly adopt the truncated estimator (3.3.13) introduced for M-dependent processes. Clearly, that would mean that the only difference between the sensitivity estimators for M-dependent processes and for our case is that in the first case the length of the vector $\underline{\boldsymbol{Y}}_t = (\boldsymbol{Y}_t, \ldots, \boldsymbol{Y}_{t-M+1})$ is fixed, while in the second it is not.

Gadrich (1989) studied numerically the issue of choosing the truncation parameter m in the $GI/G/1$ queue. She found, via extensive simulation studies, that the truncation parameter m can be chosen relatively small ($m < 100$), provided the interarrival and the service time random variables are distributed according to the exponential family, the traffic intensity $\rho \leq 0.9$, and $\gamma^1 \geq 0.01$. For $\rho > 0.95$ $(\gamma^1 \leq 0.1)$ the truncated estimators are useless (have a large variance). This is so since for large ρ, the truncation parameter m and therefore $\mathrm{Var}_{\lambda_0} \nabla\{L_t \widetilde{W}_t^{tr}(\boldsymbol{v}, m)\}$ increase rather fast in ρ.

Table 3.3.2 presents m as a function of ρ and γ^1, based on a sample of $N = 10^6$ customers. We assumed that L_t is the waiting time process in the $M/M/1$ queue with the interarrival rate $\lambda = 1$. Similar results were obtained for the $GI/G/1$ queue, provided the interarrival and the service time random variables are distributed according to the exponential family. Note that table 3.3.2 is similar to table 3.3.1 in the sense that both tables indicate that choosing m relatively small ($m < 90$), we can maintain a small bias, provided $a < 0.9$ and $\rho < 0.9$.

Table 3.3.2 m as a function of ρ and γ^1
for the the $M/M/1$ queue with interarrival rate $\lambda = 1$.

γ^1	ρ						
	0.1	0.3	0.5	0.7	0.9	0.95	0.99
0.01	3	6	11	24	84	174	894
0.5	2	4	8	18	62	132	680
0.1	1	3	7	15	54	112	579

We now consider truncated estimators for a queueing network containing r nodes. To do so we first take into account that the expected performance $\ell(\boldsymbol{v})$ can be written as $\ell(\boldsymbol{v}) = \mathbb{E}_{\boldsymbol{v}_q} \sum_{q=1}^{r} L_{qt}$ (see (3.3.1)), and then define in analogy to (3.3.3) the following aproximation of $\ell(\boldsymbol{v})$:

$$\ell^{tr}(\boldsymbol{v}) = \sum_{q=1}^{r} \mathbb{E}_g\{L_{qt}\widetilde{W}_{qt}^{tr}(m_q)\}. \tag{3.3.17}$$

Here

$$\widetilde{W}_{qt}^{tr}(m_q) = \prod_{j=t-m_q+1}^{t} W_{qj}, \quad \text{and} \quad W_{qj} = W(\boldsymbol{z}_{qj}, \boldsymbol{v}) = \frac{f(\boldsymbol{z}_{qj}, \boldsymbol{v})}{g(\boldsymbol{z}_{qj})}, \tag{3.3.18}$$

provided the FIFO discipline is used and m_q is the truncation parameter at the q-th queue. Note that for more complex models the indexing of $\widetilde{W}_{qt}^{tr}(m_q)$ must be defined in analogy to (3.2.30).

Differentiating $\ell^{tr}(\boldsymbol{v})$ next with respect to $\boldsymbol{v}$ we obtain

$$\nabla\ell^{tr}(\boldsymbol{v}) = \sum_{q=1}^{r} \mathbb{E}_g\{L_{qt}\nabla\widetilde{W}_{qt}^{tr}(m_q)\} \tag{3.3.19}$$

and similarly for higher order partial derivatives.

Based on (3.3.17) and (3.3.19) we finally estimate $\ell^{tr}(\boldsymbol{v})$ and $\nabla\ell^{tr}(\boldsymbol{v})$ as

$$\bar{\ell}_N^{tr}(\boldsymbol{v}) = N^{-1}T^{-1}\sum_{q=1}^{r}\sum_{i=1}^{N}\sum_{t=1}^{T} L_{qti}\widetilde{W}_{qti}^{tr}(m_q) \tag{3.3.20}$$

and

$$\nabla\bar{\ell}_N^{tr}(v) = N^{-1}T^{-1}\sum_{q=1}^{r}\sum_{i=1}^{N}\sum_{t=1}^{T} L_{qti}\nabla\widetilde{W}_{qti}^{tr}(m_q), \tag{3.3.21}$$

and similarly for $\nabla^k\bar{\ell}_N^{tr}(v)$, $k > 1$. Here T and N are the batch size and the number of batches, respectively.

We shall call the estimators (3.3.20) and (3.3.21) the TSF estimators.

As in the case of DSF estimators, (3.3.5) and (3.3.6), it is not difficult to see that in order to reduce the bias of the TSF estimators, (3.3.20) and (3.3.21), we can use larger modules rather than single queues, and then treat them as if they were completely independent. To do so, we extend $\widetilde{W}_{qt}^{tr}(m)$ to $\widetilde{W}_{qt}^{tr}(m_q, \tilde{m}_{q_1}, \ldots, \tilde{m}_{q_\gamma})$, where γ denotes the number of nodes in a fixed module *not including* the current node q, and $\tilde{m}_{q_k}$, $k = 1, \ldots, \gamma$, denotes the truncation parameter for the node q_k in a fixed module. We assume $\tilde{m}_{q_k} \le m_{q_k}$, $k > 1$, where m_{q_k} is the truncation parameter of the node (queue) q_k working in isolation (see also table 3.3.2). The choice of $\tilde{m}_{q_k}$ in a given module depends on how much the neighboring nodes contribute to the fixed node q or how strong the neighboring nodes are correlated with node q (see also our numerical results below). It is readily seen that for $\gamma = 0$ $\widetilde{W}_{qt}^{tr}(m_q, \tilde{m}_{q_1}, \ldots, \tilde{m}_{q_\gamma})$ reduces to $\widetilde{W}_{qt}^{tr}(m_q)$.

With $\widetilde{W}_{qt}^{tr}(m_q, \tilde{m}_{q_1}, \ldots, \tilde{m}_{q_\gamma})$ at hand we can define now in analogy to (3.3.7) and (3.3.8) the following extended version of the TSF estimators (3.3.20) and (3.3.21):

$$\bar{\ell}_N^{tr}(v, m_q, \tilde{m}_{q_1}, \ldots, \tilde{m}_{q_\gamma}) = N^{-1}T^{-1}\sum_{q=1}^{r}\sum_{i=1}^{N}\sum_{t=1}^{T} L_{qti}\widetilde{W}_{qti}^{tr}(m_q, \tilde{m}_{q_1}, \ldots, \tilde{m}_{q_\gamma}) \tag{3.3.22}$$

and

$$\nabla\bar{\ell}_N^{tr}(v, m_q, \tilde{m}_{q_1}, \ldots, \tilde{m}_{q_\gamma}) = N^{-1}T^{-1}\sum_{q=1}^{r}\sum_{i=1}^{N}\sum_{t=1}^{T} L_{qti}\nabla\widetilde{W}_{qti}^{tr}(m_q, \tilde{m}_{q_1}, \ldots, \tilde{m}_{q_\gamma}) \tag{3.3.23}$$

and similarly for $\nabla^k\bar{\ell}_N^{tr}(v, m_q, \tilde{m}_{q_1}, \ldots, \tilde{m}_{q_\gamma})$, $k > 1$.

Note again that formally speaking we whish to minimize the mean squared error of $\{\nabla^k\bar{\ell}_N^{tr}(v, m_q, \tilde{m}_{q_1}, \ldots, \tilde{m}_{q_\gamma}) - \nabla^k\ell(v)\}$, $k = 0, 1, \ldots$, with respect to $(m_q, \tilde{m}_{q_1}, \ldots, \tilde{m}_{q_\gamma})$. Since an analytic solution of such minimization problems is not feasible (even for simple queueing networks) we resorted to simulation studies.

To give some insight into this issue, consider r tandem $M/M/1$ queues. Assume for simplicity that each queue has the same traffic intensity ρ. In this case choosing $m_q = m_{q_k}$, $k = 1, \ldots, \gamma$, where m_q is given in table 3.3.2, and $\tilde{m}_k = \rho^k m_k$, $k = 1, 2, \ldots, \gamma$, we found (numerically) that the relative bias of the estimators $\overline{\ell}_N^{tr}(\boldsymbol{v}, m_q, \tilde{m}_{q_1}, \ldots, \tilde{m}_{q_\gamma})$, defined as

$$\gamma = \left| \frac{\mathbb{E}_{\boldsymbol{v}_0}\{\overline{\ell}_N^{tr}(\boldsymbol{v}, m_q, \tilde{m}_{q_1}, \ldots, \tilde{m}_{q_\gamma})\} - \ell(\boldsymbol{v})}{\ell(\boldsymbol{v})} \right|,$$

does not exceed 1% and similarly for the bias of the trace of $\nabla \overline{\ell}_N^{tr}(\boldsymbol{v}, m_q, \tilde{m}_{q_1}, \ldots, \tilde{m}_{q_\gamma})$. Our extensive simulation studies with the TSF estimators show that for typical queueing networks the bias $b \leq 0.01$, provided the following parameters are used:

$$\tilde{m}_k = \rho^k m_k, \; k = 1, 2, \ldots, \gamma, \text{ and } \gamma = 3. \tag{3.3.24}$$

Using such parameters, we obtained a substantial variance reduction relatively to the RSF counterparts (see also Sterental (1989)).

At this point we note that L'Ecuyer (1992) analyzes the truncated estimators in the replication method for the case $m = T$, i.e. when the score function process is truncated at T. In that case, increasing the number N of replications reduces the variance, while increasing T reduces the bias (typically, linearly in T), but increases the variance (again, typically, linearly in T). So, for a given computer budget C, there is a trade-off between N and T to minimize the mean-square error. L'Ecuyer (1992) shows that the optimal rates of increase of N and T as functions of C are $N = O(C^{3/4})$ and $T = O(C^{1/4})$, respectively. The mean-square error then decreases at a rate $O(C^{-1/2})$, which is very slow.

3.3.3 Network Topological Design

Let us now turn to the network topological design problem or to the so-called network "what if" design problem.

We show that the DSF estimators $\nabla^k \overline{\ell}_N^d(\boldsymbol{v}) = \nabla^k \overline{\ell}_N^d(\boldsymbol{v}, \gamma = 1, \delta_1 = 1)$ can successfully handle such "what if" problems. The treatment with the TSF estimators is similar.

We first consider a tandem queueing network containing r nodes. Assume that we want to evaluate *simultaneously from a single simulation run* both the parameters $\nabla^k \ell(\boldsymbol{v})$ in the original r node network and the perturbed one, which is obtained from the original network, say by eliminating the last node.

To do so we argue as follows. We first note that for the perturbed $(r-1)$ node network the output process L_t in (3.3.1) can be written as

$$L_t = \sum_{q=1}^{r-1} L_{qt} = \sum_{q=1}^{r} L_{qt},$$

where $L_{rt} = 0$. This clearly suggests that we use the original network as the reference case (the case to simulate) and then consider the perturbed network as a particular case, namely with the traffic intensity at the last node equal to zero ($\rho_r = 0$) while all other data remain the same.

Note that modeling of such an auxiliary queue can be done in several different ways. An appropriate alternative is to model the r-th queue in the perturbed network as an arbitrary $GI/G/1$ queue with an infinite (very large) service rate. We show now that using such an auxiliary queue, the r-th term in $\bar{\ell}_N^d(\boldsymbol{v})$ (see (3.3.5)) vanishes, that is,

$$\frac{\sum_{i=1}^{N}\sum_{t=1}^{\tau_{ri}} L_{rti}\widetilde{W}_{rti}^{d}}{\sum_{i=1}^{N}\sum_{t=1}^{\tau_{ri}} \widetilde{W}_{rti}^{d}} = \frac{0}{0}, \tag{3.3.25}$$

where we assume that $0/0 = 0$, and similarly for the r-th terms in $\nabla^k \bar{\ell}_N^d(\boldsymbol{v})$, $k > 0$. The validity of our approach will follow now from (3.3.25). To justify (3.3.25) first note that since the service rate in the auxiliary r-th tandem queue equals infinity ($\rho_r = 0$) both the values in the output processes L_{rti} and $\widetilde{W}_{rti}^{d}$ will equal zero w.p.1. Taking into account next that we assumed $0/0 = 0$, the validity of (3.3.25) follows. In practice we need to choose the service rate at the r-th queue very large (very small ρ_r) rather than infinity ($\rho_r = 0$) and also that if we need to add rather than subtract a node from the network we can use a perturbed network containing $r + 1$ nodes as the reference case and then consider the original network as a particular case of the perturbed one with the $(r+1)$-st queue having an infinite (very large) service rate.

The above ideas can be adapted to rather general queueing networks. More precisely, assume that we have several "similar" topological designs and we want to evaluate their performance parameters $\nabla^k \ell(\boldsymbol{v})$ *simultaneously from a single simulation.* The word "similar" means that all network topologies under consideration can be obtained by adding and/or eliminating at most two nodes.

The algorithm for the topological "what if" network design can be written as follows:

Algorithm 3.3.1 : Topological Design

1. *Choose the reference case by combining all "similar" networks such that any fixed network can be considered as a subnetwork of the reference case.*

2. *Add to each subnetwork the necessary auxiliary nodes (queues) having traffic intensities equal to zero, and each time obtain the reference topology.*

3. *Run the reference case and evaluate simultaneously from a single simulation all performance parameters $\nabla^k \ell(\boldsymbol{v})$ for all networks by using the DSF estimators $\nabla^k \bar{\ell}_N^d(\boldsymbol{v}) = \nabla^k \bar{\ell}_N^d(\boldsymbol{v}, \gamma = 1, \delta_1 = 1), k = 0, 1, \ldots,$.*

3.3.4 Numerical Results

We present numerical results with the DSF estimators $\nabla^k \bar{\ell}_N^d(\boldsymbol{v}, \gamma, \Delta_\gamma)$, $k = 0, 1$, and RSF estimators $\nabla^k \bar{\ell}_N(\boldsymbol{v})$, $k = 0, 1$, for two queueing models on both (a) Markovian and (b) non-Markovian levels. Our first model contains two queues in tandem while our second model is a feedback model (see fig. 3.3.1). In both cases we estimated the steady-state expected waiting time $\ell(\boldsymbol{v})$ and the associated derivatives $\nabla \ell(\boldsymbol{v})$ with respect to the parameters of the service distribution while simulating for each model $N = 1.5 \cdot 10^6$ customers. Note that in our tables 3.3.3–3.3.6 (see below) the notations Estim, $\hat{\sigma}^2$, and w_r stand for point estimators, sample variances (sample mean-squared error), and the relative width of the confidence intervals, which is defined as

$$w_r = \frac{2\hat{\sigma} z_{1-\alpha/2}}{\left|\nabla^k \bar{\ell}_N(\boldsymbol{v})\right|}, \quad k = 0, 1.$$

Note also that in tables 3.3.3–3.3.6 the DSF estimators correspond to $\gamma = 1$ and $\delta_1 = 1$ and thus are based on the totally decomposable score function estimators (3.3.5) and (3.3.6).

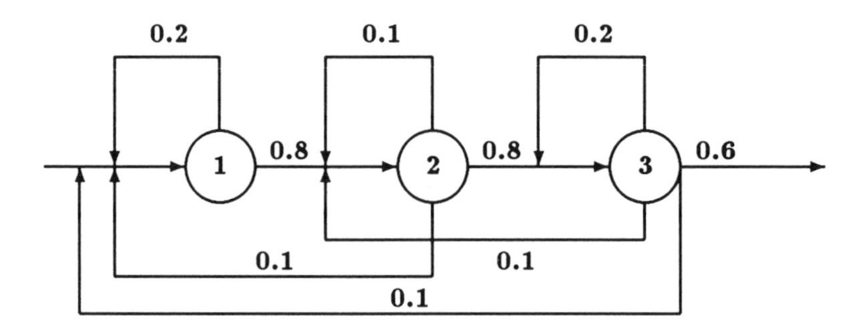

Figure 3.3.1 Three queues with feedbacks.

(a) **Markovian Models** *Two queues in tandem.*

Table 3.3.3 represents the point estimators $\bar{\ell}_N(\boldsymbol{v}), \nabla \bar{\ell}_N(\boldsymbol{v})$ and $\bar{\ell}_N^d(\boldsymbol{v})$, $\nabla \bar{\ell}_N^d(\boldsymbol{v})$ (denoted Estim) of the RSF and DSF estimators, their associated sample variances (denoted $\hat{\sigma}^2$), the relative widths of the confidence intervals (denoted w_r), and the sample efficiencies $\hat{\epsilon}^k(\boldsymbol{v})$, $k = 0, 1$, for different values of the relative perturbation, defined as $\delta_i = |(v_i - v_{0i})/v_i|$, $i = 1, 2$, where $\boldsymbol{v} = (v_1, v_2) = (\mu_1, \mu_2)$ is the vector of the service rates. We assumed that the interarrival rate is $\lambda = 1$, the reference service rates are $\mu_{01} = 2$ and $\mu_{02} = 2$ and we estimated the waiting time $\ell(\boldsymbol{v})$ and the associated gradient $\nabla\ell(\boldsymbol{v})$ for the values $\boldsymbol{v} = (v_1, v_2) = (\mu_1, \mu_2) = (2,2), (3,3), (4,4)$, which corresponds to perturbations $\delta_i = 0, 0.5, 1, \quad i = 1, 2$. Note that for $k = 0$, $\hat{\epsilon}^k(\boldsymbol{v})$ is defined as

$$\hat{\epsilon}^0(\boldsymbol{v}) = \frac{\text{sample variance of } \bar{\ell}_N(\boldsymbol{v})}{\text{sample MSE of } \bar{\ell}^d_N(\boldsymbol{v})}$$

and similarly for $k = 1$. Here MSE stands for the mean squared error and is defined as MSE $(\bar{\ell}^d_N(\boldsymbol{v})) = \mathbb{E}_{\boldsymbol{v}}\{\bar{\ell}^d_N(\boldsymbol{v}) - \bar{\ell}_N(\boldsymbol{v})\}^2$. This is needed since in general $\bar{\ell}^d_N(\boldsymbol{v})$ is a biased (inconsistent) estimator of $\ell(\boldsymbol{v})$, while its counterpart $\bar{\ell}_N(\boldsymbol{v})$ is a consistent one. The same is true for $\nabla\bar{\ell}^d_N(\boldsymbol{v})$ versus $\nabla\ell(\boldsymbol{v})$.

Table 3.3.3 Performance of the RSF and DSF estimators for two Markovian queues in tandem.

v_i/δ_i	Performance $\ell(\boldsymbol{v})$						
	Regenerative			Decomposition			
	Estim	$\hat{\sigma}^2$	w_r	Estim	$\hat{\sigma}^2$	w_r	$\hat{\epsilon}^k(\boldsymbol{v})$
2/0	0.998	1.2 E-05	0.014	0.998	1.2 E-05	0.014	1
3/0.5	0.333	1.1 E-06	0.011	0.333	5.1 E-07	0.008	2.2
4/1	0.168	8.0 E-07	0.020	0.167	1.5 E-07	0.009	5.3
	Derivative $\nabla_{v_1}\ell(\boldsymbol{v}) = \partial\ell(\boldsymbol{v})/\partial v_1$						
2/0	-0.722	1.5 E-04	0.066	-0.729	8.4 E-05	0.049	1.8
3/0.5	-0.139	1.2 E-06	0.032	-0.139	2.1 E-07	0.013	5.7
4/1	-0.049	4.3 E-07	0.053	-0.049	1.8 E-08	0.011	23.9
	Derivative $\nabla_{v_2}\ell(\boldsymbol{v}) = \partial\ell(\boldsymbol{v})/\partial v_2$						
2/0	-0.758	1.4 E-04	0.061	-0.761	9.1 E-05	0.049	1.5
3/0.5	-0.139	1.3 E-06	0.032	-0.139	2.0 E-07	0.013	6.5
4/1	-0.049	3.5 E-07	0.049	-0.048	1.7 E-08	0.011	20.6

Feedback Model

Table 3.3.4 presents data similar to that of table 3.3.3 for the feedback model given in fig.3.3.1. We assumed $\lambda = 0.8$, chose the reference service rates $\mu_{0i} = 3$, $i = 1, 2, 3$, and estimated the waiting time $\ell(\boldsymbol{v})$ and its gradient for $\mu_i = 3.0, \ 4.5, \ 6.0, \ i = 1, 2, 3$.

It follows from the results of tables 3.3.3 and 3.3.4 that the totally decomposable score function estimators $\hat{\epsilon}^k(\boldsymbol{v})$ perform reasonably well, in the sense that the relative bias

$$b = \left| \frac{\nabla^k \bar{\ell}_N^d(\boldsymbol{v}) - \nabla^k \bar{\ell}_N(\boldsymbol{v})}{\nabla^k \bar{\ell}_N(\boldsymbol{v})} \right|$$

is less than 0.15 and their relative efficiencies $\hat{\epsilon}^k(\boldsymbol{v})$ increase with the number of nodes (stations), the relative perturbation δ, and k $(k = 0, 1)$. Note again that $k = 0$ and $k = 1$ correspond to the performance $\nabla^0 \ell(\boldsymbol{v}) \equiv \ell(\boldsymbol{v})$ and to the gradient $\nabla^1 \ell(\boldsymbol{v}) \equiv \nabla \ell(\boldsymbol{v})$. It also follows from the results of tables 3.3.3 and 3.3.4 that in the first case the DSF estimators are consistent, while in the second case (feedback network) they are slightly biased $(b < 0.1)$ relative to their RSF counterparts. We also performed extensive simulation studies with the TSF estimators for more complex Markovian networks. The results are in agreement with our theoretical ones and are similar to those given in tables 3.3.3 and 3.3.4. In particular, we found that the RSF estimators become useless for $r > 3$ and $\rho_i \geq 0.5$, $i = 1, \ldots, r$. This is because the asymptotic variance of the regenerative estimators increases dramatically in r, while the asymptotic variance of the DSF estimators increases rather moderately in r.

Table 3.3.4 Performance of the RSF and DSF estimators for the Jackson network in fig. 3.3.1.

v_i/δ_i	Performance $\ell(\boldsymbol{v})$						
	Regenerative			Decomposition			
	Estim	$\hat{\sigma}^2$	w_r	Estim	$\hat{\sigma}^2$	w_r	$\hat{\epsilon}^k(\boldsymbol{v})$
3/0	1.334	1.6 E-05	0.012	1.334	1.6 E-05	0.012	1
4.5/0.5	0.485	1.3 E-04	0.091	0.488	1.2 E-06	0.009	108
6/1	0.223	4.1 E-05	0.110	0.253	3.6 E-07	0.009	114
	Derivative $\nabla_{v_1}\ell(\boldsymbol{v}) = \partial \ell(\boldsymbol{v})/\partial v_1$						
3/0	-0.396	1.0 E-04	0.103	-0.397	2.3 E-05	0.046	4.5
4.5/0.5	-0.091	2.5 E-04	0.680	-0.085	6.5 E-08	0.018	3800
6/1	-0.033	2.1 E-05	0.537	-0.031	8.5 E-09	0.011	2500
	Derivative $\nabla_{v_2}\ell(\boldsymbol{v}) = \partial \ell(\boldsymbol{v})/\partial v_2$						
3/0	-0.418	9.8 E-05	0.093	-0.398	2.4 E-05	0.048	4.08
4.5/0.5	-0.085	5.8 E-05	0.350	-0.085	6.2 E-08	0.011	940
6/1	-0.028	1.5 E-05	0.540	-0.031	6.8 E-09	0.010	2200
	Derivative $\nabla_{v_3}\ell(\boldsymbol{v}) = \partial \ell(\boldsymbol{v})/\partial v_3$						
3/0	-0.432	1.2 E-04	0.101	-0.401	2.7 E-05	0.045	4.4
4.5/0.5	-0.090	2.9 E-04	0.793	-0.084	6.7 E-08	0.012	4300
6/1	-0.037	3.8 E-05	0.650	-0.031	7.7 E-09	0.011	4900

(b) **Non-Markovian Models** *Tandem queues.*

Table 3.3.5 presents data similar to those of table 3.3.3 while choosing the following two-point interarrival and service time pdf's:
(i) *The interarrival pdf*

$$f_1(p_1) = \begin{cases} 1, & \text{with probability } p_1 = 0.5, \\ 3, & \text{with probability } 1 - p_1 = 0.5. \end{cases}$$

(ii) *The pdf of the first server*

$$f_2(p_2) = \begin{cases} 1, & \text{with probability } p_2 = 0.7, \\ 2.5, & \text{with probability } 1 - p_2 = 0.3. \end{cases}$$

(iii) *The pdf of the second server*

$$f_3(p_3) = \begin{cases} 1, & \text{with probability } p_3 = 0.7, \\ 2.5, & \text{with probability } 1 - p_3 = 0.3. \end{cases}$$

The service time pdf's f_2 and f_3 were used as the dominating pdf's to predict the expected steady-state system waiting time $\ell(\boldsymbol{v})$ and the associated gradient $\nabla \ell(\boldsymbol{v})$ for $\boldsymbol{v} = (v_1, v_2) = (p_2, p_3) = (0.7, 0.7)$, $(0.8, 0.8)$, $(0.9, 0.9)$.

Table 3.3.5 Performance of the RSF and DSF estimators for two queues in tandem with the two-point interarrival and service pdf's $f_1(p_1)$, $f_2(p_2)$ and $f_3(p_3)$.

v_i/δ_i	Performance $\ell(\boldsymbol{v})$						
	Regenerative			Decomposition			
	Estim	$\hat{\sigma}^2$	w_r	Estim	$\hat{\sigma}^2$	w_r	$\hat{\epsilon}^k(\boldsymbol{v})$
0.7/0	4.34	2.9 E-05	0.005	4.34	2.9 E-05	0.005	1
0.8/0.14	3.40	1.1 E-05	0.004	3.38	5.3 E-06	0.003	2.1
0.9/0.29	2.65	9.1 E-06	0.004	2.63	1.3 E-06	0.002	7.0
	Derivative $\nabla_{v_1}\ell(\boldsymbol{v}) = \partial\ell(\boldsymbol{v})/\partial v_1$						
0.7/0	-5.18	6.0 E-03	0.059	-5.66	2.9 E-03	0.037	2.1
0.8/0.14	-4.03	1.1 E-03	0.033	-4.23	2.8 E-04	0.015	3.9
0.9/0.29	-3.41	1.4 E-03	0.043	-3.47	1.0 E-04	0.013	14.0
	Derivative $\nabla_{v_2}\ell(\boldsymbol{v}) = \partial\ell(\boldsymbol{v})/\partial v_2$						
0.7/0	-5.48	5.7 E-03	0.054	-5.53	2.9 E-03	0.039	1.97
0.8/0.14	-4.22	9.8 E-04	0.029	-4.09	2.6 E-04	0.015	3.77
0.9/0.29	-3.56	1.1 E-03	0.037	-3.33	7.3 E-05	0.001	15.0

Feedback Model

Table 3.3.6 presents data similar to those of table 3.3.4 for the feedback model in fig.3.3.1 while assuming that the interarrival and the three service (dominating) pdf's are distributed gamma (denoted $G_i(\beta_i, \lambda_i)$, $i = 1, 2, 3, 4,$), that is $G(\beta, \lambda) = (1/\Gamma(\beta))\lambda^\beta y^{\beta-1} e^{-\lambda y}$, $y \geq 0$. More specifically, we chose the parameters of the interarrival pdf $\boldsymbol{v}_1 = (\beta_1, \lambda_1) = (0.5, 1)$ and the parameters of the service pdf's $\boldsymbol{v}_i = (\beta_i, \lambda_i)$, $\boldsymbol{v}_2 = (0.5, 2.5)$, $\boldsymbol{v}_3 = (0.4, 2.0)$ and $\boldsymbol{v}_4 = (0.25, 1.25)$. We predicted the performance $\ell(\boldsymbol{v})$ (the expected waiting time in the system) and the gradient

$$\nabla \ell(\boldsymbol{v}) = \left(\frac{\partial \ell(\boldsymbol{v})}{\partial v_1}, \frac{\partial \ell(\boldsymbol{v})}{\partial v_2}, \frac{\partial \ell(\boldsymbol{v})}{\partial v_3} \right)$$

for different values of the scale parameter vector $\boldsymbol{v} = (v_1, v_2, v_3) = (\lambda_2, \lambda_3, \lambda_4)$ while holding the shape parameter vector $(\beta_2, \beta_3, \beta_4) = (0.5, 0.4, 0.25)$. Those values are: $[\boldsymbol{v}/\delta]_1 = (2.5/0, 2/0, 1.25/0)$, $[\boldsymbol{v}/\delta]_2 = (3.125/0.25, 2.5/0.25, 1.625/3)$, and $[\boldsymbol{v}/\delta]_3 = (3.75/0.5, 3.0/0.5, 1.875/0.5)$.

Table 3.3.6 Performance of the RSF and DSF estimators for the queueing network in fig. 3.3.1 with gamma interarrival and service time pdf's.

i	Performance $\ell(\boldsymbol{v})$						
	Regenerative			Decomposition			
	Estim	$\hat{\sigma}^2$	w_r	Estim	$\hat{\sigma}^2$	w_r	$\hat{\epsilon}^k(\boldsymbol{v})$
1	1.267	1.6 E-05	0.013	1.267	1.6 E-05	0.013	1
2	0.766	7.2 E-06	0.014	0.764	3.5 E-06	0.009	2
3	0.551	1.8 E-05	0.031	0.544	1.5 E-06	0.009	12
	Derivative $\nabla_{v_1} \ell(\boldsymbol{v}) = \partial \ell(\boldsymbol{v})/\partial v_1$						
1	-0.296	6.9 E-05	0.110	-0.330	1.4 E-05	0.043	4.9
2	-0.137	1.5 E-05	0.110	-0.154	5.6 E-07	0.019	27
3	-0.075	3.4 E-05	0.314	-0.083	8.8 E-09	0.013	390
	Derivative $\nabla_{v_2} \ell(\boldsymbol{v}) = \partial \ell(\boldsymbol{v})/\partial v_2$						
1	-0.444	1.0 E-04	0.091	-0.434	2.7 E-05	0.047	3.7
2	-0.207	1.8 E-05	0.081	-0.203	1.0 E-06	0.020	18
3	-0.116	2.5 E-05	0.172	-0.113	1.6 E-07	0.014	156
	Derivative $\nabla_{v_3} \ell(\boldsymbol{v}) = \partial \ell(\boldsymbol{v})/\partial v_3$						
1	-0.912	2.4 E-04	0.067	-0.868	1.0 E-04	0.046	2.4
2	-0.370	2.7 E-05	0.055	-0.359	3.7 E-06	0.021	7.3
3	-0.239	2.5 E-05	0.083	-0.226	8.5 E-07	0.016	29

It follows that the results of tables 3.3.5 and 3.3.6 are quite similar to those in tables 3.3.3 and 3.3.4 for Markovian networks (the relative bias of

DSF estimators is still less than 0.15, the relative efficiency $\hat{\epsilon}^k(\boldsymbol{v})$ increases with the number of nodes (stations) and the relative perturbation δ and $k\,(k = 0, 1)$).

Using DSF estimators with $(\gamma, \Delta_\gamma) = (1, 1)$, we observed quite often, however, that the bias exceeds 0.1. To overcome this phenomenon we used instead DSF estimators $\nabla^k \bar{\ell}^d_N(\boldsymbol{v}, \gamma, \Delta_\gamma)$, $k = 0, 1$, and TSF estimators $\nabla^k \bar{\ell}^{tr}_N(\boldsymbol{v}, m_q, \tilde{m}_{q_1}, \ldots, \tilde{m}_{q_\gamma})$, $k = 0, 1$, while choosing the corresponding parameters (γ, Δ_γ) and $(m_q, \tilde{m}_{q_1}, \ldots, \tilde{m}_{q_\gamma})$ as suggested in (3.3.9) and (3.3.24), that is, we used larger modules (containing up to 3 nodes). By doing so we obtained a relative bias less than 0.1.

3.4 Conditional Score Function Method

In this section we present the conditional score function (CSF) method due to McLeish (1992) and McLeish and Rollans (1992) for estimating the system parameters $\nabla^k \ell(\boldsymbol{v})$, $k = 0, 1, \ldots$

We start our discussion with DESS. Consider the expected performance $\ell(\boldsymbol{v})$ in the form (2.2.26), that is

$$\ell(\boldsymbol{v}) = \mathbb{E}_{\boldsymbol{v}_0}\{L(\boldsymbol{Z}) W(\boldsymbol{Z}, \boldsymbol{v})\}, \tag{3.4.1}$$

where

$$W(\boldsymbol{Z}, \boldsymbol{v}) = f(\boldsymbol{Z}, \boldsymbol{v}) / f(\boldsymbol{Z}, \boldsymbol{v}_0),$$

$\boldsymbol{v}_0$ is a reference value of the parameter vector $\boldsymbol{v}$ and $\boldsymbol{Z} \sim f(\boldsymbol{z}, \boldsymbol{v}_0)$.

Suppose that for the considered parametric family $f(\boldsymbol{y}, \boldsymbol{v})$ we know a *sufficient statistic* $\boldsymbol{T} = \boldsymbol{T}(\boldsymbol{Y})$, i.e. the conditional distribution of $\boldsymbol{Y}$ given $\boldsymbol{T}$ is independent of $\boldsymbol{v}$. Note that $\boldsymbol{T}$ is sufficient for $\boldsymbol{v}$ if and only if (see, for example, Lehmann (1983)) $f(\boldsymbol{y}, \boldsymbol{v})$ is representable in the form

$$f(\boldsymbol{y}, \boldsymbol{v}) = q(\boldsymbol{T}(\boldsymbol{y}), \boldsymbol{v}) r(\boldsymbol{y}), \tag{3.4.2}$$

where $q(\boldsymbol{t}, \boldsymbol{v})$ and $r(\boldsymbol{y})$ are real-valued functions. It follows that for $\boldsymbol{T} = \boldsymbol{T}(\boldsymbol{Z})$,

$$W(\boldsymbol{Z}, \boldsymbol{v}) = w(\boldsymbol{T}, \boldsymbol{v}),$$

where

$$w(\boldsymbol{t}, \boldsymbol{v}) = q(\boldsymbol{t}, \boldsymbol{v}) / q(\boldsymbol{t}, \boldsymbol{v}_0).$$

We can now rewrite $\ell(\boldsymbol{v})$ as

$$\begin{aligned} \ell(\boldsymbol{v}) &= \mathbb{E}_{\boldsymbol{T}} \mathbb{E}_{\boldsymbol{Z}|\boldsymbol{T}}\{L(\boldsymbol{Z}) W(\boldsymbol{Z}, \boldsymbol{v}) | \boldsymbol{T}(\boldsymbol{Z})\} \\ &= \mathbb{E}_{\boldsymbol{T}}\{w(\boldsymbol{T}, \boldsymbol{v}) \mathbb{E}_{\boldsymbol{Z}|\boldsymbol{T}}[L(\boldsymbol{Z}) | \boldsymbol{T}]\} \\ &= \int c(\boldsymbol{t}) w(\boldsymbol{t}, \boldsymbol{v}) G_{\boldsymbol{v}_0}(\mathrm{d}\boldsymbol{t}), \end{aligned} \tag{3.4.3}$$

where $G_{\boldsymbol{v}_0}$ is the probability distribution of the random vector $\boldsymbol{T} = \boldsymbol{T}(\boldsymbol{Z})$, and

$$c(\boldsymbol{t}) = \mathbb{E}_{\boldsymbol{Z}|\boldsymbol{T}}[L(\boldsymbol{Z})|\boldsymbol{T} = \boldsymbol{t}]. \tag{3.4.4}$$

Note that by the sufficiency of $\boldsymbol{T}$ the conditional expectation $c(\boldsymbol{t})$ is independent of $\boldsymbol{v}_0$.

Assuming, as usual, the interchangeability of the expectation and differentiation operators, we obtain from (3.4.3) that

$$\begin{aligned} \nabla^k \ell(\boldsymbol{v}) &= \mathbb{E}_{\boldsymbol{T}}\{c(\boldsymbol{T})\nabla^k w(\boldsymbol{T}, \boldsymbol{v})\} = \\ &= \int c(\boldsymbol{t})\nabla^k w(\boldsymbol{t}, \boldsymbol{v}) G_{\boldsymbol{v}_0}(\mathrm{d}\boldsymbol{t}), \end{aligned} \tag{3.4.5}$$

$k = 0, 1, 2, \ldots$, where $\nabla^0 \ell(\boldsymbol{v}) \equiv \ell(\boldsymbol{v})$. In particular, by taking $\boldsymbol{v}_0 = \boldsymbol{v}$ we obtain

$$\nabla^k \ell(\boldsymbol{v}) = \int c(\boldsymbol{t}) \boldsymbol{s}^{(k)}(\boldsymbol{t}, \boldsymbol{v}) G_{\boldsymbol{v}}(\mathrm{d}\boldsymbol{t}),$$

where

$$\boldsymbol{s}^{(k)}(\boldsymbol{t}, \boldsymbol{v}) = \frac{\nabla^k q(\boldsymbol{t}, \boldsymbol{v})}{q(\boldsymbol{t}, \boldsymbol{v})} \quad .$$

Note that $\nabla^k w(\boldsymbol{t}, \boldsymbol{v}) = \boldsymbol{s}^{(k)}(\boldsymbol{t}, \boldsymbol{v}) w(\boldsymbol{t}, \boldsymbol{v}),\ k = 1, 2, \ldots$

Consider now the following exponential family:

$$f(y, \boldsymbol{v}) = a(\boldsymbol{v}) \exp\{b(\boldsymbol{v}) t(y)\} h(y) \ . \tag{3.4.6}$$

It is well known that when the input consists of an iid random sample $Y_1, \ldots, Y_m$ from such an exponential family, the random variable $T_m = \sum_{i=1}^m t(Y_i)$ has a distribution from an exponential family and is a sufficient statistic for $\boldsymbol{v}$. That is, the conditional distribution of the random vector $(Y_1, \ldots, Y_m)$ given T_m is independent of $\boldsymbol{v}$. We have in this case that the pdf $f_m(\boldsymbol{y}, \boldsymbol{v})$ of $(Y_1, \ldots, Y_m)$ is given by

$$f_m(\boldsymbol{y}, \boldsymbol{v}) = a(\boldsymbol{v})^m \exp\left\{ b(\boldsymbol{v}) \sum_{i=1}^m t(y_i) \right\} \prod_{i=1}^m h(y_i)$$

and hence

$$w(t, \boldsymbol{v}) = \frac{a(\boldsymbol{v})^m}{a(\boldsymbol{v}_0)^m} \exp\{[b(\boldsymbol{v}) - b(\boldsymbol{v}_0)]t\} \ . \tag{3.4.7}$$

Formulas (3.4.3)–(3.4.5) suggest using the following two-step algorithm for estimating $\nabla^k \ell(\boldsymbol{v})$.

Algorithm 3.4.1 :

1. *Evaluate (typically by simulation) the conditional expectation $c(\boldsymbol{t})$ defined in (3.4.4).*

2. Calculate (analytically or numerically) the integral

$$\nabla^k \ell(\boldsymbol{v}) = \int c(\boldsymbol{t}) \nabla^k w(\boldsymbol{t}, \boldsymbol{v}) G_{\boldsymbol{v}_0}(\mathrm{d}\boldsymbol{t}).$$

We show now (see McLeish (1992)) that the above conditional method *always* results in a variance reduction relative to its straightforward "what if" counterpart. Let $\boldsymbol{Z_t}$ be a random vector generated from the conditional distribution of $\boldsymbol{Z}$ given $\boldsymbol{T}(\boldsymbol{Z}) = \boldsymbol{t}$ for a fixed value $\boldsymbol{v}$ of the parameter vector. Then, based on a sample of size one, the suggested conditional estimator of $\ell(\boldsymbol{v})$ is

$$\int L(\boldsymbol{Z_t}) w(\boldsymbol{t}, \boldsymbol{v}) G_{\boldsymbol{v}_0}(\mathrm{d}\boldsymbol{t}) .$$

We clearly have that

$$\ell(\boldsymbol{v}) = \mathbb{E}\left\{\int L(\boldsymbol{Z_t}) w(\boldsymbol{t}, \boldsymbol{v}) G_{\boldsymbol{v}_0}(\mathrm{d}\boldsymbol{t})\right\} .$$

Similarly,

$$\begin{aligned}
&\mathbb{E}_{\boldsymbol{v}_0}\{L(\boldsymbol{Z})^2 W(\boldsymbol{Z}, \boldsymbol{v})^2\} \\
&= \mathbb{E}_{\boldsymbol{T}} \mathbb{E}_{\boldsymbol{Z}|\boldsymbol{T}}\{L(\boldsymbol{Z})^2 W(\boldsymbol{Z}, \boldsymbol{v})^2 | \boldsymbol{T}\} \\
&= \mathbb{E}_{\boldsymbol{T}}\{w(\boldsymbol{T}, \boldsymbol{v})^2 \mathbb{E}_{\boldsymbol{Z}|\boldsymbol{T}}[L(\boldsymbol{Z})^2 | \boldsymbol{T}]\} \\
&= \mathbb{E}\{\int L(\boldsymbol{Z_t})^2 w(\boldsymbol{t}, \boldsymbol{v})^2 G_{\boldsymbol{v}_0}(d\boldsymbol{t})\} .
\end{aligned}$$

Now by Jensen's inequality

$$\int L(\boldsymbol{Z_t})^2 w(\boldsymbol{t}, \boldsymbol{v})^2 G_{\boldsymbol{v}_0}(\mathrm{d}\boldsymbol{t}) \geq \left[\int L(\boldsymbol{Z_t}) w(\boldsymbol{t}, \boldsymbol{v}) G_{\boldsymbol{v}_0}(\mathrm{d}\boldsymbol{t})\right]^2 .$$

By taking expectations from both sides of the above inequality, we obtain

$$\mathbb{E}_{\boldsymbol{v}_0}\{L(\boldsymbol{Z})^2 W(\boldsymbol{Z}, \boldsymbol{v})^2\} \geq \mathbb{E}\left\{\left[\int L(\boldsymbol{Z_t}) w(\boldsymbol{t}, \boldsymbol{v}) G_{\boldsymbol{v}_0}(\mathrm{d}\boldsymbol{t})\right]^2\right\} .$$

Since $\mathbb{E}_{\boldsymbol{v}_0}\{LW\} = \ell(\boldsymbol{v})$ it follows that

$$\mathrm{Var}_{\boldsymbol{v}_0}\{LW\} \geq \mathrm{Var}\left\{\int L(\boldsymbol{Z_t}) w(\boldsymbol{t}, \boldsymbol{v}) G_{\boldsymbol{v}_0}(\mathrm{d}\boldsymbol{t})\right\} . \qquad (3.4.8)$$

For the corresponding estimators based on N independent simulations, the variances are as above but divided by N and so the same result holds. Similar results can be derived for the straightforward and conditional "what if" estimators of $\nabla^k \ell(\boldsymbol{v})$, $k = 1, 2, \ldots$

Note that the inequality in (3.4.8) is strict unless $L(\boldsymbol{Z}_t)w(\boldsymbol{t},\boldsymbol{v})$ is constant in $\boldsymbol{t}$ w.p.1 under the probability measure $G_{\boldsymbol{v}_0}$.

It is important to realize that algorithm 3.4.1 requires estimation of the conditional expectation $c(\boldsymbol{t})$ given in (3.4.4). This in turn requires the generation of random vectors $\boldsymbol{Z}$ *conditional* on $\boldsymbol{T} = \boldsymbol{t}$ for *different* values of $\boldsymbol{t}$. Note that in some particular cases it is possible to do so from a *single* simulation (see example 3.4.2 below), while in general, *multiple simulations* are required.

Note that in contrast to the present method, the conventional conditional Monte Carlo method (for example Ross (1989)) assumes that $c(\boldsymbol{t})$ is available analytically while only the corresponding integral, given in the right-hand side of (3.4.5), needs to be estimated from simulation. Note also that when $\boldsymbol{T}$ has more than one component, we may need to use simulation in order to evaluate the multidimensional integral at the second stage of the algorithm.

We call the estimator underlying (3.4.5), that is

$$\hat{\nabla}^k \ell_N(\boldsymbol{v}) = \mathbb{E}_{\boldsymbol{T}}\left\{\hat{c}_N(\boldsymbol{T})\nabla^k w(\boldsymbol{T},\boldsymbol{v})\right\}, \tag{3.4.9}$$

where

$$\hat{c}_N(\boldsymbol{t}) = N^{-1}\sum_{i=1}^{N} L(\boldsymbol{Z}_{\boldsymbol{t}i}) \tag{3.4.10}$$

and $\boldsymbol{Z}_{\boldsymbol{t}1},\ldots,\boldsymbol{Z}_{\boldsymbol{t}N}$ is a corresponding sample conditional on $\boldsymbol{T} = \boldsymbol{t}$, the *conditional score function* (CSF) estimator.

We present below several examples of conditional distributions based on the standard exponential families (see McLeish and Rollans (1992)). Note that in the first three examples we also present procedures for generating the associated samples, while in the other three examples this is left as an exercise. Note also that in all our examples the components $Y_1,\ldots,Y_m$ of the random vector $\boldsymbol{Y}$ are iid random variables and the sufficient statistic T is given by $T = Y_1 + \cdots + Y_m$.

Example 3.4.1 Conditional pdf's

1. *Exponential distribution.* Suppose that Y_i are independent exponentially distributed random variables with pdf $f(y,v) = ve^{-vy}$. Then given $\sum_{i=1}^m Y_i = t$ the random variables $Y_1,\ Y_1+Y_2,\ldots,\sum_{i=1}^{m-1} Y_i$, are distributed as $m-1$ uniform on the interval $(0,t)$ order statistics.

   This suggests a convenient way of generating $\boldsymbol{Y}$ conditional on $\sum_{i=1}^m Y_i = t$. First generate a sample $U_1,\ldots,U_{m-1}$ of independent uniform $U(0,1)$ variables and then arrange them in the increasing order $U_{(1)} \le \cdots \le U_{(m-1)}$ ($U_{(i)}$ are called order statistics), and finally, calculate Y_i according to $Y_i = t[U_{(i)} - U_{(i-1)}]$, $i = 1,\ldots,m$, with $U_{(0)} = 0$ and $U_{(m)} = 1$.

2. *Normal distribution.* Suppose that Y_i are distributed as independent $N(\mu, \sigma^2)$ variables. The conditional distribution of $\boldsymbol{Y} = (Y_1, \ldots, Y_m)$ given $T = t$, is multivariate normal with mean vector $tm^{-1}\boldsymbol{e}$ and covariance matrix $\sigma^2(\boldsymbol{I}_m - m^{-1}\boldsymbol{e}\boldsymbol{e}')$, where $\boldsymbol{e} = (1, \ldots, 1)'$ is the m-dimensional vector of ones and $\boldsymbol{I}_m$ is the $m \times m$ identity matrix.

   The procedure for generating random vectors $\boldsymbol{Y}$ conditional on $T = t$ can be written as follows. Generate independent variables $X_i \sim N(0, 1)$ and calculate Y_i according to

$$Y_i = \sigma X_i - \sigma m^{-1}(X_1 + \cdots + X_m) + tm^{-1}, \quad i = 1, \ldots, m.$$

3. *Poisson distribution.* Suppose that Y_i have the Poisson (v) distribution. Then given $\sum_{i=1}^m Y_i = t$, the distribution of Y_1 is binomial $(t, 1/m)$.

   The procedure for generating random vectors $\boldsymbol{Y}$ conditional on $\sum_{i=1}^m Y_i = t$ is as follows. Generate first a random variable Y_1 from the binomial $(t, 1/m)$ distribution. Generate next a random variable Y_2, conditional on $Y_2 + \cdots + Y_m = t_1$ with $t_1 = t - Y_1$, that is from the binomial $(t_1, 1/(m-1))$ distribution. Continue this until the component Y_{m-1}, conditional on $Y_{m-1} + Y_m = t - t_{m-2}$, $t_{m-2} = Y_1 + \cdots + Y_{m-2}$, is generated from the binomial $(t_{m-2}, 1/2)$ distribution. Note that the last component Y_m is determined uniquely from the previously generated sample $Y_1, \ldots, Y_{m-1}$ and is given by $Y_m = t - \sum_{i=1}^{m-1} Y_i$. Note also that in contrast to the above two procedures, where only a *single* simulation is required, here we need *multiple simulation*, namely to repeat the above procedure for different values of t.

4. *Gamma distribution.* Suppose that Y_i have a gamma distribution with parameters λ and β (see (2.2.12)). Then $T = Y_1 + \cdots + Y_m$ is a sufficient statistic for λ and the conditional distribution of $t^{-1}Y_1$ given $T = t$ is the beta $(\beta, m\beta)$ distribution.

5. *Binomial distribution.* Suppose that Y_i are distributed as binomial (n, p) variables. Then given $Y_1 + \ldots Y_m = t$, Y_1 has a hypergeometric distribution with parameters (mn, n, t).

6. *Geometric distribution.* Suppose Y_i are distributed as independent geometric variates. Then given $Y_1 + \ldots Y_m = t$, Y_1 has a negative hypergeometric distribution with parameters $\left\{ \frac{\binom{t-y-1}{m-2}}{\binom{t-1}{m-1}} \right\}$.

The following example shows how this is carried out in a simple network with exponentially distributed inputs.

Example 3.4.2 Stochastic PERT network. Consider estimation $\nabla^k \ell(\boldsymbol{v}) = \nabla^k \mathbb{E}_{\boldsymbol{v}}\{L(\boldsymbol{Y})\}$, $k = 0, 1$, in a stochastic PERT network with the sample performance

$$L(\boldsymbol{Y}) = \min[(Y_1 + Y_2), (Y_1 + Y_3 + Y_5), (Y_4 + Y_5)],$$

where $\boldsymbol{Y} = (Y_1, \ldots, Y_5)$. Assume that the rv's $Y_1, \ldots, Y_5$ are iid each distributed exponentially, that is $f(y, v) = v^{-1}e^{v^{-1}y}$, $y > 0$, and v is fixed.

The straightforward SF estimator of $\mathrm{d}\ell(v)/\mathrm{d}v$ can be written as

$$\frac{\mathrm{d}\bar{\ell}(v)}{\mathrm{d}v} = L(\boldsymbol{Y})S(\boldsymbol{Y}, v), \tag{3.4.11}$$

where

$$S(\boldsymbol{Y}, v) = \sum_{i=1}^{5} S_i(Y_i, v), \quad S_i = \frac{(Y_i - v)}{v^2}.$$

To find the CSF estimator note first that in this case conditioning on $\sum_{i=1}^m Y_i = t$ we obtain that the variates Y_i are distributed as the *spacings between uniform* $U(0, t)$ *order statistics*. In particular (see result 1 of example 3.4.1), given $\sum_{i=1}^m Y_i = t$ we obtain that $Y_i = t[U_{(i)} - U_{(i-1)}]$, where $U_{(i)}$ are the order statistics from a sample of size 4 from $U(0, 1]$, with $U_{(0)} = 0$ and $U_{(5)} = 1$. Consequently, the conditional expectation $c(t)$ reduces to

$$c(t) = tX, \tag{3.4.12}$$

where

$$X = \mathbb{E} \min[U_{(2)}, U_{(1)} + U_{(3)} - U_{(2)} + 1 - U_{(4)}, 1 - U_{(3)}].$$

Taking into account next that $s^{(1)}(t, v)$ reduces to $(t - 5v)/v^2$ and $T = \sum_{i=1}^m Y_i$ has an Erlang distribution with parameters v and m, which for $m = 5$ reduces to

$$\frac{\mathrm{d}G_{\boldsymbol{v}}}{\mathrm{d}t} = \frac{t^4 \mathrm{e}^{-t/v}}{4! v^5},$$

we readily obtain that

$$\ell(\boldsymbol{v}) = 5vX$$

and

$$\frac{\mathrm{d}\ell}{\mathrm{d}v} = \int_0^\infty \mathbb{E}_v[L(\boldsymbol{Y})|T_5 = t] \frac{t - 5v}{v^2} \frac{t^4 \mathrm{e}^{-t/v}}{4! v^5} \mathrm{d}t = 5X. \tag{3.4.13}$$

Note that $\mathrm{d}\ell/\mathrm{d}v$ is independent of v since $\ell(\boldsymbol{v})$ is linear in v. Note also that if we chose the exponential pdf in the form $f(y) = \lambda \mathrm{e}^{-\lambda y}$ rather than in the form $f(y) = v^{-1}\mathrm{e}^{-v^{-1}y}$, then $\mathrm{d}\ell/\mathrm{d}\lambda$ would be clearly dependent on λ.

The algorithm for estimating $\mathrm{d}\ell/\mathrm{d}v$ by conditioning is extremely simple and can be written as

Algorithm 3.4.2 :

1. *Estimate from simulation the conditional expectation*

$$X = \mathbb{E}\min[U_{(2)}, U_{(1)} + U_{(3)} - U_{(2)} + 1 - U_{(4)}, 1 - U_{(3)}]$$

by generating at each replication a sample $U_1, \ldots, U_4$ *from* $U(0,1)$ *and than using the order statistics* $U_{(1)}, U_{(2)}, U_{(3)}, U_{(4)}$.

2. *Estimate* $\ell(v)$ *and* $\mathrm{d}\ell/\mathrm{d}v$ *as*

$$\hat{\ell}_N(v) = 5vN^{-1}\sum_{i=1}^{N} V_i$$

and

$$\hat{\mathrm{d}\ell}_N/\mathrm{d}v = 5N^{-1}\sum_{i=1}^{N} V_i,$$

respectively, where

$$V_i = \min[U_{(2)i}, U_{(1)i} + U_{(3)i} - U_{(2)i} + 1 - U_{(4)i}, 1 - U_{(3)i}].$$

Consider now a more complex sample performance $L(\boldsymbol{Y})$, assuming again that all components of the m-dimensional vector $\boldsymbol{Y}$ are iid and distributed exponentially. In this case analogous considerations allow to estimate $\ell(v)$ and $\mathrm{d}\ell/\mathrm{d}v$ as

$$\hat{\ell}_N(v) = \int_0^\infty \hat{c}_N(t) v \mathrm{e}^{-vt}\frac{(vt)^{m-1}}{(m-1)!}\mathrm{d}t \; . \tag{3.4.14}$$

and as

$$\begin{aligned} \hat{\mathrm{d}\ell}_N/\mathrm{d}v &= \int_0^\infty \hat{c}_N(t)\frac{\mathrm{d}\ell n v^m \mathrm{e}^{-vt}}{\mathrm{d}v} G_v(\mathrm{d}t) \\ &= \int_0^\infty \hat{c}_N(t)\left(\frac{m}{v} - t\right) v\mathrm{e}^{-vt}\frac{(vt)^{m-1}}{(m-1)!}\mathrm{d}t \; , \end{aligned} \tag{3.4.15}$$

respectively. Here

$$\hat{c}_N(t) = N^{-1}\sum_{j=1}^{N} L(\boldsymbol{Z}_j) \; , \tag{3.4.16}$$

where $\boldsymbol{Z}_j = (t[U_{(1)j} - U_{(0)j}], \ldots, t[U_{(m)j} - U_{(m-1)j}])$, $j = 1, \ldots, N$. Note that $\boldsymbol{Z}_j$ and hence $L(\boldsymbol{Z}_j)$ are functions of the parameter t. Note also that

$\hat{\mathrm{d}\ell_N}/\mathrm{d}v = \mathrm{d}\hat{\ell_N}/\mathrm{d}v$ in the sense that the estimator given in (3.4.15) can be obtained by differentiation of the integral in the right-hand side of (3.4.14).

Typically, there is a very considerable variance reduction by conditioning, as well as the capability of the CSF estimators to evaluate the performance $\ell(v)$ and the derivative $\mathrm{d}\ell/\mathrm{d}v$ for *all* v.

To see this define the variance reduction (efficiency) of the conditional SF method versus the straightforward SF one as

$$\epsilon^k = \frac{\mathrm{Var}\{\bar{\mathrm{d}\ell_N}/\mathrm{d}v\}}{\mathrm{Var}\{\hat{\mathrm{d}\ell_N}/\mathrm{d}v\}}, \tag{3.4.17}$$

and consider again example 3.4.2 for $k = 1$ assuming first $v_0 = v$ (no change of measure). In this case (see McLeish and Rollans (1992)) the variance of the straightforward SF estimator $\bar{\mathrm{d}\ell_N}/\mathrm{d}v$ is approximately 16, whereas the variance of conditional estimator $\hat{\mathrm{d}\ell_N}/\mathrm{d}v = 5X$ is only around 0.3. The efficiency gain by conditioning here is about 70.

Consider now the "what if" estimator of $\ell(v)$. Suppose that we wish to estimate $\ell(v) = 5vX$ based on simulations conducted at $v_0 = 1$. The variance of the straightforward likelihood ratio estimator

$$L(\mathbf{Y})\frac{f(Y,v)}{f(Y,1)}$$

(see McLeish and Rollans (1992) is

$$\frac{30\mathbb{E}(X^2)}{v^3(2-w)^7} - 25v^2(\mathbb{E}X)^2 \approx \frac{2.1}{v^3(2-v)^7} - 1.45v^2,$$

while the variance of the conditional estimator $5vV$is approximately $0.3v^2$.

Thus, the efficiency ϵ^0 of the conditional estimator relative to the straightforward one is

$$\epsilon^0 \approx \frac{7}{v^5(2-v)^7} - 4.83.$$

It is important to note that the straightforward estimator has finite variance only for $0 < v < 2$ (see also section 2.3), and as v approaches either of these end-points, the relative efficiency of the conditional estimator approaches infinity. Table 3.4.1 shows the efficiency by conditioning for example 3.4.1.

Table 3.4.1 The efficiency of the CSF method as a function of v in example 3.4.1.

v	0	0.2	0.4	0.6	0.8	1.0	1.2	1.4	1.6	1.8	≥ 2
ϵ^0	∞	352	20.6	3.7	1.1	2.2	8.6	42	403	28937	∞

Note that the above DESS examples (with all components being iid distributed exponential family) have limited applications, and are given for illustration purposes only. They, however, shed light on and give some insight into the CSF method for more complex DES (see below).

McLeish and Rollans (1992) extended their results to distributions that are not from the exponential family. In particular, they suggested approximating the conditional expectation $\mathbb{E}[L(\boldsymbol{Y}|\boldsymbol{S}(\boldsymbol{Y},\boldsymbol{v})) = \boldsymbol{t}]$ by $\mathbb{E}[L(\boldsymbol{Y}|\hat{\boldsymbol{v}}) = \boldsymbol{t}]$, where $\hat{\boldsymbol{v}}$ is the maximum likelihood estimator of $\boldsymbol{v}$. They showed that typically the function $\mathbb{E}[L(\boldsymbol{Y}|\hat{\boldsymbol{v}}) = \boldsymbol{t}]$ varies only very slowly in $\boldsymbol{v}$. Clearly for the exponential family, $\mathbb{E}[L(\boldsymbol{Y}|\hat{\boldsymbol{v}}) = \boldsymbol{t}]$ does not depend on $\boldsymbol{v}$.

Again, the integral over the distribution of $\hat{\boldsymbol{v}}$ may either be obtained analytically, as in example 3.4.1, by crude Monte-Carlo, or by numerical integration. The latter program can exploit the asymptotic normality of the score function or maximum likelihood estimators. For more details see McLeish and Rollans (1992).

We now apply the CSF method to simple queueing models. In particular we derive CSF estimators which are based on the truncated score function (TSF) estimators (3.3.13). To do so recall first that the truncated version of $\nabla^k \ell(\boldsymbol{v})$ and the associated TSF estimator can be written (see (3.3.12) and 3.3.13)) as

$$\nabla^k \ell^{tr}(\boldsymbol{v}, m) = \mathbb{E}_g\{L_t \nabla^k \widetilde{W}_t^{tr}(\boldsymbol{v}, m)\} \tag{3.4.18}$$

and

$$\nabla^k \bar{\ell}_N^{tr}(\boldsymbol{v}, m) = N^{-1}T^{-1} \sum_{i=1}^{N} \sum_{t=1}^{T} L_{ti} \nabla^k \widetilde{W}_{ti}^{tr}(\boldsymbol{v}, m), \tag{3.4.19}$$

$k = 0, 1, \ldots$, respectively. Here

$$\widetilde{W}_t^{tr}(\boldsymbol{v}, m) = \prod_{j=t-m+1}^{t} W_j,$$

m is the truncation parameter, T is the batch size (see section 1.2), and N is the number of batches.

Assume further that $T = m$. Then one can readily define a CSF estimator underlying (3.4.18) which will be similar the DESS estimator (3.4.9). Indeed, in this case, we need only to replace

$$\hat{c}_N(\boldsymbol{t}) = N^{-1} \sum_{i=1}^{N} L(\boldsymbol{Z}_{\boldsymbol{t}_i})$$

(see (3.4.10)) by

$$\hat{c}_N(\boldsymbol{t}) = N^{-1}T^{-1} \sum_{i=1}^{N} \sum_{j=1}^{T} L(\boldsymbol{Z}_{\boldsymbol{t}_{ij}}),$$

while all other data remai the same.

As a particular example, consider the estimation $d\ell/dv$ in the $M/G/1$, where v is the expected time between (exponential) arrivals, and L_t is, say, the sojourn time. In this case (for each batch of size $T = m$) we have to estimate $c(t) = \mathbb{E}_v\{L(\boldsymbol{Z})|T_m = t\}$ by generating $m-1$ order statistics based on the uniform on the interval $(0, t)$ random variables (see 1 of example 3.4.1), which serve as the interarrival random variables, and then again evaluate numerically the integral (3.4.15).

McLeish and Rollans (1992) presented simulation results for $d\ell/dv$, where ℓ is the expected sojourn time in the $M/G/1$ queue with interarrival rates $v = 0.3$ and $v = 0.8$, respectively, while using the following two-point service time distribution:

$$f(p) = \begin{cases} 1/2, & \text{with probability } p = 0.5 \\ 3/2, & \text{with probability } 1 - p = 0.5. \end{cases}$$

They chose the batch size equal to 1786 and 3572 for $\rho = v = 0.3$ and $\rho = v = 0.8$, respectively, simulated $1.5 \cdot 10^6$ customers for each scenario, and used a simple two-point quadrature formula to evaluate the integral (3.4.5). They found that the CSF estimator outperforms the staightforward RSF (regenerative score function) estimator by a factor of 2 and 2.6 for $\rho = 0.3$ and $\rho = 0.8$, respectively. Clearly the CSF estimator also outperforms the straightforward TSF estimator, since the RSF does too. (Choosing the truncation parameter equal to 5 and 30 (see table 3.3.2) for $\rho = 0.3$ and $\rho = 0.8$, respectively we found that the RSF estimator outperforms the TSF estimator by a factor of 2.3 and 4.2).

One can easily obtain a table similar to table 3.4.1, which will clearly show the advantage of the CSF method, while comparing it, say with the RSF estimator (3.4.19). The reader is asked to perform such an experiment with the above $M/G/1$ queue, while choosing the dominating pdf $g(y) = f(y, v)$ corresponding, say, to $\rho_0 = 0.8$.

Note that the above CSF estimators are slightly biased, since they underline the truncated version of $\nabla^k \ell(\boldsymbol{v})$, that is

$$\nabla^k \ell^{tr}(\boldsymbol{v}, m) = \mathbb{E}_g\{L_t \nabla^k \widetilde{W}_t^{tr}(\boldsymbol{v}, m)\}.$$

Note also that the CSF algorithm 3.4.1 can be modified to estimate the n-dimensional gradient vector $\nabla \ell(\boldsymbol{v})$. In this case we need to replace in step 2 of algorithm 3.4.1 the one-dimensional integral (expected value) $\nabla^k \hat{\ell}(\boldsymbol{v})$, by a corresponding n-dimensional integral (expected value). For more details on evaluating multiple integrals underlying CSF estimators, see McLeish and Rollans (1992).

As an exercise the reader is asked to derive a CSF estimator for $\nabla \ell(\boldsymbol{v})$, and run a computer program, while considering two queues in tandem with

$\ell(\boldsymbol{v})$ being the sojourn time, and $\boldsymbol{v} = (v_1, v_2)$, where v_1 and v_2 correspond to the expected service times in the first and second tandem queues, having gamma and geometric service time distributions, respectively, and an arbitrary interarrival distribution, say an exponential distribution.

3.5 Optimization of DEDS

This section deals with the optimization of general open non-Markovian queueing networks where neither the objective function $\ell_0(\boldsymbol{v})$, nor the constraints, are available analytically. In this case we resort to Monte Carlo optimization methods. As in chapter 2 we deal with the program (P_0) of the form

$$(\mathrm{P}_0) \qquad \begin{array}{lll} \text{minimize} & \ell_0(\boldsymbol{v}), & \boldsymbol{v} \in V, \\ \text{subject to} & \ell_j(\boldsymbol{v}) \leq 0, & j = 1, \ldots, k, \\ & \ell_j(\boldsymbol{v}) = 0, & j = k+1, \ldots, M, \end{array} \tag{3.5.1}$$

where

$$\ell_j(\boldsymbol{v}) = \mathbb{E}_{\boldsymbol{v}}\{L_j\} = \frac{\mathbb{E}_{\boldsymbol{v}}\{\sum_{t=1}^{\tau} L_{jt}\}}{\mathbb{E}_{\boldsymbol{v}}\tau}, \quad j = 0, 1, \ldots, M, \tag{3.5.2}$$

are the steady-state expected performances corresponding to the output processes $\{L_{jt}\}$.

In this case the stochastic counterpart $(\overline{\mathrm{P}}_N)$ of DESS, that is

$$(\overline{\mathrm{P}}_N) \qquad \begin{array}{lll} \text{minimize} & \overline{\ell}_{0N}(\boldsymbol{v}), & \boldsymbol{v} \in V, \\ \text{subject to} & \overline{\ell}_{jN}(\boldsymbol{v}) \leq 0, & j = 1, \ldots, k, \\ & \overline{\ell}_{jN}(\boldsymbol{v}) = 0, & j = k+1, \ldots, M, \end{array} \tag{3.5.3}$$

(see (2.4.6)) is directly applicable to DEDS, provided $\overline{\ell}_{jN}(\boldsymbol{v})$ in (2.4.3) are replaced by

$$\overline{\ell}_{jN}(\boldsymbol{v}) = \frac{\sum_{i=1}^{N} \sum_{t=1}^{\tau_i} L_{jti} \widetilde{W}_{ti}}{\sum_{i=1}^{N} \sum_{t=1}^{\tau_i} \widetilde{W}_{ti}}. \tag{3.5.4}$$

Here

$$L_{jti} = L_{jt}(\underline{\boldsymbol{Z}}_{ti}), \quad \widetilde{W}_{ti} = \prod_{k=1}^{r} \prod_{p \in C_t^k} W_{pki}, \quad W(\boldsymbol{z}, \boldsymbol{v}) = \frac{f(\boldsymbol{z}, \boldsymbol{v})}{g(\boldsymbol{z})},$$

C_t^k, $k = 1, \ldots, r$, are the sets of indices defined in (3.2.30). Note that for r $GI/G/1$ queues in tandem with the FIFO discipline $\widetilde{W}_{ti}$ reduces to $\widetilde{W}_{ti} = \prod_{k=1}^{r} \prod_{p=1}^{t} W_{pki}$.

As soon as the processes L_{jt} are generated, the sample functions $\bar{\ell}_{jN}(\boldsymbol{v})$, $j = 0,\ldots,M$, become available analytically. This is because the processes L_{jt} do not depend on $\boldsymbol{v}$ and after simulation the likelihood ratio process $\widetilde{W}_{ti}$ is likewise available. Note that we do not even need to know explicitly the analytical expressions of the processes L_{jt}. Moreover, since for each $\boldsymbol{v}$ the functions $\bar{\ell}_{jN}(\boldsymbol{v})$ and the corresponding gradients $\nabla\bar{\ell}_{jN}(\boldsymbol{v})$ can be calculated according to (3.2.29) and (3.2.30), respectively (*all from a single simulation*), the associated optimization problem ($\overline{\text{P}}_N$) can be solved by standard analytical or numerical methods of mathematical programming. The obtained optimal value $\bar{\varphi}_N$ and the optimal solution $\bar{\boldsymbol{v}}_N$ of the program ($\overline{\text{P}}_N$) provide estimators of the optimal value φ^* and the optimal solution $\boldsymbol{v}^*$ of the program (P_0), respectively. Note again that all of this is feasible since we assumed that the output processes $\{L_{jt}(\underline{\boldsymbol{Z}}_t)\}$ *do not depend* on $\boldsymbol{v}$. Note also, that the case where $\{L_{jt}(\cdot)\}$ depends on $\boldsymbol{v}$ requires a separate treatment and is considered in chapter 5 below.

The algorithm for estimating the optimal solution $\boldsymbol{v}^*$ of the program (P_0) while using the stochastic counterpart ($\overline{\text{P}}_N$) is similar to that of algorithm 2.4.1 of chapter 2 for optimization of DESS and can be written as follows:

Algorithm 3.5.1 :

1. *Generate a random sample* $\boldsymbol{Z}_{11},\ldots,\boldsymbol{Z}_{\tau_1 1},\ \ldots,\boldsymbol{Z}_{1N},\ldots,\boldsymbol{Z}_{\tau_N N}$ *from* $g(\boldsymbol{z})$.

2. *Generate the output (sample performance) processes* $L_{jti}, j = 0,\ldots,M$, *and the likelihood ratio process* $\widetilde{W}_{ti}(\boldsymbol{v})$, $t = 1,\ldots,\tau_i$; $i = 1,\ldots,N$.

3. *Solve the program* ($\overline{\text{P}}_N$) *by methods of mathematical programming.*

4. *Deliver the solution* $\bar{\boldsymbol{v}}_N$ *of* ($\overline{\text{P}}_N$) *as an estimator of* $\boldsymbol{v}^*$.

Note that the optimal solution of the program (P_0) can be estimated (approximated) by incorporating the SF estimators into iterative algorithms of stochastic approximation (SA) type, (e.g. L'Ecuyer (1992)). A SA algorithm basically present a gradient (steepest) descent procedure in which the exact gradient $\nabla\ell(\boldsymbol{v})$ is replaced by its estimator $\nabla\bar{\ell}_N(\boldsymbol{v})$. SA can be useful if

(a) It is too expensive to store long samples $\boldsymbol{Z}_{11},\ldots,\boldsymbol{Z}_{\tau_1 1},\ldots,$ $\boldsymbol{Z}_{1N},\ldots,\boldsymbol{Z}_{\tau_N N}$ and the associated functions $\bar{\ell}_{jN}(\boldsymbol{v})$, respectively.

(b) We can not postpone optimization when a certain number of sample is gathered, say one must perform optimization after each regenerative cycle.

(c) We can not compute the functions $\bar{\ell}_{jN}(\boldsymbol{v})$ everywhere in $\boldsymbol{v}$. What we can do is to set the control vector $\boldsymbol{v}$ at any desired value, say, at $\boldsymbol{v}_t$, and then compute the functions $\bar{\ell}_{jN}(\boldsymbol{v}_t)$, and possible the gradients $\nabla\bar{\ell}_{jN}(\boldsymbol{v})$ at $\boldsymbol{v} = \boldsymbol{v}_t$.

From the viewpoint of application, the main difference between the algorithm 3.5.1 and the SA algorithm is that the first typically deal with *off-line control* models while the second deals with *on-line control* (real time) models.

Note also that SA algorithms for DEDS have been used in many references. Among them are Gaivoronski (1990, 1991); Ermoliev and Gaivoronski (1992); L'Ecuyer (1992); L'Ecuyer, Giroux, and Glynn (1991); Pflug (1992) and Rubinstein (1986). The advantage of SA is that it is a *very simple procedure*; its disadvanage is that it has *slow convergence and poor robustness* (see, for example, Rubinstein (1986), who presents various off-line and on-line stochastic optimization models). Recently Nemirovskii and Rubinstein (1992) suggested a *robust* SA algorithm for solving on-line optimization problems. They gave conditions under which their robust SA algorithm pocesses higher convergence rate than the standard SAalgorithm. The rest of this chapter deals with off-line models.

It follows from the results of the previous section that the estimator $\bar{\boldsymbol{v}}_N$ will perform well (will have a reasonably small variance) if the regenerative cycles τ_i are not too large. This actually implies that algorithm 3.4.1 will perform well *only* for queueing networks of small size, provided each queue is not too congested. For large or even moderate size networks the estimator $\bar{\boldsymbol{v}}_N$ will perform poorly for the same reasons as the RSF estimators $\nabla^k\bar{\ell}_N(\boldsymbol{v})$ perform poorly.

To overcome this difficulty we now define the so-called *decomposable* and the *truncated* stochastic counterparts of the program (P_0). They are based on the DSF estimators $\bar{\ell}^d_{jN}(\boldsymbol{v}) = \bar{\ell}^d_{jN}(\boldsymbol{v}, \gamma, \Delta_\gamma)$, $j = 0, 1, \ldots, M$, and the TSF estimators $\bar{\ell}^{tr}_{jN}(\boldsymbol{v}) = \bar{\ell}^{tr}_{jN}(\boldsymbol{v}, m_q, \tilde{m}_{q_1}, \ldots, \tilde{m}_{q_\gamma})$, $j = 0, 1, \ldots, M$, respectively. For example, the decomposable stochastic counterpart can be written in analogy to (3.5.3) as

$$(\overline{\mathrm{P}}^d_N) \qquad \begin{array}{lll} \text{minimize} & \bar{\ell}^d_{0N}(\boldsymbol{v}), & \boldsymbol{v} \in V, \\ \text{subject to} & \bar{\ell}^d_{jN}(\boldsymbol{v}) \le 0, & j = 1, \ldots, k, \\ & \bar{\ell}^d_{jN}(\boldsymbol{v}) = 0, & j = k+1, \ldots, M, \end{array} \qquad (3.5.5)$$

and similarly the truncated counterpart $(\overline{\mathrm{P}}^{tr}_N)$ (where we need to replace $\bar{\ell}^d_{jN}(\boldsymbol{v})$ by $\bar{\ell}^{tr}_{jN}(\boldsymbol{v})$ with all other data remaining the same).

It is straightforward to see that algorithm 3.4.1 can be adapted to both decomposable and truncated stochastic counterparts. To do so we need to replace $(\overline{\mathrm{P}}_N)$ in steps 3 and 4 of algorithm 3.4.1 by $(\overline{\mathrm{P}}^d_N)$ and $(\overline{\mathrm{P}}^{tr}_N)$ with all other data remaining the same. It is also important to note that unlike the stochastic counterpart $(\overline{\mathrm{P}}_N)$, which is suitable for regenerative processes *only*, the stochastic counterparts $(\overline{\mathrm{P}}_N^d)$ and $(\overline{\mathrm{P}}_N^{tr})$ are suitable for *both* regenerative and stationary processes.

Denote $\overline{\boldsymbol{v}}_N^d = \overline{\boldsymbol{v}}_N^d(\gamma, \Delta_\gamma)$, the optimal solution of the program $(\overline{\mathrm{P}}_N^d)$, and similarly $\overline{\boldsymbol{v}}_N^{tr} = \overline{\boldsymbol{v}}_N^{tr}(m_q, \tilde{m}_{q_1}, \ldots, \tilde{m}_{q_\gamma})$, the optimal solution of the program $(\overline{\mathrm{P}}_N^{tr})$. Note that differently from $\overline{\boldsymbol{v}}_N$ the estimators $\overline{\boldsymbol{v}}_N^d$ and $\overline{\boldsymbol{v}}_N^{tr}$ are typically (asymptotically) *biased*. This is because the associated estimators $\overline{\ell}_{jN}^d(\boldsymbol{v}, \gamma, \Delta_\gamma)$ and $\overline{\ell}_{jN}^{tr}(\boldsymbol{v}, m_q, \tilde{m}_{q_1}, \ldots, \tilde{m}_{q_\gamma})$ are too.

We shall discuss the statistical properties of the RSF and DSF estimators $\overline{\boldsymbol{v}}_N$ and $\overline{\boldsymbol{v}}_N^d(\gamma, \Delta_\gamma)$, respectively, for the unconstrained program (see (3.5.6) below). The statistical properties of the TSF estimator $\overline{\boldsymbol{v}}_N^{tr}(m_q, \tilde{m}_{q_1}, \ldots, \tilde{m}_{q_\gamma})$ can be derived similarly. Note also that the statistical properties of the RSF estimators $\overline{\boldsymbol{v}}_N$ and $\overline{\varphi}_N$ for the general program (3.5.1) and in particular their consistency, asymptotic normality and uniqueness will be discussed in chapter 6.

Consider the following unconstrained program (see also (2.4.6))

$$(\mathrm{P}_0) \qquad \text{minimize} \quad \ell(\boldsymbol{v}), \quad \boldsymbol{v} \in V. \tag{3.5.6}$$

The stochastic counterparts of the program (3.5.6), based on the RSF and DSF estimators, can be written as

$$(\overline{\mathrm{P}}_N) \qquad \text{minimize} \quad \overline{\ell}_N(\boldsymbol{v}), \quad \boldsymbol{v} \in V, \tag{3.5.7}$$

and

$$(\overline{\mathrm{P}}_N^d) \qquad \text{minimize} \quad \overline{\ell}_N^d(\boldsymbol{v}), \quad \boldsymbol{v} \in V, \tag{3.5.8}$$

respectively. Note that when $\tau = 1$ both (3.5.7) and (3.5.8) reduce to the program (2.4.6).

The statistical properties of the estimator $\overline{\boldsymbol{v}}_N$ are similar to those discussed in chapter 2 for DESS. In particular, recall that $\overline{\ell}_N(\boldsymbol{v})$ is given by the ratio of two "what if" estimators $\overline{\ell}_{1N}(\boldsymbol{v})$ and $\overline{\ell}_{2N}(\boldsymbol{v})$ of $\ell_1(\boldsymbol{v}) = \mathbb{E}_{\boldsymbol{v}}\{\sum_{t=1}^{\tau} L_t\}$ and $\ell_2(\boldsymbol{v}) = \mathbb{E}_{\boldsymbol{v}}\tau$, respectively (see also (3.5.4)), and we have by the Strong Law of Large Numbers that $\overline{\ell}_{1N}(\boldsymbol{v})$ and $\overline{\ell}_{2N}(\boldsymbol{v})$ converge w.p.1 to $\ell_1(\boldsymbol{v})$ and $\ell_2(\boldsymbol{v})$. Furthermore, this convergence is uniform on V, provided the set V is compact and some mild regularity conditions hold (see lemma 1 of appendix A1 in chapter 2). This implies that

1. $\overline{\ell}_N(\boldsymbol{v})$ converges to $\ell(\boldsymbol{v})$ w.p.1 uniformly on V.

2. $\overline{\boldsymbol{v}}_N$ converges to $\boldsymbol{v}^*$ w.p.1, provided V is compact and the optimal solution of the program (P_0) is unique. That is, $\overline{\boldsymbol{v}}_N$ is a consistent estimator of $\boldsymbol{v}^*$ (see theorem A1 of appendix A of chapter 2).

3. $N^{1/2}(\overline{\boldsymbol{v}}_N - \boldsymbol{v}^*)$ converges in distribution to a multivariate normal distribution with mean zero and covariance matrix $\boldsymbol{B}^{-1}\boldsymbol{\Omega}\boldsymbol{B}^{-1}$, where $\boldsymbol{\Omega}$ is the asymptotic covariance matrix of $N^{1/2}\nabla\bar{\ell}_N(\boldsymbol{v}^*)$ and $\boldsymbol{B} = \nabla^2\ell(\boldsymbol{v}^*)$. Note that by the first-order optimality conditions $\nabla\ell(\boldsymbol{v}^*) = \boldsymbol{0}$, provided $\boldsymbol{v}^*$ is an interior point of V. Note also that the covariance matrix $\boldsymbol{\Omega}$ can be written in terms of the covariances of $\sum_{t=1}^{\tau} L_t\widetilde{W}_t$, $\sum_{t=1}^{\tau} \widetilde{W}_t$, $\sum_{t=1}^{\tau} \nabla L_t\widetilde{W}_t$, and $\sum_{t=1}^{\tau} \nabla\widetilde{W}_t$. Its expression will be messy, however. Note finally that although we do not have analytical expressions for the covariance terms of $\sum_{t=1}^{\tau} L_t\widetilde{W}_t$, etc. we can still estimate *consistently* both $\boldsymbol{\Omega}$ and $\boldsymbol{B}$ from a *single sample path*.

Let us consider the statistical properties of the estimator $\overline{\boldsymbol{v}}_N^d$ defined as the optimal solution of the program (3.5.8). In this case the estimator $\bar{\ell}_N^d(\boldsymbol{v})$, given in (3.3.6), represents the sum of ratios of the corresponding "what if" estimators and by the Strong Law of Large Numbers, $\bar{\ell}_N^d(\boldsymbol{v})$ converges w.p.1 to $\ell^d(\boldsymbol{v})$, which is defined in (3.3.4). We have then that w.p.1, $\overline{\boldsymbol{v}}_N^d$ converges to the optimal solution $\boldsymbol{v}^d$ of the program:

$$(\mathrm{P_0}^d) \qquad \text{minimize} \quad \ell^d(\boldsymbol{v}), \quad \boldsymbol{v} \in V, \tag{3.5.9}$$

which is assumed to be unique.

In general, however, $\boldsymbol{v}^d$ may be different from $\boldsymbol{v}^*$, and therefore $\overline{\boldsymbol{v}}_N^d$ may have an asymptotic bias given by $|\boldsymbol{v}^d - \boldsymbol{v}^*|$. It follows from a general result on sensitivity in mathematical programming (see theorem 6.1.1 of chapter 6 and appendix A in chapter 2) that the difference between the optimal solutions of the programs $(\mathrm{P_0}^d)$ and $(\mathrm{P_0})$ can be given approximately in the form

$$\boldsymbol{v}^d - \boldsymbol{v}^* \cong -\boldsymbol{B}^{-1}\nabla\ell^d(\boldsymbol{v}^*). \tag{3.5.10}$$

Note that $\nabla\bar{\ell}_N^d(\boldsymbol{v}^*)$ converges w.p.1 to $\nabla\ell^d(\boldsymbol{v}^*)$. (Recall that $\nabla\ell(\boldsymbol{v}^*) = \boldsymbol{0}$ by the first-order optimality condition). It follows from (3.5.10) that the asymptotic bias of $\overline{\boldsymbol{v}}_N^d$ is of the same order as the asymptotic bias of the gradient $\nabla\bar{\ell}_N^d(\boldsymbol{v}^*)$, which is considered as an estimator of $\nabla\ell(\boldsymbol{v}^*)$.

It is also not difficult to show that $N^{1/2}(\overline{\boldsymbol{v}}_N^d - \boldsymbol{v}^d)$ converges in distribution to a multivariate normal random variable with mean zero and covariance matrix $\boldsymbol{B}_d^{-1}\boldsymbol{\Omega}_d\boldsymbol{B}_d^{-1}$, where $\boldsymbol{\Omega}_d$ is the asymptotic covariance matrix of $N^{1/2}\nabla\bar{\ell}_N^d(\boldsymbol{v}^d)$ and $\boldsymbol{B}_d = \nabla^2\ell^d(\boldsymbol{v}^d)$. (Note again that $\nabla\ell^d(\boldsymbol{v}^d) = \boldsymbol{0}$, provided $\boldsymbol{v}^d$ is an interior point of V). The covariance matrix $\boldsymbol{\Omega}_d$ can be written in terms of the covariances of $\sum_{t=1}^{\tau_q} L_{qt}\widetilde{W}_{qt}^d$, $\sum_{t=1}^{\tau_q} \widetilde{W}_{qt}^d$, $\sum_{t=1}^{\tau_q} \nabla L_{qt}\widetilde{W}_{qt}^d$, and $\sum_{t=1}^{\tau_q} \nabla\widetilde{W}_{qt}^d$. Note finally, that although we do not have analytical expressions for the covariance terms of $\sum_{t=1}^{\tau_q} L_{qt}\widetilde{W}_{qt}$, etc. we can still *consistently* estimate both $\boldsymbol{\Omega}_d$ and $\boldsymbol{B}_d$ using a *single sample path*.

Before turning to numerical results with both stochastic counterparts (3.5.8) and (3.5.9), we assume that

1. The parameter set V is given by

$$V = \{\boldsymbol{v} : \mathbf{0} \le \boldsymbol{\rho}(\boldsymbol{v}) \le \boldsymbol{\rho}^{\mathbf{0}}, \quad \boldsymbol{\rho} = (\rho_1, \ldots, \rho_r) \text{ and } \boldsymbol{\rho}^{\mathbf{0}} < \mathbf{1}\}, \quad (3.5.11)$$

   where $\rho_k = \rho_k(\boldsymbol{v})$, $k = 1, \ldots, r$, is the traffic intensity at the k-th queue, r is the number of nodes in the network and the inequalities between the vectors must be taken componentwise.

2. The expected performance $\ell(\boldsymbol{v})$ is given as

$$\ell(\boldsymbol{v}) = c\mathbb{E}_{\boldsymbol{v}} L_t + \sum_{k=1}^{r} b_k v_k, \quad (3.5.12)$$

   where L_t is the steady-state waiting time process, c is the cost of a waiting customer, $\boldsymbol{v} = (v_1, \ldots, v_r)$, and b_k is the cost per unit increase (decrease) of v_k. Note that in this case under some mild regularity conditions (see, e.g. Shaked and Shanthikumar (1988)) $\mathbb{E}_{\boldsymbol{v}} L_t$ is a strictly convex differentiable function with respect to $\boldsymbol{v}$. It can be also shown that convexity of $\mathbb{E}_{\boldsymbol{v}} L_t$ implies convexity of $\ell^d(\boldsymbol{v})$. Thus, both $\boldsymbol{v}^*$ and $\boldsymbol{v}^d$ are unique minimizers of (P_0) and $(\mathrm{P}_0{}^d)$ over the convex region V.

With this in mind we can solve, in analogy to (2.2.31), the stochastic counterparts $(\overline{\mathrm{P}}_N)$ and $(\overline{\mathrm{P}}_N^d)$ (see (3.5.7) and (3.5.8)) by using the following nonlinear system of equations:

$$\nabla \bar{\ell}_N(\boldsymbol{v}) = \mathbf{0}, \quad \boldsymbol{v} \in V$$

and

$$\nabla \bar{\ell}_N^d(\boldsymbol{v}) = \mathbf{0}, \quad \boldsymbol{v} \in V,$$

respectively.

To proceed with the unconstrained programs $(\overline{\mathrm{P}}_N)$ and $(\overline{\mathrm{P}}_N^d)$, we consider separately (A) a single node queueing model and (B) the general case (an r node queueing network).

(A) First consider a stable $GI/G/1$ queue with the FIFO discipline. In this case

$$\nabla \bar{\ell}_N(\boldsymbol{v}) = \mathbf{0}, \quad \boldsymbol{v} \in V,$$

reduces to

$$\begin{aligned} c^{-1}\nabla \bar{\ell}_N(\boldsymbol{v}) &= \frac{\sum_{i=1}^{N}\sum_{t=1}^{\tau_i} L_{ti}\nabla \widetilde{W}_{ti}}{\sum_{i=1}^{N}\sum_{t=1}^{\tau_i} \widetilde{W}_{ti}} - \frac{\sum_{i=1}^{N}\sum_{t=1}^{\tau_i} L_{ti}\widetilde{W}_{ti}}{\sum_{i=1}^{N}\sum_{t=1}^{\tau_i} \widetilde{W}_{ti}} \\ &\quad \frac{\sum_{i=1}^{N}\sum_{t=1}^{\tau_i} \nabla \widetilde{W}_{ti}}{\sum_{i=1}^{N}\sum_{t=1}^{\tau_i} \widetilde{W}_{ti}} + \boldsymbol{b}/c = \mathbf{0}, \quad \boldsymbol{v} \in V, \end{aligned} \quad (3.5.13)$$

where

$$\widetilde{W}_{ti} = \prod_{k=1}^{t} W(\boldsymbol{Z}_{ki}, \boldsymbol{v}) \quad \text{and} \quad W(\boldsymbol{Z}, \boldsymbol{v}) = \frac{f(\boldsymbol{Z}, \boldsymbol{v})}{g(\boldsymbol{Z})}.$$

In this case both stochastic counterparts ($\overline{\mathrm{P}}_N$) and ($\overline{\mathrm{P}}_N^d$) *coincide* since r =1. Note also that the extension of (3.5.13) to a single node queueing model with different types of service disciplines, finite buffer, etc. is similar, provided the indexing of the likelihood ratio process $\widetilde{W}_{ti}$ is defined in a more sophisticated way. It follows from (3.5.13) that we can estimate the optimal solution $\boldsymbol{v}^* = \boldsymbol{v}^*(\boldsymbol{b}, c)$ of the program (3.5.6) from a *single simulation run simultaneously* for different values of $\boldsymbol{b}$ and c. Clearly, in this case we still need to solve the system of equations (3.5.13) separately for each value of $\boldsymbol{b}$ and c.

Assume further that $g(\boldsymbol{y}) = f(\boldsymbol{y}, \boldsymbol{v}_0)$, where $\boldsymbol{v}_0$ is the reference parameter. The following remark explains how to choose a "feasible" reference traffic intensity $\rho_0 = \rho_0(\boldsymbol{v}_0)$ (see section 4.3 for the definition of a "feasible" reference traffic intensity) for the unconstrained programs ($\overline{\mathrm{P}}_N$), ($\overline{\mathrm{P}}_N^d$) and ($\overline{\mathrm{P}}_N^{tr}$).

Remark 3.5.1 It is shown in section 4.3 below that in order to choose a "feasible" reference traffic intensity $\rho_0(\boldsymbol{v}_0)$ in $\nabla\bar{\ell}_N(\boldsymbol{v})$, $\boldsymbol{v} \in V$, where

$$V = \{\boldsymbol{v} : 0 < \rho(\boldsymbol{v}) \leq \rho^0(\boldsymbol{v}) < 1\},$$

one can take ρ_0 *either equal to* ρ^0, *or moderately larger* than ρ^0. Taking into account that the asymptotic covariance matrix of $\overline{\boldsymbol{v}}_N$ is "proportional" to the covariance matrix of $\nabla\bar{\ell}_N(\boldsymbol{v}^*)$ (more precisely, $N^{1/2}(\overline{\boldsymbol{v}}_N - \boldsymbol{v}^*)$ has an asymptotic covariance matrix $\boldsymbol{B}^{-1}\boldsymbol{\Omega}\boldsymbol{B}^{-1}$, where $\boldsymbol{\Omega}$ is the covariance matrix of $N^{1/2}\nabla\bar{\ell}_N(\boldsymbol{v}^*)$ and $\boldsymbol{B} = \nabla^2\ell(\boldsymbol{v}^*)$), it follows that once chosen as a "feasible" reference traffic intensity $\rho_0(\boldsymbol{v}_0)$ of the estimator $\nabla\bar{\ell}_N(\boldsymbol{v})$, it remains a "feasible" one for the programs ($\overline{\mathrm{P}}_N$), ($\overline{\mathrm{P}}_N^d$) and ($\overline{\mathrm{P}}_N^{tr}$).

We now present numerical results, while solving the program ($\overline{\mathrm{P}}_N$) (and thus ($\overline{\mathrm{P}}_N^d$)) separately for (i) the $M/M/1$ and (ii) the $M/G/1$ queues.

(i) $M/M/1$ **queue.**

Let λ and v be the interarrival and service rates and let v be the decision parameter. Taking into account that $\mathbb{E}_v L = 1/(v - \lambda)$, it is readily seen that the optimal v^* that minimizes the performance measure given in (3.5.12) is $v^* = \lambda + (bc^{-1})^{1/2}$.

Table 3.5.1 presents theoretical values of v^*, point estimators, $\bar{v}_N$ and the 95% confidence intervals for v^* (denoted Conf. interval) as functions of b (ρ and $\alpha = [\rho_0 - \rho(b)]/\rho(b)$). We chose $\lambda = 2$, $c = 1$, the reference traffic intensity $\rho_0 = 0.8$, and ran $N = 2000$ cycles (approximately 10000 customers). Note that *all* estimators $\bar{v}_N(b)$ were obtained *simultaneously from a single simulation run* by solving the system of equations (3.5.13) for different values of b, provided the reference traffic intensity is fixed and equal to 0.8.

Table 3.5.1 Performance of the stochastic counterpart (3.5.13) for the $M/M/1$ queue with reference traffic intensity $\rho_0 = 0.8$.

ρ	b	α	v^*	$\bar{v}_N$	Conf. interval
0.88	0.074	-0.091	2.272	1.14	0.97, 1.31
0.85	0.123	-0.058	2.352	1.65	1.05 , 2.27
0.8	0.25	0.000	2.5	2.590	3.19, 1.99
0.7	0.735	0.143	2.857	2.807	2.41, 3.21
0.6	1.778	0.333	3.333	3.247	2.94 , 3.54
0.5	4.000	0.600	4.000	3.920	3.52, 4.52
0.4	9.000	1.000	5.000	5.012	4.41, 5.61
0.3	19.78	1.667	6.666	6.761	6.06, 7.46
0.2	64.00	3.000	10.00	9.785	8.88, 10.68
0.1	324	7.000	20.00	18.02	16.12, 19.92

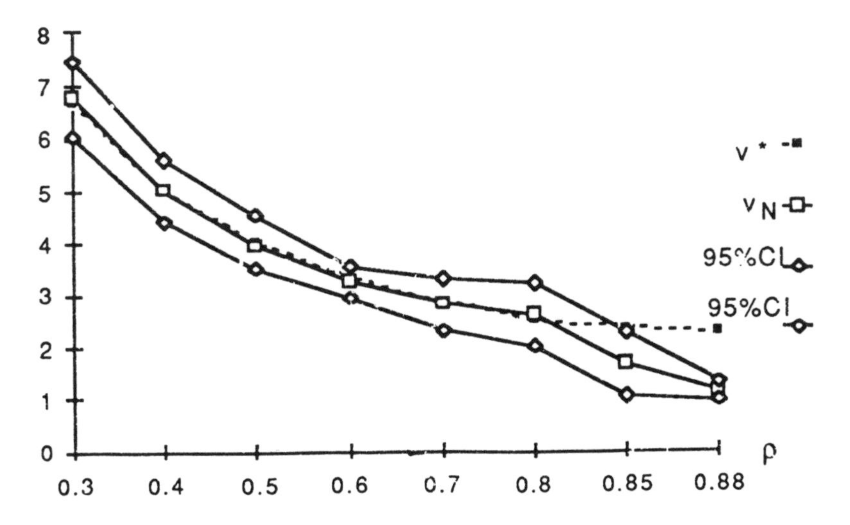

Figure 3.5.1 Performance of the estimator $\bar{v}_N$ for the $M/M/1$ queue with reference traffic intensity $\rho_0 = 0.8$.

Based on the results of table 3.5.1, fig. 3.5.1 depicts theoretical values of v^*, point estimators, $\bar{v}_N$ and the 95% confidence intervals for v^* as functions of ρ, provided $0.3 < \rho < 0.85$.

It is readily seen that the estimator $\bar{v}_N$ performs reasonably well for $\rho \in (0.2, 0.8)$. Note that poor perfomance of the estimator $\bar{v}_N$ for
(a) $\rho > 0.8$ is the result of violation of the requirement of remark 3.5.1. (According to remark 3.5.1 we must have $\rho \leq \rho_0 = 0.8$, while in fact we have $0.88 = \rho \geq \rho_0 = 0.8$.)
(b) $\rho < 0.2$ is the result of very large relative perturbations ($\alpha > 3$) in the likelihood ratio process $\widetilde{W}_t$.

(ii) $M/G/1$ **queue.**
Assume that the service time pdf is given as

$$f(y,p) = \begin{cases} p, & \text{if } y = a_1, \\ 1-p, & \text{if } y = a_2, \end{cases}$$

where $a_1 > 0,\ a_2 > 0$ and $0 < p < 1$.

By the Pollaczek-Khinchin formula (e.g. Gross and Harris (1985)) the expected sojourn time can be written as

$$\ell(p) = \mathbb{E}_p L = \beta_1 + \frac{\lambda \beta_2}{2(1-\lambda\beta_1)},$$

where

$$\beta_1 = \mathbb{E}Y = pa_1 + (1-p)a_2, \quad \beta_2 = \mathbb{E}Y^2 = pa_1^2 + (1-p)a_2^2,$$

and Y is the service time random variable. Note that in this case the likelihood ratio W in (3.5.13) reduces to

$$W(Z,p) = \left(\frac{p}{p_0}\right)^{\frac{a_2 - Z}{a_1 - a_2}} \left(\frac{1-p}{1-p_0}\right)^{\frac{Z - a_1}{a_2 - a_1}}.$$

Table 3.5.2 presents data similar to those of table 3.5.1. We chose $\lambda = 0.6$, $p_0 = 0.5$, $a_1 = 0.5$, $a_2 = 1.2$ ($\rho_0 = 0.85$) and $c = 1$, assumed $\rho^0 = 0.7$, and ran the $M/G/1$ queue for $N = 2000$ cycles (approximately 12500 customers). Following remark 3.5.1 and taking into account that $\rho^0 = 0.7$, we chose the reference traffic intensity moderately larger than ρ^0, namely $\rho_0 = 0.85$.

Table 3.5.2 Performance of the stochastic counterpart (3.5.13) for the $M/G/1$ queue with reference traffic intensity $\rho_0 = 0.85$.

ρ	b	α	p^*	$\bar{p}_N$	Conf. interval
0.70	5.013	0.212	0.712	0.688	0.65, 0.73
0.65	3.820	0.310	0.787	0.777	0.74, 0.81
0.60	3.167	0.405	0.850	0.852	0.82, 0.88
0.55	2.622	0.538	0.925	0.951	0.91, 0.99
0.50	2.240	0.700	1.000	1.000	0.94, 1.06

(B) Consider now the unconstrained program (P_0) assuming that the parameter set V and the expected performance $\ell(\boldsymbol{v})$ are given by (3.5.11) and (3.5.12), respectively. In this case the stochastic counterpart

$$\nabla \bar{\ell}_N(\boldsymbol{v}) = \mathbf{0}, \quad \boldsymbol{v} \in V,$$

reduces to (3.5.13), while the the stochastic counterpart

$$\nabla \bar{\ell}_N^d(\boldsymbol{v}) = \mathbf{0}, \quad \boldsymbol{v} \in V,$$

can be written as

$$\begin{aligned} c^{-1}\nabla \bar{\ell}_N^d(\boldsymbol{v}) &= \sum_{q=1}^{r} \left(\frac{\sum_{i=1}^N \sum_{t=1}^{\tau_{qi}} L_{qti} \nabla \widetilde{W}_{qti}^d(\boldsymbol{v}, \gamma, \Delta_\gamma)}{\sum_{i=1}^N \sum_{t=1}^{\tau_{qi}} \widetilde{W}_{qti}^d(\boldsymbol{v}, \gamma, \Delta_\gamma)} \right. \\ &- \frac{\sum_{i=1}^N \sum_{t=1}^{\tau_{qi}} L_{qti} \widetilde{W}_{qti}^d(\boldsymbol{v}, \gamma, \Delta_\gamma)}{\sum_{i=1}^N \sum_{t=1}^{\tau_{qi}} \widetilde{W}_{qti}^d(\boldsymbol{v}, \gamma, \Delta_\gamma)} \\ &\times \left. \frac{\sum_{i=1}^N \sum_{t=1}^{\tau_{qi}} \nabla \widetilde{W}_{qti}^d(\boldsymbol{v}, \gamma, \Delta_\gamma)}{\sum_{i=1}^N \sum_{t=1}^{\tau_{qi}} \widetilde{W}_{qti}^d(\boldsymbol{v}, \gamma, \Delta_\gamma)} \right) \\ &+ \mathbf{b}/c = \mathbf{0}, \quad \boldsymbol{v} \in V \end{aligned} \tag{3.5.14}$$

The stochastic counterpart

$$\nabla \bar{\ell}_N^{tr}(\boldsymbol{v}) = \mathbf{0}, \quad \boldsymbol{v} \in V,$$

of the program ($\overline{\mathrm{P}}_N^{tr}$), which is based on the TSF estimators (3.3.23), can be written as

$$c^{-1}\nabla \bar{\ell}_N^{tr}(\boldsymbol{v}) + N^{-1}T^{-1} \sum_{q=1}^{r} \sum_{i=1}^{N} \sum_{t=1}^{T_{qi}} L_{qti} \nabla \widetilde{W}_{qti}^{tr}(\boldsymbol{v}, m_q, \tilde{m}_{q_1}, \ldots, \tilde{m}_{q_\gamma}) + \boldsymbol{b}/c = \mathbf{0}.$$

Assume again that $g(y) = f(y, v_0)$, where v_0 is the reference parameter. The following remark discusses the choice of a "feasible" vector of traffic intensities $\rho_0 = \rho_0(v_0)$ in the unconstrained programs $(\overline{\mathrm{P}}_N)$, $(\overline{\mathrm{P}}_N^d)$ and $(\overline{\mathrm{P}}_N^{tr})$.

Remark 3.5.2 Arguing as in remark 3.5.1 we obtain that in order to choose a "feasible" vector of reference traffic intensities $\rho - 0(v_0)$, while solving the unconstrained programs $(\overline{\mathrm{P}}_N)$, $(\overline{\mathrm{P}}_N^d)$, and $(\overline{\mathrm{P}}_N^{tr})$, where V is given by(3.5.11), one can assume that ρ_0 is either equal to ρ^o, or moderately larger than ρ^0 componentwise.

We now show how to calculate the components of the reference service rate vector $v_0 = (v_{01}, \ldots, v_{0r})$.

Let λ_q be the mean arrival rate to the node q in a stable open Jackson network, and let p_{qj} be the elements of the routing probability matrix, which are assumed to be independent of the system state. Then it is well known (see Gross and Harris (1985)) that the total mean flow rate Λ_q into the node q (from outside and from other nodes) satisfies the following system of equations:

$$\Lambda_q = \lambda_q + \sum_{j=1}^{k} p_{jq}\Lambda_j, \quad q = 1, \ldots, r. \tag{3.5.15}$$

We can now calculate the elements of the vector v_0 as

$$v_{0q} = \Lambda_q / \rho_q^0, \quad q = 1, \ldots, r.$$

In order to approximate the components of the reference service rate vector v_0 for a nonproduct open queueing network we can solve in addition to (3.5.15) the so-called *traffic variability equation* due to Whitt (1983), which takes into account the squared coefficients of variation of the arrival and service processes.

We evaluated the optimal solution v^* of the unconstrained program (P_0) for different network topologies, different dimensionalities n $(1 \leq n \leq 100)$ of the decision vector v, different traffic intensities ρ $(0.1 \leq \rho \leq 0.85)$, and different sample sizes N $(10^2 \leq N \leq 10^5)$ assuming as before that v represents the service rate vector, the performance measure $\ell(v)$ is given in (3.5.12), and the parameter set V satisfies

$$V = \{\rho : 0 < \rho \leq \rho^0 < 0.9, \ \rho = (\rho_1, \ldots, \rho_r)\},$$

componentwise. We found that for a fixed sample size N, the asymptotic variance (the asymptotic mean squared errors) of the optimal solutions of these stochastic counterparts increase moderately in n and rather quickly in

$\boldsymbol{\rho}$. We also found that, although the DSF and TSF estimators of the optimal solution of $\boldsymbol{v}^*$ contain some moderate bias, they are typically much more accurate than their RSF counterparts, which is virtually useless, provided $n > 3$ and $\boldsymbol{\rho} \geq 0.5$.

Let us turn to the evaluation of the optimal solution of the constrained program (P_0) with the stochastic counterparts ($\overline{\mathrm{P}}_N$) and ($\overline{\mathrm{P}}_N^d$) (see (3.5.3) and (3.5.5)). Assume as before that the sample performance L_0 of the objective function $\ell_0(\boldsymbol{v}) = \mathbb{E}_{\boldsymbol{v}} L_0$ represents the steady-state sojourn time in the network and consider the following constraints:

1. $\ell_1(\boldsymbol{v}) = \sum_{k=1}^{r} b_k v_k - c$ is the "deterministic" constraint function with b_k being the cost per unit increase in v_k, where $v_k = 1/\mathbb{E}Y_k$ and Y_k is the service time random variable at the k-th queue.

2. $\ell_2(\boldsymbol{v}) = P\{L_2 \geq x\} \leq \beta$, where L_2 is the number of customers in the network, I is the indicator of the event $\{L_2 \geq x\}$, and x and β are fixed numbers.

3. $V = \{\boldsymbol{\rho} : 0 < \boldsymbol{\rho} \leq \boldsymbol{\rho}^0 < 0.7, \quad \boldsymbol{\rho} = (\rho_1, \ldots, \rho_r)\}$.

The function $\ell_0(\boldsymbol{v})$ in the stochastic counterparts ($\overline{\mathrm{P}}_N$), ($\overline{\mathrm{P}}_N^d$), and ($\overline{\mathrm{P}}_N^{tr}$) must be replaced by

$$\bar{\ell}_{0N}(\boldsymbol{v}) = \frac{\sum\limits_{i=1}^{N} \sum\limits_{t=1}^{\tau_i} L_{0ti} \widetilde{W}_{ti}}{\sum\limits_{i=1}^{N} \sum\limits_{t=1}^{\tau_i} \widetilde{W}_{ti}},$$

$$\bar{\ell}_{0N}^d(\boldsymbol{v}) = \sum_{q=1}^{r} \frac{\sum_{i=1}^{N} \sum_{t=1}^{\tau_{qi}} L_{0qti} \widetilde{W}^d_{qti}(\boldsymbol{v}, \gamma, \Delta_\gamma)}{\sum_{i=1}^{N} \sum_{t=1}^{\tau_{qi}} \widetilde{W}^d_{qti}(\boldsymbol{v}, \gamma, \Delta_\gamma)},$$

and

$$\bar{\ell}_{0N}^{tr}(\boldsymbol{v}) = N^{-1} T^{-1} \sum_{q=1}^{r} \sum_{i=1}^{N} \sum_{t=1}^{T_{qi}} L_{0qti} \widetilde{W}^{tr}_{qti}(\boldsymbol{v}, m_q, \tilde{m}_{q1}, \ldots, \tilde{m}_{q\gamma}),$$

respectively, and similarly for the function $\ell_2(\boldsymbol{v})$. Note that the "deterministic" function $\ell_1(\boldsymbol{v})$ remains the same.

We evaluated the optimal solution $\boldsymbol{v}^*$ of the above constrained program (P_0) with our QNSO package for different network topologies, different dimensionalities n ($1 \leq n \leq 100$) of the decision vector $\boldsymbol{v}$, different traffic intensities $\boldsymbol{\rho}$, and different sample sizes N ($10^2 \leq N \leq 10^5$), while using the stochastic counterparts ($\overline{\mathrm{P}}_N^d$) and ($\overline{\mathrm{P}}_N^{tr}$). We found that the DSF and TSF estimators perform reasonably well; their accuracies (asymptotic variances) are similar to those for the unconstrained problems, and the accuracies of the DSF and TSF estimators for both the unconstrained and constrained problems are similar. More details on the computational aspects of the above can be found in Izxaki (1992).

3.6 Sensitivity Analysis and Robustness of the Optimal Solutions

In this section we discuss (a) sensitivity analysis of the optimal solution $\boldsymbol{v}^*$ of the program (P_0) and of its associated stochastic counterpart ($\tilde{\mathrm{P}}_N$) with respect to the parameters of the system, say $\boldsymbol{r}$, while considering the model

$$\ell(\boldsymbol{\theta}) = \mathbb{E}_{\boldsymbol{\theta}} L(\boldsymbol{Y}, \boldsymbol{\theta}),$$

where $\boldsymbol{\theta} = (\boldsymbol{v}, \boldsymbol{r})$, and (b) robustness of the optimal solution $\boldsymbol{v}^*$ with respect to contamination of the underlying pdf $f(\boldsymbol{y}, \boldsymbol{v})$.

(a) Let $\boldsymbol{v}^*(\boldsymbol{r})$ be the minimizer of $\ell(\cdot, \boldsymbol{r})$ of the program (P_0) over the parameter set V for fixed $\boldsymbol{r}$. For example, $\boldsymbol{v}^*(\boldsymbol{r})$ can represent the optimal service rates in a queueing network, while $\boldsymbol{r}$ can represent the corresponding interarrival rates, the reference parameter $\boldsymbol{v}_0$, and the correlation between the interarrival random variables.

Suppose that we want to evaluate the sensitivity of $\boldsymbol{v}^*(\boldsymbol{r})$ with respect to changes (perturbations) in $\boldsymbol{r}$, that is to evaluate $\nabla \boldsymbol{v}^*(\boldsymbol{r}_0)$. By the Implicit Function Theorem (see appendix A of chapter 2), we have that the Jacobian matrix of $\boldsymbol{v}^*(\boldsymbol{r})$ at $\boldsymbol{r} = \boldsymbol{r}_0$ is given by

$$\nabla \boldsymbol{v}^*(\boldsymbol{r}_0) = -\left[\nabla^2_{\boldsymbol{v}\boldsymbol{v}} \ell(\boldsymbol{v}^*, \boldsymbol{r}_0)\right]^{-1} \nabla^2_{\boldsymbol{v}\boldsymbol{r}} \ell(\boldsymbol{v}^*, \boldsymbol{r}_0)\,, \tag{3.6.1}$$

where $\boldsymbol{v}^* = \boldsymbol{v}^*(\boldsymbol{r}_0)$. We can then estimate $\nabla \boldsymbol{v}^*(\boldsymbol{r}_0)$ as

$$\bar{\nabla} \boldsymbol{v}^*_N(\boldsymbol{r}_0) = -\left[\nabla^2_{\boldsymbol{v}\boldsymbol{v}} \bar{\ell}_N(\bar{\boldsymbol{v}}_N, \boldsymbol{r}_0)\right]^{-1} \nabla^2_{\boldsymbol{v}\boldsymbol{r}} \bar{\ell}_N(\bar{\boldsymbol{v}}_N, \boldsymbol{r}_0)\,, \tag{3.6.2}$$

where $\bar{\boldsymbol{v}}_N$ is the minimizer of $\bar{\ell}_N(\cdot, \boldsymbol{r}_0)$ over V derived from the solution of the stochastic counterpart ($\tilde{\mathrm{P}}_N$).

The obtained estimator $\bar{\nabla} \boldsymbol{v}^*_N(\boldsymbol{r}_0)$ of $\nabla \boldsymbol{v}^*(\boldsymbol{r}_0)$ is consistent provided $\bar{\boldsymbol{v}}_N$ is a consistent estimator of $\boldsymbol{v}^*$. (Consistency of $\bar{\boldsymbol{v}}_N$ holds under mild regularity conditions, see appendix A of chapter 2.) It can also be shown (by the delta method) that the estimator $\bar{\nabla} \boldsymbol{v}^*_N(\boldsymbol{r}_0)$ is asymptotically normal.

(b) Consider now robustness of the optimal solution $\boldsymbol{v}^*$ of the unconstrained optimization problem (P_0) with respect to small perturbations of the pdf $f(\boldsymbol{y}, \boldsymbol{v})$. A discussion of the robustness for the general case will be postponed until chapter 6. More definitely, suppose that $f(\boldsymbol{y}, \boldsymbol{v})$ is perturbed by a pdf $h(\boldsymbol{y}, \boldsymbol{v})$ which depends on the same parameter vector $\boldsymbol{v} \in V$. The perturbed pdf $\tilde{f}(\boldsymbol{y}, \boldsymbol{v}, r)$ is then given as

$$\tilde{f}(\boldsymbol{y}, \boldsymbol{v}, r) = (1 - r) f(\boldsymbol{y}, \boldsymbol{v}) + r h(\boldsymbol{y}, \boldsymbol{v}) , \tag{3.6.3}$$

where the parameter $r \in [0, 1]$ represents the percentage of contamination of $f(\boldsymbol{y}, \boldsymbol{v})$ by $h(\boldsymbol{y}, \boldsymbol{v})$. We can view r as the additional parameter in the parametric family $\tilde{f}(\boldsymbol{y}, \boldsymbol{\theta})$, $\boldsymbol{\theta} = (\boldsymbol{v}, r)$, and consider the corresponding expected performance

$$\ell(\boldsymbol{\theta}) = \mathbb{E}_{\boldsymbol{\theta}}\{L(\boldsymbol{Y}, \boldsymbol{\theta})\},$$

with respect to $\tilde{f}(\boldsymbol{y}, \boldsymbol{\theta})$. Of course, for $r = 0$ the expected performance $\ell(\boldsymbol{v}, 0)$ is given by $\mathbb{E}_{\boldsymbol{v}} L$.

We show now how to estimate the sensitivity of the optimal solution of the program (P_0) with respect to the considered contamination, while considering for simplicity a static queueing model, like the $GI/G/\infty$ queue. The extension of our results to more complex queueing models is straightforward and is left as an exercise.

To do so consider the expected performance

$$\ell(\boldsymbol{v}, r) = \int L(\boldsymbol{y}, \boldsymbol{v}) \tilde{f}(\boldsymbol{y}, \boldsymbol{v}, r) \mathrm{d}\boldsymbol{y} = \mathbb{E}_{\tilde{f}}\{L(\boldsymbol{Y}, \boldsymbol{v})\} . \tag{3.6.4}$$

We have that

$$\begin{aligned} \ell(\boldsymbol{v}, r) &= (1 - r) \textstyle\int L(\boldsymbol{y}, \boldsymbol{v}) f(\boldsymbol{y}, \boldsymbol{v}) d\boldsymbol{y} + r \textstyle\int L(\boldsymbol{y}, \boldsymbol{v}) h(\boldsymbol{y}, \boldsymbol{v}) \mathrm{d}\boldsymbol{y} \\ &= (1 - r)\ell(\boldsymbol{v}) + r\eta(\boldsymbol{v}) = \ell(\boldsymbol{v}) + r[\eta(\boldsymbol{v}) - \ell(\boldsymbol{v})] , \end{aligned}$$

where

$$\eta(\boldsymbol{v}) = \int L(\boldsymbol{y}, \boldsymbol{v}) h(\boldsymbol{y}, \boldsymbol{v}) \mathrm{d}\boldsymbol{y} .$$

Let for a given $r \in [0, 1]$, $\tilde{\boldsymbol{v}}(r)$ be the minimizer of $\ell(\boldsymbol{v}, r)$ over V. Clearly for $r = 0$ we have that $\ell(\boldsymbol{v}, 0) = \ell(\boldsymbol{v})$ and hence $\tilde{\boldsymbol{v}}(0)$ reduces to the minimizer $\boldsymbol{v}^*$ of $\ell(\boldsymbol{v})$ over V.

The minimizer $\tilde{\boldsymbol{v}}(r)$ is a function of the parameter r. Its derivative

$$\left.\frac{\mathrm{d}\tilde{\boldsymbol{v}}(r)}{\mathrm{d}r}\right|_{r=0} = \lim_{r \to 0^+} \frac{\tilde{\boldsymbol{v}}(r) - \boldsymbol{v}^*}{r} , \tag{3.6.5}$$

at $r = 0$, represents the sensitivity of the optimal value of the parameter vector with respect to the contamination of $f(\boldsymbol{y}, \boldsymbol{v})$ by $h(\boldsymbol{y}, \boldsymbol{v})$.

We show now how to calculate and estimate this derivative. To do so note first that the difference $\ell(\boldsymbol{v}, r) - \ell(\boldsymbol{v})$ between the objective functions of the perturbed and unperturbed minimization problems is equal to $r[\eta(\boldsymbol{v}) - \ell(\boldsymbol{v})]$

and tends to zero as $r \to 0^+$ for any fixed $\boldsymbol{v}$. Suppose next that the functions $\ell(\boldsymbol{v})$ and $\eta(\boldsymbol{v})$ are continuous and that the set V is compact. Then the function $\eta(\boldsymbol{v}) - \ell(\boldsymbol{v})$ is bounded on V and hence $\ell(\boldsymbol{v}, r)$ tends to $\ell(\boldsymbol{v})$ as $r \to 0^+$ uniformly on V. This implies that $\tilde{\boldsymbol{v}}(r) \to \boldsymbol{v}^*$ as $r \to 0^+$ provided that the minimizer $\boldsymbol{v}^*$ is unique (see the proof of theorem A1 in appendix A of chapter 2). Assuming further that $\boldsymbol{v}^*$ is an interior point of V, that $\ell(\boldsymbol{v})$ and $\eta(\boldsymbol{v})$ are twice continuously differentiable and that the matrix $\boldsymbol{B} = \nabla^2 \ell(\boldsymbol{v}^*)$ is nonsingular, we can finally write

$$\left.\frac{\mathrm{d}\tilde{\boldsymbol{v}}(r)}{\mathrm{d}r}\right|_{r=0} = -\boldsymbol{B}^{-1} \nabla \eta(\boldsymbol{v}^*). \tag{3.6.6}$$

Note that formula (3.6.6) follows from the Implicit Function Theorem applied to the system of equations representing the first-order necessary optimality conditions for the minimizer $\tilde{\boldsymbol{v}}(t)$ (see equations (2.6.11) and (2.6.12) in appendix A of chapter 2).

It follows from (3.6.6) that

$$\tilde{\boldsymbol{v}}(r) = \boldsymbol{v}^* - r\boldsymbol{B}^{-1} \nabla \eta(\boldsymbol{v}^*) + o(r). \tag{3.6.7}$$

For small values of r the term $-r\boldsymbol{B}^{-1}\nabla\eta(\boldsymbol{v}^*)$ gives an approximation of a possible error in $\boldsymbol{v}^*$ owing to contamination of the model $f(\boldsymbol{y}, \boldsymbol{v})$. The sensitivity of $\boldsymbol{v}^*$ depends on the contaminating pdf $h(\boldsymbol{y}, \boldsymbol{v})$ through the gradient

$$\begin{aligned} \nabla \eta(\boldsymbol{v}^*) &= \int \nabla [L(\boldsymbol{y}, \boldsymbol{v}^*) h(\boldsymbol{y}, \boldsymbol{v}^*)] \mathrm{d}\boldsymbol{y} \\ &= \int [h(\boldsymbol{y}, \boldsymbol{v}^*) \nabla L(\boldsymbol{y}, \boldsymbol{v}^*) + L(\boldsymbol{y}, \boldsymbol{v}^*) \nabla h(\boldsymbol{y}, \boldsymbol{v}^*)] \mathrm{d}\boldsymbol{y}. \end{aligned}$$

We show now how to estimate $\mathrm{d}\tilde{\boldsymbol{v}}(r)/\mathrm{d}r$ from a simulation. Let $\boldsymbol{Z}_1, \ldots, \boldsymbol{Z}_N$ be a sample from a dominating pdf $g(\boldsymbol{z})$. Then the "what if" estimator of $\ell(\boldsymbol{v}, r)$ can be written as

$$\bar{\ell}_N(\boldsymbol{v}, r) = N^{-1} \sum_{i=1}^{N} L(\boldsymbol{Z}_i, \boldsymbol{v}) W(\boldsymbol{Z}_i, \boldsymbol{v}, r) = (1-r)\bar{\ell}_N(\boldsymbol{v}) + r\bar{\eta}_N(\boldsymbol{v}), \tag{3.6.8}$$

where

$$W(\boldsymbol{z}, \boldsymbol{v}, r) = \frac{(1-r) f(\boldsymbol{z}, \boldsymbol{v}) + r h(\boldsymbol{z}, \boldsymbol{v})}{g(\boldsymbol{z})}$$

and

$$\bar{\eta}_N(\boldsymbol{v}) = N^{-1} \sum_{i=1}^{N} L(\boldsymbol{Z}_i, \boldsymbol{v}) \frac{h(\boldsymbol{Z}_i, \boldsymbol{v})}{g(\boldsymbol{Z}_i)} .$$

Let $\bar{\boldsymbol{v}}_N(r)$ be the minimizer of $\bar{\ell}_N(\cdot, r)$ over V. Note that for $r = 0$, $W(\boldsymbol{z}, \boldsymbol{v}, 0)$ reduces to the standard likelihood ratio $W(\boldsymbol{z}, \boldsymbol{v}) = f(\boldsymbol{z}, \boldsymbol{v})/g(\boldsymbol{z})$ and $\bar{\boldsymbol{v}}_N(0) =$

$\bar{\boldsymbol{v}}_N$ becomes the standard SF estimator of $\boldsymbol{v}^*$. It follows from the above that the derivative of $\bar{\boldsymbol{v}}_N(r)$ at $r = 0$ can be written in analogy to (3.6.6) as

$$\left.\frac{\mathrm{d}\bar{\boldsymbol{v}}_N(r)}{\mathrm{d}r}\right|_{r=0} = -\bar{\boldsymbol{B}}_N^{-1}\nabla\bar{\eta}_N(\bar{\boldsymbol{v}}_N), \tag{3.6.9}$$

where $\bar{\boldsymbol{B}}_N = \nabla^2\bar{\ell}_N(\bar{\boldsymbol{v}}_N)$.

The sample equivalent of (3.6.7) can be written as

$$\bar{\boldsymbol{v}}_N(r) = \bar{\boldsymbol{v}}_N - r\bar{\boldsymbol{B}}_N^{-1}\nabla\bar{\eta}_N(\bar{\boldsymbol{v}}_N) + o(r).$$

The matrix $\bar{\boldsymbol{B}}_N$ and the gradient $\nabla\bar{\eta}_N(\bar{\boldsymbol{v}}_N)$, calculated from the sample $\boldsymbol{Z}_1, \ldots, \boldsymbol{Z}_N$, provide consistent estimators of the matrix $\boldsymbol{B}$ and the gradient $\nabla\eta(\boldsymbol{v}^*)$, respectively. The derivative $\mathrm{d}\bar{\boldsymbol{v}}_N(r)/\mathrm{d}r$, $r = 0$, which can also be calculated from the same sample $\boldsymbol{Z}_1, \ldots, \boldsymbol{Z}_N$, serves a double purpose: it provides a consistent estimator of the derivative $\mathrm{d}\tilde{\boldsymbol{v}}(r)/\mathrm{d}r$, $r = 0$, and at the same time gives the sensitivity of the estimator $\bar{\boldsymbol{v}}_N$ with respect to the considered contamination.

Example 3.6.1 Let $f(y, \lambda) = f(y, \lambda, \beta_0)$ be the gamma pdf with $\beta_0 = 1$ and consider $L(y, \lambda) = y + \lambda$. The corresponding expected performance is $\ell(\lambda) = \lambda^{-1} + \lambda$ and its minimizer, over $\mathbb{R}_+$, is $\lambda^* = 1$. Suppose now that $f(y, \lambda)$ is contaminated by the gamma distribution with pdf $h(y, \lambda) = f(y, \lambda, \beta_1)$. The corresponding minimizer $\tilde{\lambda}(r)$ is then given by

$$\tilde{\lambda}(r) = [1 + r(\beta_1 - 1)]^{1/2}.$$

In this case $B = -2$, $\eta'(\lambda^*) = \beta_1 - 1$ and hence, by (3.6.6), we have

$$\tilde{\lambda}'(0) = \tfrac{1}{2}(\beta_1 - 1),$$

which can be easily checked by direct calculation. Consequently

$$\tilde{\lambda}(r) = 1 + \tfrac{1}{2}r(\beta_1 - 1) + o(r).$$

Let us take, for example, $\beta_1 = 10$ and $\beta_1 = 100$. Then 1% contamination of $f(y, \lambda)$ by $h(y, \lambda)$ gives about 4.5% and 45% errors, respectively, in the optimal value of the parameter λ.

3.7 Exercises

1. *M/G/1* **queue.** Let $\ell(\boldsymbol{v}) = \mathbb{E}_{\boldsymbol{v}} L_t$, $\boldsymbol{v} = (\lambda, \mu)$, where λ and μ are the interarrival and service rates, respectively. Find analytical expressions for $\nabla\ell(\boldsymbol{v}) = \nabla\mathbb{E}_{\boldsymbol{v}} L_t$, where $\nabla\ell(\boldsymbol{v}) = (\partial\ell(\boldsymbol{v})/\partial\lambda,\ \partial\ell(\boldsymbol{v})/\partial\mu)$ assuming that L_t is

(i) the waiting time in the queue,

(ii) the sojourn time,

(iii) the number of the customer in the $M/G/1$ queue, respectively.

Hint. Use the Pollaczek Khinchin formula, which for the sojourn time, for example, can be written as

$$\ell(\boldsymbol{v}) = \mathbb{E}S + \frac{\lambda \mathbb{E}S^2}{2(1-\lambda \mathbb{E}S)},$$

where S denotes the service time random variable.

2. Consider a cyclic network containing m $GI/G/1$ queues. Assuming that there is only a *single customer* in the network, find

   (i) $\ell(\boldsymbol{v}) = \mathbb{E}_{\boldsymbol{v}} L$, where $L = \sum_{j=1}^{m} L_j$, and L_j is the sojourn time at the j-th queue. In this case $L_j = Y_j$, where Y_j is the service time at the j-th queue,

   (ii) $\nabla l(\boldsymbol{v}) = \nabla \mathbb{E}_{\boldsymbol{v}} L = \mathbb{E}_{\boldsymbol{v}_0} L \nabla W$,

   (iii) $\mathrm{Var}_{\boldsymbol{v}_0}\{L \nabla W\}$, where $W = f(\boldsymbol{Z}, \boldsymbol{v})/f(\boldsymbol{Z}, \boldsymbol{v}_0)$, $\boldsymbol{Z} \sim f(\boldsymbol{z}, \boldsymbol{v}_0)$, $f(\boldsymbol{z},\ \boldsymbol{v}) = \prod_{j=1}^{m} f_j(z_j, \boldsymbol{v}_j)$, and the components of the random vector $\boldsymbol{Z} = (Z_1, \ldots, Z_m)$ are distributed according to the canonical exponential family of distributions, given in (2.2.19)

3. $GI/G/1$ **queue**. Write expressions for the score function processes $\widetilde{S}_t^{(1)}(\boldsymbol{v})$ and $\widetilde{S}_t^{(2)}(\boldsymbol{v})$ for the $GI/G/1$ queue, while considering the pdf's given in examples 2.2.2–2.2.5 of chapter 2.

4. $GI/G/\infty$ **queues in tandem** . Consider m queues in tandem. Assuming that each queue contains an *infinite number* of servers and the servers' time distributions are $G(\lambda_0, \beta)$, where G denotes gamma distributions, show that

   (i) $\mathrm{Var}_{\boldsymbol{v}_0}\{L\widetilde{W}\} = (1+\alpha^2/(1+2\alpha))^m \, \mathbb{E}_{2\boldsymbol{v}_0-\boldsymbol{v}}\{L^2(Z_1, \ldots, Z_m)\} - \ell^2(\boldsymbol{v})$,
   provided $\boldsymbol{v}_0 = (\lambda_0, \beta)$, that is only λ is perturbed,
   $\alpha = (\lambda - \lambda_0)/\lambda_0$, $\widetilde{W} = \prod_{j=1}^{m} W_j$, $W_j = f_j(Z_j, \boldsymbol{v})/f_j(Z_j, \boldsymbol{v}_0)$

   (ii) Find $\mathrm{Var}_{\boldsymbol{v}_0}\{L \nabla \tilde{W}\}$, provided $\boldsymbol{v}_0 = (\lambda_0, \beta)$.

   (iii) Let $L(Z_{1,\ldots,}Z_m) = \sum_{j=1}^{m} Z_j$. Show that for any fixed λ there exist λ_0 such that if $\lambda_0 < \lambda$, then $\mathrm{Var}_{\lambda_0}\{L\tilde{W}\} \leq \mathrm{Var}_{\lambda} L$. That is, the "what if" estimator is more accurate than the straightforward one.

(iv) Similarly to (iii) show that there exist $\tilde{\lambda}_0$ such that if $\tilde{\lambda}_0 < \lambda$, then
$\text{Var}_{\tilde{\lambda}_0}\{L\nabla\widetilde{W}\} \le \text{Var}_{\lambda}\{L\widetilde{S}\}$, where $\widetilde{S} = \sum_{j=1}^{m} \nabla \log f(Y_j, \lambda)$ and $Y_j \sim G(\lambda, \beta)$.

(v) Let $m = 1$, $\beta = 1$ and $L(Y) = I_{-\infty,a}(Y)$. Assuming that $Y \sim \exp(v)$ and a is fixed find the solution of the following optimization problem:

$$\min_{v_0} \text{Var}_{v_0}\{LW(v)\}.$$

(vi) Repeat (v), replacing LW by $L\nabla W$.

5. $GI/G/1/b$ **queue** (b is the buffer size). Present an algorithm for computing the likelihood process $\widetilde{W}_t$ while estimating the performance parameters

$$\nabla^k \ell(\boldsymbol{v}) = \mathbb{E}_{\boldsymbol{v}_0}\left\{\sum_{t=1}^{\tau} L_t \nabla^k W_t(\boldsymbol{v})\right\}, \quad k = 0, 1, \ldots,$$

under the FIFO and LIFO disciplines, provided $\boldsymbol{v} = (\lambda, \mu)$.

Hint: Note that for the $GI/G/1$ queue we have

$$\begin{aligned} W_t &= \prod_{j=1}^{t} \widetilde{W}_j; \quad \widetilde{W}_j = \widetilde{W}_{j1} \cdot \widetilde{W}_{j2}, \quad \widetilde{W}_{ji} = \frac{f_i(Z_{ji}, \boldsymbol{v}_i)}{f_i(Z_{ji}, \boldsymbol{v}_{0i})}, \\ Z_{ji} &\sim f_i(z, \boldsymbol{v}_i), \quad i = 1, 2. \end{aligned}$$

6. Consider the $GI/G/1$ queue with a feedback and repeat example 5.

7. Simulate the queueing models in examples 5–6 and estimate the system parameters

$$\nabla^k \ell(\boldsymbol{v}) = \mathbb{E}_{\boldsymbol{v}_0}\left\{\sum_{t=1}^{\tau} L_t \nabla^k W_t(\boldsymbol{v})\right\}, \quad k = 0, 1,$$

for the $M/M/1/m$ queue and the $M/M/1$ queue with a feedback, respectively, while using the SF method.

8. Consider an open (closed) queueing network with a routing probabilities matrix $P(v) = \{p_{ij}(v)\}$ depending on the parameter v. Find an expression for the SF estimator of

$$\nabla^k \ell(\boldsymbol{v}) = \mathbb{E}_{\boldsymbol{v}_0}\left\{\sum_{t=1}^{\tau} L_t \nabla^k W_t(\boldsymbol{v})\right\}, \quad k = 0, 1, \ldots$$

9. Find confidence intervals (regions) for the DSF estimators (3.3.8) and (3.3.9).

10. Write an expression for the asymptotic covariance matrix Ω of $N^{1/2}\nabla\bar{\ell}_N(\boldsymbol{v}_0^*)$ and its sample equivalent in terms of the covariances of $\sum_{i=1}^{\tau} L_t\widetilde{W}_t$, $\sum_{i=1}^{\tau}\widetilde{W}_t$, $\sum_{i=1}^{\tau}\nabla L_t\widetilde{W}_t$, and $\sum_{i=1}^{\tau}\nabla\widetilde{W}_t$. Its expression will be messy, however.

11. Let $b^1(M,m)$ be the bias of the estimator $\nabla\bar{\ell}_N^{\,t}(\boldsymbol{v},m)$ (see (3.3.13)). Show (3.3.14), i.e. the bias

$$b^1 = \begin{cases} R_{\{L\nabla\widetilde{W}^{tr}\}}(m), & \text{if } m \le M, \\ 0, & \text{otherwise,} \end{cases}$$

where $R_{\{L\widetilde{W}^{tr}\}}(h) = \mathrm{Cov}\{L_t, \nabla\widetilde{W}^{tr}{}_{t+h}(M-m)\}$. That is, the bias introduced by using the truncated estimator $\nabla\bar{\ell}_N^{\,t}(\boldsymbol{v},m)$ instead of the untruncated one $\nabla\bar{\ell}_t(\boldsymbol{v},M)$ equals the value of the *cross-covariance* function $R_{\{L\widetilde{W}^{tr}\}}(h)$ between the processes L_t and $\widetilde{W}^{tr}{}_t(M)$ at the point m.

12. Consider the expected performance $\ell(v) = \mathbb{E}_v\{L(\boldsymbol{Y}\}$, $v \in \mathbb{R}$, and the corresponding efficient score function $S^{(1)}(\boldsymbol{y},v)$. Let $T(\boldsymbol{y},v)$ be a function such that the equality

$$\frac{\mathrm{d}\ell(v)}{\mathrm{d}v} = \mathbb{E}_v\{L(\boldsymbol{Y})T(\boldsymbol{Y},v)\} \tag{3.6.10}$$

holds for any sample performance function $L(\boldsymbol{y})$ having a finite variance (compare with (2.2.3)). Show then that

$$\mathrm{Var}_v\{T(\boldsymbol{Y},v)\} \ge \mathrm{Var}_v\{S^{(1)}(\boldsymbol{Y},v)\}\,. \tag{3.6.11}$$

This inequality shows that the efficient score function has the smallest possible variance among all functions satisfying (3.6.10) (McLeish and Rollans (1992)).

<u>Hint</u> Take $L(\boldsymbol{y}) = S^{(1)}(\boldsymbol{y},v_0)$ and, use (3.6.10) to show that

$$\mathbb{E}_{v_0}\{S^{(1)}(\boldsymbol{Y},v_0)T(\boldsymbol{Y},v_0)\} = \mathbb{E}_{v_0}\{S^{(1)}(\boldsymbol{Y},v_0)^2\}\,.$$

Then apply the Cauchy Schwarz inequality to the left-hand side of (3.6.11).

Show that if $\boldsymbol{v}$ is a vector, then a similar result holds for the corresponding covariance matrices in the Loewner sense of inequality.

13. Derive expressions for the CSF estimator $\hat{\mathrm{d}\ell_N}/\mathrm{d}v$, similar to those given in example 3.4.2, while using normal and Poisson distributions.

14. Prove the validity of the procedure for generating random vectors $\boldsymbol{Y}$ conditional on $\sum_{i=1}^{m} Y_i = t$ for the Poisson distribution in example 3.4.1 (see Weisman (1992)).

15. Consider two queues in tandem with $\ell(\boldsymbol{v})$ being the sojourn time, and $\boldsymbol{v} = (v_1, v_2)$, where v_1 and v_2 correspond to the expected service times in the first and the second tandem queues, having gamma and geometric service time distributions, respectively, and an arbitrary interarrival distribution, say exponential distribution. Run a computer program and estimate $\nabla\ell(\boldsymbol{v})$, while using the CSF method.

16. Using (3.1.3) modify algorithm 2.4.1 and theorem 2.4.1 by making them suitable for the finite-horizon regime.

Chapter 4

What is a "Good" Reference System to Simulate

4.1 Introduction

In this chapter we discuss how to choose a "good" parameter vector $\boldsymbol{v}_0$, which we call the *reference parameter vector*, in order to obtain a variance reduction while estimating $\nabla^k \ell(\boldsymbol{v})$. In particular we consider the following problems:

(i) Given a stable queueing model, that is assuming that the traffic intensity ρ is less than unity, what is a "good" reference system (if any) to simulate in order to obtain reasonably good SF estimators? What is the optimal reference traffic intensity ρ_0^* of such a system?

(ii) For a given reference value ρ_0, how do the SF estimators perform in different ranges of ρ? How does one choose ρ_0 in order to increase the range of ρ without affecting substantially the performance of the SF estimators?

We also show that for the exponential families the variance of the SF estimators (see, for example, (3.2.13)) is a convex function with respect to $\boldsymbol{v}_0$ and that typically there exists a unique optimal reference parameter, say $\boldsymbol{v}_0^*$, which minimizes this variance. We then show numerically that the optimal reference parameter $\boldsymbol{v}_0^*$ can be approximated rather well if we minimize the variance with respect to the parameters of the *bottleneck* queue alone, provided such a bottleneck queue exists. We also show that the SF method is a fairly efficient tool for handling rare events. Here we deal first with the choice of a "good" vector of reference parameters in a single-node queueing model, in the sense that the associated rare event SF estimators are unbiased and have a manageable variance. We then proceed to queueing networks again showing that minimum variance of these rare-event estimators is essentially insensitive to the parameters of non bottleneck queues and, thus a "good" reference vector $\boldsymbol{v}_0$ can be derived from a bottleneck queue alone.

Assume therefore that the dominating pdf $g(\boldsymbol{z})$ is taken in the form $g(\boldsymbol{z}) = f(\boldsymbol{z}, \boldsymbol{v}_0)$, where $\boldsymbol{v}_0$ is the reference parameter vector or simply the *reference parameter.* Note that whenever it is more convenient we use $\nabla^k \bar{\ell}_N(\boldsymbol{v}, \boldsymbol{v}_0, p)$ instead of $\nabla^k \bar{\ell}_N(\boldsymbol{v})$ assuming further that $\ell(\boldsymbol{v}) = \mathbb{E}_{\boldsymbol{v}} L^p$ and $\nabla^k \bar{\ell}_N(\boldsymbol{v}, \boldsymbol{v}_0, p)$ is the corresponding SF estimator. Our goal here is twofold:

(a) To show how to choose a "good" vector of reference parameters $\boldsymbol{v}_0$ in order to obtain a variance reduction in the sense that

$$\mathrm{Var}_{\boldsymbol{v}_0}\{\nabla^k \bar{\ell}_N(\boldsymbol{v}, \boldsymbol{v}_0, p)\} < \mathrm{Var}_{\boldsymbol{v}}\{\nabla^k \bar{\ell}_N(\boldsymbol{v}, \boldsymbol{v}, p)\}, \tag{4.1.1}$$

where $k = 0,\ 1$. For $k = 1$ we define $\mathrm{Var}_{\boldsymbol{v}_0}\{\nabla \bar{\ell}_N(\boldsymbol{v}, \boldsymbol{v}_0, p)\}$ as the trace of the covariance matrix of $\nabla \bar{\ell}_N(\boldsymbol{v}, \boldsymbol{v}_0, p)$. That is

$$\mathrm{Var}_{\boldsymbol{v}_0}\{\nabla \bar{\ell}_N(\boldsymbol{v}, \boldsymbol{v}_0, p)\} \equiv \mathrm{tr}[\mathrm{Cov}_{\boldsymbol{v}_0}\{\nabla \bar{\ell}_N(\boldsymbol{v}, \boldsymbol{v}_0, p\}]. \tag{4.1.2}$$

(b) To establish conditions under which $\mathrm{Var}_{\boldsymbol{v}_0}\{\nabla^k \bar{\ell}_N(\boldsymbol{v}, \boldsymbol{v}_0, p)\}$ is a convex (strictly convex) function of $\boldsymbol{v}_0$ and to evaluate the optimal solution, say, $\boldsymbol{v}_0^*$ of the corresponding mathematical programming problem

$$\min_{\boldsymbol{v}_0} \mathrm{Var}_{\boldsymbol{v}_0}\{\nabla^k \bar{\ell}_N(\boldsymbol{v}, \boldsymbol{v}_0, p)\}, \quad \boldsymbol{v}_0 \in V, \tag{4.1.3}$$

$k = 0, 1.$

Let $\rho = \rho(\boldsymbol{v})$ and $\rho_0 = \rho_0(\boldsymbol{v}_0)$ be the nominal and the reference traffic intensities of the underlying queueing model, where, say, $\boldsymbol{v}$ is the parameter vector of interarrival and service time distributions in the $GI/G/1$ queue. We shall show that for typical output processes, like the waiting time process, a "good" reference traffic intensity ρ_0 must satisfy

$$\rho \le \rho_0 \le \rho_n, \tag{4.1.4}$$

where $\rho_n = \rho_n(\boldsymbol{v}_n, k, p)$ satisfies

$$\mathrm{Var}_{\rho_n} \nabla^k \bar{\ell}_N(\rho, \rho_n, p) = \mathrm{Var}_\rho \nabla^k \bar{\ell}_N(\rho, \rho, p). \tag{4.1.5}$$

Here $k = 0, 1$, $p = 1, 2 \ldots$, $\nabla^k \bar{\ell}_N(\rho, \rho_n, p) = \nabla^k \bar{\ell}_N(\rho_n)$ and $\nabla^k \bar{\ell}_N(\rho, \rho, p) = \nabla^k \bar{\ell}_N(\rho)$ are the estimators of $\nabla^k \ell(\rho)$ corresponding to the pdf's $f(y, \boldsymbol{v}_n)$ and $f(y, \boldsymbol{v})$. Note that (4.1.4) and (4.1.5) mean that for the "what if " estimator $\nabla^k \bar{\ell}_N(\rho_0)$ to perform "well" in the sense of (4.1.1), the reference traffic intensity ρ_0 must belong to the interval (ρ, ρ_n), that is $\rho_0 \in (\rho, \rho_n)$.

We also show how to choose a "good" vector of reference parameters for non-Markovian queueing networks. In particular we extend the results (4.1.1)–(4.1.5) in the sense that we show that in this case a "good" vector of reference parameters $\boldsymbol{\rho}_o = (\rho_{01}, \ldots, \rho_{0r})$ must satisfy (4.1.4) componentwise (separately for each node). We also discuss how to choose a "good" vector $\boldsymbol{\rho}_0$ while estimating $\nabla^k \ell(\boldsymbol{v})$ *simultaneously* for several values (scenarios) $\boldsymbol{v}$ and $k = 0, 1; p = 1, 2, \ldots$. We finally show how to estimate rare events in a queueing network while using the SF method.

To proceed further let

$$\epsilon^k(\boldsymbol{v}, \boldsymbol{v}_0, p) = \frac{\text{Var}_{\boldsymbol{v}_0} \; \nabla^k \bar{\ell}_N(\boldsymbol{v}, \boldsymbol{v}_0, p) t(\boldsymbol{v}_0)}{\text{Var}_{\boldsymbol{v}} \; \nabla^k \bar{\ell}_N(\boldsymbol{v}, \boldsymbol{v}, p) t(\boldsymbol{v})}, \quad k = 0, 1, \tag{4.1.6}$$

be the efficiency of the "what if" estimator $\nabla^k \bar{\ell}_N(\boldsymbol{v}, \boldsymbol{v}_0, p)$, $k = 0, 1$, relative to its standard counterpart $\nabla^k \bar{\ell}_N(\boldsymbol{v}, \boldsymbol{v}, p)$. Here $t(\boldsymbol{v}_0)$ and $t(\boldsymbol{v})$ represent the CPU time needed by the "what if" and standard estimators, respectively. We assume further for simplicity that $t(\boldsymbol{v}_0) = t(\boldsymbol{v})$. Note that the variances of $\nabla^k \bar{\ell}_N(\boldsymbol{v}, \boldsymbol{v}_0, p)$ and $\nabla^k \bar{\ell}_N(\boldsymbol{v}, \boldsymbol{v}, p)$ are measured in terms of *variance/customer* rather than *variance/cycle*.

We say that the "what if" estimator is the more efficient of the two, or that a variance reduction is obtained, if

$$\epsilon^k(\boldsymbol{v}, \boldsymbol{v}_0, p) < 1, \quad k = 0, 1.$$

Note also that if we replace in (4.1.6) $\boldsymbol{v}_0$ by $\boldsymbol{v}_0^*$ with all other data remaining the same, then $\epsilon^k(\boldsymbol{v}, \boldsymbol{v}_0^*, p)$ corresponds to the efficiency of the "what if" estimator $\nabla^k \bar{\ell}_N(\boldsymbol{v}, \boldsymbol{v}_0^*, p)$, where $\boldsymbol{v}_0^*(k, p)$ is the optimal solution of (4.1.3). We shall call $\boldsymbol{v}_0^*(k, p)$ the *optimal reference parameter vector.*

To provide better insight into the subsequent material and, in particular, into the formulas (4.1.1)–(4.1.6), consider the $M/G/1$ queue with service time distributed gamma (denoted $\Gamma(\lambda, \beta)$), the interarrival rate equal unity and $\rho = 0.6$.

Table 4.1.1 and fig. 4.1.1 display how different values of ρ_0 perform for the selected value $\rho = 0.6$. In particular table 4.1.1 displays the point estimators $\nabla^k \bar{\ell}_N(\rho, \rho_0, p)$, the sample variance of $\nabla^k \bar{\ell}_N(\rho, \rho_0, p)$, denoted $\hat{\sigma}^2(\rho, \rho_0, k, p)$, the sample efficiency $\bar{\epsilon}^k(\rho, \rho_0, p)$ as a function of ρ_0 (and $\Delta\delta = (\rho_0 - \rho)/\rho$ –the relative perturbation interval) for the chosen $\rho = 0.6$ and $k = 0, 1$; $p = 1, 3$, where $\bar{\epsilon}^k(\rho, \rho_0, p)$ is defined as

$$\bar{\epsilon}^k(\rho, \rho_0, p) = \frac{\hat{\sigma}^2(\rho, \rho_0, k, p)}{\hat{\sigma}^2(\rho, \rho, k, p)}. \tag{4.1.7}$$

Table 4.1.1 $\nabla^k \bar{\ell}_N(\rho,\rho_0,p)$, $\hat{\sigma}^2(\rho,\rho_0,k,p)$ and $\bar{\epsilon}^k(\rho,\rho_0,p)$ as functions of ρ_0 (and $\Delta\delta$) for the $M/G/1$ queue with $\rho = 0.6$, where $G = \Gamma(\lambda,\beta)$, $\beta = 2$.

ρ_0	$\Delta\delta$	$\bar{\ell}_N(\rho,\rho_0,1)$ $\hat{\sigma}^2(\rho,\rho_0,0,1)$ $\bar{\epsilon}^0(\rho,\rho_0,1)$	$\bar{\ell}_N(\rho,\rho_0,3)$ $\hat{\sigma}^2(\rho,\rho_0,0,3)$ $\bar{\epsilon}^0(\rho,\rho_0,3)$	$\nabla\bar{\ell}_N(\rho,\rho_0,1)$ $\hat{\sigma}^2(\rho,\rho_0,1,1)$ $\bar{\epsilon}^1(\rho,\rho_0,1)$	$\nabla\bar{\ell}_N(\rho,\rho_0,3)$ $\hat{\sigma}^2(\rho,\rho_0,1,3)$ $\bar{\epsilon}^1(\rho,\rho_0,3)$
0.58	-0.03	6.710E-01 2.790E-05 2.011E+00	4.440E+00 3.667E-02 2.560E+00	-7.018E-01 5.333E-04 2.780E+00	-1.117E+01 1.910E-02 3.480E+00
0.60	0.00	6.790E-01 1.390E-05 1.000E+00	4.820E+00 1.430E-02 1.000E+00	-7.230E-01 1.920E-04 1.000E+00	-1.290E+01 5.490E-01 1.000E+00
0.66	0.10	6.730E-01 8.010E-06 5.760E-01	4.580E+00 4.380E-03 3.060E-01	-7.030E-01 4.660E-05 2.430E-01	-1.180E+01 8.510E-02 1.550E-01
0.70	0.17	6.730E-01 7.640E-06 5.500E-01*	4.590E+00 3.890E-03 2.720E-01	-7.080E-01 3.260E-05 1.700E-01*	-1.200E+01 5.160E-02 9.400E-02
0.72	0.20	6.750E-01 8.760E-06 6.300E-01	4.620E+00 3.40E-03 2.470E-01*	- 7.090E-01 3.310E-05 1.730E-01	-1.200E+01 4.610E-02 8.410E-02*
0.76	0.27	6.740E-01 9.910E-06 7.130E-01	4.620E+00 5.030E-03 3.520E-01	-7.140E-01 3.330E-05 1.740E-01	-1.200E+01 4.940E-02 8.990E-02
0.80	0.33	6.700E-01 1.250E-05 8.990E-01	4.490E+00 4.420E-03 3.090E-01	-7.040E-01 4.220E-05 2.200E-01	-1.190E+01 6.510E-02 1.180E-01
0.90	0.50	6.940E-01 3.330E-05 2.390E+00	5.300E+00 4.030E-02 2.820E+00	-7.330E-01 4.160E-04 2.150E+00	-1.290E+01 7.170E-01 1.310E+00

Note that the values of $\bar{\epsilon}^k(\rho, \rho_0, p)$, $k = 0, 1$ marked with the upperscript $*$ correspond to the maximal efficiency (maximal variance reduction).

Figure 4.1.1 displays $\bar{\epsilon}^k(\rho, \rho_0, p)$ as a function of ρ_0 for $k = 0, 1$; $p = 1, 3$ on the basis of data of table 4.1.1. We assumed that $\ell(v)$ is the steady-state expected waiting time of a customer in the $M/G/1$ queue, took $\lambda_0 = v_0$ as the reference parameter, that is assumed $\rho_0 = \lambda_0 \cdot \beta$, chose the interarrival rate equal to 1, the scale $\lambda = v = 0.3$, the shape $\beta = 2$ $[\rho = \mathbb{E}(Y) = \lambda \cdot \beta = 0.6]$, and simulated $N = 10^6$ customers. Note that the number $N = 10^6$ was chosen with the view to obtaining very high precision (small variance) of the estimators $\bar{\epsilon}^k(\rho, \rho_0, p)$.

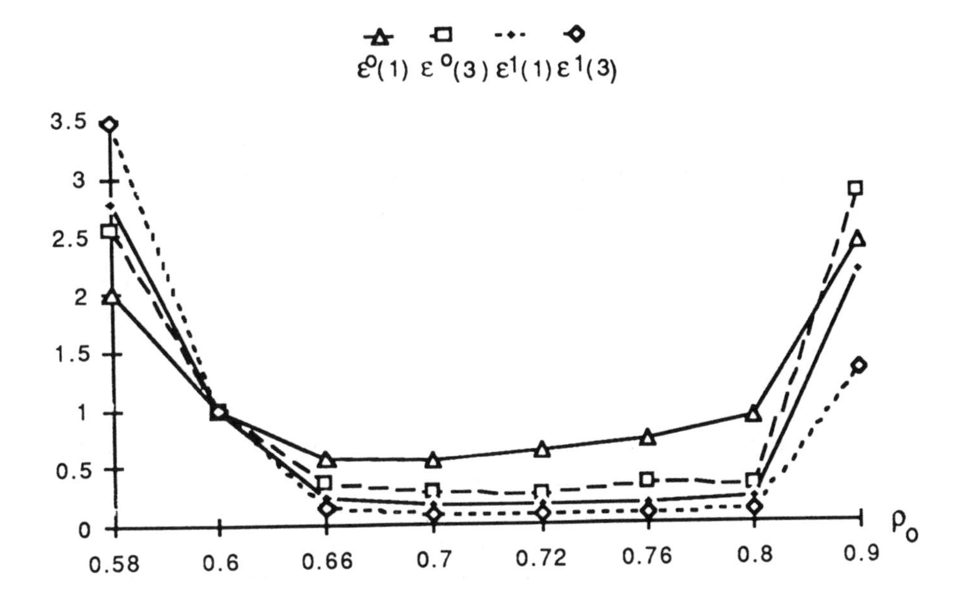

Figure 4.1.1 The relative efficiency $\bar{\epsilon}^k(\rho, \rho_0, p)$ (denoted $\epsilon^k(p)$, $k = 0, 1$, $p = 1, 3$) as a function of ρ_0 for the $M/G/1$ queue with $\rho = 0.6$.

It follows from table 4.1.1 and fig. 4.1.1 that

(i) For fixed ρ_0 the sample variance $\hat{\sigma}^2(\rho, \rho_0, k, p)$ increases in both k and p.

(ii) For fixed p and k there exists $\rho_n = \rho_n(v, v_n, k, p)$ satisfying (4.1.4) and (4.1.5). In particular we found for the above $M/G/1$ queue that $\Delta\rho(\rho_0, k, p) = \rho_n(k, p) - \rho$, called the *variance reduction intervals*, are

$$\Delta\rho(\rho_0, k = 0, p = 1) \approx 0.21,$$
$$\Delta\rho(\rho_0, k = 0, p = 3) \approx 0.22,$$
$$\Delta\rho(\rho_1, k = 1, p = 1) \approx 0.23,$$
$$\Delta\rho(\rho_0, k = 1, p = 3) \approx 0.28.$$

(iii) If ρ_0 satisfies (4.1.4), then (4.1.1) holds and thus a variance reduction, is achieved. It also follows from (4.1.4) that we are on the safe side if we choose the reference parameter $\boldsymbol{v}_0$, corresponding to the traffic intensity $\rho_0 = \rho_0(\boldsymbol{v}_0)$, *moderately larger* than the nominal traffic intensity $\rho = 0.6$.

We also found that for $\rho_0 < \rho$ the relative efficiency $\bar{\epsilon}^k(\rho, \rho_0, p)$ decreases fast in ρ_0 and for $\rho_0 < 0.5$ the SF estimator $\nabla^k \bar{\ell}_N(\rho, \rho_0, p)$ has a very large variance (see proposition 4.3.3 below). With this in mind we shall further assume that $\rho_0 > \rho$.

The rest of this chapter is organized as follows. Section 4.2, which is based on Rubinstein and Shapiro (1992), deals with convexity and uniqueness of the optimal solution $\boldsymbol{v}_0^*$ of the mathematical programming problem (4.1.3) for DESS. It also discusses how to choose a "good" vector of reference parameters $\boldsymbol{v}_0$ (in the sense of (4.1.1)) for the SF estimators $\nabla^k \bar{\ell}_N(\boldsymbol{v}, \boldsymbol{v}_0, p)$ with particular emphasis on static queues, like the $GI/G/\infty$ queue. In section 4.3 we extend the main results of section 4.2 on convexity and on the choice of a "good" $\boldsymbol{v}_0$ to queueing models. We start with the $GI/G/1$ queue and then proceed to non-Markovian queueing networks. Here we show that the optimal solution $\boldsymbol{v}_0^*$ of the program (4.1.3) can be approximated rather well by solving (4.1.3) with respect to the parameters of the bottleneck queue alone, provided such a queue exists (see also Asmussen and Rubinstein (1992b)). In other words, we show that the optimal solution $\boldsymbol{v}_0^*$ of the program (4.1.3) is essentially insensitive to the parameters of the other (non bottleneck) queues. Section 4.4, which is based on Asmussen, Rubinstein and Wang (1992), deals with the estimation of rare events. Here we show that the SF method can handle rare events rather efficiently. We again start with a single-node queueing model by showing how to choose a "good" vector $\boldsymbol{v}_0$ in the sense that the associated rare-event estimators $\nabla^k \bar{\ell}_N(\boldsymbol{v}), k = 0, 1, \ldots$, are both unbiased and have a reasonable variance. We then proceed to queueing networks, by showing that the bottleneck phenomenon is even more prevalent here. We also discuss how to choose a "good" $\boldsymbol{v}_0$ while estimating *simultaneously from a single run* both the standard performance parameters (like the expected waiting time in a queueing network and its associated sensitivities) and rare events. We finally discuss the relationship between importance sampling, rare events and the score function method. Finally, in the appendix we present some auxiliary results borrowed from Rubinstein and Shapiro (1992).

4.2 How to Choose "Good" Reference Parameters for DESS

Bearing in mind that DESS can be considered as a particular case of DEDS with $\tau = 1$, we have from chapter 2 that

$$\nabla^k \ell(\boldsymbol{v}) = \mathbb{E}_{\boldsymbol{v}_0}\{L \nabla^k W(\boldsymbol{v})\}, \tag{4.2.1}$$

where

$$W(\boldsymbol{v}) = W(\boldsymbol{Z}, \boldsymbol{v}) = \frac{f(\boldsymbol{Z}, \boldsymbol{v})}{f(\boldsymbol{Z}, \boldsymbol{v}_0)}, \quad \boldsymbol{Z} \sim f(\boldsymbol{z}, \boldsymbol{v}_0).$$

The SF estimator of $\nabla^k \ell(\boldsymbol{v})$ is

$$\nabla^k \bar{\ell}_N(\boldsymbol{v}) = N^{-1} \sum_{i=1}^{N} L(\boldsymbol{Z}_i) \nabla^k W(\boldsymbol{Z}_i, \boldsymbol{v}). \tag{4.2.2}$$

Consider the program (4.1.3), which for $k = 0, 1$ is equivalent to

$$\min_{\boldsymbol{v}_0} \mathcal{L}^k(\boldsymbol{v}, \boldsymbol{v}_0), \quad \boldsymbol{v}_0 \in V, \tag{4.2.3}$$

where

$$\mathcal{L}^0(\boldsymbol{v}, \boldsymbol{v}_0) = \mathbb{E}_{\boldsymbol{v}_0}\{L^2 W(\boldsymbol{v})^2\} \tag{4.2.4}$$

and

$$\mathcal{L}^1(\boldsymbol{v}, \boldsymbol{v}_0) \equiv \operatorname{tr} \mathbb{E}_{\boldsymbol{v}_0}\{L^2 \nabla W(\boldsymbol{v}) \nabla W(\boldsymbol{v})'\}. \tag{4.2.5}$$

Clearly, the optimal solution $\boldsymbol{v}_0^*$ of (4.2.3) depends on the function $L(\boldsymbol{y})$ and the considered parametric family $f(\boldsymbol{y}, \boldsymbol{v})$. The above optimization problems cannot be typically solved in a closed form since we assume that $L(\boldsymbol{y})$ is not available. They can however, be estimated from simulation by solving the stochastic counterpart of (4.2.3). To demonstrate this, consider first (4.2.4) in more detail. For a fixed $\boldsymbol{v}$, the gradient of $\mathcal{L}^0(\boldsymbol{v}_0) = \mathcal{L}^0(\boldsymbol{v}, \boldsymbol{v}_0)$ with respect to $\boldsymbol{v}_0$ can be written as

$$\nabla \mathcal{L}^0(\boldsymbol{v}_0) = -\mathbb{E}_{\boldsymbol{v}_0}\{L(\boldsymbol{Z})^2 W(\boldsymbol{Z}, \boldsymbol{v})^2 \boldsymbol{S}^{(1)}(\boldsymbol{Z}, \boldsymbol{v}_0)\}, \tag{4.2.6}$$

provided the corresponding derivatives can be taken inside the expected value. In order to find a minimizer $\boldsymbol{v}_0^*$ of the function $\mathcal{L}^0(\boldsymbol{v}_0)$ we need to solve the associated system of equations $\nabla \mathcal{L}^0(\boldsymbol{v}_0) = \boldsymbol{0}$. To do so we may perform the first step in Newton's algorithm starting from a given point $\boldsymbol{v}$. The point

$$\boldsymbol{v}_0^{(1)} = \boldsymbol{v} - [\nabla^2 \mathcal{L}^0(\boldsymbol{v})]^{-1} \nabla \mathcal{L}^0(\boldsymbol{v}) \tag{4.2.7}$$

gives an approximation (based on the second-order Taylor expansion of $\mathcal{L}^0(\boldsymbol{v}_0)$ at $\boldsymbol{v}_0 = \boldsymbol{v}$) of the minimizer $\boldsymbol{v}_0^*$. Here the gradient $\nabla\mathcal{L}^0(\boldsymbol{v})$ is obtained from (4.2.6) by taking $\boldsymbol{v}_0 = \boldsymbol{v}$, that is

$$\nabla\mathcal{L}^0(\boldsymbol{v}) = -\mathbb{E}_{\boldsymbol{v}}\{L(\boldsymbol{Y})^2\boldsymbol{S}^{(1)}(\boldsymbol{Y},\boldsymbol{v})\} . \tag{4.2.8}$$

It is not difficult to see that

$$\nabla^2\mathcal{L}^0(\boldsymbol{v}) = \mathbb{E}_{\boldsymbol{v}}\{L(\boldsymbol{Y})^2[2\boldsymbol{S}^{(1)}(\boldsymbol{Y},\boldsymbol{v})\boldsymbol{S}^{(1)}(\boldsymbol{Y},\boldsymbol{v})' - \boldsymbol{S}^{(2)}(\boldsymbol{Y},\boldsymbol{v})]\} . \tag{4.2.9}$$

Typically the point $\boldsymbol{v}_0^{(1)}$ gives a reasonably good approximation of $\boldsymbol{v}_0^*$ if $\boldsymbol{v}$ and $\boldsymbol{v}_0^*$ lie not "too far" apart. Note also that $-\nabla\mathcal{L}^0(\boldsymbol{v})$ gives the direction of the steepest descent for the function $\mathcal{L}^0(\boldsymbol{v}_0)$ at the point $\boldsymbol{v}_0 = \boldsymbol{v}$.

To estimate now $\boldsymbol{v}_0^*$ from simulation we first take a sample $\boldsymbol{Z}_1, \ldots, \boldsymbol{Z}_N$ from $f(\boldsymbol{z}, \bar{\boldsymbol{v}})$, $\bar{\boldsymbol{v}} \in V$, which dominates the pdf $f(\boldsymbol{z}, \boldsymbol{v}_0)$. We consider next the stochastic counterpart of the system $\nabla\mathcal{L}^0(\boldsymbol{v}_0) = \boldsymbol{0}$, that is

$$\sum_{i=1}^{N} L(\boldsymbol{Z}_i)^2 W(\boldsymbol{Z}_i, \boldsymbol{v})^2 \frac{\nabla f(\boldsymbol{Z}_i, \boldsymbol{v}_0)}{f(\boldsymbol{Z}_i, \bar{\boldsymbol{v}})} = \boldsymbol{0} . \tag{4.2.10}$$

We estimate finally $\boldsymbol{v}_0^*$ by solving the system of equations (4.2.10) and taking the calculated root of the system as an estimator of $\boldsymbol{v}_0^*$.

Let us consider now an exponential family $f(\boldsymbol{y}, \boldsymbol{v})$ in the canonical form, that is

$$f(\boldsymbol{y}, \boldsymbol{v}) = c(\boldsymbol{v}) \exp\left\{\sum_{k=1}^{n} v_k t_k(\boldsymbol{y})\right\} h(\boldsymbol{y}). \tag{4.2.11}$$

Proposition 4.2.1 *Let* $\boldsymbol{Y}$ *be a random vector distributed according to an exponential family in the canonical form (4.2.11). Then* $\mathcal{L}^k(\boldsymbol{v}, \boldsymbol{v}_0)$, $k = 0, 1$, *is a convex function with respect to* $\boldsymbol{v}_0 = (v_{01}, \ldots, v_{0n})$.

The proof of proposition 4.2.1 is given in the appendix (section 4.5).

Note that if V is a convex set, then for a given $\boldsymbol{v}$, $\mathcal{L}^k(\boldsymbol{v}, \boldsymbol{v}_0)$ *is finite-valued on a convex subset* of V, and *strictly convex* if $L(\boldsymbol{y}) \neq 0$. Among other things this implies that the function $\mathcal{L}^k(\boldsymbol{v}, \cdot)$ may have at most a *unique* minimizer $\boldsymbol{v}_0^*$ over V. (For example, one can often verify that $\mathcal{L}^k(\boldsymbol{v}, \boldsymbol{v}_0)$ is continuous and tends to infinity as $\boldsymbol{v}_0$ approaches the boundaries of the domain, where $\mathcal{L}^k(\boldsymbol{v}, \boldsymbol{v}_0)$ is finite valued. It will follow then that the minimizer $\boldsymbol{v}_0^*$ does exist.) We discuss these issues in detail for the gamma distribution, that is we assume that $Y \sim \Gamma(\lambda, \beta)$, where the density of $\Gamma(\lambda, \beta)$ is

$$f(y, \lambda, \beta) = \Gamma(\beta)^{-1}\lambda^{\beta}\mathrm{e}^{-y\lambda}y^{\beta-1}, \quad y > 0, \quad \lambda > 0, \quad \beta > 0.$$

Note that the gamma distribution represents a particular case of the two-parameter exponential family in the canonical form with

$$v_1 = \lambda,\ v_2 = \beta,\ t_1(y) = -y, \text{ and } t_2(y) = \log y.$$

We shall study the properties of $\mathcal{L}^k(\boldsymbol{v}, \boldsymbol{v}_0)$ with respect to the parameter λ, assuming that β is fixed and known. It follows from the results of chapter 2 that in this case the corresponding function $\mathcal{L}^k(\lambda, \lambda_0)$ is typically finite-valued for all $\lambda_0 \in (0, 2\lambda)$. It also follows from the previous discussion that $\mathcal{L}^k(\lambda, \lambda_0)$ is convex on the interval $(0, 2\lambda)$. Typically $\mathcal{L}^k(\lambda, \lambda_0)$ tends to infinity as λ_0 approaches zero or 2λ in which case we obtain that $\mathcal{L}^k(\lambda, \lambda_0)$ has a unique minimizer λ_0^* on the interval $(0, 2\lambda)$.

Proposition 4.2.2 *Let* $Y \sim \Gamma(\lambda, \beta)$, *and assume that* $L(y)^2$ *monotonically increases on the interval* $[0, \infty)$. *Then the optimal value* $\lambda_0^* = \lambda_0^*(k, p)$ *of the reference parameter is less than* λ, *that is*

$$\lambda_0^*(k,p) < \lambda \quad \text{for all } p > 0 \text{ and } k = 0, 1. \tag{4.2.12}$$

The proof of proposition 4.2.2 is given in the appendix (section 4.5).

Consider now the multivariate situation where components $Y_1, \ldots, Y_m$ of the random vector $\boldsymbol{Y}$ are independent, each distributed gamma with the respective parameters λ_k and β_k, $k = 1, \ldots, m$, and let $\boldsymbol{\lambda} = (\lambda_1, \ldots, \lambda_m)'$. Consider the corresponding function

$$\begin{aligned}\mathcal{L}^0(\boldsymbol{\lambda}_0) &= \mathbb{E}_{\boldsymbol{\lambda}_0}\{L^2W^2\} \\ &= \int_0^\infty \cdots \int_0^\infty L(y_1, \ldots, y_m)^2 \prod_{k=1}^m \frac{f_k(y_k, \lambda_k, \beta_k)^2}{f_k(y_k, \lambda_{0k}, \beta_k)}\, dy_1, \ldots, dy_m\ .\end{aligned}$$

The partial derivative of $\mathcal{L}^0(\boldsymbol{\lambda}_0)$ with respect to the first component λ_{01} can be written as

$$\begin{aligned}&\frac{\partial \mathcal{L}^0(\boldsymbol{\lambda}_0)}{\partial \lambda_{01}} \\ &= \int_0^\infty \cdots \int_0^\infty \left\{ \int_0^\infty -L(y_1, \ldots, y_m)^2 \frac{f_1(y_1, \lambda_1, \beta_1)^2}{f_1(y_1, \lambda_{01}, \beta_1)^2} \frac{\partial f_1(y_1, \lambda_{01}, \beta_1)}{\partial \lambda_{01}}\, dy_1 \right\} \\ &\quad \times \prod_{k=2}^m \frac{f_k(y_k, \lambda_k, \beta_k)^2}{f_k(y_k, \lambda_{0k}, \beta_k)}\, dy_2 \cdots dy_m\ .\end{aligned} \tag{4.2.13}$$

In particular if we calculate this derivative at $\boldsymbol{\lambda}_0$ such that $\lambda_{01} = \lambda_1$, then the term $f_1(y_1, \lambda_1, \beta_1)^2 / f_1(y_1, \lambda_{01}, \beta_1)^2$ inside the integral equals one. Assuming furthermore that for all values of $y_2, \ldots, y_m$ the function

$L(y_1, \ldots, y_m)^2$ is monotonically increasing in y_1, it can be shown in the same way as in proposition 4.2.2 (see the proof in the appendix) that the one-dimensional integral inside (4.2.13) is positive and thus the corresponding multi-dimensional integral and hence $\partial \mathcal{L}^0(\boldsymbol{\lambda}_0)/\partial \lambda_{01}$ are positive. Note that we established positivity of $\partial \mathcal{L}^0(\boldsymbol{\lambda}_0)/\partial \lambda_{01}$ at points $\boldsymbol{\lambda}_0$ such that $\lambda_{01} = \lambda_1$ regardless of the values of other components λ_{0k}, $k = 2, \ldots, m$, of $\boldsymbol{\lambda}_0$. Similar arguments can be applied to the partial derivatives $\partial \mathcal{L}^0(\boldsymbol{\lambda}_0)/\partial \lambda_{0k}$, $k = 2, \ldots, m$, as well. In this case we obtain that if $L(\boldsymbol{y})^2$ is *monotonically increasing in every component* y_k of $\boldsymbol{y}$, then the partial derivatives $\partial \mathcal{L}^0(\boldsymbol{\lambda}_0)/\partial \lambda_{0k}$, calculated at points $\boldsymbol{\lambda}_0$ such that $\lambda_{0k} = \lambda_k$, are positive. Together with the result of convexity of $\mathcal{L}(\boldsymbol{\lambda}_0)$ this implies that *every* component of the optimal value $\boldsymbol{\lambda}_0^*$ of the reference parameter vector is *less* than that of the corresponding component of $\boldsymbol{\lambda}$, that is $\boldsymbol{\lambda}_0^* < \boldsymbol{\lambda}$.

Example 4.2.1 Let $L(y) = y^p$. Note that this simple example can model certain static queues, like the $GI/G/\infty$ queue, where, for example, $L(Y) = Y^p$ is the p-th moment of the sojourn time and Y is the service time random variable. Assume further that $Y \sim \Gamma(\lambda, \beta)$.

We have for $k = 0$

$$\mathcal{L}^0(\lambda, \lambda_0, p) = \frac{\lambda^{2\beta}}{\lambda_0^\beta (2\lambda - \lambda_0)^\beta} \mathbb{E}_{2\lambda - \lambda_0}\{Y^{2p}\} = \frac{\lambda^{2\beta}\Gamma(\beta + 2p)}{\lambda_0^\beta (2\lambda - \lambda_0)^{\beta + 2p}\Gamma(\beta)}$$

and similarly for $\mathcal{L}^1(\lambda, \lambda_0, p)$.

The minimizer $\lambda_0^*(k = 0, p)$ of $\mathcal{L}^0(\lambda, \lambda_0, p)$ on the interval $(0, 2\lambda)$ is then given by

$$\lambda_0^*(k = 0, p) = \frac{\lambda\beta}{\beta + p}.$$

It is readily seen that in this case $\lambda_0^*(p, k = 0)$ monotonically decreases in p and tends to zero as $p \to \infty$.

Consider the relative efficiency $\epsilon^k(\lambda, \lambda_0, p)$. We have for $k = 0$

$$\epsilon^0(\lambda, \lambda_0, p) = \frac{\lambda^{2\beta + 2p}}{\lambda_0^\beta (2\lambda - \lambda_0)^{\beta + 2p}}.$$

Also we have that $\hat{\lambda}$ satisfying

$$\epsilon^0(\lambda, \lambda, p) = \epsilon^0(\lambda, \hat{\lambda}, p) = 1$$

equals

$$\hat{\lambda} \approx \frac{\lambda^{(2 - 2p/\beta)}}{(2\lambda - \lambda/2 \cdot 2^{2p/\beta}) \cdot (2\lambda - \lambda/2^{(1 - 2p/\beta)})^{2p/\beta}}.$$

Take, for example, $\lambda = 2$ and $\beta = 0.5$. In this case $\lambda_0^*(k = 0,p) = 1/(0.5+p)$, and the variance reduction interval is

$$(\hat{\lambda},\lambda) \approx \left(\frac{\lambda^{(2-2p/\beta)}}{(2\lambda - \lambda/2 \cdot 2^{2p/\beta}) \cdot (2\lambda - \lambda/2^{(1-2p/\beta)})^{2p/\beta}}, 2 \right).$$

Table 4.2.1 presents $\mathcal{L}^k(v, v_0, p)$ and $\epsilon^k(v, v_0, p)$ as a function of $v_0 = \lambda_0$ for the $M/G/\infty$ queue where $G = \Gamma(v,\beta)$, $v = 2$, $\beta = 0.5$.

Table 4.2.1 $\mathcal{L}^k(v, v_0, p)$ and $\epsilon^k(v, v_0, p)$ as a function of $v_0 = \lambda_0$ for the $M/G/\infty$ queue where $G = \Gamma(v,\beta)$, $v = 2$, $\beta = 0.5$.

v_0	$\Delta\delta$	$\mathcal{L}^0(v,v_0,1)$ $\epsilon^0(v,v_0,1)$	$\mathcal{L}^0(v,v_0,3)$ $\epsilon^0(v,v_0,3)$	$\mathcal{L}^1(v,v_0,1)$ $\epsilon^1(v,v_0,1)$	$\mathcal{L}^1(v,v_0,3)$ $\epsilon^1(v,v_0,3)$
0.1	-0.95	0.158E+00 0.763E+00	0.148E+00 0.374E-01	0.501E-01 0.119E+00	0.360E+00 0.880E-02
0.2	-0.90	0.119E+00 0.453E+00	0.124E+00 0.277E-01	0.405E-01 0.859E-01	0.320E+00 0.731E-02*
0.3	-0.85	0.104E+00 0.332E+00	0.120E+00 0.263E-01*	0.378E-01 0.768E-01	0.330E+00 0.769E-02
0.4	-0.80	0.965E-01 0.272E+00	0.124E+00 0.280E-01	0.377E-01 0.762E-01*	0.363E+00 0.893E-02
0.6	-0.70	0.908E-01 0.227E+00*	0.147E+00 0.372E-01	0.410E-01 0.879E-01	0.489E+00 0.136E-01
0.8	-0.60	0.916E-01 0.232E+00	0.189E+00 0.540E-01	0.482E-01 0.113E+00	0.720E+00 0.222E-01
2.00	0.00	0.188E+00 1.000E+00	0.254E+01 1.000E+00	0.305E+00 1.000E+00	0.270E+02 1.000E+00

4.3 How to Choose "Good" Reference Parameters for DEDS

In this section we show that our basic results derived for the static queues, in particular those related to the choice of the reference vector $\boldsymbol{v}_0$ and convexity of $\mathcal{L}^k(\boldsymbol{v}, \boldsymbol{v}_0, \boldsymbol{p})$ with respect to $\boldsymbol{v}_0$, hold for non-Markovian queueing models. In addition we discuss the following issues:

(i) Suppose we are only interested in a single value of $\boldsymbol{v}$, or in a rather narrow range. Which values of $\boldsymbol{v}_0$ are then reasonable? Which one is optimal?

(ii) For a given reference value $\boldsymbol{v}_0$, how does $\nabla^k \bar{\ell}_N(\boldsymbol{v}, \boldsymbol{v}_0, \boldsymbol{p})$ perform in various ranges of $\boldsymbol{v}$?

In particular:

- For which values of $\boldsymbol{v}$ is the variance of $\nabla^k \bar{\ell}_N(\boldsymbol{v}, \boldsymbol{v}_0, \boldsymbol{p})$ finite?
- For which of the remaining values of $\boldsymbol{v}$ is the variance of $\nabla^k \bar{\ell}_N(\boldsymbol{v}, \boldsymbol{v}_0, \boldsymbol{p})$ too large for $\nabla^k \bar{\ell}_N(\boldsymbol{v}, \boldsymbol{v}_0, \boldsymbol{p})$ to be useful, and for which values of $\boldsymbol{v}$ is it reasonably small?
- For which values of $\boldsymbol{v}$ (if any) do the "what if" estimators $\nabla^k \bar{\ell}_N(\boldsymbol{v}, \boldsymbol{v}_0, \boldsymbol{p})$ lead to variance reduction?

To proceed further, consider separately (I) a single queue and (II) a queueing network.

(I) **Single queue** Here we treat separately estimators (a) $\nabla^k \bar{\ell}_{1N}(\boldsymbol{v}, \boldsymbol{v}_0, \boldsymbol{p})$ and (b) $\nabla^k \bar{\ell}_N(\boldsymbol{v}, \boldsymbol{v}_0, \boldsymbol{p})$.

(a) Assume as before that $g(\boldsymbol{z}) = f(\boldsymbol{z}, \boldsymbol{v}_0)$ and $\ell(\boldsymbol{v}) = \mathbb{E}_{\boldsymbol{v}} L^p$, where L is, say, the steady-state waiting time in the $GI/G/1$ queue. In this case (see also (3.2.25) and (3.2.26)) we have

$$\nabla^k \ell_1(\boldsymbol{v}, \boldsymbol{v}_0, \boldsymbol{p}) = \mathbb{E}_{\boldsymbol{v}_0} \left\{ \sum_{t=1}^{\tau} L_t^p \nabla^k \widetilde{W}_t(\boldsymbol{v}) \right\} \tag{4.3.1}$$

and

$$\nabla^k \bar{\ell}_{1N}(\boldsymbol{v}, \boldsymbol{v}_0, \boldsymbol{p}) = N^{-1} \sum_{i=1}^{N} \sum_{t=1}^{\tau_i} L_{ti}^p \nabla^k \widetilde{W}_{ti}(\boldsymbol{v}), \tag{4.3.2}$$

respectively. Note that formulas (4.2.1) and (4.2.2) represent particular cases of (4.3.1) and (4.3.2) with $\tau = 1$.

Consider now the program

$$\min_{\boldsymbol{v}_0} \mathrm{Var}\{\nabla^k \bar{\ell}_{1N}(\boldsymbol{v}, \boldsymbol{v}_0, p)\}, \quad \boldsymbol{v}_0 \in V, \tag{4.3.3}$$

$k = 0, 1$. Arguing similarly to (4.1.3) and (4.2.3), it is readily seen that the program (4.3.3) is equivalent to

$$\min_{\boldsymbol{v}_0} \mathcal{L}_1^k(\boldsymbol{v}, \boldsymbol{v}_0, p), \ \boldsymbol{v}_0 \in V, \tag{4.3.4}$$

where

$$\mathcal{L}_1^0(\boldsymbol{v}, \boldsymbol{v}_0, p) = \mathbb{E}_{\boldsymbol{v}_0} \left\{ \left[\sum_{t=1}^{\tau} L_t^p \widetilde{W}_t(\boldsymbol{v}) \right]^2 \right\}$$

and

$$\mathcal{L}_1^1(\boldsymbol{v}, \boldsymbol{v}_0, p) = \mathrm{tr}\ \mathbb{E}_{\boldsymbol{v}_0} \left\{ \left[\sum_{t=1}^{\tau} L_t^p \nabla \widetilde{W}_t(\boldsymbol{v}) \right] \left[\sum_{t=1}^{\tau} L_t^p \nabla \widetilde{W}_t(\boldsymbol{v}) \right]' \right\}$$

Note that for $\tau = 1$, (4.3.4) reduces to (4.2.3).

The following proposition extends the results of proposition 4.2.1.

Proposition 4.3.1 *Consider a stable queueing model. Suppose that the input sequence $\boldsymbol{Y}_1, \boldsymbol{Y}_2, \ldots$ is distributed according to an exponential family in the canonical form (4.2.11). Then $\mathcal{L}_1^k(\boldsymbol{v}, \boldsymbol{v}_0)$, $k = 0, 1$, is a convex function with respect to $\boldsymbol{v}_0$, provided the output process $\{L_t : t > 0\}$ is non-negative-valued.*

The proof of proposition 4.3.1 is given in the appendix (section 4.5).

Proposition 4.3.2 *Let $L_t = L_t(\underline{\boldsymbol{Y}}_t)$ be the steady-state waiting (sojourn) time process in a stable queueing model. Suppose that the vectors $\boldsymbol{Y}_i = (Y_{1i}, Y_{i2})$ of the service and the interarrival times have independent components each distributed gamma, that is $Y_{ji} \sim \Gamma(\lambda_i, \beta_i)$, $j = 1, 2$, $i = 1, 2, \ldots$. Let β_1, β_2 be fixed and consider $\mathcal{L}_1^{k=0}(\boldsymbol{\lambda}, \boldsymbol{\lambda}_0)$ as a function of the reference parameter $\boldsymbol{\lambda}_0$ (here $\boldsymbol{\lambda} = (\lambda_1, \lambda_2)$ and $\boldsymbol{\lambda}_0 = (\lambda_{01}, \lambda_{02})$). Then the minimizer $\boldsymbol{\lambda}_0^*$ of $\mathcal{L}_1^0(\boldsymbol{\lambda}, \cdot)$ satisfies $\lambda_{01}^* < \lambda_1$ and $\lambda_{02}^* > \lambda_2$, implying*

$$\rho_0^*(k, p) > \rho \quad \textit{for all} \quad k \geq 0 \textit{ and } p > 0. \tag{4.3.5}$$

The proof of proposition 4.3.2 is given in the appendix (section 4.5).

It follows from proposition 4.3.2 that for any fixed traffic intensity ρ there exists $\rho_n = \rho_n(k,p)$ satisfying

$$\rho_n(k,p) > \rho_0^*(k,p) > \rho, \tag{4.3.6}$$

such that

$$\mathcal{L}_1^k(\rho,\rho_n,p) = \mathcal{L}_1^k(\rho,\rho,p), \tag{4.3.7}$$

and

$$\mathcal{L}_1^k(\rho,\rho_0,p) < \mathcal{L}_1^k(\rho,\rho,p), \tag{4.3.8}$$

provided $\rho_0 \in (\rho,\rho_n)$.

Formulas (4.3.6) and (4.3.8) are directly related to the choice of a "good" reference parameter ρ_0. In particular (4.3.8) means that if $\rho_0 \in (\rho,\rho_n)$, then the "what if" estimator $\nabla^k \bar{\ell}_{1N}(\rho,\rho_0,p)$ is more efficient than its standard counterpart $\nabla^k \bar{\ell}_{1N}(\rho,\rho,p)$ in the sense of (4.1.1).

(b) Consider the program (4.1.3) where $\bar{\ell}_N(\boldsymbol{v})$ is given in (3.2.29). Although we found that extension of proposition 4.3.1 to the program (4.1.3) is not an easy task our simulation studies (see also below) indicate, however, that the variance of the regenerative estimator $\nabla^k \bar{\ell}_N(\rho,\rho_0,p)$ is a convex function with respect to $\boldsymbol{v}_0$, provided the process $\{L_t : t > 0\}$ is non-negative-valued. Explicit calculation of the variance $\sigma^2(\boldsymbol{v},\boldsymbol{v}_0,k=0,p=1) \equiv \sigma^2(\rho,\rho_0)$ of the estimator $\bar{\ell}_N(\boldsymbol{v})$ of the waiting time in the $M/M/1$ queue, is given in Asmussen and Rubinstein (1992b).

It is not difficult to see that under the conditions of proposition 4.3.1 the variance of the truncated estimator (3.3.22) is a convex function with respect to $\boldsymbol{v}_0$. Also, extension of Prorositions 4.3.1 and 4.3.2 to the truncated estimators (3.3.22) and (3.3.23) for queueing networks is not difficult.

After this remark let us turn to numerical results. Tables 4.3.1 and 4.3.2 display data similar to that of table 4.1.1 for $\rho = 0.3$ and 0.9, respectively. As in table 4.1.1 we assumed that $\ell(\boldsymbol{v})$ is the steady-state expected waiting time of a customer in the $M/G/1$ queue, chose the interarrival rate equal to 1, the shape $\beta = 2$, took $\lambda_0 = v_0$ as the reference parameter, that is assumed $\rho_0 = \lambda_0 \cdot \beta$, and simulated $N = 10^6$ customers.

Table 4.3.1 $\nabla^k \bar{\ell}_N(\rho, \rho_0, p)$, $\hat{\sigma}^2(\rho, \rho_0, k, p)$ and $\bar{\epsilon}^k(\rho, \rho_0, p)$ as functions of ρ_0 (and $\Delta\delta$) for the $M/G/1$ queue with $\rho = 0.3$, where $G = \Gamma(\lambda, \beta), \beta = 2$.

ρ_0	$\Delta\delta$	$\bar{\ell}_N(\rho, \rho_0, 1)$ $\hat{\sigma}^2(\rho, \rho_0, 0, 1)$ $\bar{\epsilon}^0(\rho, \rho_0, 1)$	$\bar{\ell}_N(\rho, \rho_0, 3)$ $\hat{\sigma}^2(\rho, \rho_0, 0, 3)$ $\bar{\epsilon}^0(\rho, \rho_0, 3)$	$\nabla\bar{\ell}_N(\rho, \rho_0, 1)$ $\hat{\sigma}^2(\rho, \rho_0, 1, 1)$ $\bar{\epsilon}^1(\rho, \rho_0, 1)$	$\nabla\bar{\ell}_N(\rho, \rho_0, 3)$ $\hat{\sigma}^2(\rho, \rho_0, 1, 3)$ $\bar{\epsilon}^1(\rho, \rho_0, 3)$
0.30	0.00	9.650E-02 1.490E-07 1.000E+00	4.880E-02 4.960E-07 1.000E+00	-3.500E-02 1.450E-07 1.000E+00	-3.920E-02 1.680E-06 1.000E+00
0.38	0.27	9.580E-02 8.680E-08 5.830E-01*	4.920E-02 1.380E-07 2.790E-01	- 3.510E-02 3.230E-08 2.230E-01	-3.990E-02 1.880E-07 1.120E-01
0.42	0.40	9.620E-02 9.270E-08 6.220E-01	4.930E-02 1.300E-07 2.630E-01	-3.500E-02 2.660E-08 1.840E-01*	-3.970E-02 1.360E-07 8.110E-02
0.46	0.53	9.650E-02 1.140E-07 7.680E-01	4.890E-02 1.290E-07 2.610E-01*	-3.480E-02 2.710E-08 1.870E-01	-3.920E-02 1.090E-07 6.500E-02*
0.50	0.67	9.650E-02 1.620E-07 1.090E+00	4.980E-02 1.910E-07 3.850E-01	-3.540E-02 3.270E-08 2.260E-01	-4.030E-02 1.370E-07 8.160E-02
0.80	1.67	9.730E-02 3.450E-06 2.310E+01	4.930E-02 2.010E-06 4.050E+00	-3.490E-02 1.100E-06 7.580E+00	-3.930E-02 1.240E-06 7.350E-01

Table 4.3.2 $\nabla^k \bar{\ell}_N(\rho, \rho_0, p)$, $\hat{\sigma}^2(\rho, \rho_0, k, p)$ and $\bar{\epsilon}^k(\rho, \rho_0, p)$ as functions of ρ_0 (and $\Delta\delta$) for the $M/G/1$ queue with $\rho = 0.8$, where $G = \Gamma(\lambda_0, \beta), \beta = 2$.

ρ_0	$\Delta\delta$	$\bar{\ell}_N(\rho,\rho_0,1)$ $\hat{\sigma}^2(\rho,\rho_0,0,1)$ $\bar{\epsilon}^0(\rho,\rho_0,1)$	$\bar{\ell}_N(\rho,\rho_0,3)$ $\hat{\sigma}^2(\rho,\rho_0,0,3)$ $\bar{\epsilon}^0(\rho,\rho_0,3)$	$\nabla\bar{\ell}_N(\rho,\rho_0,1)$ $\hat{\sigma}^2(\rho,\rho_0,1,1)$ $\bar{\epsilon}^1(\rho,\rho_0,1)$	$\nabla\bar{\ell}_N(\rho,\rho_0,3)$ $\hat{\sigma}^2(\rho,\rho_0,1,3)$ $\bar{\epsilon}^1(\rho,\rho_0,3)$
0.80	0.00	2.360E+00 4.880E-04 1.000E+00	1.180E+02 2.720E+01 1.000E+00	- 5.530E+00 3.660E-02 1.000E+00	-7.260E+02 5.650E+03 1.000E+00
0.84	0.05	2.410E+00 3.350E-04 6.860E-01*	1.250E+02 1.180E+01 4.340E-01	- 5.830E+00 1.380E-02 3.770E-01	-8.240E+02 1.480E+03 2.620E-01
0.86	0.07	2.420E+00 3.640E-04 7.470E-01	1.280E+02 1.320E+01 4.840E-01	- 5.830E+00 1.040E-02 2.850E-01	-8.250E+02 1.030E+03 1.830E-01
0.88	0.10	2.380E+00 3.380E-04 6.920E-01	1.190E+02 9.370E+00 3.440E-01*	- 5.600E+00 8.980E-03 2.450E-01*	-7.650E+02 7.880E+02 1.390E-01
0.92	0.15	2.390E+00 6.250E-04 1.280E+00	1.200E+02 2.220E+01 8.150E-01	- 5.630E+00 9.840E-03 2.690E-01	-7.360E+02 7.050E+02 1.250E-01*
0.94	0.17	2.390E+00 1.270E-03 2.600E+00	1.220E+02 4.150E+01 1.530E+00	- 5.730E+00 1.860E-02 5.090E-01	-8.240E+02 2.370E+03 4.190E-01
0.98	0.22	2.290E+00 2.090E-03 4.280E+00	1.040E+02 3.700E+01 1.360E+00	- 5.760E+00 1.210E-01 3.300E+00	-7.100E+02 3.740E+03 6.620E-01

It seems reasonable to ask whether there is a simple intuitive explanation why, for a given ρ, one can obtain a variance reduction by choosing ρ_0 moderately larger than ρ. We answer this in terms of importance sampling as follows. When estimating $\ell(\rho)$, the main contributors typically come from the cycles which are somewhat larger than average (for example in heavy traffic, essentially only cycles of order $O(1-\rho)^{-2}$ matter, while the cycles themselves are of order $O(1)$ in distribution; (see the discussion by Asmussen (1987)); and in choosing a reference parameter $\rho_0 > \rho$ a greater weight is assigned to the large cycles.

In tables 4.3.3–4.3.5 the question has been turned around, in the sense that they display how $\nabla^k \bar{\ell}_N(\rho)$ performs in the $M/G/1$ queue for different values of ρ for fixed ρ_0, namely for $\rho_0 = 0.3, 0.6$ and 0.8 . In particular, they present the sample relative efficiency $\bar{\epsilon}^k(\rho, p|\rho_0)$, $k = 0, 1$; $p = 1, 3$ defined in analogy to $\bar{\epsilon}^k(\rho, \rho_0, p)$ (see also (4.1.7)) as

$$\bar{\epsilon}^k(\rho, p|\rho_0) = \frac{\hat{\sigma}^2(\rho, k, p|\rho_0)}{\hat{\sigma}^2(\rho_0, k, p|\rho_0)} \tag{4.3.9}$$

along with the sample performance $\nabla^k \bar{\ell}_N(\rho, p|\rho_0)$ and the sample variance $\hat{\sigma}^2(\rho, k, p|\rho_0)$ as a function of ρ for the $M/G/1$ queue, with $G = \Gamma(\lambda_0, \beta)$, $\beta = 2$ and $\rho_0 = 0.3, 0.6$ and 0.8 respectively. We chose the interarrival rate equal to 1, $\lambda_0 = v_0$ as the reference parameter, and in order to obtain high precision of $\bar{\epsilon}^k(\rho, p|\rho_0)$, we simulated 10^6 customers.

Table 4.3.3 $\nabla^k \bar{\ell}_N(\rho, p|\rho_0)$, $\hat{\sigma}^2(\rho, k, p|\rho_0)$ and $\bar{\epsilon}^k(\rho, p|\rho_0)$ as functions of ρ for the $M/G/1$ queue with $\rho_0 = 0.3$, where $G = \Gamma(\lambda_0, \beta)$, $\beta = 2$.

ρ	$\Delta\delta$	$\bar{\ell}_N(\rho,1\|\rho_0)$ $\hat{\sigma}^2(\rho,0,1\|\rho_0)$ $\bar{\epsilon}^0(\rho,1\|\rho_0)$	$\bar{\ell}_N(\rho,3\|\rho_0)$ $\hat{\sigma}^2(\rho,0,3\|\rho_0)$ $\bar{\epsilon}^0(\rho,3\|\rho_0)$	$\nabla\bar{\ell}_N(\rho,1\|\rho_0)$ $\hat{\sigma}^2(\rho,1,1\|\rho_0)$ $\bar{\epsilon}^1(\rho,1\|\rho_0)$	$\nabla\bar{\ell}_N(\rho,3\|\rho_0)$ $\hat{\sigma}^2(\rho,1,3\|\rho_0)$ $\bar{\epsilon}^1(\rho,3\|\rho_0)$
0.10	2.000	8.400E-03 1.100E-08 6.510E+00	2.660E-04 1.280E-11 8.850E-01	-8.670E-04 1.190E-10 1.310E+00	-5.800E-05 4.200E-13 1.780E-01
0.16	0.875	2.280E-02 9.640E-09 9.450E-01	2.190E-03 2.190E-10 2.170E-01	-4.000E-03 3.620E-10 2.000E-01	-8.130E-04 3.190E-11 6.270E-02
0.20	0.500	3.750E-02 1.550E-08 6.330E-01	6.300E-03 1.570E-09 1.530E-01*	-8.450E-03 1.370E-09 1.720E-01*	-3.040E-03 5.800E-10 4.050E-02*
0.22	0.364	4.650E-02 2.130E-08 5.750E-01*	1.000E-02 4.490E-09 2.130E-01	-1.170E-02 2.950E-09 1.960E-01	-5.430E-03 2.520E-09 1.040E-01
0.26	0.154	6.860E-02 4.850E-08 6.280E-01	2.320E-02 4.130E-08 3.680E-01	-2.100E-02 1.670E-08 3.420E- 01	-1.540E-02 4.900E-08 2.210E-01
0.30	0.000	9.650E-02 1.490E-07 1.000E+00	4.880E-02 4.960E-07 1.000E+00	- 3.500E-02 1.450E-07 1.000E+00	-3.920E-02 1.680E-06 1.000E+00

Table 4.3.4 $\nabla^k \bar{\ell}_N(\rho, p|\rho_0)$, $\hat{\sigma}^2(\rho, k, p|\rho_0)$ and $\bar{\epsilon}^k(\rho, p|\rho_0)$ as functions of ρ for the $M/G/1$ queue with $\rho_0 = 0.6$, where $G = \Gamma(\lambda_0, \beta)$, $\beta = 2$.

ρ	$\Delta\delta$	$\bar{\ell}_N(\rho,1\|\rho_0)$ $\hat{\sigma}^2(\rho,0,1\|\rho_0)$ $\bar{\epsilon}^0(\rho,1\|\rho_0)$	$\bar{\ell}_N(\rho,3\|\rho_0)$ $\hat{\sigma}^2(\rho,0,3\|\rho_0)$ $\bar{\epsilon}^0(\rho,3\|\rho_0)$	$\nabla\bar{\ell}_N(\rho,1\|\rho_0)$ $\hat{\sigma}^2(\rho,1,1\|\rho_0)$ $\bar{\epsilon}^1(\rho,1\|\rho_0)$	$\nabla\bar{\ell}_N(\rho,3\|\rho_0)$ $\hat{\sigma}^2(\rho,1,3\|\rho_0)$ $\bar{\epsilon}^1(\rho,3\|\rho_0)$
0.10	5.00	9.000E-03 2.110E-07 1.250E+02	3.050E-04 1.300E-09 8.940E+01	- 9.510E-04 4.940E-09 5.440E+01	-6.830E-05 6.010E-11 2.550E+01
0.30	1.00	9.720E-02 4.850E-07 3.260E+00	4.960E-02 4.680E-07 9.440E-01	- 3.450E-02 1.450E-07 9.990E-01	-3.980E-02 2.830E-07 1.680E-01
0.48	0.25	3.320E-01 1.380E-06 6.030E-01	8.500E-01 7.710E-05 2.570E-01	- 2.350E-01 2.110E-06 1.680E-01	-1.430E+00 4.160E-04 6.500E-02
0.50	0.20	3.760E-01 1.800E-06 5.450E-01*	1.140E+00 1.560E-04 1.150E-01*	- 2.840E-01 3.620E-06 1.510E-01*	-2.050E+00 1.100E-03 2.900E-02*
0.52	0.15	4.240E-01 2.440E-06 6.190E-01	1.520E+00 3.310E-04 3.630E-01	-3.420E-01 6.500E-06 2.730E-01	-2.940E+00 3.020E-03 1.890E-01
0.60	0.00	6.790E-01 1.390E-05 1.000E+00	4.820E+00 1.430E-02 1.000E+00	-7.230E-01 1.920E-04 1.000E+00	-1.290E+01 5.490E-01 1.000E+00

Table 4.3.5 $\nabla^k \bar{\ell}_N(\rho, p|\rho_0)$, $\hat{\sigma}^2(\rho, k, p|\rho_0)$ and $\bar{\epsilon}^k(\rho, p|\rho_0)$ as functions of ρ for the $M/G/1$ queue with $\rho_0 = 0.8$, where $G = \Gamma(\lambda_0, \beta)$, $\beta = 2$.

ρ	$\Delta\delta$	$\bar{\ell}_N(\rho,1\|\rho_0)$ $\hat{\sigma}^2(\rho,0,1\|\rho_0)$ $\bar{\epsilon}^0(\rho,1\|\rho_0)$	$\bar{\ell}_N(\rho,3\|\rho_0)$ $\hat{\sigma}^2(\rho,0,3\|\rho_0)$ $\bar{\epsilon}^0(\rho,3\|\rho_0)$	$\nabla\bar{\ell}_N(\rho,1\|\rho_0)$ $\hat{\sigma}^2(\rho,1,1\|\rho_0)$ $\bar{\epsilon}^1(\rho,1\|\rho_0)$	$\nabla\bar{\ell}_N(\rho,3\|\rho_0)$ $\hat{\sigma}^2(\rho,1,3\|\rho_0)$ $\bar{\epsilon}^1(\rho,3\|\rho_0)$
0.200	3.000	3.960E-02 9.430E-06 3.860E+02	6.910E-03 1.240E-06 1.200E+02	-8.480E-03 5.370E-08 6.750E+00	-2.930E-03 1.020E-08 7.130E-01
0.300	1.667	9.730E-02 3.450E-06 2.310E+01	4.930E-02 2.010E-06 4.050E+00	-3.490E-02 1.100E-06 7.580E+00	-3.930E-02 1.240E-06 7.350E-01
0.400	1.000	2.000E-01 3.350E-06 2.730E+00	2.480E-01 1.760E-05 1.040E+00	-1.050E-01 1.430E-06 8.740E-01	-2.990E-01 2.690E-05 2.190E-01
0.500	0.600	3.730E-01 5.670E-06 1.710E+00	1.070E+00 2.320E-04 1.710E-01*	-2.790E- 01 6.730E-06 2.820E- 01	-1.890E+00 8.880E-04 2.340E-02*
0.600	0.333	6.700E-01 1.250E-05 8.990E-01	4.490E+00 4.420E-03 3.090E-01	-7.040E- 01 4.220E-05 2.200E- 01*	-1.190E+01 6.510E-02 1.180E-01
0.700	0.143	1.220E+00 4.460E-05 7.060E-01	2.060E+01 1.790E-01 6.250E-01	- 1.840E+00 4.200E-04 2.550E-01	-8.050E+01 6.160E+00 2.350E-01
0.760	0.053	1.790E+00 1.380E-04 6.280E-01*	5.600E+01 2.100E+00 5.520E-01	- 3.460E+00 3.560E-03 2.510E-01	-2.850E+02 1.520E+02 1.870E-01
0.800	0.000	2.360E+00 4.880E-04 1.000E+00	1.180E+02 2.720E+01 1.000E+00	- 5.530E+00 3.660E-02 1.000E+00	-7.260E+02 5.650E+03 1.000E+00

The results of tables 4.3.3-4.3.5 are self-explanatory. Based on the results of table 4.3.5, fig. 4.3.1 presents the relative efficiency $\bar{\epsilon}^k(\rho, p|\rho_0)$, $k = 0, 1$; $p = 1, 3$ as a function of ρ with $\rho_0 = 0.8$.

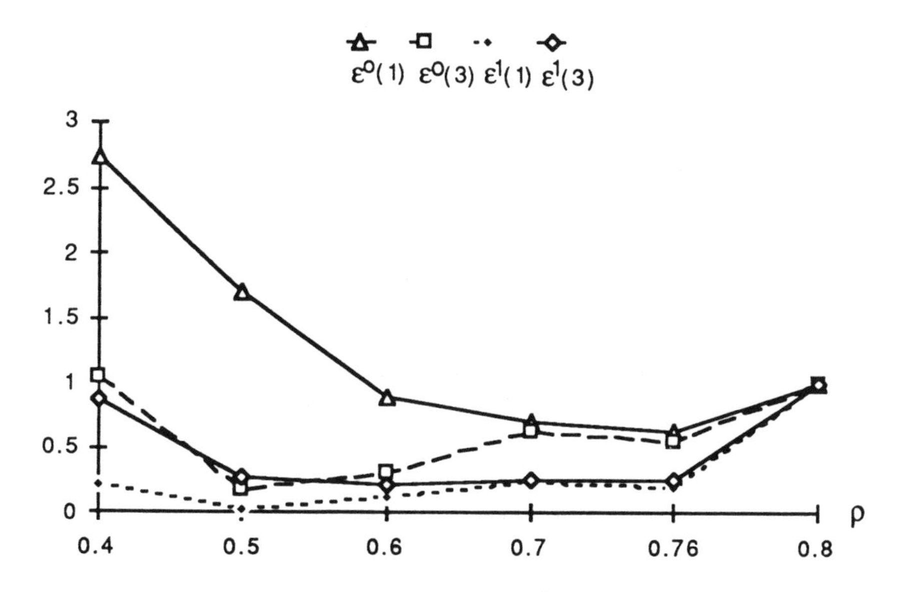

Figure 4.3.1 The relative efficiency $\bar{\epsilon}^k(\rho, p|\rho_0)$ (denoted $\epsilon^k(p)$, $k = 0, 1$, $p = 1, 3$) as a function of ρ for the $M/G/1$ queue with $\rho_0 = 0.8$.

Our extensive simulation studies with single-node queues, like the $GI/G/1$ queue, show that

- There exists a rather broad interval (ρ_n, ρ_0) where some moderate variance reduction is achieved, that is, $\bar{\epsilon}^k(\rho, p|\rho_0) < 1$. Here ρ_n is the solution of the equation

$$\hat{\sigma}^2(\rho, k, p|\rho_0) = \hat{\sigma}^2(\rho_0, k, p|\rho_0),$$

  with respect to ρ, provided ρ_0, k and p are fixed. Note that the interval $\Delta\rho_n = \{(\rho_n, \rho_0) : \rho_n(\rho) < \rho < \rho_0\}$, called the *variance reduction interval*, increases in k and p. Consider, for example, table 4.3.4 with $\rho_0 = 0.6$. We have
  $\Delta\rho_n(\rho, k = 0, p = 1) \approx 0.20$,
  $\Delta\rho_n(\rho, k = 0, p = 3) \approx 0.30$,
  $\Delta\rho_n(\rho, k = 1, p = 1) \approx 0.30$,
  $\Delta\rho_n(\rho, k = 1, p = 3) \approx 0.45$.

- The efficiency $\bar{\epsilon}^k(\rho, p|\rho_0)$ decreases moderately as we move down from ρ_0 to zero, and decreases rapidly as we move above ρ_0.

The following proposition shows how far we can move *above* ρ_0 in the $M/M/1$ queue.

Proposition 4.3.3 *Let $\sigma^2(\rho|\rho_0)$ be the variance of the estimator $\bar{\ell}_N(\rho|\rho_0) = \bar{\ell}_N(\rho, 0, 1|\rho_0)$ of the waiting time in the $M/M/1$ queue. Then $\sigma^2(\rho|\rho_0)$ is finite for*

$$\rho \leq \rho_c = \rho_c(\rho_0) = \frac{2\rho_0}{1+\sqrt{\rho_0}} \tag{4.3.10}$$

and is infinite for $\rho > \rho_c$.

The proof of this proposition is given in Asmussen and Rubinstein (1992b).

Table 4.3.6 presents ρ_c as a function of ρ_0 calculated according to (4.3.10).

Table 4.3.6 ρ_c as function of ρ_0.

ρ_0	0.10	0.20	0.30	0.40	0.50	0.60	0.70	0.80	90	0.95	0.99
ρ_c	0.15	0.28	0.39	0.49	0.59	0.66	0.76	0.84	0.92	0.96	0.99

It follows from (4.3.10) that $\rho_0 < \rho_c < 2\rho_0$. Note also that in light and heavy traffic we have $\rho_c \cong 2\rho_0$ and $\rho_c \cong \rho_0$, respectively. Take, for example, $\rho_0 = 0.1$ and 0.9. We have $\rho_c = 0.152$ and 0.924, respectively.

Figure 4.3.2 displays the curves ρ_c (the critical value), ρ_n (the neutral value), and ρ_0^* (the optimal value) as functions of ρ_0 for the expected waiting time $\ell(\boldsymbol{v})$ in the $M/G/1$ queue, with the interarrival rate equal to 1, $G = \Gamma(\lambda_0, \beta)$, $\beta = 2$, and $\lambda_0 = v_0$ being the reference parameter. Figure 4.3.3 displays similar curves for the derivative $\partial\ell(\boldsymbol{v})/\partial\lambda$. For both cases we performed separate simulations at $\rho_0 = 0.1, 0.2, \ldots, 0.9$, each of size 10^6 customers.

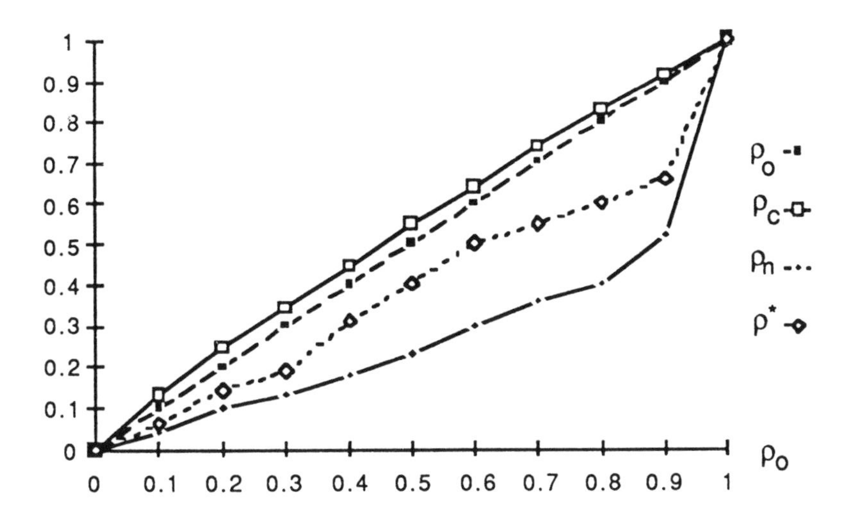

Figure 4.3.2 The curves ρ_c, ρ_n, and ρ_0^* as functions of ρ_0 for the expected waiting time $\ell(v)$ in the $M/G/1$ queue.

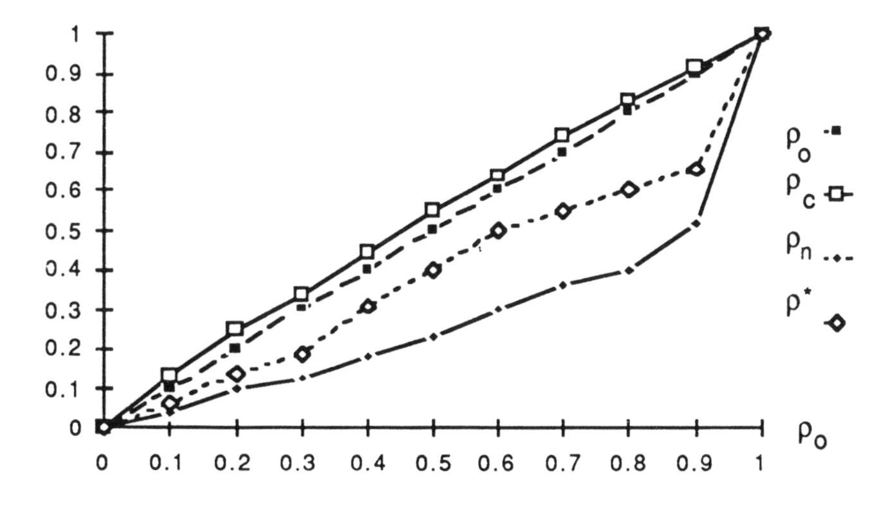

Figure 4.3.3 The curves ρ_c, ρ_n, and ρ_0^* as functions of ρ_0 for the derivative $\partial\ell(v)/\partial\lambda$ of the expected waiting time $\ell(v)$ in the $M/G/1$ queue.

It is readily seen that $\rho_0^*(\rho_0)$ is located inside the variance reduction region $\Delta\rho_n = \{(\rho_n, \rho_0) : \rho_n(\rho_0) < \rho \leq \rho_0\}$, which in turn is quite large. This means that we have good performance over a reasonably large neighborhood. Similar results were obtained for the $GI/G/1$ queue, provided the interarrival and service times are distributed according to the exponential family

Remark 4.3.1 Suppose that we want to estimate $\nabla^k \ell(\boldsymbol{v})$ *simultaneously* for several values of $\boldsymbol{v}$, say, for $\boldsymbol{v} = \boldsymbol{v}_1, \ldots, \boldsymbol{v}_s$, and $k = 0, 1$. We can consider in this case the following extended version of the program (4.1.3):

$$\min_{\boldsymbol{v}_0} \psi^k(\boldsymbol{v}_0, k, p), \tag{4.3.11}$$

where

$$\psi^k(\boldsymbol{v}_0, k, p) = \sum_{j=1}^{s} \mathrm{Var}_{\boldsymbol{v}_0}\{\nabla^k \bar{\ell}_N(\boldsymbol{v}_j, k, p)\} \tag{4.3.12}$$

or

$$\psi^k(\boldsymbol{v}_0, k, p) = \max_{j=1,\ldots,s} \mathrm{Var}_{\boldsymbol{v}_0}\{\nabla^k \bar{\ell}_N(\boldsymbol{v}_j, k, p)\}. \tag{4.3.13}$$

Arguing as before it seems *natural* to choose the reference parameter $\boldsymbol{v}_0$ either according to

$$\rho_0(\boldsymbol{v}_0) = \max_{j=1,\ldots,s} \rho(\boldsymbol{v}_j) \tag{4.3.14}$$

or moderately larger than $\rho_0(\boldsymbol{v}_0)$.

Here (4.3.14) means that the reference parameter $\boldsymbol{v}_0$ must correspond to the *highest traffic intensity* among all traffic intensities associated with the selected values $\boldsymbol{v}_1, \ldots, \boldsymbol{v}_s$.

Consider, for example, the data in table 4.1.1 for the $M/G/1$ queue and assume that we seek to estimate $\nabla^k \ell(\rho)$ for $\rho = 0.4, 0.5$ and 0.6 and $k = 0, 1$, $p = 1, 3$ *simultaneously from a single simulation.* It follows from table 4.1.1 that we can do this efficiently in the sense that $\bar{\epsilon}^k(\rho, p|\rho_0) < 1$ and simultaneously for all $\boldsymbol{v}, k$ and p by simulating a single scenario with the reference traffic intensity $\rho_0 = 0.6$. This becomes feasible, since in this case, the variance reduction region is $\Delta\rho_n = \{(\rho_n, \rho_0) : \rho_n(\rho) < \rho < \rho_0\} \geq 0.2$. Note that if we need to estimate $\ell(\boldsymbol{v})$, say, for $\rho \in (0.3, 0.6)$, we could still choose $\rho_0 = 0.6$ as a reference parameter.

In this case (see table 4.1.1) we would obtain a variance reduction approximately in the interval (0.4, 0.6), while we obtain a moderate increase in the variance (compared with the CMC method) in the remaining interval (0.3, 0.4).

With $\sigma^2(\rho, k, p|\rho_0)$ or its estimate $\hat{\sigma}^2(\rho, k, p|\rho_0)$ at hand, we can depict various confidence intervals for $\ell(\rho)$ with ρ fixed and satisfying (4.3.10). By doing so we can see how well the "what if" estimators perform for different values of ρ, provided ρ_0 is fixed.

Figure 4.3.4 depicts the curve $\bar{\ell}_N(\rho|\rho_0)$ (denoted ℓ_N) along with curves

$$J_1 = \{\bar{\ell}_N(\rho|\rho_0) - w_r\},\ J_2 = \{\bar{\ell}_N(\rho|\rho_0) + w_r\},\ \text{where } w_r = \frac{1.96\hat{\sigma}(\rho|\rho_0)}{\bar{\ell}_N(\rho|\rho_0)} \tag{4.3.15}$$

(denoted 95%CI) as a functions of ρ for the $M/G/1$ queue with $\rho_0 = 0.8$. Here ω_r represents the half width of the 95% (relative) confidence interval. Note that $\bar{\ell}_N(\rho|\rho_0)$ and ω_r in J_1, J_2 are given in different scales. Fig. 4.3.5 depicts similar data for the derivative of the expected waiting time in the $M/G/1$ queue with respect to λ. We used the same $M/G/1$ queue as for table 4.3.5, that is we chose the interarrival rate equal to 1, $\beta = 2$, assumed that $\lambda_0 = v_0$ as the reference parameter and estimated from a *single simulation run* the performance $\ell(v)$ and its derivative *simultaneously* for $\rho = 0.2, 0.3, \ldots, 0.8$, while simulating *only* $N = 10^5$ customers. It is readily seen that the "what if" estimators $\bar{\ell}_N(\rho|\rho_0)$ and $\nabla\bar{\ell}_N(\rho|\rho_0)$ perform reasonably well in the range $\rho \in (0.4, 0.8)$, that is when the relative perturbation in ρ does not exceed 100%. For large relative perturbations ($\geq 100\%$) the SF process $\nabla^k \widetilde{W}$ causes substantial increase in the variance of the estimators $\bar{\ell}_N(\rho|\rho_0)$ and $\nabla\bar{\ell}_N(\rho|\rho_0)$.

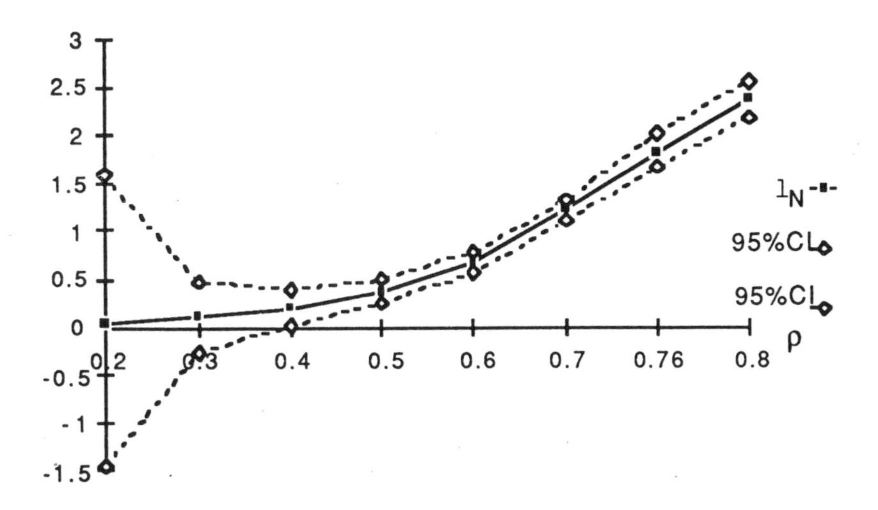

Figure 4.3.4 The performance of the "what if" estimators $\bar{\ell}_N(\rho|\rho_0)$ for different values of ρ in the $M/G/1$ queue with the reference traffic intensity $\rho_0 = 0.8$.

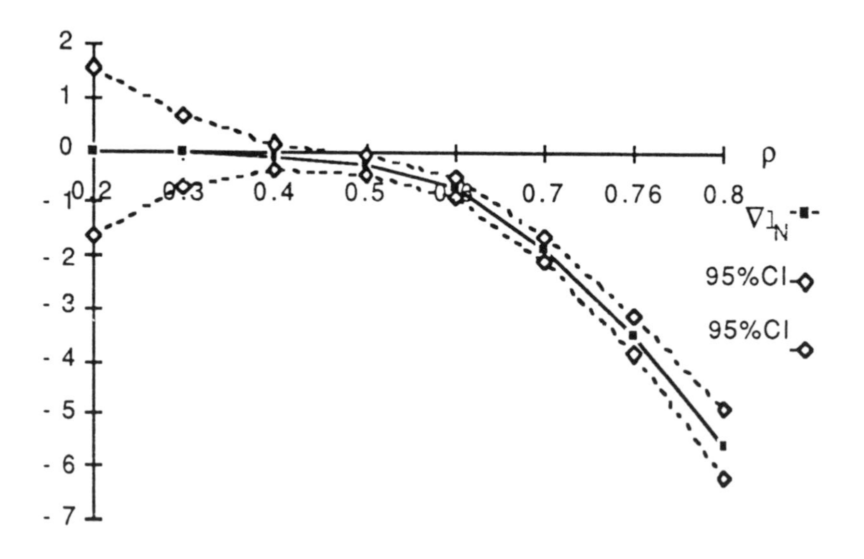

Figure 4.3.5 The performance of the "what if" estimators $\nabla\bar{\ell}_N(\rho|\rho_0)$ for different values of ρ in the $M/G/1$ queue with the reference traffic intensity $\rho_0 = 0.8$.

(II) **Queueing networks** We now give guidelines on how to choose a "good" vector of reference parameters $\boldsymbol{v}_0$ (and $\boldsymbol{\rho}_o$) while using the SF method to estimate the performance parameters $\nabla^k\ell(\boldsymbol{v}) = \nabla^k\mathbb{E}_{\boldsymbol{v}}L_t$, $k = 0, 1$, in a non-Markovian queueing network containing r nodes. If not stated otherwise we assume below that L_t is the steady-state sojourn time in the network.

We start with simulation results for the following two simple queueing models: two queues in tandem and two queues with a feedback on the second queue. Note that in both queueing models we chose one of the queues more congested (working under heavier traffic) than the other. According to common terminology, we shall call the congested queue the *bottleneck queue*. As we shall see below, the bottleneck queue plays a crucial role when choosing a "good" (optimal) vector of reference parameters.

(i) **Two queues in tandem** Table 4.3.7 presents the sample relative efficiency $\bar{\epsilon}^k(\boldsymbol{\rho}, \boldsymbol{\rho}_o)$ (the latter being the sample equivalent of $\epsilon^k(\boldsymbol{\rho}, \boldsymbol{\rho}_o)$, where $\boldsymbol{\rho} = (\rho_{01}, \rho_{02})$; $k = 0, 1$; $p = 1$) along with the sample variance and the trace of the sample variance of $\nabla^k\bar{\ell}_N(\boldsymbol{\rho})$, $k = 0, 1$, denoted $\hat{\sigma}^2(\boldsymbol{\rho}, \boldsymbol{\rho}_o, k)$, $k = 0, 1$, as a function of ρ_{01} and ρ_{02} for two $GI/G/1$ queues in tandem with $\rho_1 = 0.3$ and $\rho_2 = 0.6$. We assumed that the interarrival and two service times distributions are gamma (denoted $\Gamma_i(\beta_i, \lambda_i)$, $i = 1, 2, 3$). In particular we chose $(\beta_1, \lambda_1) =(0.5, 3)$;

(β_2, λ_2) =(0.5,10); and (β_3, λ_3) =(0.5, 5), took $\boldsymbol{v}_0 = \boldsymbol{\lambda}_0 = (\lambda_{20}, \lambda_{30})$ as the reference vector and simulated 5×10^4 customers. Note that $\bar{\epsilon}^k(\boldsymbol{\rho}_o{}^*)$, $k = 0, 1$, and $\bar{\epsilon}^k(\boldsymbol{\rho}_o{}^\diamond)$, $k = 0, 1$, correspond to the optimal reference vector $\boldsymbol{\rho}_o{}^* = (\rho_{01}^*, \rho_{02}^*)$ and to $\boldsymbol{\rho}_o{}^\diamond = (\rho_1, \rho_{02}^\diamond)$ (that is, where ρ_1 is fixed, namely $\rho_1 = 0.3$), respectively.

Table 4.3.7 $\hat{\sigma}^2(\boldsymbol{\rho}, \boldsymbol{\rho}_o, k)$ and $\bar{\epsilon}^k(\boldsymbol{\rho}, \boldsymbol{\rho}_o)$ as functions of ρ_{01} and ρ_{02} for two $GI/G/1$ queues in tandem with $\boldsymbol{\rho} = (\rho_1, \rho_2) = (0.3, 0.6)$.

ρ_{01}	ρ_{02}	$\hat{\sigma}^2(\boldsymbol{\rho}, \boldsymbol{\rho}_o, 0)$	$\bar{\epsilon}^0(\boldsymbol{\rho}, \boldsymbol{\rho}_o)$	$\hat{\sigma}^2(\boldsymbol{\rho}, \boldsymbol{\rho}_o, 1)$	$\bar{\epsilon}^1(\boldsymbol{\rho}, \boldsymbol{\rho}_o)$
0.300	0.600	3.386E-04	1.000	1.850E-01	1.000
0.300	0.620	2.861E-04	0.845	1.654E-01	0.894
0.300	0.640	2,305E-04	0.681	1.095E-01	0.592
0.300	0.660	2.168E-04	0.640	8.916E-02	0.482
0.300	0.680	1.902E-04	0.562	8.008E-02	0.433
0.300	0.700	1.732E-04	$0.511^\diamond$	6.275E-02	$0.329^\diamond$
0.300	0.720	2.006E-04	0.592	8.527E-02	0.461
0.310	0.600	2.880E-04	0.851	1.457E-01	0.787
0.310	0.620	2.840E-04	0.839	1.461E-01	0.790
0.310	0.640	2.330E-04	0.688	9.753E-02	0.527
0.310	0.660	2.183E-04	0.645	9.023E-02	0.488
0.310	0.680	1.867E-04	0.551	7.468E-02	0.404
0.310	0.700	1.734E-04	0.512	6.394E-02	0.346
0.310	0.720	1.682E-04	0.497^*	5.268E-02	0.285^*
0.320	0.600	3.664E-04	1.082	2.091E-01	1.130
0.320	0.620	2.805E-04	0.828	1.587E-01	0.858
0.320	0.640	2.748E-04	0.812	1.651E-01	0.893
0.320	0.660	1.928E-04	0.569	7.565E-02	0.409
0.320	0.680	1.837E-04	0.543	8.015E-02	0.433
0.320	0.700	1.750E-04	0.517	6.974E-02	0.377
0.320	0.720	1.728E-04	0.510	6.151E-02	0.333

[i] **Feedback model** Table 4.3.8 presents data similar to those of table 4.3.7 for the two-node queueing model with the feedback on the second queue. We chose the feedback probability $p = 0.3$; assumed

that the interarrival and two service times distributions are gamma ($\Gamma_i(\beta_i, \lambda_i)$, $i = 1, 2, 3$); chose (β_1, λ_1) =(2, 2); (β_2, λ_2) = (2, 6.666); (β_3, λ_3) =(2, 6.666); and took $\boldsymbol{v}_0 = \boldsymbol{\lambda}_0 = (\lambda_{02}, \lambda_{03})$ as the reference vector.

Table 4.3.8 $\hat{\sigma}^2(\boldsymbol{\rho}, \boldsymbol{\rho}_o, k)$ and $\bar{\epsilon}^k(\boldsymbol{\rho}, \boldsymbol{\rho}_o)$ as functions of λ_{02} and λ_{03} for two $GI/G/1$ queues with feedback on the second queue.

λ_{02}	λ_{03}	$\hat{\sigma}^2(\boldsymbol{\rho}, \boldsymbol{\rho}_o, 0)$	$\bar{\epsilon}^0(\boldsymbol{\rho}, \boldsymbol{\rho}_o)$	$\hat{\sigma}^2(\boldsymbol{\rho}, \boldsymbol{\rho}_o, 1)$	$\bar{\epsilon}^1(\boldsymbol{\rho}, \boldsymbol{\rho}_o)$
6.66	6.66	5.2 E-6	1.00	3.5E-5	1
6.66	5.5	4.11 E-6	0.79$^\diamond$	1.37E-5	0.39
6.66	5.0	4.33 E-6	0.83	1.08E-5	0.31$^\diamond$
6.66	4.5	8.3 E-6	1.59	2.4E-5	0.68
6.66	4.0	1.1 E-5	2.11	7.4E-5	2.11
6.66	3.6	4.6 E-5	8.84	3.6E-4	10.3
6.66	3.2	5.1 E-5	9.81	4.1E-4	11.7
5.5	6.66	4.9E-6	0.94	3.1E-5	0.88
5.5	5.5	4.03E-6	0.77	1.8E-5	0.51
5.5	5.0	3.8E-6	0.73*	1.6E-5	0.48
5.5	4.5	4.9E-6	0.94	8.2E-6	0.23*
5.5	4.0	8.3E-6	1.59	2.1E-5	0.60
5.5	3.6	2.1E-5	4.03	8.1E-5	2.31
5.5	3.2	4.7E-5	9.03	3.4E-4	9.70
5.0	6.66	4.6E-6	0.88	2.6E-5	0.74
5.0	5.5	4.4E-6	0.84	1.4E-5	0.40
5.0	5.0	4.2E-6	0.81	1.8E-5	0.51
5.0	4.5	6.1E-6	1.17	1.3E-5	0.37
5.0	4.0	9.3E-6	1.78	7.9E-5	2.25
5.0	3.6	3.2E-5	6.15	2.5E-4	7.14
5.0	3.2	5.3E-5	10.2	4.4E-4	12.5

It follows from these tables that the optimal vector of traffic intensities $\boldsymbol{\rho}_o{}^* = (\rho_{01}^*, \rho_{02}^*)$ and the minimal value $\sigma^2(\boldsymbol{\rho}, \boldsymbol{\rho}_o^*, k)$ (and thus the minimal value of the relative efficiency $\epsilon^k(\boldsymbol{\rho}, \boldsymbol{\rho}_o{}^*)$) are essentially *insensitive* to the parameters of the non bottleneck queues in the sense that both $\boldsymbol{\rho}_o^* = (\rho_{01}^*, \rho_{02}^*)$ and $\sigma^2((\boldsymbol{\rho}, \boldsymbol{\rho}_o{}^*, k)$ can be *approximated rather well* if we minimize the objective function $\sigma^2(\boldsymbol{\rho}, \boldsymbol{\rho}_o, k)$ with respect to the reference traffic intensity of the bottleneck queue alone, that is $\boldsymbol{\rho}_o^* = (\rho_{01}^*, \rho_{02}^*)$ can be approximated rather well by $\boldsymbol{\rho}_o{}^\diamond = (\rho_1, \rho_{02}^\diamond)$. Notice that $\boldsymbol{\rho}_o{}^\diamond$ corresponds to the optimal solution of (4.1.3) with respect to $\boldsymbol{v}_{02}$ *alone*, that is when $\boldsymbol{v}_1$ (and thus ρ_1) are fixed.

Consider, for example, the results of table 4.3.7. We have respectively

$$\bar{\epsilon}^0(\boldsymbol{\rho}_o^*) = 0.497, \quad \boldsymbol{\rho}_o^* = (\rho_{01}^*, \rho_{02}^*) = (0.31,\ 0.72);$$

$$\bar{\epsilon}^1(\boldsymbol{\rho}_o^*) = 0.285, \quad \boldsymbol{\rho}_o^* = (\rho_{01}^*, \rho_{02}^*) = (0.31,\ 0.72);$$

and

$$\bar{\epsilon}^0(\boldsymbol{\rho}_o{}^\diamond) = 0.511, \quad \boldsymbol{\rho}_o{}^\diamond = (\rho_1, \rho_{02}^\diamond) = (0.30,\ 0.70);$$

$$\bar{\epsilon}^1(\boldsymbol{\rho}_o{}^\diamond) = 0.329, \quad \boldsymbol{\rho}_o{}^\diamond = (\rho_1, \rho_{02}^\diamond) = (0.30,\ 0.70).$$

Note that the first case represents the relative efficiencies $\bar{\epsilon}^k(\boldsymbol{\rho}_o^*)$, $k = 0, 1$, for the optimal reference vector $\boldsymbol{\rho}_o^* = (\rho_{01}^*, \rho_{02}^*)$, while the second represents the relative efficiencies $\bar{\epsilon}^k(\boldsymbol{\rho}_o{}^\diamond)$, $k = 0, 1$, for $\boldsymbol{\rho}_o{}^\diamond = (\rho_1, \rho_{02}^\diamond)$, that is, where ρ_1 is fixed, namely $\rho_1 = 0.3$. (The third case corresponding to $\rho_{01} = 0.32$ is given just for comparison with $\bar{\epsilon}^k(\boldsymbol{\rho}_o^*)$ and $\bar{\epsilon}^k(\boldsymbol{\rho}_o{}^\diamond)$.) It follows from the above that $\bar{\epsilon}^k(\boldsymbol{\rho}_o{}^\diamond)$ approximates $\bar{\epsilon}^k(\boldsymbol{\rho}_o^*)$ rather well, and thus we can adopt $\boldsymbol{\rho}_o{}^\diamond = (\rho_1, \rho_{02}^\diamond)$ as a "good" vector of reference parameters.

Consider now a general r-node queueing network. Assume for simplicity that the first node is the bottleneck one and our goal is again to choose a "good" vector of reference parameters while estimating the performance parameters $\nabla^k \ell(\boldsymbol{v})$. Arguing as before we can solve the program (4.1.3) with respect to the parameters of the bottleneck queue alone, that is with respect to $\boldsymbol{v}_{01}$ and then adopt the vector $\boldsymbol{v}_o{}^\diamond = (\boldsymbol{v}_{01}^\diamond, \boldsymbol{v}_2 \ldots, \boldsymbol{v}_r)$ as the vector of reference parameters. Here $\boldsymbol{v}_{01}^\diamond$ corresponds to the optimal solution of (4.1.3) with respect to $\boldsymbol{v}_{01}$ *alone* and $(\boldsymbol{v}_2, \ldots, \boldsymbol{v}_r)$ is the vector of underlying (nominal) parameters of the remaining $r-1$ non bottleneck queues. Our extensive simulation studies support the above so-called *bottleneck phenomenon*. Notice that if we have more than one bottleneck queue we must increase the dimensionality of the program (4.1.3) accordingly, that is to solve it with respect to all bottleneck queues.

We shall now explain why the optimal vector $\boldsymbol{\rho}_o^*$ and the optimal value function $\epsilon^k(\boldsymbol{\rho}_o^*)$ are typically insensitive with respect to the parameters $(\rho_{02} \ldots, \rho_{0r})$ of the non bottleneck queues. To see this assume, for simplicity, that $k = 0$ and we solve the program (4.1.3) by using the following system of equations:

$$\nabla \sigma^2(\boldsymbol{\rho}, \boldsymbol{\rho}_o) = \boldsymbol{0}. \tag{4.3.16}$$

Since the first component of $\nabla \sigma^2(\boldsymbol{\rho}, \boldsymbol{\rho}_o)$ (the bottleneck component) is *substantially larger* than any of the remaining $r-1$ (non bottleneck) components (see the results of Asmussen (1992), Asmussen and Rubinstein (1992b), Shalmon and Rubinstein (1990) and subsection 3.3.3 on the variance of performance measures and their heavy traffic approximations) the *insensitivity* result follows. Note that since $\sigma^2(\rho, \rho_o)$ is typically analytically not available, we can consider instead of $\nabla\sigma^2(\rho, \rho_o) = \mathbf{0}$ its stochastic counterpart, that is

$$\nabla\hat{\sigma}^2(\rho, \rho_o) = \mathbf{0}, \tag{4.3.17}$$

where $\hat{\sigma}^2(\rho, \rho_o)$ is the estimator of $\sigma^2(\rho, \rho_o)$ and similarly for $k = 1$.

4.4 Estimating Rare Events

We open this section by noting that the conventional approach for estimating rare events in queueing networks (see, for example, Anantharam, Heidelberger and Tsoucas (1990), Bucklew, Ney and Sadowsky (1990), Frater, Lennon and Anderson (1989), Parekh and Walrand (1988), Goyal et al (1992), Shwartz and Weiss (1991), and Walrand (1987)) is similar to our approach in the sense that it also based on changes of the probability measure, but differs from our in the sense that the problem under their study being *transient behavior*, say, mean time to buffer overflow, rather than *steady-state behavior*. Exceptions are Siegmund (1976) and Asmussen (1985), (1990), but the technique there relies on random walk methods which do not generalize to queueing networks. In this literature much attention is given to the optimal choice of the probability measure by referring to the theory of large deviations and exponential change of measure. Our intention here is not to discuss such optimality issues. Rather, we show that the *standard* LR (SF) method can handle rare events rather efficiently by using the dominating pdf in the form $g(\boldsymbol{y}) = f(\boldsymbol{y}, \boldsymbol{v}_0)$, that is without resorting to the exponential change of the probability measure. Before we start with the standard LR (SF) method for estimating rare events the following remark (related to estimation of rare events in the transient regime) is in order. Consider, for example, the probability of buffer overflow within a busy cycle of the $GI/G/1$ queue, that is $P(T_x < \tau)$, where $T_x = \inf\{t : L_t > x\}$. It follows from Theorem 5.1 of Asmussen (1982) that given $T_x < \tau$, the conditional distribution of the interarrival times and service times prior to T_x is asymptotically given by an exponential change of measure which hence is asymptotically optimal for the simulation.

Let now $\{L_t : t > 0\}$ be a steady-state output process, say, the steady-state waiting time process of a customer in a $GI/G/1$ queue, and let the quantity x be chosen such that the probability $P_{\boldsymbol{v}}\{L \geq x\} = \ell(\boldsymbol{v})$ represents a rare event, say, $\ell(\boldsymbol{v})$ is of order 10^{-4} and less. Note that in this case our basic formulas (3.2.27), (3.2.28), (3.2.29) and (3.2.30), originally derived to estimate the standard performance parameters like the expected waiting time $\ell(\boldsymbol{v})$ and the associated sensitivities $\nabla^k\ell(\boldsymbol{v})$, can be modified in a straightforward way to estimate rare events and their associated sensitivities. To do so, we have to replace the process $\{L_t : t > 0\}$ in (3.2.27), (3.2.28), (3.2.29) and (3.2.30) by an associated indicator process $\{I_t : t > 0\}$, where $I_t = I_{\{L_t \geq x\}}$, and I_A denotes the indicator function on the event A, while all other data remain the same. In such case, formulas (3.2.27) and (3.2.29) can be rewritten as

$$\ell(\boldsymbol{v}, \boldsymbol{v}_0, x) = \frac{\mathbb{E}_{\boldsymbol{v}_0}\{\sum_{t=1}^{\tau} I_t \widetilde{W}_t\}}{\mathbb{E}_{\boldsymbol{v}_0}\{\sum_{t=1}^{\tau} \widetilde{W}_t\}} \tag{4.4.1}$$

and

$$\bar{\ell}_N(\boldsymbol{v}, \boldsymbol{v}_0, x) = \frac{\sum_{i=1}^{N} \sum_{t=1}^{\tau_i} I_{ti} \widetilde{W}_{ti}}{\sum_{i=1}^{N} \sum_{t=1}^{\tau_i} \widetilde{W}_{ti}}, \tag{4.4.2}$$

respectively, and similarly $\nabla^k \ell(\boldsymbol{v}, \boldsymbol{v}_0, x)$ and $\nabla^k \bar{\ell}_N(\boldsymbol{v}, \boldsymbol{v}_0, x)$.

Note that in the particular case $f(\boldsymbol{y}, \boldsymbol{v}_0) = f(\boldsymbol{y}, \boldsymbol{v})$, (4.4.1) and (4.4.2) reduce to

$$\ell(\boldsymbol{v}, \boldsymbol{v}, x) = \frac{\mathbb{E}_{\boldsymbol{v}}\{\sum_{t=1}^{\tau} I_t\}}{\mathbb{E}_{\boldsymbol{v}}\{\sum_{t=1}^{\tau} 1\}} \tag{4.4.3}$$

and

$$\bar{\ell}_N(\boldsymbol{v}, \boldsymbol{v}, x) = \frac{\sum_{i=1}^{N} \sum_{t=1}^{\tau_i} I_{ti}}{\sum_{i=1}^{N} \sum_{t=1}^{\tau_i} 1}, \tag{4.4.4}$$

respectively. Note also that for static queueing models like $GI/G/\infty$, (4.4.1) and (4.4.2) reduce to

$$\ell(\boldsymbol{v}, \boldsymbol{v}_0, x) = \mathbb{E}_{\boldsymbol{v}_0}\{I_{\{L \geq x\}}(\boldsymbol{Y}) W(\boldsymbol{Y}, \boldsymbol{v})\} \tag{4.4.5}$$

and

$$\bar{\ell}_N(\boldsymbol{v}, \boldsymbol{v}_0, x) = N^{-1} \sum_{i=1}^{N} I_{\{L \geq x\}}(\boldsymbol{Y}_i) W(\boldsymbol{Y}_i, \boldsymbol{v}), \tag{4.4.6}$$

respectively.

At this point it is crucial to understand that under the *original* pdf $f(\boldsymbol{y}, \boldsymbol{v})$ the rare-event estimators $\nabla^k \bar{\ell}_N(\boldsymbol{v}, \boldsymbol{v}, x)$, $k = 0, 1, \ldots$, typically have a large *run-length ratio* defined (see Whitt (1989a)) as $\sigma_{\boldsymbol{v}}^2 / \ell^2$, where

$$\sigma_{\boldsymbol{v}}^2 = \operatorname{Var}_{\boldsymbol{v}} \left\{ \frac{\sum_{t=1}^{\tau} I_t}{\sum_{t=1}^{\tau} 1} \right\},$$

and thus large associated relative confidence intervals (see (4.3.15)). This means that in order to obtain a meaningful rare-event estimator for $\mathbb{E}_{\boldsymbol{v}} I_{\{L \geq x\}} = P_{\boldsymbol{v}}\{L \geq x\}$ we need an enormous sample. To see this, assume that $\tau = 1$. In this case $L_1, L_2, \ldots$ are iid with the same distribution as L. By the standard formula for Bernoulli sampling, the variance of the Crude Monte Carlo estimator $1/N \sum_{t=1}^{N} I_{\{L \geq x\}}$ is

$$\frac{\ell(\boldsymbol{v}, \boldsymbol{x})(1 - \ell(\boldsymbol{v}, \boldsymbol{x}))}{N} \approx \frac{\ell(\boldsymbol{v}, \boldsymbol{x})}{N}.$$

Assume next that $\ell(\boldsymbol{v}, \boldsymbol{x}) = 10^{-6}$ and that we want to have at most a 25% error on $\ell(\boldsymbol{v}, \boldsymbol{x})$ with 95% confidence. This means that we must have

$$P\{|\ell(\boldsymbol{v}, \boldsymbol{x}) - \bar{\ell}_N(\boldsymbol{v}, \boldsymbol{v}, \boldsymbol{x})| \leq 0.25\ell(\boldsymbol{v}, \boldsymbol{x})\} = 0.95.$$

Using the Central Limit Theorem we obtain that the minimal sample size N is determined by

$$0.25\ell(\boldsymbol{v}, \boldsymbol{x}) \quad \approx \quad 1.96\sqrt{\frac{\ell(\boldsymbol{v}, \boldsymbol{x})}{N}}, \quad \text{i.e. } N \approx \frac{1.96^2}{\ell(\boldsymbol{v}, \boldsymbol{x})0.25^2} \approx 1.3 \cdot 10^7.$$

Hence, even in this simple case the direct simulation would require a very large amount of computation. Another difficulty with rare-event estimators is their bias: if the sample size is not sufficiently large, not only the relative precision will be small, but also the steady-state estimates of $\ell(\boldsymbol{v}, \boldsymbol{x})$ will be typically on the low side (see Asmussen (1992)).

The main idea of estimating rare-events by the SF method is based on the fact that in most queueing applications there exists a subset $V_0 \subset V$, large enough such that $\mathbb{E}_{\boldsymbol{v}_0} I_{\{L \geq x\}} >> \mathbb{E}_{\boldsymbol{v}} I_{\{L \geq x\}}$, provided $\boldsymbol{v}_0 \in V_0$ and $\boldsymbol{v} \in V$ (see also tables 3.5.12–3.5.17 below). This implies that under $f(\boldsymbol{y}, \boldsymbol{v}_0)$, the event $\{L \geq x\}$ is *not rare*. We shall show below that using the "what if" estimator $\nabla^k \bar{\ell}_N(\boldsymbol{v}, \boldsymbol{v}_0, \boldsymbol{x})$, $k = 0, 1$, instead of the standard one $\nabla^k \bar{\ell}_N(\boldsymbol{v}, \boldsymbol{v}, \boldsymbol{x})$, $k = 0, 1$, leads to a dramatic decrease in both the run-length ratio $\sigma^2_{\boldsymbol{v}_0}/\ell^2$, where

$$\sigma^2_{\boldsymbol{v}_0} = \mathrm{Var}_{\boldsymbol{v}_0} \left\{ \frac{\sum_{t=1}^{\tau} I_t \widetilde{W}_t}{\sum_{t=1}^{\tau} \widetilde{W}_t} \right\},$$

and the associated relative efficiency, which can be written in analogy to (4.1.6) as

$$\epsilon^k(\boldsymbol{v}, \boldsymbol{v}_0, \boldsymbol{x}) = \frac{\mathrm{Var}_{\boldsymbol{v}_0} \ \nabla^k \bar{\ell}_N(\boldsymbol{v}, \boldsymbol{v}_0, \boldsymbol{x}) t(\boldsymbol{v}_0)}{\mathrm{Var}_{\boldsymbol{v}} \ \nabla^k \bar{\ell}_N(\boldsymbol{v}, \boldsymbol{v}, \boldsymbol{x}) t(\boldsymbol{v})}, \quad k = 0, 1. \tag{4.4.7}$$

Here, as before, $t(\boldsymbol{v}_0)$ and $t(\boldsymbol{v})$ are the CPU times needed by the "what if" and the standard estimator. As before we assume that $t(\boldsymbol{v}_0) = t(\boldsymbol{v})$.

To provide a better insight into how much variance reduction one can get using the optimal value of the reference parameter vector $\boldsymbol{v}_0^*$ instead of the nominal one consider the following simple example.

Example 4.4.1 $M/M/\infty$ **queue** Let L be the expected sojourn time in the $M/M/\infty$ queue with service rate μ. In this case $\ell(\mu) = \mathrm{P}_\mu\{L > x\} = \mathbb{E}_\mu I_{\{L>x\}} = \mathbb{E}_\mu\{\varphi(L)\} = \mathrm{e}^{-\mu x}$, where $\varphi(\cdot) = I_{(x,\infty)}(\cdot)$, and we have

$$\mathcal{L}^0(\mu, \mu_0) = \mathbb{E}_{\mu_0}\{\varphi(L)^2 W(\mu)^2\} = \frac{\mu^2 \mathrm{e}^{-(2\mu-\mu_0)x}}{\mu_0(2\mu - \mu_0)}.$$

The optimal value of the reference parameter $\mu_0^* = \mu_0^*(x)$ which minimizes $\mathcal{L}^0(\mu, \cdot)$ is

$$\mu_0^*(x) = \mu + x^{-1} - (\mu^2 + x^{-2})^{1/2}.$$

Suppose that $x >> \mu$ and hence $\{L > x\}$ is a rare-event, that is the probability $\mathrm{P}_\mu\{L > x\}$ is small, say, less than 10^{-6}. In this case we have that $\mu_0^*(x) \approx x^{-1}$ and $\mathbb{E}_{\mu_0^*} I_{\{L>x\}} \approx \mathrm{e}^{-1}$.

Consider the relative efficiency $\epsilon(\mu, \mu_0, x)$, defined as

$$\epsilon(\mu, \mu_0, x) = \frac{\mathrm{Var}_{\mu_0}\ \bar{\ell}_N(\mu, \mu_0, x)}{\mathrm{Var}_\mu\ \bar{\ell}_N(\mu, \mu, x)}.$$

For $\mu_0 = \mu_0^*$ its value is given by

$$\epsilon^* = \epsilon(\mu, \mu_0^*, x) \approx 0.5x\mu \mathrm{e}^{1-\mu x}.$$

Let, for example, $\mu = 1$ and $x = 12$. We have then

$$\mathbb{E}_{\mu=1} I_{\{L>12\}} = P_{\mu=1}\{L > 12\} = \mathrm{e}^{-12} \approx 10^{-6},$$

$\mu_0^*(x) \approx 1/12$ and $\epsilon^* = \epsilon(1, \mu_0^*, 12) \approx 10^{-4}$. Thus, using the optimal value $\mu_0^* \approx 1/12$ we obtain dramatic a variance reduction, namely of the order of 10^4.

Consider now the squared coefficient of variation of the LR estimators

$$\kappa(\mu, \mu_0, x) = \frac{N\ Var_{\mu_0} \bar{\ell}_N(\mu, \mu_0, x)}{\ell(\mu)^2}.$$

It is not difficult to calculate that in our case

$$\kappa(\mu, \mu_0, x) = \frac{\mu^2 \mathrm{e}^{\mu_0 x}}{\mu_0(2\mu - \mu_0)} - 1.$$

Therefore for large x the asymptotic behavior of $\kappa(\mu, \mu_0, x)$ for the crude Monte Carlo estimators with $\mu_0 = \mu$ and the optimal LR estimators with $\mu_0 = \mu_0^*$ is given as

$$\kappa(\mu, \mu, x) \approx \mathrm{e}^{\mu x}$$

and

$$\kappa(\mu, \mu_0^*, x) \approx 0.5x\mu \mathrm{e},$$

respectively. That is, for large x, the squared coefficient of relative variation of the crude Monte Carlo and the optimal LR estimators increases in x *exponentially* and *linearly*, respectively.

It is also not difficult to see that if we choose $\mu_0 = k\mu_0^*$ (instead of the optimal $\mu_0 = \mu_0^*$), then for large x we obtain

$$\kappa(\mu, \mu_0, x) \approx 0.5k^{-1}x\mu e^k.$$

For example, if $k = 2$, that is $\mu_0 = 2\mu_0^* \approx 2/x$, we obtain

$$\kappa(\mu, \mu_0, x) \approx 0.25x\mu e^2.$$

In this case the relative efficiency

$$\epsilon(\mu_0^*, \mu_0, x) = \frac{\mathrm{Var}_{\mu_0}\, \bar{\ell}_N(\mu, \mu_0, x)}{\mathrm{Var}_{\mu_0^*}\, \bar{\ell}_N(\mu, \mu_0^*, x)} \approx 2e.$$

That is, perturbing μ_0^* by 100% we increase the variance approximately only 2e times.

To proceed further we consider (as in section 3.5.2) separately (I) a single-node queue and (II) a queueing network.

(I) **Single-node queue.** Asmussen, Rubinstein and Wang (1992) derived an explicit expression for the variance of

$$\bar{\ell}_N(v, v_0, x) = \frac{\sum_{t=1}^{\tau} I_t \widetilde{W}_t}{\sum_{t=1}^{\tau} \widetilde{W}_t},$$

where $I_t = I_{\{L_t \geq x\}}$ and $\{L_t\}$ is the steady-state waiting time process in the $M/G/1$ queue with constant interarrival rate λ and $\boldsymbol{v}_0$ being the reference parameter vector in the service time distribution ($\rho_0 = \mathbb{E}_{\boldsymbol{v}_0} Y$, where Y is the service time random variable).

Table 4.4.1, taken from Asmussen, Rubinstein and Wang (1992), displays the optimal value of the reference parameter ρ_0^* and the optimal relative efficiency ϵ^* as functions of $\ell(v) = P_v\{L > x\}$ for $\rho = 0.3, 0.6$ and 0.9.

Table 4.4.1 The optimal value of the reference parameter ρ_0^* and the optimal relative efficiency ϵ^* as functions of $\ell(v) = P_v\{L > x\}$ for $\rho = 0.3, 0.6$ and 0.9.

ρ	$P_v\{L > x\}$	ρ_0^*	ϵ^*
0.3	10^{-1}	0.395	7.22E-01
	10^{-3}	0.52	9.51E-02
	10^{-5}	0.56	9.82E-03
	10^{-7}	0.58	9.66E-04
	10^{-9}	0.591	9.30E-05
	10^{-11}	0.600	8.84E-06
	10^{-13}	0.606	8.35E-07
	10^{-15}	0.610	7.86E-08
0.6	10^{-1}	0.711	5.78E-01
	10^{-3}	0.777	9.84E-02
	10^{-5}	0.795	1.57E-02
	10^{-7}	0.804	2.46E-03
	10^{-9}	0.811	3.81E-04
	10^{-11}	0.815	5.84E-05
	10^{-13}	0.818	8.91E-06
	10^{-15}	0.820	1.36E-06
0.9	10^{-1}	0.937	5.36E-01
	10^{-3}	0.951	1.22E-01
	10^{-5}	0.955	2.58E-02
	10^{-7}	0.958	5.34E-03
	10^{-9}	0.959	1.09E-03
	10^{-11}	0.960	2.22E-04
	10^{-13}	0.9605	4.50E-05
	10^{-15}	0.961	9.09E-06

Let, for example, $P_v\{L > x\} = 10^{-9}$. We have then $\epsilon^* = 9.3 \cdot 10^{-5}$, $3.81 \cdot 10^{-4}$, $1.09 \cdot 10^{-3}$ for $\rho = 0.3$, 0.6 and 0.9, respectively. The resulting variance reduction is therefore $9.3 \cdot 10^5$, $3.81 \cdot 10^4$, $1.09 \cdot 10^3$. This implies that for the above selected values of ρ and any fixed value of the confidence interval, the LR estimator $\bar{\ell}_N(v, v_0, x)$ is faster than the crude Monte Carlo estimator by the factor of $(9.3 \cdot 10^5)^{1/2}$, $(3.81 \cdot 10^4)^{1/2}$ and $(1.09 \cdot 10^3)^{1/2}$, respectively. It is also important to note that the amount of variance reduction (the reciprical of ϵ^*) increases as $P_v\{L > x\}$ (the value of the rare-event) decreases. Take, for example, $\rho = 0.6$. We have $\epsilon^* = 2.46 \cdot 10^{-3}$, $5.84 \cdot 10^{-5}$, $1.36 \cdot 10^{-6}$ for $P_v\{L > x\} = 10^{-7}$, 10^{-11}, 10^{-15}, respectively.

Although the $M/G/1$ setup is quite specialized, we found (via simulation) that the above properties of the LR estimators hold for more general models, allowing estimation of rare-events up to the order of 10^{-10} in a reasonable amount of time. We show below that the results of table 4.4.1 shed some light on and provide a basic insight into the more complicated queueing models, and in particular how to choose a "good" set of reference parameters.

Table 4.4.2 displays point estimators $\bar{\ell}_N(v, v_0, x)$, sample variances $\hat{\sigma}^2(v, v_0, x)$ and the relative widths w_r of the 95% confidence intervals as a function of x for the $M/M/1$ queue with $\rho = 1/\mu = 0.6$. We chose $\mu_0 = 1.21$, $(\rho_0 = \lambda/\mu_0 = 0.826)$ and simulated $N = 10^7$ customers.

Table 4.4.2 $\bar{\ell}_N(v, v_0, x)$ and $\hat{\sigma}^2(v, v_0, x)$, as functions of x for the $M/M/1$ queue with $\rho = 0.6$.

x	$\bar{\ell}_N(v, v_0, x)$	$\hat{\sigma}^2(v, v_0, x)$	$w_r(\%)$
15	2.73E-05	1.52E-12	17.6
20	8.01E-07	2.69E-15	25.3
25	2.60E-08	1.11E-17	50.3
31	6.62E-10	7.09E-20	157.4

It is readily seen that the results of this table are in agreement with those of table 4.4.1 (see $\rho = 0.6$). Similar results were obtained for the $M/M/1$ queue with $\rho = 0.3$ and $\rho = 0.9$ and for other single-node queues.

Table 4.4.3 displays point estimators $\nabla^k \bar{\ell}_N(\rho, \rho_0, x)$, $k = 0, 1$, the associated sample variance denoted $\hat{\sigma}^2(\rho, \rho_0, x)$, along with the point estimator $\bar{P}_{\rho_0}\{L > x\}$ of $\ell(\rho_0) = P_{\rho_0}\{L > x\}$ as functions of $\rho_0 = \lambda/\mu_0$ for the $M/M/1$ queue with $\rho = 1/\mu = 0.3$, $(\lambda = 1)$ and $x = 3$, while simulating $N = 4 \cdot 10^6$ customers. Table 4.4.4 displays similar data for $x = 4$.

Table 4.4.3 $\nabla \bar{\ell}_N(\rho, \rho_0, x)$, $k = 0, 1$, and $\hat{\sigma}^2(\rho, \rho_0, x)$, $k = 0, 1$, as functions of ρ_0 for the $M/M/1$ queue with $\rho = 0.3$ and $x = 3$.

ρ_0	$\bar{P}_{\rho_0}(L > 3)$	$\bar{\ell}_N(\rho, \rho_0, x)$	$\hat{\sigma}^2(\rho, \rho_0, 0, x)$	$\nabla \bar{\ell}_N(\rho, \rho_0, x)$	$\hat{\sigma}^2(\rho, \rho_0, 1, x)$
0.3	0.00	0.00	0.00	0.00	0.00
0.4	0.011	2.3E-4	4.8E-11	-8.57E-4	7.8.7E-11
0.5	0.025	2.7E-4	9.9E-12	-9.0 E-4	7.5E-11
0.55	0.042	2.6E-4	7.2E-12*	-8.9E-4	5.7E-11*
0.58	0.076	2.7E-4	1.3E-11	-9.1E-4	6.1E-11
0.625	0.115	2.7E-4	9.9E-12	-9.0E-4	6.5E-11
0.715	0.22	2.7E-4	1.8E-11	-9.1E-4	1.1E-10
0.8	0.37	2.8E-4	2.2E-11	-9.2E-4	1.7E-10
0.9	0.64	2.7E-4	7.5E-11	-9.2E-4	6.7E-10

Table 4.4.4 $\nabla \bar{\ell}_N(\rho, \rho_0, x)$, $k = 0, 1$, and $\hat{\sigma}^2(\rho, \rho_0, x)$, $k = 0, 1$, as functions of ρ_0 for the $M/M/1$ queue with $\rho = 0.3$ and $x = 4$.

ρ_0	$\bar{P}_{\rho_0}(L > 4)$	$\bar{\ell}_N(\rho, \rho_0, x)$	$\hat{\sigma}^2(\rho, \rho_0, 0, x)$	$\nabla \bar{\ell}_N(\rho, \rho_0, x)$	$\hat{\sigma}^2(\rho, \rho_0, 1, x)$
0.3	0.00	0.00	0.00	0.00	0.00
0.4	0.003	2.3E-5	4.2E-12	-0.09E-4	5.2E-12
0.5	0.01	2.7E-5	3.7E-13	-1.1E-4	3.2E-12
0.55	0.021	2.5E-5	2.5E-13	-1.1E-4	3.1E-12
0.58	0.030	2.6E-4	2.3E-13*	-1.0 E-4	3.0E-12*
0.625	0.043	2.5E-5	2.4E-13	-1.1E-4	3.2E-12
0.715	0.16	2.6E-5	4.1E-13	-1.1E-4	4.4E-12
0.8	0.29	2.6E-5	4.9E-13	-1.1E-4	6.2E-11
0.9	0.48	2.5E-5	6.4E-12	-1.0E-4	7.8E-11

Table 4.4.5 displays data similar to those of table 4.4.3 for the $M/G/1$ queue with $\rho = 0.6$ and $x = 12$. We assumed that the interarrival rate equals 1, the service time distribution is gamma (denoted $\Gamma(\beta, \lambda)$), chose $(\beta, \lambda) =(2, 3.333)$, took the scale parameter $\lambda_0 = v_0$ as the reference parameter, and simulated as before $4 \cdot 10^6$ customers.

Table 4.4.5 $\nabla^k \bar{\ell}_N(\rho, \rho_0, x)$, $k = 0, 1$, and $\hat{\sigma}^2(\rho, \rho_0, x)$, $k = 0, 1$, as functions of ρ_0 for the $M/G/1$ queue with $\rho = 0.6$ and $x = 12$.

ρ_0	$\bar{P}_{\rho_0}(L > 12)$	$\bar{\ell}_N(\rho, \rho_0, x)$	$\hat{\sigma}^2(\rho, \rho_0, 0, x)$	$\nabla \bar{\ell}_N(\rho, \rho_0, x)$	$\hat{\sigma}^2(\rho, \rho_0, 1, x)$
0.60	0.00	0.00	0.00	0.00	0.00
0.65	0.001	3.7E-6	3.6E-13	-3.87E-5	6.8E-11
0.70	0.002	6.4E-6	3.3E-13	-8.6E-5	4.7E-11
0.75	0.003	6.8E-6	2.1E-13	-9.0E-5	4.9E-11
0.77	0.005	6.9E-6	1.1E-13	-9.1E-5	3.9E-11
0.80	0.013	6.7E-6	3.6E-14*	-9.0E-5	5.2E-12*
0.82	0.023	7.1E-6	5.6E-14	-9.5E-5	1.0E-11
0.85	0.045	7.5E-6	1.4E-13	-9.4E-5	1.4E-11
0.90	0.15	6.5E-6	2.2E-13	-8.9E-5	2.1E-11

Note that an additional amount of variance reduction can be obtained by using both λ_0 and β_0 as reference parameters. For example, consider the above $M/G/1$ queue with $\rho = 0.6$. Choosing λ_0 as in table 4.4.5, that is $\lambda_0 = \beta/\rho_0 = 2/\rho_0$, and $2.2 \leq \beta_0 \leq 2.45$ (instead of $\beta = 2$) we obtained a variance reduction at least 10 times.

Based on the results of table 4.4.5, fig. 4.4.1 depicts $\bar{\ell}_N(\rho, \rho_0, x)$ (denoted ℓ_N) along with the 95% (relative) confidence intervals (denoted 95%CI) as functions of ρ_0.

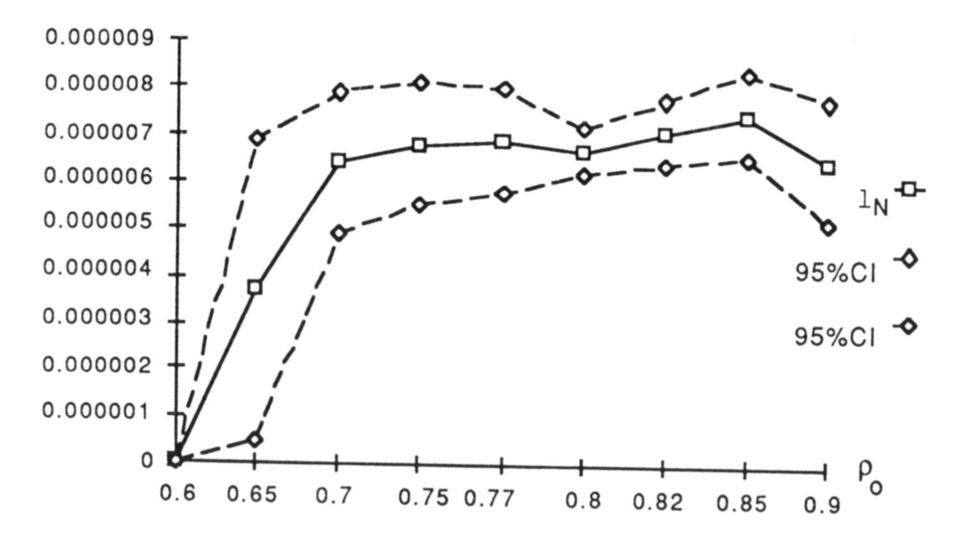

Figure 4.4.1 Performance of the "what if" estimator $\bar{\ell}_N(\rho, \rho_0, x)$ as a function of ρ_0 for the $M/G/1$ queue with $\rho = 0.6$ and $x = 12$.

The results of these tables and figures are self-explanatory. It is important to note that under the original pdf $f(\boldsymbol{y}, \boldsymbol{v})$, all estimators in tables 4.4.3 –4.4.5 result in zero.

Based on the above results and our extensive simulation studies with simple queues, like the $GI/G/1$ queue, we found that

(i) Under the original pdf $f(\boldsymbol{y}, \boldsymbol{v})$, the standard regenerative estimator $\bar{\ell}_N(\boldsymbol{v}, \boldsymbol{v}, x)$ typically *underestimates* the underlying quantity $\mathbb{E}_\rho I_{\{L \geq x\}}$, provided $\mathbb{E}_\rho I_{\{L \geq x\}} \leq 10^{-2}$, and when estimating rare-events, say, when $\mathbb{E}_\rho I_{\{L \geq x\}} \leq 10^{-4}$, the estimator $\bar{\ell}_N(\boldsymbol{v}, \boldsymbol{v}, x)$ is useless (in all cases $\bar{\ell}_N(\boldsymbol{v}, \boldsymbol{v}, x)$ resulted in zero, provided the sample size $N \leq 4 \cdot 10^6$).

(ii) There exists a rather broad interval in the form

$$\Delta\rho^c = \{(\rho^c, \rho_n) : \rho \leq \rho^c \leq \rho_0 \leq \rho_n \leq 1\}, \tag{4.4.8}$$

where the estimator $\bar{\ell}_N(\rho, \rho_0, x)$ consistently estimates the rare-event $\ell(\rho) = \mathbb{E}_\rho I_{\{L \geq x\}}$. Here $\rho^c = \rho^c(\boldsymbol{v}_n, x)$ and $\rho_n = \rho_n(\boldsymbol{v}_n, x)$ are chosen such that

$$\mathrm{Var}_{\rho_n} \bar{\ell}_N(\rho, \rho_n, x) = \mathrm{Var}_{\rho^c} \bar{\ell}_N(\rho, \rho^c, x), \tag{4.4.9}$$

and as mentioned in the introduction, they are called the *critical* and *neutral* traffic intensity parameters, respectively.

(iii) In order to be on the safe side, it is desirable to choose the reference traffic intensity ρ_0 moderately larger than ρ^c, where ρ^c by itself must be chosen moderately larger than ρ.

We also found that once chosen a "good" reference traffic intensity ρ_0 for estimating rare-events can serve *simultaneously* as a "good" reference traffic intensity for estimating standard performance parameters, like the waiting time. For example, using $\rho_0^* = 0.80$ as a reference traffic intensity for estimating a rare-event in the $M/G/1$ queue with $\rho = 0.6$ (see table 4.4.5), we can, at the same time, improve the efficiency (accuracy) of its LR (SF) estimators for estimating the expected waiting time (see table 4.1.1).

It again seems reasonable to ask why, for a given traffic intensity ρ, one obtains a variance reduction by choosing ρ_0 moderately larger than ρ^c ($\rho^c > \rho$). The answer is in terms of importance sampling. Namely, when estimating a rare-event $\ell(\rho)$, the main contributors typically come from the cycles which are much longer than average, and when choosing a reference parameter $\rho_0 > \rho^c$ greater weight is assigned to large cycles. The reason that the optimal ρ_0 is not substantially larger than ρ is that a too large ρ_0 will increase the variance of the estimator of $\ell(\rho)$.

Note that if the process $\{L_t\}$ is stationary and ergodic rather than regenerative, then we can estimate rare-events by using the truncated estimators discussed in subsection 3.3.2.

(II) **Queueing networks** We shall show now numerically that the SF method can handle rare-events rather efficiently while estimating rare-events for queueing networks and in particular that the "insensitivity" phenomenon observed earlier for the standard performance parameters $\nabla^k \ell(\boldsymbol{v})$, where $\ell(\boldsymbol{v})$, is say, the expected steady-state waiting time, is even more prevalent here. As before we consider separately two queues in tandem and two queues with a feedback on the second queue.

Two queues in tandem Table 4.4.6 presents point estimators $\bar{\ell}_N(\boldsymbol{\rho}, \boldsymbol{\rho}_o, \boldsymbol{x})$ of $\ell(\boldsymbol{\rho}, \boldsymbol{\rho}_o, \boldsymbol{x}) = \mathbb{E}_{\boldsymbol{\rho}} I_{\{L \geq \boldsymbol{x}\}}$, the associated sample variance denoted $\hat{\sigma}^2(\boldsymbol{\rho}, \boldsymbol{\rho}_o, \boldsymbol{x})$ along with the point estimators $\bar{P}_{\boldsymbol{\rho}_o}(L > \boldsymbol{x})$ of $\ell(\boldsymbol{\rho}_o, \boldsymbol{x}) = \mathbb{E}_{\boldsymbol{\rho}_o} I_{\{L \geq \boldsymbol{x}\}}$ as functions of λ_{02} and λ_{03} for two $GI/G/1$ queues in tandem with $\rho_1 = 0.2$ and $\rho_2 = 0.3$. We assumed that $\boldsymbol{x} = 3.5$, the interarrival and two service times distributions are gamma (denoted $\Gamma_i(\beta_i, \lambda_i)$, $i = 1, 2, 3$). We chose $(\beta_1, \lambda_1) =(4, 1)$; $(\beta_2, \lambda_2) = (4, 3.333)$; and $(\beta_3, \lambda_3) =(4, 1.667)$; took $\boldsymbol{v}_0 = \boldsymbol{\lambda}_o = (\lambda_{02}, \lambda_{03})$ as the reference vector; and simulated 4×10^6 customers.

Table 4.4.6 $\bar{\ell}_N(\rho, \rho_o, x)$ and $\hat{\sigma}^2(\rho, \rho_o, x)$ as functions of λ_{02} and λ_{03} for two $GI/G/1$ queues in tandem with $\rho_1 = 0.2$ and $\rho_2 = 0.3$.

ρ_{01}	ρ_{02}	$\bar{P}_{\rho_o}(L > 3.5)$	$\bar{\ell}_N(\rho, \rho_o, x)$	$\hat{\sigma}^2(\rho, \rho_o, x)$
0.2	0.6	0.03	7.7E-5	2.7E-11
0.2	0.66	0.08	7.3E-5	2.4E-11
0.2	0.71	0.13	7.5E-5	2.3E-11$^\diamond$
0.2	0.77	0.22	7.1E-5	7.6E-11
0.2	0.85	0.35	7.2E-5	9.2E-11
0.2	0.9	0.56	6.9E-5	2.7E-10
0.22	0.6	0.03	7.6E-5	2.6E-11
0.22	0.66	0.08	7.3E-5	2.2E-11*
0.22	0.71	0.14	7.5E-5	2.3E-11
0.22	0.77	0.23	7.2E-5	7.8E-11
0.22	0.85	0.36	7.2E-5	9.4E-11
0.22	0.9	0.58	7.0E-5	2.9E-10
0.24	0.6	0.03	7.5E-5	2.6E-11
0.24	0.66	0.09	7.2E-5	2.4E-11
0.24	0.71	0.15	7.4E-5	2.5E-11
0.24	0.77	0.24	7.2E-5	7.8E-11
0.24	0.81	0.37	7.1E-5	9.5E-11
0.24	0.9	0.59	6.8E-5	3.7E-10

Feedback model Table 4.4.7 presents data similar to those of table 4.4.6 for the two-node queueing model with feedback on the second queue. We chose the feedback probability $p = 0.3$; assumed that the interarrival and two service times distributions are gamma $(\Gamma_i(\beta_i, \lambda_i),\ \ i = 1, 2, 3)$; chose (β_1, λ_1) =(2, 2); (β_2, λ_2) = (2, 6.666); (β_3, λ_3) =(2, 6.666); took $\boldsymbol{v}_0 = \boldsymbol{\lambda}_o = (\lambda_{02}, \lambda_{03})$ as the reference vector; and simulated $N = 4 \cdot 10^6$ customers with $x = 5$ and $x = 6$, respectively.

Table 4.4.7 $\bar{\ell}_N(\rho, \rho_o, x)$ and $\hat{\sigma}^2(\rho, \rho_o, x)$ as functions of λ_{02} and λ_{03} for two $GI/G/1$ queues with the feedback on the second queue.

λ_{02}	λ_{03}	$\bar{P}(L > 5)$	$\bar{\ell}_N(\rho_o, x)$	$\hat{\sigma}^2(\rho, \rho_o, x)$	$\bar{P}(L > 6)$	$\bar{\ell}_N(\rho_o, x)$	$\hat{\sigma}^2(\rho, \rho_o,$
6.66	4.5	0.008	8.4E-5	1.7E-10	0.005	2.8E-5	4.5E-11
6.66	4.2	0.016	6.8E-5	8.7E-11	0.011	1.9E-5	3.3E-11
6.66	4.0	0.026	6.6E-5	7.7E-11$^\diamond$	0.017	2.1E-5	1.8E-11
6.66	3.8	0.045	6.6E-5	2.1E-10	0.029	2.0E-5	4.6E-12$^\diamond$
6.66	3.6	0.078	6.8E-5	4.3E-10	0.055	5.4E-5	8.7E-12
6.66	3.4	0.140	5.1E-5	6.2E-10	0.112	8.3E-5	1.1E-11
6.66	3.2	0.270	4.7E-5	7.2E-10	0.220	1.5E-6	1.8E-11
6.5	4.5	0.008	8.2E-5	1.9E-10	0.005	2.6E-5	4.2E-11
6.5	4.2	0.017	7.1E-5	8.5E-11	0.012	1.9E-5	3.1E-11
6.5	4.0	0.027	6.9E-5	7.5E-11*	0.019	2.3E-5	1.7E-11
6.5	3.8	0.047	6.8E-5	1.3E-10	0.032	2.1E-5	4.4E-12*
6.5	3.6	0.081	6.6E-5	3.2E-10	0.059	5.5E-5	6.4E-12
6.5	3.4	0.146	5.4E-5	5.7E-10	0.120	8.6E-5	9.2E-12
6.5	3.2	0.282	4.9E-5	7.1E-10	0.231	1.4E-6	1.9E-11
6.4	4.5	0.008	8.5E-5	2.4E-10	0.005	2.4E-5	4.7E-11
6.4	4.2	0.019	6.9E-5	8.9E-11	0.013	2.0E-5	3.2E-11
6.4	4.0	0.028	6.6E-5	8.2E-11	0.021	2.0E-5	1.7E-11
6.4	3.8	0.049	6.7E-5	9.8E-11	0.035	3.2E-5	5.8E-12
6.4	3.6	0.084	6.7E-5	2.4E-10	0.063	4.7E-5	8.4E-12
6.4	3.4	0.151	5.7E-5	4.8E-10	0.127	9.1E-5	1.3E-11
6.4	3.2	0.293	5.1E-5	6.3E-10	0.240	1.6E-6	2.4E-12

It is readily seen from the results of table 4.4.6 and table 4.4.7 that they are similar to those of table 4.3.7 and table 4.3.8 in the sense that the optimal vector of traffic intensities $\rho_o^* = (\rho_{01}^*, \rho_{02}^*)$ and the optimal value of $\sigma^2(\rho, \rho_o^*, x)$ are essentially *insensitive* to the parameters of the non bottleneck queues in the sense that both $\rho_o^* = (\rho_{01}^*, \rho_{02}^*)$ and $\sigma^2(\rho, \rho_o^*, x)$ can be *approximated rather well* if we minimize $\sigma^2(\rho, \rho_o, k, x)$ with respect to the reference traffic intensity of the bottleneck queue alone (see the values of $\hat{\sigma}^2(\rho, \rho_o, x)$ marked with $\diamond$) provided, of course, such a bottleneck queue exists.

We also performed extensive simulation studies with different queueing networks and found that the bottleneck phenomenon is even more prevalent for rare-events. This is because the relationship between the bottleneck component of $\nabla \sigma^2(\rho, \rho_o, x)$ and any of the remaining $r - 1$ (non bottleneck) components is more prevalent in this case (see also (4.3.16) and (4.3.17)).

Note finally that our tables clearly indicate that one can estimate *simultaneously and rather efficiently from a single run both* the standard performance $\nabla^k \ell(\boldsymbol{v})$, where $\ell(\boldsymbol{v})$ is, say, the expected waiting time and the associated rare-event. Consider, for example, tables 4.1.1 and 4.4.5, corresponding to the $M/G/1$ queue with $\rho = 0.6$. It is readily seen that the expected waiting time, the rare-event (of order 10^{-5}) and their associated sensitivities can be estimated simultaneously by simulating the $M/G/1$ queue *only* once, say, with the reference traffic intensity $\rho = 0.8$.

4.5 Extensions

We introduce now two simple modifications for the choice of the dominating pdf

$$g_t(\underline{\boldsymbol{z}}_t) = f_t(\underline{\boldsymbol{z}}_t, \boldsymbol{v}_0) = \prod_{j=1}^{t} f(\boldsymbol{z}_j, \boldsymbol{v}_0)$$

and the associated likelihood ratios $\widetilde{W}_t$. Our first modification is suitable for any standard performance measure of the form $\nabla^k \ell(\boldsymbol{v}) = \nabla^k \mathbb{E}_{\boldsymbol{v}}\{\varphi(L)\}$ (including rare-events), while the second one involves some additional random variables associated with level crossing, and is suitable for specific performance measures like rare-events (see below). The idea behind both modifications is to introduce more congestion at the beginning of the cycle by making the corresponding servise or arrival rates (or both) controllable.

Our first modification of $g_t(\underline{\boldsymbol{z}}_t)$ and $\widetilde{W}_t$ can be written as

$$g_t(\underline{\boldsymbol{z}}_t) = f_t(\underline{\boldsymbol{z}}_t, \underline{\boldsymbol{v}}_{0t}) = \prod_{j=1}^{t} f(\boldsymbol{z}_j, \boldsymbol{v}_{0j}), \tag{4.5.1}$$

and

$$\widetilde{W}_t = \widetilde{W}_t(\underline{\boldsymbol{Z}}_t, \boldsymbol{v}, \underline{\boldsymbol{v}}_{0t}) = \prod_{j=1}^{t} W_j(\boldsymbol{Z}_j, \boldsymbol{v}, \boldsymbol{v}_{0j}), \tag{4.5.2}$$

respectively. Here the LR function

$$W_j(\boldsymbol{Z}_j, \boldsymbol{v}, \boldsymbol{v}_{0j}) = f(\boldsymbol{Z}_j, \boldsymbol{v}) / f(\boldsymbol{Z}_j, \boldsymbol{v}_{0j}),$$

depends on the reference parameter vector $\boldsymbol{v}_{0j}$ which can *vary* for different values of $j = 1, \ldots$. The vector $\underline{\boldsymbol{v}}_{0t} = (\boldsymbol{v}_{01}, \boldsymbol{v}_{02}, \ldots, \boldsymbol{v}_{0t})$ represents the history of the chosen reference parameters up to the time t, and $\underline{\boldsymbol{Z}}_t = (\boldsymbol{Z}_1, \ldots, \boldsymbol{Z}_t) \sim f_t(\underline{\boldsymbol{z}}_t, \underline{\boldsymbol{v}}_{0t})$.

For example, let Y_1 and Y_2 be the service and interarrival time random variables in a $GI/G/1$ queue depending on the parameters $v_0^{(1)} = \mathbb{E}Y_1$ and

$v_0^{(2)} = \mathbb{E}Y_2$, respectively. We argue heuristically that there exist sequences $v_{01}^{(1)} \geq v_{02}^{(1)} \geq \cdots \geq v_{0,\tau}^{(1)}$ and $v_{01}^{(2)} \leq v_{02}^{(2)} \leq \cdots \leq v_{0,\tau}^{(2)}$ such that the SF estimator $\nabla^k \bar{\ell}_N(\boldsymbol{v})$, with $\widetilde{W}_t$ given in (4.5.2), will outperform the standard SF estimator $\nabla^k \bar{\ell}_N(\boldsymbol{v})$ with

$$\widetilde{W}_t = \widetilde{W}_t(\boldsymbol{v}, \boldsymbol{v}_0) = \prod_{j=1}^{t} W_j(\boldsymbol{v}, \boldsymbol{v}_0).$$

As a particular case of (4.5.1) consider the following so-called *switching* pdf

$$g_t(\underline{\boldsymbol{z}}_t, \alpha) = \prod_{j=1}^{\alpha} f(\boldsymbol{z}_j, \boldsymbol{v}_{01}) \prod_{j=\alpha+1}^{t} f(\boldsymbol{z}_j, \boldsymbol{v}_{02}), \tag{4.5.3}$$

where α is a *fixed* integer.

In our second modification, $g_t(\underline{\boldsymbol{z}}_t)$ and $\widetilde{W}_t$ are chosen as

$$g_t(\underline{\boldsymbol{z}}_t, \zeta) = \prod_{j=1}^{\zeta} f(\boldsymbol{z}_j, \boldsymbol{v}_{01}) \prod_{j=\zeta+1}^{t} f(\boldsymbol{z}_j, \boldsymbol{v}_{02}), \tag{4.5.4}$$

and

$$\widetilde{W}_t(\zeta) = \begin{cases} \prod_{j=1}^{t} W(\boldsymbol{z}_j, \boldsymbol{v}_{01}), & \text{if } t \leq \zeta, \\ \prod_{j=1}^{\zeta} W(\boldsymbol{z}_j, \boldsymbol{v}_{01}) \prod_{j=\zeta+1}^{t} W(\boldsymbol{z}_j, \boldsymbol{v}_{02}), & \text{if } \zeta < t \leq \tau, \end{cases} \tag{4.5.5}$$

respectively. Here ζ is a *random variable*, defined as

$$\zeta = \zeta^{\delta} = \inf\{t : \sum_{j=1}^{t} I_{\{L_t = x\}} = \delta\},$$

where δ is a fixed integer. For example, think of L_t and x as the queue length just prior customer arrivals and the buffer size in the $GI/G/1/b$ queue, wehere $b = x$. Note that the swithing case (4.5.3) differs from that of (4.5.4) since in the first case α is fixed, while in the second case ζ is a random variable (a stopping time for the process L_t). In the particular case where $\delta = 1$ and $\boldsymbol{v}_{02} = \boldsymbol{v}$, (4.5.5) reduces to

$$\widetilde{W}_t = \begin{cases} \prod_{j=1}^{t} W(\boldsymbol{z}_j, \boldsymbol{v}_{01}), & \text{if } t \leq \zeta^1, \\ \prod_{j=1}^{\zeta^1} W(\boldsymbol{z}_j, \boldsymbol{v}_{01}), & \text{if } \zeta^1 < t \leq \tau, \end{cases} \tag{4.5.6}$$

where

$$\zeta^1 = \inf\{t : L_t = x\}.$$

We again argue heuristically that the SF estimator with $\widetilde{W}_t$ given in (4.5.5), which will be denoted $\nabla^k \bar{\ell}_N^s(\boldsymbol{v}, \boldsymbol{v}_0, x)$ and called the *switching* SF estimator, will typically outperform the standard SF estimator $\nabla^k \bar{\ell}_N(\boldsymbol{v}, \boldsymbol{v}_0, \boldsymbol{x})$, provided $v_{01}^{(1)} \geq v_{02}^{(1)}$ and $v_{01}^{(2)} \leq v_{02}^{(2)}$, and that v_{01} and v_{02} are properly chosen. Note that justification for using switching estimators can be found in Asmussen, Rubinstein and Wang (1992).

Table 4.5.1 represents the performance of the standard estimator $\bar{\ell}_N(\boldsymbol{v}, \boldsymbol{v}_0, \boldsymbol{x})$ and the switching estimator $\bar{\ell}_N^s(\boldsymbol{v}, \boldsymbol{v}_0, \boldsymbol{x})$ with $\widetilde{W}_t$ given in (4.5.6), while estimating the probability of the buffer overflow $\ell(\boldsymbol{v}) = P_{\boldsymbol{v}}\{L = x\}$ in the $M/M/1/b$ queue, where $\boldsymbol{v} = (\lambda, \mu) = (1, 1.667)$ and $b = x$ is the buffer size. More precisely, it presents the theoretical values $\ell(\boldsymbol{v}) = P_{\boldsymbol{v}}\{L = x\}$, point estimators $\bar{\ell}_N(\boldsymbol{v}, \boldsymbol{v}_0, \boldsymbol{x})$ and $\bar{\ell}_N^s(\boldsymbol{v}, \boldsymbol{v}_0, \boldsymbol{x})$ (denoted Estim), their associated sample variances (denoted $\hat{\sigma}^2$), the relative widths of the confidence intervals in % (denoted $w_r(\%)$), and the sample efficiency

$$\hat{\epsilon}^0(\boldsymbol{v}, \boldsymbol{v}_0, \boldsymbol{x}) = \frac{\hat{\sigma}^2\{\bar{\ell}_N(\boldsymbol{v}, \boldsymbol{v}_0, \boldsymbol{x})\}}{\hat{\sigma}^2\{\bar{\ell}_N^s(\boldsymbol{v}, \boldsymbol{v}_{01}, \boldsymbol{x})\}},$$

as functions of x. We chose $\boldsymbol{v}_0 = (\lambda, \mu_0) = (1, 1.18)$ and $\boldsymbol{v}_{01} = (\lambda_{01}, \mu_{01}) = (1.667, 1)$ for the standard and the switching estimators, respectively, and simulated $N = 10^6$ customers for both cases. Note that such choice of $\boldsymbol{v}_0$ corresponds to the reference traffic intensities $\rho_0 = \lambda_0/\mu_0 = 0.85$ and

$$\rho_0^s(t) = \begin{cases} \rho_{01} = \lambda_{01}/\mu_{01} = \mu/\lambda = 1.667, & \text{if } t \leq \zeta^1, \\ \rho_{02} = \rho = \lambda_{02}/\mu_{02} = \lambda/\mu = 0.6, & \text{if } \zeta^1 < t \leq \tau, \end{cases} \quad (4.5.7)$$

for the standard and switching estimators, respectively. Here the superscript s in ρ_0^s stands to indicate the switching policy. Notice also that $\rho_0 = 0.85$ is chosen with the view of obtaining maximum variance reduction for the standard estimator. Notice finally that for the switching estimator we start each regenerative cycle with a high utilization factor ($\rho_{01} = 1.667$), that is by *speeding up* the system. After crossing the level ζ^1 (buffer overflow) we *slow down* the system and finish each cycle with the original utilization $\rho = 0.6$.

Table 4.5.1 Performance of the standard and switching estimators for the $M/M/1/b$ queue.

x	Theor	Rare event $\ell(\boldsymbol{v}) = P_{\boldsymbol{v}}\{L_t = x\}$						
		Switching			Standard			
		Estim	$\hat{\sigma}^2$	$w_r(\%)$	Estim	$\hat{\sigma}^2$	w_r	$\hat{\epsilon}(\boldsymbol{v}, \boldsymbol{v}_0, \boldsymbol{x})$
19	1.5E-05	1.5E-05	1.7E-14	3.5	1.6E-05	2.1E-12	31.5	1.26E+02
28	1.5E-07	1.5E-07	2.9E-18	4.6	9.6E-08	2.3E-16	81.1	3.17E+02
37	1.5E-09	1.6E-09	4.8E-22	5.5	8.6E-10	3.5E-19	267	6.29E+02
37	1.5E-11	1.5E-11	4.9E-26	5.7	failed	–	–	–
46	9.1E-14	8.8E-14	2.0E-30	6.4	failed	–	–	–
64	1.5E-15	1.5E-15	4.9E-34	5.9	failed	–	–	–

It is readily seen that for $P_{\boldsymbol{v}}\{L_t > x\} = 1.46 \cdot 10^{-5}$ the variance reduction of two orders of magnitude ($\hat{\epsilon}(\boldsymbol{v}, \boldsymbol{v}_0, x) = 1.26 \cdot 10^{+02}$) and this variance reduction increases as $P_{\boldsymbol{v}}\{L_t > x\}$ decreases. We also found that for $P_{\boldsymbol{v}}\{L_t = x\}$ less than of order 10^{-10} and for sample size of the order of 10^6 customers the standard estimator failes to work (see table 4.5.1) in the sense that its relative width $w_r(\%)$ is of order of thousands and and even more. Similar results were obtained for other single-node queues and simple queueing models. More research on switching policy estimators is underway.

Summary We showed

(i) How to estimate *simultaneously from a single simulation* the standard performance parameters $\nabla^k \ell(\boldsymbol{v})$, like the expected waiting time and the associated sensitivities along with the rare-events $\nabla^k \mathbb{E}_{\boldsymbol{\rho}} I_{\{L \geq x\}} = \nabla^k P_{\boldsymbol{\rho}}\{L \geq x\}$.

(ii) That both the optimal vector of traffic intensities $\boldsymbol{\rho}_\text{o}^* = (\rho_1^*, \rho_2^*)$ and the optimal relative efficiency $\epsilon^k(\boldsymbol{\rho}_\text{o}^*)$ (see program (4.1.3)) are virtually insensitive to the parameters of the non bottleneck queues.

(iii) In order to be on the safe side (obtain a variance reduction) one has to choose ρ_0 moderately larger than ρ and ρ^c when estimating standard performance measures and rare-events, respectively. The reason for this can be explained in terms of importance sampling as follows: when estimating $\nabla^k \ell(\rho)$, $k \geq 0$, the main contribution is typically from cycles which are somewhat larger than average. As we mentioned, in heavy traffic, essentially only the cycles of the order of magnitude $(1 - \rho)^{-2}$ matter, whereas the typical cycles are of the order of magnitude $O(1)$, and thus choosing a reference parameter $\rho_0 > \rho$ (and $\rho_0 > \rho^c$) we place a greater weight on large cycles.

[i] The SF estimators with switching policies have high potentials for estimating rare-events in queueing models.

4.6 Appendix. Convexity Results

Proposition 4.2.1 *Let* $\boldsymbol{Y}$ *be distributed according to an exponential family in the canonical form (4.2.11). Then* $\mathcal{L}^k(\boldsymbol{v}, \boldsymbol{v}_0)$, $k = 0, 1$, *(see 4.2.4) and (4.2.5)) is a convex function with respect to* $\boldsymbol{v}_0$.

Proof First consider the case $k = 0$. Assume without loss of generality that $p = 1$ in $L(\boldsymbol{y})^p$. We have

$$\mathcal{L}^0(\boldsymbol{v}, \boldsymbol{v}_0) = c(\boldsymbol{v})^2 \int L(\boldsymbol{y})^2 c(\boldsymbol{v}_0)^{-1} \exp\left\{\sum_{k=1}^{n} (2v_k - v_{0k}) t_k(\boldsymbol{y})\right\} h(\boldsymbol{y}) \mathrm{d}\boldsymbol{y}, \tag{4.6.1}$$

where

$$c(\boldsymbol{v}_0)^{-1} = \int \exp\left\{\sum_{k=1}^{n} v_{0k} t_k(\boldsymbol{z})\right\} h(\boldsymbol{z}) \mathrm{d}\boldsymbol{z}.$$

Putting this together we obtain

$$\mathcal{L}^0(\boldsymbol{v}, \boldsymbol{v}_0) = c(\boldsymbol{v})^2 \int\int L(\boldsymbol{y})^2 \cdot \exp\left\{\sum_{k=1}^{n} [2v_k t_k(\boldsymbol{y}) + v_{0k}(t_k(\boldsymbol{z}) - t_k(\boldsymbol{y}))]\right\} h(\boldsymbol{y}) h(\boldsymbol{z}) \mathrm{d}\boldsymbol{y} \mathrm{d}\boldsymbol{z}. \tag{4.6.2}$$

Now, for any linear function $a(\boldsymbol{x})$ of $\boldsymbol{x}$, the function $\mathrm{e}^{a(\boldsymbol{x})}$ is convex. Therefore, since $L(\boldsymbol{y})^2$ is non-negative, for any fixed $\boldsymbol{v}, \boldsymbol{y}$ and $\boldsymbol{z}$ the function under the second integral sign in (4.6.2) is convex in $\boldsymbol{v}_0$. This implies convexity of $\mathcal{L}^0(\boldsymbol{v}, \cdot)$.

Let now $k = 1$. In this case $\mathcal{L}^1(\boldsymbol{v}, \boldsymbol{v}_0)$ is given by the trace of the covariance matrix. We have

$$\mathrm{tr}\ \mathbb{E}_{\boldsymbol{v}_0}\{L^2(\nabla W)(\nabla W)'\} = \mathbb{E}_{\boldsymbol{v}_0}\{L^2 W^2 \mathrm{tr}\ \mathbf{S}^{(1)} \mathbf{S}^{(1)'}\}. \tag{4.6.3}$$

Taking into account that $\mathrm{tr}\ [\mathbf{S}^{(1)}(\boldsymbol{y}, \boldsymbol{v}) \mathbf{S}^{(1)}(\boldsymbol{y}, \boldsymbol{v})']$ is non-negative for any $\boldsymbol{y}$ and $\boldsymbol{v}$, we conclude as before that $\mathrm{tr}\ \mathbb{E}_{\boldsymbol{v}_0}\{L^2(\nabla W)(\nabla W)'\}$ is a convex function of $\boldsymbol{v}_0$. □

Remark 4.6.1 Note that proposition 4.2.1 can be extended in the following way. Let $\varphi(x_1, x_2)$ be a real-valued differentiable function of two variables x_1 and x_2. Consider the corresponding expected value function

$$\ell(\boldsymbol{v}) = \varphi(\ell_1(\boldsymbol{v}), \ell_2(\boldsymbol{v})),$$

with

$$\ell_i(\boldsymbol{v}) = \mathbb{E}_{\boldsymbol{v}}\{L_i(\boldsymbol{Y})\} = \mathbb{E}_{\boldsymbol{v}_0}\{L_i(\boldsymbol{Y}) W(\boldsymbol{Y}, \boldsymbol{v})\}, \quad i = 1, 2,$$

where $L_1(\boldsymbol{Y})$ and $L_2(\boldsymbol{Y})$ are two sample functions associated with the same random vector $\boldsymbol{Y}$. In this case the "what if" estimators of $\ell(\boldsymbol{v})$ can be written as

$$\bar{\ell}_N(\boldsymbol{v}, \boldsymbol{v}_0) = \varphi(\bar{\ell}_{1N}(\boldsymbol{v}, \boldsymbol{v}_0), \bar{\ell}_{2N}(\boldsymbol{v}, \boldsymbol{v}_0)),$$

where $\bar{\ell}_{1N}(\boldsymbol{v}, \boldsymbol{v}_0)$ and $\bar{\ell}_{2N}(\boldsymbol{v}, \boldsymbol{v}_0)$ are the "what if" estimators of $\ell_1(\boldsymbol{v})$ and $\ell_2(\boldsymbol{v})$, respectively.

By the delta theorem, we have that $N^{1/2}(\bar{\ell}_N(\boldsymbol{v}) - \ell(\boldsymbol{v}))$ is distributed asymptotically normal with mean zero and variance

$$\begin{aligned}\sigma^2(\boldsymbol{v}, \boldsymbol{v}_0) &= a^2 \mathrm{Var}_{\boldsymbol{v}_0}\{L_1 W\} + b^2 \mathrm{Var}_{\boldsymbol{v}_0}\{L_2 W\} + 2ab\mathrm{Cov}_{\boldsymbol{v}_0}\{L_1 W, L_2 W\} \\ &= \mathbb{E}_{\boldsymbol{v}_0}\{[aL_1 + bL_2]^2 W^2\} + \text{ terms independent of } \boldsymbol{v}_0. \quad (4.6.4)\end{aligned}$$

Here $a = \partial\varphi(x_1, x_2)/\partial x_1$ and $b = \partial\varphi(x_1, x_2)/\partial x_2$ at $(x_1, x_2) = (\ell_1(\boldsymbol{v}), \ell_2(\boldsymbol{v}))$. For example, for $\varphi(x_1, x_2) = x_1/x_2$ we have $a = 1/\ell_2(\boldsymbol{v})$ and $b = -\ell_1(\boldsymbol{v})/\ell_2(\boldsymbol{v})^2$.

Convexity of $\sigma^2(\boldsymbol{v}, \boldsymbol{v}_0)$ with respect to $\boldsymbol{v}_0$ follows in a way similar to the proof of proposition 4.2.1. (This result can be extended further by considering a function $\varphi(x_1, \ldots, x_r)$ of r variables.)

Proposition 4.2.2 *Let* $Y \sim \Gamma(\lambda, \beta)$. *Suppose that* $L(y)^2$ *is a monotonically increasing function on the interval* $[0, \infty)$. *Then*

$$\lambda_0^*(k, p) < \lambda \quad \text{for all } p > 0 \text{ and } k = 0, 1. \tag{4.6.5}$$

Proof The proof will be given for $k = 0$ and $p = 1$ only. The proof for $k = 1$ using the trace operator and for $p > 1$ is similar.

Taking into account that $\mathcal{L}^k(\lambda_0)$ is convex it suffices to prove that the derivative of $\mathcal{L}^k(\lambda_0)$ with respect to λ_0 is positive at $\lambda_0 = \lambda$. To do so we present first $\mathcal{L}^0(\lambda_0)$ as

$$\mathcal{L}^0(\lambda_0) = c\int_0^\infty \lambda_0^{-\beta} L(y)^2 y^{\beta-1} \exp\{-(2\lambda - \lambda_0)y\} \mathrm{d}y,$$

where the constant $c = \lambda^{2\beta}\Gamma(\beta)^{-1}$ is independent of λ_0.

Differentiating now $\mathcal{L}^0(\lambda_0)$ with respect to λ_0 at $\lambda_0 = \lambda$ we obtain

$$(\mathcal{L}^0)'(\lambda_0)|_{\lambda_0=\lambda} = c\int_0^\infty (y - \beta\lambda^{-1})\lambda^{-\beta} L(y)^2 y^{\beta-1} \exp(-\lambda y) \mathrm{d}y.$$

Integrating $\int_0^\infty L(y)^2 y^\beta \exp(-\lambda y)\mathrm{d}y$ by parts, we obtain

$$\begin{aligned}\int_0^\infty L(y)^2 y^\beta \exp(-\lambda y)\mathrm{d}y &= \lambda^{-1}\beta\int_0^\infty L(y)^2 y^{\beta-1}\exp(-\lambda y)\mathrm{d}y \\ &+ \lambda^{-1}\int_0^\infty y^\beta \exp(-\lambda y)\mathrm{d}L(y)^2,\end{aligned}$$

provided $L(y)^2 y^\beta \exp(-\lambda y)$ tends to zero as $y \to +\infty$.

Putting all this together we obtain

$$(\mathcal{L}^0)'(\lambda) = c\lambda^{-\beta-1}\int_0^\infty y^\beta \exp(-\lambda y)\mathrm{d}L(y)^2. \tag{4.6.6}$$

Finally, taking into account that $L(y)^2$ monotonically increases in y, we obtain that the integral in the right-hand side of (4.6.6) is positive and therefore $(\mathcal{L}^0)'(\lambda) > 0$. This, and the convexity of $(\mathcal{L}^0)(\lambda_0)$ imply that $\lambda_0^*(k=0) < \lambda$, and thus proposition 4.2.1 is proved. □

Proposition 4.2.1 can be extended to the multidimensional case of gamma distribution as follows:

Proposition 4.2.1(a) *Suppose that the components $Y_1, \ldots, Y_m$ of the random vector $\boldsymbol{Y}$ are iid each distributed $\Gamma(\lambda, \beta)$ and $L(y_1, \ldots, y_m)^2$ is a monotonically increasing function in every component y_j, $j = 1, \ldots, m$. Then*

$$\lambda_0^*(k,p) < \lambda \textit{ for all } p > 0 \textit{ and } k = 0, 1 \tag{4.6.7}$$

Proof Convexity of the corresponding variance follows here from the general result of proposition 4.2.1. The proof can be completed then by calculations similar to those in proposition 4.2.2. □

Proposition 4.3.1 *Consider a stable queueing model, say, a stable $GI/G/1$ queue. Let the input sequence $\boldsymbol{Y}_1, \boldsymbol{Y}_2, \ldots$ be distributed according to the exponential family given in the canonical form (4.2.11). Then $\mathcal{L}_1^k(\boldsymbol{v}, \boldsymbol{v}_0)$ is a convex function with respect to $\boldsymbol{v}_0$, provided the output process $\{L_t : t > 0\}$ is non-negative-valued.*

Proof Let us consider the random variable

$$U = \sum_{t=1}^{\tau} L_t(\underline{\boldsymbol{Z}}_t)\widetilde{W}_t(\underline{\boldsymbol{Z}}_t, \boldsymbol{v}). \tag{4.6.8}$$

Note that since $\ell_1(\boldsymbol{v}) = \mathbb{E}_{\boldsymbol{v}_0}\{U\}$ we have

$$\mathrm{Var}_{\boldsymbol{v}_0} U = \mathbb{E}_{\boldsymbol{v}_0}\{U^2\} - \ell_1(\boldsymbol{v})^2,$$

and hence the variance $\mathrm{Var}_{\boldsymbol{v}_0} U$ depends on $\boldsymbol{v}_0$ only through the second moment $\mathbb{E}_{\boldsymbol{v}_0}\{U^2\}$. We show now that for the exponential family pdf $f(\boldsymbol{y}, \boldsymbol{v})$, given in the canonical form (4.2.11), $\mathbb{E}_{\boldsymbol{v}_0}\{U^2\}$ is a convex function of $\boldsymbol{v}_0$, provided all $L_t(\underline{\boldsymbol{y}}_t)$, $t = 1, 2, \ldots$, are *non-negative-valued.* To do so we note first that

$$\begin{aligned} U^2 &= \sum_{r=1}^{\infty} \left(\sum_{t=1}^{r} L_t \widetilde{W}_t \right)^2 I_{[\tau = r]} \\ &= \sum_{r=1}^{\infty} \left(\sum_{s,t \le r} L_s L_t \widetilde{W}_s \widetilde{W}_t \right) I_{[\tau = r]} \end{aligned}$$

$$= \sum_{s,t=1}^{\infty} \sum_{r \geq \max(s,t)} L_s L_t \widetilde{W}_s \widetilde{W}_t I_{[\tau=r]}$$

$$= \sum_{s,t=1}^{\infty} L_s L_t \widetilde{W}_s \widetilde{W}_t I_{[\tau \geq \max(s,t)]} \cdot \tag{4.6.9}$$

Therefore it suffices to show that

$$\mathbb{E}_{\boldsymbol{v}_0}\{L_s L_t \widetilde{W}_s \widetilde{W}_t I_{[\tau \geq t]}\}$$

is a convex function of $\boldsymbol{v}_0$ for any $s \leq t$.

Now, for $s \leq t$ we have

$$\begin{aligned} & \mathbb{E}_{\boldsymbol{v}_0}\{L_s L_t \widetilde{W}_s \widetilde{W}_t I_{[\tau \geq t]}\} \\ = & \int H(\underline{\boldsymbol{z}}_t) \widetilde{W}_s(\underline{\boldsymbol{z}}_s, \boldsymbol{v}) \widetilde{W}_t(\underline{\boldsymbol{z}}_t, \boldsymbol{v}) f_t(\underline{\boldsymbol{z}}_t, \boldsymbol{v}_0) \mathrm{d}\underline{\boldsymbol{z}}_t \\ = & \int H(\underline{\boldsymbol{z}}_t) \frac{f_s(\underline{\boldsymbol{z}}_s, \boldsymbol{v}) f_t(\underline{\boldsymbol{z}}_t, \boldsymbol{v})}{f_s(\underline{\boldsymbol{z}}_s, \boldsymbol{v}_0)} \mathrm{d}\underline{\boldsymbol{z}}_t, \end{aligned} \tag{4.6.10}$$

where

$$H(\underline{\boldsymbol{z}}_t) = L_s(\underline{\boldsymbol{z}}_s) L_t(\underline{\boldsymbol{z}}_t) I_{[\tau \geq t]}(\underline{\boldsymbol{z}}_t)$$

is a non-negative-valued function. From (4.2.11) we have that

$$f_s(\underline{\boldsymbol{z}}_s, \boldsymbol{v}_0)^{-1} = c(\boldsymbol{v}_0)^{-1} \exp\left\{ \sum_{j=1}^{s} \sum_{k=1}^{m} -v_{0k} t_k(\boldsymbol{z}_j) \right\} h_s(\underline{\boldsymbol{z}}_s)^{-1},$$

where $h_s(\underline{\boldsymbol{z}}_s) = \prod_{j=1}^{s} h(\boldsymbol{z}_j)$ and $c(\boldsymbol{v}_0)$ is the normalization factor, that is

$$c(\boldsymbol{v}_0)^{-1} = \int \exp\left\{ \sum_{j=1}^{s} \sum_{k=1}^{m} v_{0k} t_k(\boldsymbol{x}_j) \right\} h_s(\underline{\boldsymbol{x}}_s) \mathrm{d}\underline{\boldsymbol{x}}_s.$$

Putting all this together we have

$$\begin{aligned} & \mathbb{E}_{\boldsymbol{v}_0}\left\{L_s L_t \widetilde{W}_s \widetilde{W}_t I_{[\tau \geq t]}\right\} \\ = & \int\int A(\underline{\boldsymbol{z}}_t, \underline{\boldsymbol{x}}_s, \boldsymbol{v}) \exp\{a(\boldsymbol{v}_0, \underline{\boldsymbol{z}}_t, \underline{\boldsymbol{x}}_s)\} \mathrm{d}\underline{\boldsymbol{z}}_t \mathrm{d}\underline{\boldsymbol{x}}_s, \end{aligned} \tag{4.6.11}$$

where $A(\underline{\boldsymbol{z}}_t, \underline{\boldsymbol{x}}_s, \boldsymbol{v})$ is a non-negative-valued function independent of $\boldsymbol{v}_0$ and $a(\boldsymbol{v}_0, \underline{\boldsymbol{z}}_t, \underline{\boldsymbol{x}}_s)$ is a function linear in $\boldsymbol{v}_0$ for every fixed $\underline{\boldsymbol{z}}_t$ and $\underline{\boldsymbol{x}}_s$. It follows that the function under the integral sign on the right-hand side of (4.6.11) is convex in $\boldsymbol{v}_0$. Consequently the expectation in (4.6.11) is a convex function of $\boldsymbol{v}_0$, and so $\mathbb{E}_{\boldsymbol{v}_0}\{U^2\}$ too. □

Note that in the proof we only used tye non-negativeness of the products $L_s(\underline{y}_s)L_t(\underline{y}_t)$ for any $\underline{y}_s$ and $\underline{y}_t$. It follows that $\mathbb{E}_{\boldsymbol{v}_0}\left\{\left(\sum_{t=1}^{\tau}\widetilde{W}_t\right)^2\right\}$ is a convex function of $\boldsymbol{v}_0$. Note also that since $\nabla^k\widetilde{W}_t(\underline{z}_t,\boldsymbol{v}) = \widetilde{W}_t(\underline{z}_t,\boldsymbol{v})\tilde{S}_t^{(k)}(\underline{z}_t,\boldsymbol{v})$ the convexity result can be extended to the trace of $\mathbb{E}_{\boldsymbol{v}_0}\{[\nabla\bar{\ell}_{1N}(\boldsymbol{v})][\nabla\bar{\ell}_{1N}(\boldsymbol{v})]'\}$.

Proposition 4.3.2 *Let $L_t = L_t(\underline{\mathbf{Y}}_t)$ be the steady-state waiting (sojourn) time process in a stable queueing model. Suppose that the vectors $\mathbf{Y}_i = (Y_{1i}, Y_{i2})$ of the service and the interarrival times have independent components each distributed gamma, that is $Y_{ji} \sim \Gamma(\lambda_i, \beta_i)$, $j = 1,2$, $i = 1,2,\ldots$. Let β_1 and β_2 be fixed and consider $\mathcal{L}_1^0(\boldsymbol{\lambda},\boldsymbol{\lambda}_0)$ as a function of the reference parameter $\boldsymbol{\lambda}_0$ (here $\boldsymbol{\lambda} = (\lambda_1,\lambda_2)$ and $\boldsymbol{\lambda}_0 = (\lambda_{01},\lambda_{02})$). Then the minimizer $\boldsymbol{\lambda}_0^*$ of $\mathcal{L}_1^0(\boldsymbol{\lambda},\cdot)$ satisfies $\lambda_{01}^* < \lambda_1$ and $\lambda_{02}^* > \lambda_2$, implying*

$$\rho_0^*(k,p) > \rho \quad \textit{for all} \quad k \geq 0 \textit{ and } p > 0.$$

Lemma 4.6.1 *Consider an output process $L_t = L_t(\underline{Y}_t)$ with $Y_i \sim \Gamma(\lambda,\beta)$. Suppose that $L_t(\underline{y}_t)$, $t = 1,2,\ldots$, are non-negative-valued and monotonically increasing in every component y_i, $i = 1,\ldots,t$, of $\underline{y}_t$, and that also the indicator function $I_{[\tau\geq t]}(\underline{y}_t)$ is non-decreasing in every component of $\underline{y}_t$. Then the minimizer λ_0^* of the function*

$$\mathcal{L}(\lambda_0) = \mathbb{E}_{\lambda_0}\left\{\left[\sum_{t=1}^{\tau} L_t(\underline{Z}_t)\tilde{W}_t(\underline{Z}_t,\lambda)\right]^2\right\}$$

satisfies $\lambda_0^* < \lambda$.

Proof Since the function $\mathcal{L}(\lambda_0)$ is convex, it will be sufficient to show that $\mathrm{d}\mathcal{L}(\lambda_0)/\mathrm{d}\lambda_0|_{\lambda_0=\lambda}$ is positive. By (4.6.10) we have

$$\mathbb{E}_{\lambda_0}\left\{L_s L_t \tilde{W}_s \tilde{W}_t I_{[\tau\geq t]}\right\} = \int H(\underline{z}_t)\frac{f_s(\underline{z}_s,\lambda)f_t(\underline{z}_t,\lambda)}{f_s(\underline{z}_s,\lambda_0)}\mathrm{d}\underline{z}_t,$$

where

$$H(\underline{z}_t) = L_s(\underline{z}_s)L_t(\underline{z}_t)I_{[\tau\geq t]}(\underline{z}_t)$$

is monotonically increasing in every component of $\underline{z}_t$. Then

$$\begin{aligned}&\frac{\mathrm{d}}{\mathrm{d}\lambda_0}\left[\mathbb{E}_{\lambda_0}\left\{L_s L_t \tilde{W}_s \tilde{W}_t I_{[\tau\geq t]}\right\}\right]\Big|_{\lambda_0=\lambda}\\ &= -\int H(\underline{z}_t)\tilde{S}_s^{(1)}(\underline{z}_s,\lambda)f_t(\underline{z}_t,\lambda)\mathrm{d}\underline{z}_t.\end{aligned} \tag{4.6.12}$$

By (4.6.9) it will be sufficient to show that for $s \leq t$ the expression given in the right-hand side of (4.6.12) is positive. Now (4.6.12) can be written in the form

$$\frac{1}{\Gamma(\beta)^t} \int \left(z_1 + \cdots + z_s - s\frac{\beta}{\lambda} \right) H(\underline{z}_t) \lambda^{t\beta} e^{(-\lambda_1 z_1 - \cdots - \lambda_t z_t)} \prod_{j=1}^{t} z_j^{\beta-1} d\underline{z}_t$$

$$= \frac{\lambda^{t\beta}}{\Gamma(\beta)^t} \sum_{i=1}^{s} \int \left(z_i - \frac{\beta}{\lambda} \right) H(\underline{z}_t) \prod_{j=1}^{t} (e^{-\lambda z_j} z_j^{\beta-1}) d\underline{z}_t.$$

Therefore it will be sufficient to show that each integral in the sum is positive for $i = 1, \ldots, s$. We prove this for $i = 1$, the proof being similar for $i > 1$. We have

$$\int \left(z_1 - \frac{\beta}{\lambda} \right) H(\underline{z}_t) \prod_{j=1}^{t} (e^{-\lambda z_j} z_j^{\beta-1}) d\underline{z}_t$$

$$= \int \prod_{j=2}^{t} (e^{-\lambda z_j} z_j^{\beta-1}) \left(\int_0^\infty (z_1 - \frac{\beta}{\lambda}) H(\underline{z}_t) e^{-\lambda z_1} z_1^{\beta-1} dz_1 \right) dz_2 \ldots dz_t.$$

Now, for fixed $z_2 \ldots z_t$ we have that the inner integral is

$$\frac{1}{\lambda} \int_0^\infty z_1^\beta e^{-\lambda z_1} dH(z_1, z_2, \ldots, z_t).$$

Since for fixed $z_2, \ldots, z_t$ the function $H(z_1, z_2, \ldots, z_t)$ is monotonically increasing in z_1, this integral is positive. Thus also the corresponding multiple integral is positive, and the proof is complete. □

Note that if $L_t(\underline{y}_t)$ is non-negative and monotonically decreasing in every component of $\underline{y}_t$ and $I_{[\tau \geq t]}(\underline{y}_t)$ is monotonically nonincreasing in every component of $\underline{y}_t$, then similar arguments show that the corresponding minimizer λ_0^* is greater than λ.

Now let $\boldsymbol{Y}_i = (Y_{1i}, Y_{2i})$, where $Y_{1i} \sim \Gamma(\lambda_1, \beta_1)$ and $Y_{2i} \sim \Gamma(\lambda_2, \beta_2)$. Suppose that Y_{1i} and Y_{2i} are independent, $L_t(\underline{y}_t)$ is non-negative and that for any y_{2i}, $i = 1, \ldots, t$, $L_t(\underline{y}_t)$ is monotonically increasing in every y_{1i}, $i = 1, \ldots, t$, and that for any y_{1i}, $i = 1, \ldots, t$, $L_t(\underline{y}_t)$ is monotonically decreasing in every y_{2i}, $i = 1, \ldots, t$. Suppose also that for any y_{2i}, $i = 1, \ldots, t$, $I_{[\tau \geq t]}(\underline{y}_t)$ is monotonically nondecreasing in every y_{1i}, $i = 1, \ldots, t$, and for any y_{1i}, $i = 1, \ldots, t$, $I_{[\tau \geq t]}(\underline{y}_t)$ is monotonically nonincreasing in every y_{2i}, $i = 1, \ldots, t$. It then follows from lemma 4.5.1 that for any λ_{02},

$$\left. \frac{\partial \mathcal{L}_1^0(\boldsymbol{\lambda}, \boldsymbol{\lambda}_0)}{\partial \lambda_{01}} \right|_{\lambda_{01} = \lambda_1} > 0,$$

and that for any λ_{01},

$$\left.\frac{\partial \mathcal{L}_1^0(\boldsymbol{\lambda}, \boldsymbol{\lambda}_0)}{\partial \lambda_{02}}\right|_{\lambda_{02}=\lambda_2} < 0.$$

Together with the convexity of $\mathcal{L}^0(\boldsymbol{\lambda}, \cdot)$, these inequalities imply that $\lambda_{01}^* < \lambda_1$ and $\lambda_{02}^* > \lambda_2$. It only remains to note that for the output processes, like the waiting time processes, the above assumptions hold. □

4.7 Exercises

1. Consider example 4.4.1. Find the optimal value of the reference parameter $\mu_0^*(x, k = 1)$ which minimizes $\mathcal{L}^1(\mu_0, x)$. Prove that
   (i) $\mu_0^*(x, k)$, $k = 0, 1$, decreases in both x and k,
   (ii) as $x \to \infty$, $\mu_0^*(x, k)$ tends to zero,
   (iii) $\mu_0^*(x, k = 1) < \mu_0^*(x, k = 0)$.

2. Consider the $M/M/\infty$ queue in example 4.4.1 with $P_\mu\{L \geq x\} = \mathrm{e}^{-\mu x}$, where μ is the service rate. Find an importance sampling density $g(\boldsymbol{z})$ such that
$$\mathrm{Var}_g\left\{\frac{\sum_{t=1}^{\tau} I_t \widetilde{W}_t}{\sum_{t=1}^{\tau} \widetilde{W}_t}\right\} = 0.$$

3. Let Y be a random variable distributed normally with mean μ and variance σ.

   (a) Suppose that σ is known and fixed and consider, for a given μ, the function (see section 2.5)
   $$\mathcal{L}^0(\mu_0) = \mathbb{E}_{\mu_0}\{L^2 W^2\}.$$

   (i) Show that if for all $\mu \in \mathbb{R}$ the expectation $\mathbb{E}_\mu\{L^2\}$ is finite, then $\mathcal{L}^0(\mu_0)$ is convex and continuous on $\mathbb{R}$ and if, in addition, $\mathbb{E}_{\mu_n}\{L^2\}$ does not tend to zero for any sequence $\mu_n \to \infty$, then $\mathcal{L}^0(\mu_0)$ has a unique minimizer μ_0^* over $\mathbb{R}$.

   (ii) Show that if $L(y)^2$ is monotonically increasing on $\mathbb{R}$, then $\mu_0^* < \mu$.

   (b) Suppose now that μ is known and consider the parameter σ. Note that the obtained exponential family does not have the canonical form. Therefore let us parameterize it by $\xi = \sigma^{-2}$. The family is parametrized then in the canonical form with $t(y) = -(y - \mu)^2/2$ and $c(\xi) = (2\pi)^{-1/2}\xi^{1/2}$.

(i) Show that

$$\mathbb{E}_{\xi_0}\{W^2\} = \frac{\xi^{1/2}}{\xi_0^{1/2}(2\xi - \xi_0)^{1/2}},$$

provided $0 < \xi_0 < 2\xi$.

(ii) Show that, for a given ξ, the function

$$\mathcal{L}^0(\xi_0) = \mathbb{E}_{\xi_0}\{L^2 W^2\}$$

has a unique minimizer ξ_0^* on the interval $(0, 2\xi)$ provided the expectation $\mathbb{E}_{\xi'}\{L^2\}$ is finite for all $\xi' \in (0, 2\xi)$ and does not tend to zero as ξ' approaches 0 or 2ξ. (Notice that this implies that the corresponding optimal value $\sigma_0^* = \xi_0^{*-1/2}$ of the reference parameter σ_0 is also unique.)

(iii) Show that if $L(y)^2$ is monotonically increasing on $\mathbb{R}$, then $\xi_0^* < \xi$. (Notice that this implies that $\sigma_0^* > \sigma$.)

4. Let L_t be the waiting time process in an $M/G/c$ queue (c denotes the number of servers) with arrival rate λ, and let τ be the number of customers served during a busy period. Consider the corresponding expected value functions

$$\ell_1(\lambda) = \mathbb{E}_\lambda\left\{\sum_{t=1}^{\tau} L_t\right\}$$

and

$$\mathcal{L}^0(\lambda, \lambda_0) = \mathbb{E}_{\lambda_0}\left\{\left[\sum_{t=1}^{\tau} L_t \tilde{W}_t(\lambda)\right]^2\right\}.$$

Show that the optimal value λ_0^* of the reference parameter (given by the minimizer of the function $\mathcal{L}^0(\lambda, \cdot)$) satisfies $\lambda_0^* > \lambda$.

Chapter 5

Extensions of the SF Method

5.1 Introduction

This chapter deals with extensions of the SF method for sensitivity analysis and stochastic optimization to the following model of DES:

$$\ell(\boldsymbol{v}) = \mathbb{E}_{\boldsymbol{v}_1}\{L(\underline{\boldsymbol{Y}}_t, \boldsymbol{v}_2)\}. \tag{5.1.1}$$

Here $L(\underline{\boldsymbol{Y}}_t)$, $(\underline{\boldsymbol{Y}}_t = (\boldsymbol{Y}_1, \ldots, \boldsymbol{Y}_t)$ is the sample performance driven by an input sequence $\boldsymbol{Y}_1, \ldots, \boldsymbol{Y}_t$ of iid random vectors with a common pdf $f(\boldsymbol{y}, \boldsymbol{v}_1)$. The combined vector of parameters is given here by $\boldsymbol{v} = (\boldsymbol{v}_1, \boldsymbol{v}_2)$, and the subscript $\boldsymbol{v}_1$ in $\mathbb{E}_{\boldsymbol{v}_1} L$ indicates that the expectation is taken with respect to the pdf $f(\boldsymbol{y}, \boldsymbol{v}_1)$. Note that we assume that f depends on the parameter vector $\boldsymbol{v}_1$ and not on $\boldsymbol{v}_2$, and vice versa for L. We shall call $\boldsymbol{v}_1$ and $\boldsymbol{v}_2$ the *distribution* and the *structural* parameter vectors, respectively. Note also that the model $\ell(\boldsymbol{v}) = \mathbb{E}_{\boldsymbol{v}}\{L(\underline{\boldsymbol{Y}}_t)\}$ of chapter 3 can be considered as a particular case of the model (5.1.1) when L does not dependent on $\boldsymbol{v}_2$ and $\boldsymbol{v} = \boldsymbol{v}_1$.

As in chapters 2 and 3 we suppose that $\ell(\boldsymbol{v})$ is not available analytically and we want to evaluate it, as well as the associated sensitivities $\nabla^k \ell(\boldsymbol{v})$, $k = 1, 2, \ldots$, for different values of $\boldsymbol{v} = (\boldsymbol{v}_1, \boldsymbol{v}_2)$ via simulation.

Consider the following motivating examples.

1. **Stochastic PERT network**. Consider the stochastic PERT network (2.1.2) with the sample performance

$$L(\boldsymbol{Y}) = \min_{j=1\ldots,p} \left\{ \sum_{i \in \mathcal{L}_j} Y_i \right\}, \tag{5.1.2}$$

where $\mathcal{L}_j$ is the j-th complete path from a source to a sink in the system, $\boldsymbol{Y} = (Y_1, \ldots, Y_m)$ is the vector of durations of the components, and p is the number of complete paths in the system. Suppose that the vector $\boldsymbol{Y}$ can be decomposed into two parts, stochastic and deterministic. That is, $\boldsymbol{Y} = (\boldsymbol{Y}_1, \boldsymbol{Y}_2)$, where $\boldsymbol{Y}_1 \sim f_1(\boldsymbol{y}_1, \boldsymbol{v}_1)$ and $\boldsymbol{Y}_2$ is the deterministic part, say $\boldsymbol{Y}_2 = \boldsymbol{v}_2$. Then $L(\boldsymbol{Y}) = L(\boldsymbol{Y}_1, \boldsymbol{v}_2)$ and the corresponding expected performance takes the form (5.1.1).

2. $GI/D/1$ **and** $D/G/1$ **queues.** Suppose that we want to evaluate the steady-state performance of a stable $GI/D/1$ or $D/G/1$ queue, say the steady-state expected waiting time $\ell(\boldsymbol{v}) = \mathbb{E}_{\boldsymbol{v}_1}\{L(\boldsymbol{Y}, \boldsymbol{v}_2)\}$ for different values of $\boldsymbol{v}_1$ and $\boldsymbol{v}_2$. Here for the $GI/D/1$ queues, $Y \sim f_1(y_1, v_1)$ represents the interarrival random variable with v_1 being the interarrival rate and v_2 being the length of the service time, and similarly for the $D/G/1$ queue.

3. $GI/G/1$ **queue.** Suppose that we want to evaluate the cdf

$$H_L(x, \boldsymbol{v}) = P_{\boldsymbol{v}}\{L \le x\} \tag{5.1.3}$$

of the steady-state sample performance L and the associated derivative $\partial H_L(x, \boldsymbol{v})/\partial x$ *simultaneously* for different values of $\boldsymbol{v}$ and x. In this case, we can represent $H_L(x, \boldsymbol{v})$ as

$$H_L(x, \boldsymbol{v}) = \mathbb{E}_{\boldsymbol{v}_1}\{I_{(-\infty,0]}(L - x)\}\ , \tag{5.1.4}$$

where $I_{(-\infty,0]}(\cdot)$ is the indicator function of the interval $(-\infty, 0]$.

4. $GI/G/1/b$ **queue.** Suppose that we want to evaluate the steady-state expected waiting time $\ell(\boldsymbol{v}) = \mathbb{E}_{\boldsymbol{v}_1}\{L(\boldsymbol{Y}, v_2)\}$ of a stable $GI/G/1/b$ queue for different values of $\boldsymbol{v}_1$ and $v_2 = b$, where b denotes the buffer size.

5. **DES with autocorrelated inputs.** Consider an output process $\{L_t : t > 0\}$ driven by an autocorrelated input sequence $\{\boldsymbol{X}_t : t > 0\}$, that is $L_t(\cdot) = L_t(\underline{\boldsymbol{X}}_t)$, where $\underline{\boldsymbol{X}}_t = (\boldsymbol{X}_1, \boldsymbol{X}_2, \ldots, \boldsymbol{X}_t)$, and $L_t(\cdot)$ is a sequence of real-valued functions. As an example one may think of autocorrelated interarrival and service time random variables in a $GI/G/1$ queue with $\{L_t : t > 0\}$ being the waiting time process. We assume, for simplicity, that $\{X_t : t > 0\}$ is a scalar process and it can be written recursively as

$$X_t(\boldsymbol{v}_2) = h(X_{t-1}, Y_t, \boldsymbol{v}_2), \quad t > 0, \tag{5.1.5}$$

where $h(x, y, \boldsymbol{v}_2)$ is a real valued function, and, as before, $\{Y_t : t > 0\}$ is an iid sequence. This implies that $\{X_t : t > 0\}$ is a function of the sequence of an iid rv's $\{Y_t : t > 0\}$, that is, $X_t = X_t(\boldsymbol{v}_2) = \boldsymbol{x}_t(\underline{Y}_t, \boldsymbol{v}_2)$, where $\boldsymbol{x}_t(\cdot)$ is real-valued function calculated from $h(\boldsymbol{x}, \boldsymbol{y}, \boldsymbol{v}_2)$, and $\boldsymbol{v}_2$ is the parameter vector associated with the autocorrelation of the rv's $X_1, X_2, \ldots, X_t$, which will be called the *autocorrelation parameter vector.* We also assume that as $t \to \infty$, both processes $\{X_t : t > 0\}$ and $\{L_t : t > 0\}$ become stationary and ergodic.

As examples of $\{X_t : t > 0\}$ consider:

- **First order autoregressive (AR(1)) processes.**

$$X_t(v_2) = \begin{cases} X_1, & \text{if } t = 1, \\ v_2 X_{t-1} + Y_t, & \text{if } t > 1, \end{cases} \tag{5.1.6}$$

  where $-1 < v_2 < 1$. One can also consider higher order autoregressive processes.

- **Transform-Expand-Sample (TES) processes.**

$$X_t(\boldsymbol{v}_2) = F^{-1}[\eta_t^+(\boldsymbol{v}_2)], \tag{5.1.7}$$

  where F^{-1} is the inverse of a cdf $F(y, \boldsymbol{v})$. For example, if $Y \sim \exp(v)$, then $X_t = (-1/v)\ln(1 - \eta_t^+)$, and $\eta_t^+(\boldsymbol{v}_2)$ is given recursively (see Jagerman and Melamed (1992a,b) and Melamed (1991)) as

$$\eta_t^+(\boldsymbol{v}_2) = \begin{cases} U_1, & \text{if } t = 1, \\ \langle \eta_{t-1}^+ + \theta_1 + (\theta_2 - \theta_1) U_t \rangle, & \text{if } t > 1. \end{cases} \tag{5.1.8}$$

  Here $\boldsymbol{v}_2 = (\theta_1, \theta_2)$, $-1/2 \leq \theta_1 < \theta_2 \leq 1/2$, $\langle x \rangle = x - \lfloor x \rfloor$ is the fractional part of x, $\lfloor x \rfloor = \max\{n \text{ integer} : n \leq x\}$ is the integral part of x, and $\{U_t\}$ is the uniform (0,1) sequence of (iid) random numbers. It can be shown (Melamed (1991)) that the sequence η_t^+ covers all positive lag-1 autocorrelations in the range [0,1). To produce negative autocorrelations one can use the following (antithetic) formula (see Melamed (1991)):

$$\eta_t^- = \begin{cases} \eta_t^+, & \text{if } t \text{ even}, \\ 1 - \eta_t^+, & \text{if } t \text{ odd}. \end{cases} \tag{5.1.9}$$

  Notice that when $\theta_2 - \theta_1 = 1$, the autocorrelated sequence $\{X_t : t > 0\}$ reduces to the independent one $\{Y_t : t > 0\}$. Note also that for all $t \geq 1$ the marginals of both the η_t^+ and η_t^- sequences are uniformly distributed on the interval $[0, 1)$. For more details on TES sequences see Melamed (1991) and Jagerman and Melamed (1992a,b) .

- **Minification/maxification sequence** (see Lewis and McKenzie (1991)).

$$X_t(\boldsymbol{v}_2) = F^{-1}[\eta_t^{\min}(\boldsymbol{v}_2)], \tag{5.1.10}$$

where $\{\eta_t^{\min}(\boldsymbol{v}_2)\}$ is a minification sequence, defined by

$$\eta_t^{\min}(\boldsymbol{v}_2) = \begin{cases} U_1, & \text{if } t = 1, \\ v_2 \min[\eta_{t-1}^{\min}, U_{t-1}/(v_2 - 1 + U_{t-1})], & \text{if } t > 1. \end{cases} \tag{5.1.11}$$

Note that for $v_2 = 1$, the autocorrelated sequence $\{X_t : t > 0\}$ reduces to the independent one $\{Y_t : t > 0\}$. In analogy to (5.1.10) we can define

$$X_t(\boldsymbol{v}_2) = F^{-1}[\eta_t^{\max}(\boldsymbol{v}_2)], \tag{5.1.12}$$

where $\{\eta_t^{\max}(\boldsymbol{v}_2)\}$ is a maxification sequence, defined by

$$\eta_t^{\max}(\boldsymbol{v}_2) = \begin{cases} U_1, & \text{if } t = 1, \\ \max[(\eta_{t-1}^{\max})^{v_2}, U_{t-1}^{v_2/(v_2-1)}], & \text{if } t > 1. \end{cases} \tag{5.1.13}$$

Note that in this case as $\boldsymbol{v}_2 \to \infty$, the autocorrelated sequence $\{X_t : t > 0\}$ approaches the independent one $\{Y_t : t > 0\}$.

Section 5.2 deals with sensitivity analysis of the model (5.1.1). Particular emphasis is placed on smoothness of the sample function $L(\boldsymbol{y}, \boldsymbol{v}_2)$ and variance reduction. Here we present two techniques, based on transformation of random variables, which will be called the "push out" and "push in" techniques, respectively. The terms "push out" and "push in" derive from the fact (see below) that in the first case we "push out" the parameter vector $\boldsymbol{v}_2$ from the original sample performance $L(\boldsymbol{Y}, \boldsymbol{v}_2)$ into an auxiliary pdf by a suitable transformation and then apply the standard SF method; while in the second case we act the other way around, namely we first "push in" (by a suitable transformation) the parameter vector $\boldsymbol{v}_1$ into the sample performance $L(\boldsymbol{Y}, \boldsymbol{v}_2)$ and then differentiate the resulting (auxiliary) sample performance with respect to $\boldsymbol{v} = (\boldsymbol{v}_1, \boldsymbol{v}_2)$. We discuss conditions under which such transformations are useful in the sense that they either generate *smooth sample performances* or lead to a *variance reduction.* We also show how the *infinitesimal perturbation analysis* (IPA) method introduced by Ho and his co-workers corresponds to the "push in" technique; the latter can be viewed as a dual of the "push out" technique. Finally, section 5.3 deals with autocorrelated input sequences. More precisely we extend some of our earlier results on sensitivity analysis of DES for the independent input sequences to correlated ones. We conclude this section by making the following

Remark 5.1.1 Consider the model $\ell(\boldsymbol{v}) = \mathbb{E}_{\boldsymbol{v}}\{L(\underline{X}_t)\}$. Assume that $\underline{X}_t$ can be written as

$$\underline{X}_t = \varphi_k(\varphi_{k-1}(\ldots\varphi_1(\underline{U}_t)\ldots)), \tag{5.1.14}$$

where φ_r, $r = 1,\ldots,k$, are real-valued functions calculated by using $h(x, y, \boldsymbol{v}_2)$ (see (5.1.5)), $\underline{U}_t = (U_1,\ldots,U_t)$ is the underlying random number stream, i.e. U_j, $j = 1,\ldots,t$ are iid uniform on the interval [0,1) rv's. Denote

$$\underline{X}_t{}^{(0)} \equiv \underline{U}_t;\ \underline{X}_t{}^{(1)} = \varphi_1(\underline{U}_t);\ldots;$$
$$\underline{X}_t{}^{(k-1)} = \varphi_{k-1}(\ldots\varphi_1(\underline{U}_t)\ldots);\ \underline{X}_t{}^{(k)} \equiv \underline{X}_t \tag{5.1.15}$$

and assume that $\underline{X}_t{}^{(j)} \sim H_t^{(j)}(\underline{x}_t, \boldsymbol{v}^{(j)})$, $j = 0, 1,\ldots,k$, and $\boldsymbol{v}^{(k)} \equiv \boldsymbol{v} = (\boldsymbol{v}_1, \boldsymbol{v}_2)$.

Using (5.1.14) and (5.1.15), we can write $\ell(\boldsymbol{v})$ as

$$\ell(\boldsymbol{v}) = \mathbb{E}_{H_t^{(j)}}\{L(\underline{X}_t{}^{(j)}, \boldsymbol{v}^{(j)})\}, \tag{5.1.16}$$

where $H_t^{(j)}(\underline{x}_t, \boldsymbol{v}^{(j)})$ is the cdf of $\underline{X}_t{}^{(j)}$. Note that in (5.1.16) the *same* performance $\ell(\boldsymbol{v})$ is expressed as the expectation taken with respect to different probability measures (cdf's) $H_t^{(j)}(\underline{x}_t, \boldsymbol{v}^{(j)})$, which we assume to be given. We shall call the index j in (5.1.16) the *j-th simulation level.* The two extremes, namely $j = 0$ and $j = k$, correspond to $\underline{X}_t{}^{(0)} = \underline{U}_t$ and $\underline{X}_t = \underline{X}_t{}^{(k)} = \varphi_k(\varphi_{k-1}(\ldots\varphi_1(\underline{U}_t)\ldots))$, respectively. (Clearly if we know the cdf of the random variable $L(\underline{X}_t)$ there is no need at all for simulation.) We shall see below that, typically, the higher the simulation level j, the more *knowledgeable* we are about the system, and in particular about the sample performance $L(\underline{X}_t{}^{(j)}, \boldsymbol{v}^{(j)})$. We shall also see below that this knowledge is not completely free, in the sense that the higher the index j, the more complicated is the associated pdf $H_t^{(j)}$. Thus, there exists a tradeoff between the level j and the simulation complexity, say, in terms of CPU time. We shall show below that in typical situations, a CPU investment is beneficial since smoothness and some other nice properties of the sample performance $L(\underline{X}_t{}^{(j)}, \boldsymbol{v}^{(j)})$ are brought about by increasing j.

For the model $\ell(\boldsymbol{v}) = \mathbb{E}_{\boldsymbol{v}}\{L(\underline{X}_t)\}$ with the autocorrelated sequence $X_t = x_t(\underline{Y}_t, \boldsymbol{v}_2,)$ we have three simulation levels $j = 0$, $j = 1$ and $j = 2$ corresponding to

$$\underline{X}_t{}^{(0)} = \underline{U}_t;\ \underline{X}_t{}^{(1)} = \underline{Y}_t;\ \underline{X}_t{}^{(2)} = \underline{X}_t,$$

which are associated with the so-called "push in", direct and "push out" estimators, respectively (see section 5.3 below). Here

$$\underline{Y}_t = (Y_1,\ldots,Y_t),\ Y_j \sim F(y, \boldsymbol{v}_1)$$

and

$$\underline{X}_t = (X_1,\ldots,X_t),\ X_j \sim \tilde{F}(y, \boldsymbol{v}_2),\quad j = 1,\ldots,t.$$

5.2 Direct, "Push out", and "Push in" Estimators

5.2.1 Direct Estimators

Let us extend first our basic formulas (2.2.27)–(2.2.29) and (3.2.28)–(3.2.33) to the model (5.1.1) while considering separately DESS and DEDS models.

(a) DESS models

Let G again be a probability measure having a density function $g(\boldsymbol{y})$, that is $\mathrm{d}G(\boldsymbol{y}) = g(\boldsymbol{y})\mathrm{d}\boldsymbol{y}$. Suppose that for every permissible value of the parameter vector $\boldsymbol{v}_1$ the support of $f(\boldsymbol{y}, \boldsymbol{v}_1)$ lies within that of $g(\boldsymbol{y})$, that is

$$\operatorname{supp}\{f(\boldsymbol{y}, \boldsymbol{v}_1)\} \subset \operatorname{supp}\{g(\boldsymbol{y})\}.$$

Then we can represent $\ell(\boldsymbol{v})$ as

$$\ell(\boldsymbol{v}) = \mathbb{E}_g\{L(\boldsymbol{Z}, \boldsymbol{v}_2)W(\boldsymbol{Z}, \boldsymbol{v}_1)\}, \tag{5.2.1}$$

where

$$W(\boldsymbol{z}, \boldsymbol{v}_1) = f(\boldsymbol{z}, \boldsymbol{v}_1)/g(\boldsymbol{z}) \tag{5.2.2}$$

is the likelihood ratio, $\boldsymbol{Z} \sim g(\boldsymbol{z})$, and the subscript g indicates that the expectation is taken with respect to the dominating pdf $g(\boldsymbol{y})$. Under suitable regularity conditions (see appendix A of chapter 2), we have

$$\nabla^k_{\boldsymbol{v}_1}\ell(\boldsymbol{v}) = \mathbb{E}_g\{L(\boldsymbol{Z}, \boldsymbol{v}_2)\nabla^k_{\boldsymbol{v}_1}W(\boldsymbol{Z}, \boldsymbol{v}_1)\} \tag{5.2.3}$$

and

$$\nabla^k_{\boldsymbol{v}_2}\ell(\boldsymbol{v}) = \mathbb{E}_g\{\nabla^k_{\boldsymbol{v}_2}L(\boldsymbol{Z}, \boldsymbol{v}_2)W(\boldsymbol{Z}, \boldsymbol{v}_1)\}, \tag{5.2.4}$$

$k = 1, 2, \ldots$.

Consider formula (5.2.3). For fixed $\boldsymbol{v}_2$ it allows us to estimate the performance $\ell(\boldsymbol{v})$ and *all* sensitivities $\nabla^k_{\boldsymbol{v}_1}\ell(\boldsymbol{v})$, $k = 1, 2, \ldots$, from a *single simulation, simultaneously* for different values of $\boldsymbol{v}_1$. The corresponding SF estimator can be written as

$$\nabla^k_{\boldsymbol{v}_1}\bar{\ell}_N(\boldsymbol{v}) = N^{-1}\sum_{i=1}^{N} L(\boldsymbol{Z}_i, \boldsymbol{v}_2)\nabla^k_{\boldsymbol{v}_1}W(\boldsymbol{Z}_i, \boldsymbol{v}_1), \tag{5.2.5}$$

and similarly for the estimator of (5.2.4). Here $\boldsymbol{Z}_1, \ldots, \boldsymbol{Z}_N$ is a random sample from the dominating pdf $g(\boldsymbol{z})$.

Example 5.2.1 Let $L(\boldsymbol{Y}, v_1) = \max\{(v_1 + Y_2), Y_3\}$, where $\boldsymbol{Y} = (Y_2, Y_3)$, $Y_i \sim F_i(y, v_i)$, $i = 2, 3$, Y_2 and Y_3 are independent. We have

$$\ell(\boldsymbol{v}) = \mathbb{E}_g\{L(\boldsymbol{Z}, v_1)W(\boldsymbol{Z}, \boldsymbol{v}_1)\},$$

$$\frac{\partial \ell(\boldsymbol{v})}{\partial v_1} = \mathbb{E}_g\left\{\frac{\partial L(\boldsymbol{Z}, v_1)}{\partial v_1} W(\boldsymbol{Z}, \boldsymbol{v}_1)\right\},$$

$$\frac{\partial \ell(\boldsymbol{v})}{\partial v_2} = \mathbb{E}_g\left\{L(\boldsymbol{Z}, v_1)\frac{\partial W(\boldsymbol{Z}, \boldsymbol{v}_1)}{\partial v_2}\right\},$$

and similarly for $\partial \ell(\boldsymbol{v})/\partial v_3$. Here

$$\boldsymbol{v} = (v_1, v_2, v_3), \; \boldsymbol{v}_1 = (v_2, v_3), \; L(\boldsymbol{Z}, v_1) = \max\{(v_1 + Z_2), Z_3\},$$

$$W(\boldsymbol{Z}, \boldsymbol{v}_1) = \frac{f_2(Z_2, v_2)}{g_2(Z_2)} \frac{f_3(Z_3, v_3)}{g_3(Z_3)},$$

$Z_i \sim g_i(z_i)$, $i = 2, 3$ and

$$\frac{\partial L(\boldsymbol{Z}, v_1)}{\partial v_1} = \begin{cases} 1, & \text{if } v_1 + Z_2 > Z_3, \\ 0, & \text{otherwise.} \end{cases}$$

Note that in this case, using an appropriate dominating pdf $g(\boldsymbol{z}) = g_2(z_2)g_3(z_3)$, allows estimation of $\nabla \ell(\boldsymbol{v})$ for an entire region of permissible values of $\boldsymbol{v}$. If, however, the sample performance $L(\boldsymbol{y}, \boldsymbol{v}_2)$ in (5.2.1) has a complex form, then estimation of $\nabla_{\boldsymbol{v}_2} \ell(\boldsymbol{v})$ (see (5.2.4)) might become time consuming.

(b) DEDS models

Consider (3.2.27). Differentiating $\ell(\boldsymbol{v})$ with respect to $\boldsymbol{v}_1$ and $\boldsymbol{v}_2$ we obtain

$$\begin{aligned} \nabla_{\boldsymbol{v}_1} \ell(\boldsymbol{v}) &= \frac{\mathbb{E}_g\{\sum_{t=1}^{\tau} L_t \nabla \widetilde{W}_t(\boldsymbol{v}_1)\}}{\mathbb{E}_g\{\sum_{t=1}^{\tau} \widetilde{W}_t(\boldsymbol{v}_1)\}} - \frac{\mathbb{E}_g\{\sum_{t=1}^{\tau} L_t \widetilde{W}_t(\boldsymbol{v}_1)\}}{\mathbb{E}_g\{\sum_{t=1}^{\tau} \widetilde{W}_t(\boldsymbol{v}_1)\}} \\ &\times \frac{\mathbb{E}_g\{\sum_{t=1}^{\tau} \nabla \widetilde{W}_t(\boldsymbol{v}_1)\}}{\mathbb{E}_g\{\sum_{t=1}^{\tau} \widetilde{W}_t(\boldsymbol{v}_1)\}}, \end{aligned} \tag{5.2.6}$$

and

$$\nabla_{\boldsymbol{v}_2} \ell(\boldsymbol{v}) = \frac{\mathbb{E}_g\{\sum_{t=1}^{\tau} \nabla L_t(\boldsymbol{v}_2) \widetilde{W}_t(\boldsymbol{v}_1)\}}{\mathbb{E}_g\{\sum_{t=1}^{\tau} \widetilde{W}_t(\boldsymbol{v}_1)\}}, \tag{5.2.7}$$

provided the interchangeability conditions hold.

With (5.2.6) and (5.2.7) at hand we can derive statistical estimators of $\nabla_{\boldsymbol{v}_1} \ell(\boldsymbol{v})$ and $\nabla_{\boldsymbol{v}_2} \ell(\boldsymbol{v})$ and similarly for higher order partial and mixed derivatives.

Example 5.2.2 *$GI/D/1$* **queue.** Let L_t be the steady-state waiting time of a customer in the $GI/D/1$ queue. In this case (see also example 3.2.2) the expected steady-state performance $\ell(\boldsymbol{v}) = \mathbb{E}_{\boldsymbol{v}_1} L_t$ can be written as

$$\ell(\boldsymbol{v}) = \frac{\mathbb{E}_{\boldsymbol{v}_1}\{\sum_{t=1}^{\tau} L_t(\boldsymbol{v}_2)\}}{\mathbb{E}_{\boldsymbol{v}_1}\tau} = \frac{\mathbb{E}_{\boldsymbol{v}_1}\left\{\sum_{t=1}^{\tau}\sum_{j=1}^{t-1} Z_j\right\}}{\mathbb{E}_{\boldsymbol{v}_1}\tau}, \tag{5.2.8}$$

where $Z_j = v_2 - Y_{1(j+1)}$, v_1 is the constant service time of the customer, $Y_{1j} = A_j - A_{j-1}$, A_j is the time of arrival of the j-th customer, $A_0 = 0$, $\tau = \min\{t : L_t \leq 0\}$, and $Y_{1j} \sim f_1(y_1, \boldsymbol{v}_1)$.

In order to estimate the gradient $\nabla\ell(\boldsymbol{v})$, $\boldsymbol{v} = (\boldsymbol{v}_1, v_2)$, we can first differentiate $\ell(\boldsymbol{v})$ with respect to $\boldsymbol{v}_1$ and v_2 by taking the derivative inside the expected value (why?), and then, based on N regenerative cycles, derive a consistent estimator of $\nabla\ell(\boldsymbol{v})$.

5.2.2 Smoothness and Variance Reduction

We now describe two techniques, called (a) the "push out" and (b) the "push in" techniques. We shall show below that the first technique typically smooths out the sample performance function $L(\boldsymbol{y}, \boldsymbol{v}_2)$ with respect to $\boldsymbol{v}_2$ by making it *independent* of $\boldsymbol{v}_2$, while the second typically leads to a variance reduction. Both techniques are based on the standard change of variables method.

(a) To demonstrate the idea of the "push out" technique consider a simple DESS and suppose that there exist a vector-valued function $\boldsymbol{x} = \boldsymbol{x}(\boldsymbol{y}, \boldsymbol{v}_2)$ and a real-valued function $\tilde{L}(\boldsymbol{x})$ independent of $\boldsymbol{v}_2$, such that $L(\boldsymbol{y}, \boldsymbol{v}_2)$ can be represented as

$$L(\boldsymbol{y}, \boldsymbol{v}_2) = \tilde{L}[\boldsymbol{x}(\boldsymbol{y}, \boldsymbol{v}_2)]. \tag{5.2.9}$$

Furthermore, suppose that for $\boldsymbol{Y} \sim f(\boldsymbol{y}, \boldsymbol{v}_1)$, the corresponding random vector $\boldsymbol{X} = \boldsymbol{x}(\boldsymbol{Y}, \boldsymbol{v}_2)$ has a known pdf $\tilde{f}(\boldsymbol{x}, \boldsymbol{v}_1, \boldsymbol{v}_2)$. Then we can write $\ell(\boldsymbol{v})$ as

$$\ell(\boldsymbol{v}) = \int \tilde{L}(\boldsymbol{x})\tilde{f}(\boldsymbol{x}, \boldsymbol{v})\mathrm{d}\boldsymbol{x} = \mathbb{E}_{\tilde{f}}\{\tilde{L}(\boldsymbol{X})\}, \tag{5.2.10}$$

where the expectation is now taken with respect to the pdf $\tilde{f}(\boldsymbol{x}, \boldsymbol{v}_1, \boldsymbol{v}_2)$. As we mentioned in section 4.1 the term "push out" derives from the fact that the parameter vector $\boldsymbol{v}_2$ is "pushed out" from $L(\boldsymbol{y}, \boldsymbol{v}_2)$ to an auxiliary pdf $\tilde{f}(\boldsymbol{x}, \boldsymbol{v}_1, \boldsymbol{v}_2)$.

It is important to understand that the representation of $L(\boldsymbol{y}, \boldsymbol{v}_2)$ in the form (5.2.9) and the consequent transformation (5.2.10) is not always available, and if available, the corresponding random vector $\boldsymbol{X}$ may not have a density function (with respect to the Lebesgue measure). Also, even if $\tilde{f}(\boldsymbol{x}, \boldsymbol{v}_1, \boldsymbol{v}_2)$ exists, in some cases it may be not easy to calculate it.

Suppose now that for every $\boldsymbol{v}_2$ the function $\boldsymbol{x} = \boldsymbol{x}(\boldsymbol{y}, \boldsymbol{v}_2)$ is one-to-one and thus has an inverse $\boldsymbol{y} = \boldsymbol{y}(\boldsymbol{x}, \boldsymbol{v}_2)$, which is assumed to be continuously differentiable in $\boldsymbol{x}$. In this case we have

$$\tilde{f}(\boldsymbol{x}, \boldsymbol{v}_1, \boldsymbol{v}_2) = f[\boldsymbol{y}(\boldsymbol{x}, \boldsymbol{v}_2), \boldsymbol{v}_1] \left| \frac{\partial \boldsymbol{y}(\boldsymbol{x}, \boldsymbol{v}_2)}{\partial \boldsymbol{x}} \right|, \tag{5.2.11}$$

where $|\partial \boldsymbol{y}/\partial \boldsymbol{x}|$ denotes the absolute value of the determinant of the Jacobian matrix of $\boldsymbol{y}(\boldsymbol{x}, \boldsymbol{v}_2)$ with respect to $\boldsymbol{x}$.

For example, suppose that $\boldsymbol{v}_2$ has the same dimensionality as $\boldsymbol{y}$ and that the function $L(\boldsymbol{y}, \boldsymbol{v}_2)$ can be represented as $L(\boldsymbol{y}, \boldsymbol{v}_2) = \tilde{L}(\boldsymbol{y} + \boldsymbol{v}_2)$. Then we can define $\boldsymbol{x} = \boldsymbol{y} + \boldsymbol{v}_2$, and thus obtain

$$\tilde{f}(\boldsymbol{x}, \boldsymbol{v}) = f(\boldsymbol{x} - \boldsymbol{v}_2, \boldsymbol{v}_1). \tag{5.2.12}$$

The following simple example demonstrates the advantages of using the transformation $\boldsymbol{x} = \boldsymbol{x}(\boldsymbol{y}, \boldsymbol{v}_2)$.

Example 5.2.3 Consider the sample performance function $L(\boldsymbol{Y}, v_1)$ given in example 5.2.1. The corresponding transformation can be written then as $x_1 = v_1 + y_2$ and $x_2 = y_3$. It follows that $\tilde{L}(x) = \max\{x_1, x_2\}$ and $\tilde{f}(x, v_1, v_2, v_3) = f_2(x_1 - v_1, v_2) f_3(x_2, v_3)$. Note that here $\partial L(y, v_1)/\partial v_1$ is a piecewise constant function and that $\partial^2 L(\boldsymbol{y}, v_1)/\partial v_1^2$ is zero almost everywhere. Therefore, in this case, the second-order derivatives cannot be taken inside the expected value. On the other hand, after the transformation the obtained functions are everywhere differentiable in $\boldsymbol{v}$, provided $f_2(x_1 - v_1, v_2)$ is smooth.

In subsection 5.2.4 we discuss the "push out" technique in more detail, and in particular we show how to construct (for typical DES) the auxiliary functions $\tilde{L}$, and $\tilde{f}$ which allow us to estimate from *a single simulation run* the system parameters $\nabla^k \ell(\boldsymbol{v})$, $\boldsymbol{v} = (\boldsymbol{v}_1, \boldsymbol{v}_2)$, $k = 0, 1, \ldots$ (the expected performance and the associated sensitivities) *everywhere* in $\boldsymbol{v} \in V$.

(b) The "push in" technique can be considered as a dual to "push out" in the sense that one searches for a transformation $\boldsymbol{y} = \boldsymbol{y}(\boldsymbol{x}, \boldsymbol{v}_1)$ such that the distribution of the corresponding random vector $\boldsymbol{x} = \boldsymbol{x}(\boldsymbol{Y}, \boldsymbol{v}_1)$ will be independent of $\boldsymbol{v}_1$. In this case, $\ell(\boldsymbol{v})$ can be represented as

$$\ell(\boldsymbol{v}) = \int \hat{L}(\boldsymbol{x}, \boldsymbol{v}) \hat{f}(\boldsymbol{x}) \mathrm{d}\boldsymbol{x} = \mathbb{E}_{\hat{f}}\{\hat{L}(\boldsymbol{X}, \boldsymbol{v})\}, \tag{5.2.13}$$

where $\hat{L}(\boldsymbol{x}, \boldsymbol{v})) = L[\boldsymbol{y}(\boldsymbol{x}, \boldsymbol{v}_1), \boldsymbol{v}_2]$ and $\hat{f}(\boldsymbol{x})$ is the pdf of $\boldsymbol{x}$, (assumed to be independent of $\boldsymbol{v}_1$ and $\boldsymbol{v}_2$ and $\boldsymbol{v} = (\boldsymbol{v}_1, \boldsymbol{v}_2)$) .

5.2.3 Sensitivity Analysis via the IPA Technique

In this subsection we consider the *infinitesimal perturbation analysis* (IPA) method, pioneered by Ho and his co-workers (see Glasserman (1991c), and Ho and Cao). We observe that IPA corresponds to the "push in" technique. Furthemore, the case where $\hat{f} = 1$, on the interval $(0, 1)$, i.e. X has uniform $[0, 1)$ distribution, is of particular interest. In this case the transformation $\boldsymbol{x} = \boldsymbol{x}(\boldsymbol{y}, \boldsymbol{v}_1)$ reduces to

$$x = x(y, \boldsymbol{v}_1) = F(y, \boldsymbol{v}_1) \tag{5.2.14}$$

or

$$y = F^{-1}(x, \boldsymbol{v}_1). \tag{5.2.15}$$

Here $F(y, \boldsymbol{v}_1)$ is the cumulative distribution function (cdf) of the random variable Y. Similar derivations can be applied to DEDS models and have been treated rigorously in Glasserman (1991c), L'Ecuyer (1990a), Ho and Cao (1991), and Suri and Zazanis (1988).

We point out that the original IPA approach is constrained by the fact that the transformation (5.2.14) typically leads to a sample performance function $\hat{L}(\boldsymbol{x}, \boldsymbol{v})$ which is *nondifferentiable* in $\boldsymbol{v}$ (see, for example, Heidelberger et al. (1988) and L'Ecuyer (1990a)). As a result, the required interchangeability conditions for expectation and differentiation do not hold. Furthermore, when $\hat{L}(\boldsymbol{v})$ is not available analytically, typically it allows estimation of the sensitivities at a fixed $\boldsymbol{v}$ *only*, while the SF method allows estimation of $\nabla^k \ell(\boldsymbol{v})$ for an entire region of permissible values of $\boldsymbol{v}$. Consequently, when dealing with optimization problems, the IPA approach must rely on iterative algorithms of the stochastic approximation type, and thus on multiple simulations; (e.g. Kushner and Clark (1978) and L'Ecuyer et al. (1991)). In contrast, its SF method counterpart solves an entire constrained optimization problem from a *single sample path.* Comparison of the efficiency of SA and stochastic counterparts is not an easy task and needs further investigation.

Example 5.2.4 Indicator functions. Suppose that we want to estimate $\nabla \ell(\boldsymbol{v})$, where $\ell(\boldsymbol{v}) = P_{\boldsymbol{v}_1}\{L(Y, \boldsymbol{v}_2) \leq \alpha\}$, and L is a sample performance, say the waiting time in a $GI/G/1$ queue. The IPA approach will not work, since the inverse transformation (5.2.15) leads to a piecewise constant sample function (taking only 0 and 1 values). However, if we represent $\ell(\boldsymbol{v})$ as $\ell(\boldsymbol{v}) = \mathbb{E}_{\boldsymbol{v}_1}\{I_{(-\infty,0)}[L(\boldsymbol{v}_2) - \alpha]\}$, by combining the standard SF approach with the "push out" technique (see subsection 5.3.4 below), we can estimate $\nabla \ell(\boldsymbol{v})$, $\boldsymbol{v} = (\boldsymbol{v}_1, \boldsymbol{v}_2)$, for an entire region of permissible values of $\boldsymbol{v}$ from a single simulation run.

Now we present several cases where the IPA approach is useful in the sense that subject to the interchangeability conditions, its sensitivity estimators typically have *smaller variance* than the standard SF estimators. For more useful IPA examples see Glasserman (1991c), Gong and Ho (1987), Ho and Cao (1991), L'Ecuyer (1990a) and Wardi et al. (1992).

Example 5.2.5 Consider example 5.2.1, that is the sample performance

$$L(\boldsymbol{Y}, v_1) = \max\{(v_1 + Y_2), Y_3\},$$

where $\boldsymbol{Y} = (Y_2, Y_3)$, $Y_i \sim F_i(y, v_i)$, $i = 2, 3$. Using the inverse transformations $Y_i = F_i^{-1}(U_i, v_i), i = 2, 3$, we obtain

$$\hat{L}(\boldsymbol{U}, v_1) = \max\{(v_1 + F_2^{-1}(U_2, v_2)), F_3^{-1}(U_3, v_3)\},$$

where $\boldsymbol{U} = (U_2, U_3)$, U_2 and U_3 are independent uniform (0,1) random variables, and $\boldsymbol{v} = (v_1, v_2, v_3)$. In this case the interchangeability conditions for the gradient (but not for the Hessian, why?) hold, and we obtain

$$\nabla \ell(\boldsymbol{v}) = \mathbb{E}\{\nabla \hat{L}(\boldsymbol{U}, \boldsymbol{v})\} .$$

The IPA estimator can then be written as

$$\nabla \hat{\ell}_N(\boldsymbol{v}) = N^{-1} \sum_{i=1}^{N} \nabla \hat{L}(\boldsymbol{U}_i, \boldsymbol{v}). \tag{5.2.16}$$

Note that the first-order derivatives of $\hat{L}(\boldsymbol{U}, \cdot)$ are piecewise continuous functions with discontinuities at the points where $v_1 + F_2^{-1}(U_2, v_2) = F_3^{-1}(U_3, v_3)$.

Consider now more complex DESS models, say those given in formulas (2.1.1) and (2.1.2). Although in this case computation of the sample performance $\hat{L}(\boldsymbol{U}_i, \boldsymbol{v})$ while using the IPA estimator (5.2.16) might be time consuming, we still can obtain a variance reduction compared with the standard SF estimator (2.2.9) (see also (2.2.22)), provided the dimension of the vector $\boldsymbol{v} = (v_1, \ldots, v_n)$ is large.

As for another situation where the IPA estimators may outperform their standard SF counterparts, consider regenerative simulation with long regenerative cycles. The RSF (regenerative score function) estimators are useless here since the variance of the associated likelihood ratio process $\widetilde{W}_t$ (see section 3.3) is large. The IPA approach will produce estimators which have lower variance (see Glasserman (1991c), Ho and Cao (1991), and L'Ecuyer (1990a, 1992)), provided the interchangeability conditions hold. Note also that in this case we may use the following alternatives to IPA: the CSF (conditional score function) estimators of McLeish and Rollans (1992), the DSF (decomposable score function) and TSF (truncated score function) estimators (see section 3.3). It is also important to note that the IPA and CSF estimators are *consistent*, while their DSF and TSF estimators are *asymptotically biased.* Note finally that when the output process is not regenerative we can still use the CSF, DSF,TSF, and IPA estimators, but not the RSF one.

To provide more insight into the IPA and SF approaches, we consider separately (a) the regenerative and (b) the nonregenerative estimators of $\nabla \ell(\boldsymbol{v})$, assuming that $\ell(\boldsymbol{v})$ is the steady-state expected waiting time in a $GI/G/1$ queue.

(a) The regenerative IPA sensitivity estimators

Example 5.2.6 Let L_t be the steady-state waiting time of a customer in the $GI/G/1$ queue. In this case (see also example 3.2.2) $\ell(\boldsymbol{v}) = \mathbb{E}_{\boldsymbol{v}} L_t$ can be written as

$$\ell(\boldsymbol{v}) = \frac{\mathbb{E}_{\boldsymbol{v}}\{\sum_{t=1}^{\tau} L_t\}}{\mathbb{E}_{\boldsymbol{v}}\tau} = \frac{\mathbb{E}_{\boldsymbol{v}}\left\{\sum_{t=1}^{\tau} \sum_{j=1}^{t-1} Z_j\right\}}{\mathbb{E}_{\boldsymbol{v}}\tau}, \tag{5.2.17}$$

where $Z_j = Y_{1j} - Y_{2(j+1)}$, Y_{1j} is the service time of the j-th customer, $Y_{2j} = A_j - A_{j-1}$, A_j is the time of arrival of the j-th customer, $A_0 = 0$ and $\tau = \min\{t : L_t \leq 0\}$.

The RSF "what if" estimators of $\ell(\boldsymbol{v})$ and $\nabla \ell(\boldsymbol{v})$ are given by (3.2.30) and (3.2.31), respectively. To derive the IPA estimator of $\nabla \ell(\boldsymbol{v})$ we may use the following algorithm:

Algorithm 5.2.1 :

1. *Generate the output process* $\hat{L}_t(\boldsymbol{v})$ *as*

$$\hat{L}_t(\boldsymbol{v}) = \sum_{j=1}^{t-1} \hat{Z}_j(\boldsymbol{v}) = \sum_{j=1}^{t-1}[F_1^{-1}(U_{1j}, v_1) - F_2^{-1}(U_{2(j+1)}, v_2)], \tag{5.2.18}$$

where $F_1(Y_1, v_1)$ *and* $F_2(Y_2, v_2)$ *are the cdf's of the rv's* Y_1 *and* Y_2, *respectively.*

2. *Represent* $\ell(\boldsymbol{v})$ *in the ratio form (5.2.17) with* L_t *replaced by* $\hat{L}_t(\boldsymbol{v})$, *and calculate the corresponding derivatives.*

3. *Run the* $GI/G/1$ *for* N *regenerative cycles and estimate* $\nabla \ell(\boldsymbol{v})$ *accordingly.*

(b) The nonregenerative IPA sensitivity estimators

Consider Lindley's recursive (sample path) equation for the waiting time process L_t in the $GI/G/1$ queue, that is (e.g. Kleinrock (1975), p. 277),

$$L_{t+1} = \max\{0, L_t + Z_t\}, \quad t \geq 1, \; L_1 = 0, \tag{5.2.19}$$

where $Z_t = Y_{1t} - Y_{2(t+1)}$.

Consider separately the SF and the IPA approaches. The straightforward SF approach based on the representation

$$\nabla^k \ell(\boldsymbol{v}) = \mathbb{E}_g\{L_t \nabla^k \widetilde{W}_t\}, \tag{5.2.20}$$

(see section 3.3) *fails* since for large t the variance of the associated SF estimators (see, for example, (2.2.29)) is very large. Here $\widetilde{W}_t = \prod_{j=1}^t W_j$, $W_j = f(\boldsymbol{Z}, \boldsymbol{v})/g(\boldsymbol{Z})$, and $\boldsymbol{Z} \sim g(\boldsymbol{z})$. Its counterpart, the IPA approach, produces, however, an unbiased estimator of $\nabla \ell(\boldsymbol{v})$ with a manageable variance. The corresponding IPA algorithm for estimation of $\nabla \ell(\boldsymbol{v})$ is similar to algorithm 5.2.1 and can can be written as follows:

Algorithm 5.2.2 :

1. *Generate the output process $\hat{L}_t(\boldsymbol{v})$ by using Lindley's recursive equation*

$$\hat{L}_{t+1}(\boldsymbol{v}) = \max\{0, \hat{L}_t + \hat{Z}_t(\boldsymbol{v})\}, \quad t \geq 1, \; L_1 = 0. \tag{5.2.21}$$

   where $\hat{Z}_t(\boldsymbol{v}) = F_1^{-1}(U_{1t}, v_1) - F_2^{-1}(U_{2(t+1)}, v_2)$.

2. *Differentiate $\ell(\boldsymbol{v}) = \mathbb{E}\hat{L}_t(\boldsymbol{v})$ with respect to $\boldsymbol{v}$ (by taking the derivative inside the expected value), where $\hat{L}_t$ is given in (5.2.21) and derive* $\nabla \ell(\boldsymbol{v}) = \mathbb{E}\nabla \hat{L}_t(\boldsymbol{v})$.

3. *Run the $GI/G/1$ for N customers and estimate $\nabla \ell(\boldsymbol{v})$ as*

$$\nabla \hat{\ell}_N(\boldsymbol{v}) = N^{-1} \sum_{t=1}^{N} \nabla \hat{L}_t(\boldsymbol{v}). \tag{5.2.22}$$

5.2.4 Sensitivity Analysis by the "Push out" Technique

This subsection is based on Rubinstein (1992a) and is organized as follows. Part (A) deals with DESS. In particular we show how to construct the auxiliary functions $\tilde{L}$ and $\tilde{f}$ (see (5.2.9), (5.2.11) and (5.2.11)) in order to estimate the system parameters $\nabla^k \ell(\boldsymbol{v})$, $\boldsymbol{v} = (\boldsymbol{v}_1, \boldsymbol{v}_2)$, by the "push out" technique for several typical DESS. We shall put emphasis on the case where the sample performance function $L(\boldsymbol{y}, \boldsymbol{v}_2)$ is not everywhere differentiable in $\boldsymbol{v}_2$.

Additionally, we show (see example 5.2.8 below) that using the "push out" technique one can easily estimate the distribution and density functions of the sample performance $L(\boldsymbol{Y})$. Part (B) deals with application of the "push out" technique to DEDS and in particular extends the results of part (A) to queueing models. Here we also discuss how to estimate the system parameters $\ell(\boldsymbol{v})$ and $\nabla^k \ell(\boldsymbol{v})$, $k = 1, \ldots,$ provided the sample performance process $\{L_t : t > 0\}$, say the sojourn time process in a queueing model, is not available analytically and can be written as $L_t = L_t(\underline{\boldsymbol{Y}}_t, \boldsymbol{v}_2)$, where $\underline{\boldsymbol{Y}}_t = (\boldsymbol{Y}_1, \boldsymbol{Y}_2, \ldots, \boldsymbol{Y}_t)$ is the input sequence of iid (independent identically distributed) random vectors generated from the pdf $f(\boldsymbol{y}, \boldsymbol{v}_1)$. Finally, in part (C) numerical examples supporting the "push out" approach are given.

(A) The "push out" technique for DESS

Let $g(\boldsymbol{x})$ be a pdf dominating the pdf $\tilde{f}(\boldsymbol{x}, \boldsymbol{v})$. Then in analogy to (5.2.1)–(5.2.3) we have

$$\nabla^k \ell(\boldsymbol{v}) = \mathbb{E}_g\{\tilde{L}(\boldsymbol{Z}) \nabla^k \widehat{W}(\boldsymbol{Z}, \boldsymbol{v})\}, \tag{5.2.23}$$

where

$$\widehat{W}(\boldsymbol{z}, \boldsymbol{v}) = \tilde{f}(\boldsymbol{z}, \boldsymbol{v}) / g(\boldsymbol{z}) \tag{5.2.24}$$

and $\boldsymbol{Z} \sim g(\boldsymbol{z})$. Consequently, for a given random sample $\boldsymbol{Z}_1, \ldots, \boldsymbol{Z}_N$ from $g(\boldsymbol{z})$ we can estimate $\nabla^k \ell(\boldsymbol{v})$ by

$$\nabla^k \tilde{\ell}_N(\boldsymbol{v}) = N^{-1} \sum_{i=1}^{N} \tilde{L}(\boldsymbol{Z}_i) \nabla^k \widehat{W}(\boldsymbol{Z}_i, \boldsymbol{v}), \quad k = 0, 1, \ldots, \tag{5.2.25}$$

where by convention

$$\nabla^0 \ell_N(\boldsymbol{v}) \equiv \ell_N(\boldsymbol{v}).$$

We shall call the estimator $\nabla^k \tilde{\ell}_N(\boldsymbol{v})$ the POSF *(push out score function)* estimator.

We now present two examples of the application of the "push out" technique. The first example deals with the case where the sample function $L(\boldsymbol{y}, \boldsymbol{v}_2)$ is not necessarily everywhere differentiable with respect to $\boldsymbol{v}_2$, while the second example shows how the method can yield a convenient parametric description of the distribution and the density function of the rv $L(\boldsymbol{Y})$ (and of the associated estimators), provided $L(\boldsymbol{y})$ can be represented as $L(\boldsymbol{y}) = L(\boldsymbol{y}_1) + L(\boldsymbol{y}_2)$. We shall call such $L(\boldsymbol{y})$ an *additive sample function.*

Example 5.2.7 Estimating $\nabla^k \ell(\boldsymbol{v})$ in a stochastic-PERT network. Consider the following particular case of (5.1.2):

$$L(\boldsymbol{Y}) = \min(Y_1 + Y_3,\ Y_2 + Y_4) \tag{5.2.26}$$

with Y_3 and Y_4 being deterministic, say $Y_3 = v_3$ and $Y_4 = v_4$.

In this case, in order to construct the POSF estimator $\nabla^k \tilde{\ell}_N(\boldsymbol{v})$ we consider the following steps:

(i) Define $x_1 = y_1 + v_3$, $x_2 = y_2 + v_4$ and the auxiliary sample performance $\tilde{L}(\boldsymbol{x}) = \min(x_1, x_2)$, with $\boldsymbol{x} = (x_1, x_2)$,

(ii) Calculate the auxiliary pdf $\tilde{f}(\boldsymbol{x}, \boldsymbol{v})$ of the random vector $\boldsymbol{X} = (X_1, X_2)$, where $X_1 = Y_1 + v_3$ and $X_2 = Y_2 + v_4$, by using formula (5.2.12).

(iii) Choose a dominating pdf $g(\boldsymbol{z})$ for $\tilde{f}(\boldsymbol{z}, \boldsymbol{v})$ and finally estimate $\nabla^k \ell(\boldsymbol{v})$ according to (5.2.25).

It is interesting to note that the sample function

$$L(y_1, y_2, v_3, v_4) = \min(y_1 + v_3,\ y_2 + v_4) \tag{5.2.27}$$

is not everywhere differentiable. In particular the derivatives $\partial L / \partial v_3$ and $\partial L / \partial v_4$ are discontinuous at the points where $y_1 + v_3 = y_2 + v_4$. For example, $\partial L / \partial v_3$ equals one if $y_1 + v_3 < y_2 + v_4$ and zero if $y_1 + v_3 > y_2 + v_4$. Therefore the associated second-order derivatives cannot be taken inside the expected value in (5.2.4). On the other hand, formula (5.2.23), and therefore the "push out" technique, are applicable here, provided the corresponding pdf $f(\boldsymbol{y}, \boldsymbol{v}_1)$ is sufficiently smooth. Thus, under smoothness of $f(\boldsymbol{y}, \boldsymbol{v}_1)$, the "push out" technique *smooths out* the original *nonsmooth* performance measures.

Note also that if we consider, instead of (5.2.27), say

$$L(y_1, y_2, v_3, v_4) = \min\{\max(y_1, v_3), \max(y_2, v_4)\},$$

then we could still define $x_1 = \max(y_1, v_3)$, $x_2 = \max(y_2, v_4)$ and $\tilde{L}(\boldsymbol{x}) = \min(x_1, x_2)$. In this case, however, the corresponding random variables X_1 and X_2 would take values v_3 and v_4 simultaneously with nonzero probability and hence $\boldsymbol{X} = (X_1, X_2)$ would not have a density function. Consequently, in such a case the SF method would need to be applied to distributions representing a mixture of continuous and discrete types. This problem requires, however, further investigations.

Example 5.2.8 Estimating the cdf of a DESS for the case where $L(\boldsymbol{y})$ is an additive function. Suppose that we want to estimate the cdf of $L(\boldsymbol{Y})$, that is

$$H_L(x, \boldsymbol{v}) = P_{\boldsymbol{v}}\{L(\boldsymbol{Y}) \le x\} \tag{5.2.28}$$

for different values of $x \le x_0$ and $\boldsymbol{v}$, where $\boldsymbol{Y} = (Y_1, \ldots, Y_n)$, $\boldsymbol{Y} \sim f(\boldsymbol{y}, \boldsymbol{v})$ and x_0 is a fixed number. Assume that the function $L(\boldsymbol{y})$ can be represented as $L(\boldsymbol{y}) = L_1(\boldsymbol{y}_1) + L_2(y_2)$, and that the cdf of the random variable $L_2(Y_2)$ is available. In particular, assume without loss of generality that $L_2(y_2) = y_n$. Then the procedure of estimating $H_L(x, \boldsymbol{v})$ can be written as follows:

(i) Write H_L as

$$H_L(\boldsymbol{x}, \boldsymbol{v}) = \mathbb{E}_{\boldsymbol{v}}\{I_{(-\infty,0]}(L - x)\}. \tag{5.2.29}$$

(ii) Change the vector of variables $\boldsymbol{y} = (y_1, \ldots, y_n)$ to $\boldsymbol{z} = (z_1, \ldots, z_n)$, where $z_1 = y_1, \ldots, z_{n-1} = y_{n-1}, z_n = y_n - x$, and write the cdf H_L as

$$H_L(x, \boldsymbol{v}) = \mathbb{E}_{\boldsymbol{u}}\{\tilde{I}(\boldsymbol{Z})\}, \tag{5.2.30}$$

where $\boldsymbol{u} = (x, \boldsymbol{v})$, $\boldsymbol{z}_1 = (z_1, \ldots, z_{n-1})$, $\tilde{I}(\boldsymbol{z}) = I_{(-\infty,0]}(L(\boldsymbol{z}_1) + z_n)$ and the expectation is taken with respect to the pdf
$\tilde{f}(\boldsymbol{z}, \boldsymbol{u}) = f(\boldsymbol{z}_1, z_n + x, \boldsymbol{v})$.

(iii) Represent the cdf H_L as

$$H_L(x, \boldsymbol{v}) = \mathbb{E}_g\{\tilde{I}(\boldsymbol{Z})\widehat{W}(\boldsymbol{Z}, \boldsymbol{u})\}, \tag{5.2.31}$$

where $g(\boldsymbol{z})$ is a pdf dominating the pdf $\tilde{f}(\boldsymbol{z}, \boldsymbol{u})$, $\widehat{W}(\boldsymbol{z}, \boldsymbol{u}) = \tilde{f}(\boldsymbol{z}, \boldsymbol{u})/g(\boldsymbol{z})$, and $\boldsymbol{Z} \sim g(\boldsymbol{z})$.

(iv) Take a random sample $\boldsymbol{Z}_1, \ldots, \boldsymbol{Z}_N$ from $g(\boldsymbol{z})$ and estimate $H_L(x, \boldsymbol{v})$ as

$$\tilde{H}_{LN}(x, \boldsymbol{v}) = N^{-1} \sum_{i=1}^{N} \tilde{I}(\boldsymbol{Z}_i)\widehat{W}(\boldsymbol{Z}_i, \boldsymbol{u}). \tag{5.2.32}$$

If we choose $g(\boldsymbol{z}) = f(\boldsymbol{z}, \boldsymbol{u}_0)$, where $\boldsymbol{u}_0 = (x_0, \boldsymbol{v}_0)$ and $x_0 \geq x$, and if the components Y_k, $k = 1, \ldots, n$, of the vector $\boldsymbol{Y}$ are independent, distributed according to the pdf's $f_k(z_k, v_k)$, then the likelihood ratio $\hat{W}$ can be written in the form

$$\widehat{W}(\boldsymbol{z}, \boldsymbol{u}) = \frac{f_n(z_n + x, v_n)}{f_n(z_n + x_0, v_{0n})} \cdot \prod_{k=1}^{n-1} \frac{f_k(z_k, v_k)}{f_k(z_k, v_{0k})}.$$

Extension to Sensitivity Analysis

Suppose that the pdf $\tilde{f}(\boldsymbol{z}, \boldsymbol{u})$ is p times continuously differentiable with respect to both x and $\boldsymbol{v}$. Then under standard regularity conditions the sensitivities $\nabla^k H_L(x, \boldsymbol{v})$ and their associated estimators $\nabla^k \tilde{H}_{LN}(x, \boldsymbol{v})$ can be written as

$$\nabla^k H_L(x, \boldsymbol{v}) = \mathbb{E}_g\{\tilde{I}(\boldsymbol{Z})\nabla^k \widehat{W}(\boldsymbol{Z}, \boldsymbol{u})\} \tag{5.2.33}$$

and

$$\nabla^k \tilde{H}_{LN}(x, \boldsymbol{v}) = N^{-1} \sum_{i=1}^{N} \tilde{I}(\boldsymbol{Z}_i)\nabla^k \widehat{W}(\boldsymbol{Z}_i, \boldsymbol{u}), \tag{5.2.34}$$

respectively, where $\nabla \widehat{W}$ is called the *score function.* Note that in particular

$$h_L(\boldsymbol{x}, \boldsymbol{v}) = \frac{\partial H_L(\boldsymbol{x}, \boldsymbol{v})}{\partial \boldsymbol{x}}$$

corresponds to the pdf of the random variable L and can be estimated as

$$\tilde{h}_{LN}(\boldsymbol{x}, \boldsymbol{v}) = \frac{\partial \tilde{H}_{LN}(\boldsymbol{x}, \boldsymbol{v})}{\partial \boldsymbol{x}} = N^{-1} \sum_{i=1}^{N} \tilde{I}(\boldsymbol{Z}_i) \frac{\partial \widehat{W}(\boldsymbol{Z}_i, \boldsymbol{x}, \boldsymbol{v})}{\partial \boldsymbol{x}}.$$

Denoting $\nabla^0 H \equiv H$ and $\nabla^0 W \equiv W$ the algorithm for estimating the quantities $\nabla^k H_L(\boldsymbol{x}, \boldsymbol{v})$, $k = 0, 1, \ldots, p$ (the cdf H_L and the sensitivities $\nabla^k H_L$, $k = 1, 2, \ldots, p$) can be written as follows:

Algorithm 5.2.3 :

1. *Given* $\boldsymbol{Y} \sim f(\boldsymbol{y}, \boldsymbol{v})$ *and* $L(\boldsymbol{Y}) = L_1(Y_1, \ldots, Y_{n-1}) + Y_n$ *define the new pdf* $\tilde{f}(\boldsymbol{y}, \boldsymbol{u}), \boldsymbol{u} = (\boldsymbol{x}, \boldsymbol{v})$.
2. *Choose a pdf* $g(\boldsymbol{z})$ *dominating the pdf* $\tilde{f}(\boldsymbol{z}, \boldsymbol{u})$ *and take a sample* $\boldsymbol{Z}_1, \ldots, \boldsymbol{Z}_N$ *from* $g(\boldsymbol{z})$.
3. *Estimate the cdf* $H_L(\boldsymbol{x}, \boldsymbol{v})$ *and the associated sensitivities* $\nabla^k H_L(\boldsymbol{x}, \boldsymbol{v})$, $k = 1, 2, \ldots,$ *by using the SF estimator (5.2.34).*

Note that deriving the *smooth estimators* of pdf's of discontinuous random variables is one of the *major problems in the area of density estimation* (e.g. Topia and Thompson (1978)). The advantage of the POSF estimators $\nabla^k \tilde{H}_{LN}(\boldsymbol{x}, \boldsymbol{v})$ is that typically they are *smooth* in $\boldsymbol{x}$, when the underlying pdf $\tilde{f}(\boldsymbol{z}, \boldsymbol{u})$ is too (e.g. the exponential family).

(B) The "push out" technique for queueing models

In this subsection we show how to estimate the cdf of the steady-state output processes in a queueing model, like the cdf of the steady-state sojourn time, the cdf of the steady-state number of customers in the system, etc. and some associated quantities, while using the "push out" technique, provided the underlying output processes are regenerative. Although our main emphasis will be on the $GI/G/1$ queue, most of our results can be easily adapted to other queueing models, like the $GI/D/1$ and $D/G/1$ and to rather general queueing networks. Here we also show that in order to estimate the system parameters $\nabla^k \ell(\boldsymbol{v})$, $k = 0, 1, \ldots,$ we do not need to know the analytical expressions for the underlying output processes, say the sojourn time process.

Before proceeding further, we shall introduce the following notations:

(i) Y_n and A_n denote the service time of the n-th customer and the time between the arrivals of the n-th and $(n-1)$-st customer, respectively, where $Y_0 = A_0 = 0$.

(ii) $L_n, \mathcal{L}_n, V_t, R_n$ and M_t denote the sojourn time, the waiting time in the queue, the virtual waiting time, the queue length just prior to the n-th arrival and the queue length, respectively.

(iii) X denotes a random variable having the steady-state distribution H of X_n, that is

$$H_X(x, \boldsymbol{v}) = P_{\boldsymbol{v}}\{X \leq x\} = \lim_{n\to\infty} P_{\boldsymbol{v}}\{X_n \leq x\}.$$

With these notations at hand let us show how the "push out" technique makes it possible to estimate (for different values of x and $\boldsymbol{v}$) the cdf $H_X(x, \boldsymbol{v})$ and the pdf $h_X(x, \boldsymbol{v}) = \partial H_X(x, \boldsymbol{v})/\partial x$ of the random variables $X = \{L, \mathcal{L}, V, R \text{ and } M\}$ and some related quantities using a *single simulation run.*

(a) We start with the cdf of the steady-state random variable (rv)

$$X = \sum_{t=1}^{\tau} L_t\,,$$

where τ is the number of customers served during the busy period in the $GI/G/1$ queue with the FIFO discipline, assuming that $\boldsymbol{v}$ is a vector of parameters in the service time distribution $f(y, \boldsymbol{v})$. We note first that using the standard likelihood ratios (see chapters 2 and 3), we can represent H_X as

$$H_X(x, \boldsymbol{v}) = \mathbb{E}_g\{I_{(-\infty,0]}(X - x)\widetilde{W}_\tau(\underline{Z}_\tau, \boldsymbol{v})\}, \tag{5.2.35}$$

where

$$\widetilde{W}_t = \prod_{j=1}^{t} W_j, \;\; W_j = \frac{f(Z_j, \boldsymbol{v})}{g(Z_j)}\;, \;\; Z_j \sim g(z)$$

$\underline{Z}_t = (Z_1, \ldots, Z_t)$ and similarly for the higher order derivatives of H_X. Note that the standard likelihood estimators of $H_X(x, \boldsymbol{v})$ based on (5.2.35) require calculation of the indicator function $I_{(-\infty,0]}(X - x)$ for each value of x separately.

The following procedure, based on the "push out" technique, permits estimation of $H_X(x, \boldsymbol{v})$ from a *single simulation simultaneously* for different values of x and $\boldsymbol{v}$ and can be written as follows:

(i) Represent $H_X(x, \boldsymbol{v})$ as

$$H_X(x, \boldsymbol{v}) = \mathbb{E}_{\boldsymbol{v}}\{I_{(-\infty,0]}(X - x)\}.$$

(ii) Taking into account that

$$L_n = \mathcal{L}_n + Y_n, \quad n = 1, 2, \ldots, \tag{5.2.36}$$

where L_n and $\mathcal{L}_n$ are the sojourn and the waiting time of a customer in the $GI/G/1$ queue, rewrite H_X as

$$H_X(\boldsymbol{x}, \boldsymbol{v}) = \mathbb{E}_{\boldsymbol{u}}\{I_{(-\infty,0]}(\tilde{X})\}, \quad \boldsymbol{u} = (\boldsymbol{x}, \boldsymbol{v}), \tag{5.2.37}$$

where

$$\tilde{X} = \sum_{n=1}^{\tau-1} L_n + \mathcal{L}_\tau + \tilde{Y}_\tau, \tag{5.2.38}$$

$\tilde{Y}_\tau = Y_\tau - x$ and $\tilde{Y}_\tau \sim \tilde{f}(y, \boldsymbol{u})$. Note that $\tilde{f}(y, 0, \boldsymbol{v}) = f(y, \boldsymbol{v})$.

(iii) With (5.2.37) and (5.2.38) in hand, represent the cdf $H_X(\boldsymbol{x}, \boldsymbol{v})$ as

$$H_X(\boldsymbol{x}, \boldsymbol{v}) = \mathbb{E}_g\{\tilde{I}_\tau \widetilde{W}_\tau(\boldsymbol{u})\}, \tag{5.2.39}$$

where

$$\begin{aligned} \widetilde{W}_\tau(\boldsymbol{u}) &= W_\tau(\boldsymbol{u}) \prod_{j=1}^{\tau-1} W_j(\boldsymbol{v}),\ W_j(\boldsymbol{v}) = \frac{f(Z_j, \boldsymbol{v})}{g(Z_j)},\ j = 1, \ldots, \tau - 1, \\ W_\tau(\boldsymbol{u}) &= \tilde{f}(Z_\tau, \boldsymbol{u})/g(Z_\tau),\ Z_k \sim g(z),\ k = 1, \ldots, \tau, \end{aligned} \tag{5.2.40}$$

and

$$\tilde{I}_\tau = I_{(-\infty,0]}\left(\sum_{n=1}^{\tau-1} L_n + \mathcal{L}_\tau + Z_\tau\right). \tag{5.2.41}$$

(iv) Estimate $H_X(\boldsymbol{x}, \boldsymbol{v})$ as

$$\tilde{H}_{XN}(\boldsymbol{x}, \boldsymbol{v}) = N^{-1} \sum_{i=1}^{N} \tilde{I}(\underline{\boldsymbol{Z}}_{\tau i}) \widetilde{W}(\underline{\boldsymbol{Z}}_{\tau i}, \boldsymbol{u}), \tag{5.2.42}$$

where $\underline{\boldsymbol{Z}}_{\tau i} = (Z_{1i}, \ldots, Z_{\tau i})$, $i = 1, \ldots, N$, is a random sample from $g(z)$ and similarly the estimators of the higher order derivatives of $H_X(\boldsymbol{x}, \boldsymbol{v})$. For example, let

(i) $f(y, v) = v \exp(-vy)$. Then $\tilde{f}(y, x, v) = v \exp(-v(y + x))$, $y \geq -x$. In this case the dominating pdf can be chosen as

$$g(y) = v_0 \exp(-v_0(y + x_0)), \tag{5.2.43}$$

where $x < x_0$ and $y > -x_0$. Note that we could also choose $\tilde{f}(y, x, v) = f(y, v)$ for $n = 1, \ldots, \tau - 1$ and $\tilde{f}(y, x, v) = v \exp(-(y + x))$ for $n = \tau$, and similarly $g(y)$.

(ii) $f(y, v) = v \exp(-v(y - x_0), y > x_0 > 0$. Then $\tilde{f}(y, v, x) = v \exp(-v(y - x_0 + x)),\ y > x_0 - x$, and the dominating pdf can be chosen as $g(y) = v_0 \exp(-v_0 y),\ y > 0$.

In the special case of the $GI/G/\infty$ queue, $\tau = 1$ and (5.2.42) reduces to (5.2.32).

Remark 5.2.1 Let M_t be the steady-state number of customers in the $GI/G/1$ queue. Assume that we want to estimate the cdf

$$H_U(x, \boldsymbol{v}) = P_{\boldsymbol{v}}\{U_\eta \leq x\}$$

of the rv $U_\eta = \int_0^\eta M_t dt$ representing the total number of customers served in a regenerative cycle of length η. Taking into account that for a fixed sample path $U_{\eta_i} = X_{\tau_i},\ i = 1, \ldots, N$, we immediately obtain $H_U = H_X$.

It is important to note that the above procedure can be adapted to general queueing networks and in particular to those where neither L_n nor $\mathcal{L}_\tau$ are analytically available. This is because both rv's $\sum_{n=1}^{\tau-1} L_n$ and $\mathcal{L}_\tau$ of the sample performance

$$\tilde{I}_\tau = I_{(-\infty,0]}\left(\sum_{n=1}^{\tau-1} L_n + \mathcal{L}_\tau + Z_\tau\right)$$

are obtained from simulation, and the rv Z_τ is generated from the dominating pdf $g(z)$.

(b) We now present a procedure for estimating the cdf

$$H_L(x, \boldsymbol{v}) = P_{\boldsymbol{v}}(L \leq x) \tag{5.2.44}$$

of the steady-state sojourn time rv L in the $GI/G/1$ queue. To do so, we argue similarly to (5.2.36)–(5.2.42), namely

(i) Write $H_L(x, \boldsymbol{v})$ as

$$H_L(x, \boldsymbol{v}) = \mathbb{E}_{\boldsymbol{v}}\{I_{(-\infty,0]}(L - x)\}. \tag{5.2.45}$$

(ii) Represent H_L as

$$H_L(x, \boldsymbol{v}) = \mathbb{E}_{\boldsymbol{u}}\{I_{(-\infty,0]}(\tilde{L}_n)\}\,, \tag{5.2.46}$$

where $\tilde{L}_n = \mathcal{L}_n + \tilde{Y}_n,\ Y_n \sim f(y, \boldsymbol{v}),\ \tilde{Y}_n = Y_n - x$ and $\tilde{Y}_n \sim \tilde{f}(y, \boldsymbol{u})$.

(iii) Taking into account that $I_n = I_{(-\infty,0]}(\tilde{L}_n)$ is a regenerative process, write H_L as

$$H_L(\boldsymbol{x}, \boldsymbol{v}) = \frac{\mathbb{E}_{\boldsymbol{u}}\{\sum_{n=1}^{\tau} I_n\}}{\mathbb{E}_{\boldsymbol{u}}\tau}. \tag{5.2.47}$$

(iv) With any dominating pdf $g(z)$ at hand we can rewrite the cdf H_L in (5.2.47) as

$$H_L(\boldsymbol{x}, \boldsymbol{v}) = \frac{\mathbb{E}_g \sum_{n=1}^{\tau} \tilde{I}_n \widetilde{W}_n(\boldsymbol{u})}{\mathbb{E}_g \sum_{n=1}^{\tau} \widetilde{W}_n(\boldsymbol{u})}, \tag{5.2.48}$$

where

$$\tilde{I}_n = I_{(-\infty,0]}(\mathcal{L}_n + Z_n), \tag{5.2.49}$$

$$\widetilde{W}_n(\boldsymbol{u}) = W_n(\boldsymbol{u}) \prod_{j=1}^{n-1} W_j(\boldsymbol{v}), \tag{5.2.50}$$

$$W_j(\boldsymbol{v}) = \frac{f(Z_j, \boldsymbol{v})}{g(Z_j)}, \quad W_n(\boldsymbol{u}) = \frac{\tilde{f}(Z_n, \boldsymbol{u})}{g(Z_n)}, \quad Z_n \sim g(z), \ n = 1, \dots, \tau.$$

It is not difficult to see that since $\tilde{Y} = Y + \boldsymbol{x}$, we can choose the dominating pdf $g(y)$ equal to $\tilde{f}(y, \boldsymbol{x}_0, \boldsymbol{v}_0)$, provided $\boldsymbol{x}_0 \geq \boldsymbol{x}$ and $y \geq 0$.

(v) Taking into account that (5.2.48) is presented as the ratio of two expected values (both taken with respect to the same probability measure $g(z)$), we can estimate $H_L(\boldsymbol{x}, \boldsymbol{v})$ as

$$\tilde{H}_{LN}(\boldsymbol{x}, \boldsymbol{v}) = \frac{\sum_{i=1}^{N} \sum_{n=1}^{\tau_i} \tilde{I}_{ni}(\underline{\boldsymbol{Z}}_{ni}) \widetilde{W}_{ni}(\underline{\boldsymbol{Z}}_{ni}, \boldsymbol{u})}{\sum_{i=1}^{N} \sum_{n=1}^{\tau_i} \widetilde{W}_{ni}(\underline{\boldsymbol{Z}}_{ni}, \boldsymbol{u})}, \tag{5.2.51}$$

where $\underline{\boldsymbol{Z}}_{ni} = (Z_{1i}, \dots, Z_{ni})$, $n = 1, \dots, \tau_i$; $i = 1, \dots, N$, is a random sample from $g(z)$.

In order to estimate the pdf h of the steady-state sojourn time process L_n we have to differentiate (with respect to $\boldsymbol{x}$) the ratio formula (5.2.51) and similarly for the estimators of the higher order derivatives of $H_L(\boldsymbol{x}, \boldsymbol{v})$.

It is important to note again that the above procedure can be adapted to *rather general queueing networks* and in particular to those where the underlying process $\mathcal{L}_n$ is analytically not available. This is because both rv's $\mathcal{L}_n$ and Z_n associated with the sample performance

$$\tilde{I}_n = I_{(-\infty,0]}(\mathcal{L}_n + Z_n)$$

can be obtained from simulation and the random variable $\tilde{Y}$ can be generated from the dominating pdf $g(y) = \tilde{f}(y, \boldsymbol{x}_0, \boldsymbol{v}_0)$.

We shall use next the "push out" technique to estimate the cdf's of the virtual time V_t, and the standard SF method to estimate both the queue length just prior to the n-th arrival R_n and the queue length M_t in the $GI/G/1$ queue. Before doing so, let us note that there are many useful (sample path) relationships between L_n, V_t, R_n and M_t which can serve our purpose. We shall adapt some of these from chapter 8 of Asmussen (1987). More specifically, we shall use Asmussen's formulas: (3.6), (4.10)–(4.11) and (4.4)–(4.7) which in our notation can be written as

$$(c)\ H_V(x, \boldsymbol{v}) = 1 - \frac{\mathbb{E}_{\boldsymbol{v}} \sum_{n=1}^{\tau} A_n \Lambda (L_n + Y_n - x)^+}{\mathbb{E}_{\boldsymbol{v}} \tau}, \tag{5.2.52}$$

$$(d)\ H_R(r, \boldsymbol{v}) = P_{\boldsymbol{v}}\{R_n \leq r\} \tag{5.2.53}$$

$$= P_{\boldsymbol{v}} \left\{ L_{n-r} \leq \sum_{j=1}^{r} A_{n-j} \right\},$$

$$(e)\ H_M(r, \boldsymbol{v}) = 1 - \frac{1}{\mathbb{E}_{\boldsymbol{v}} \tau} \sum_{n=1}^{\tau} \{(L_n + Y_n - A_n^{(r)})^+ \tag{5.2.54}$$

$$- (L_n - A_n^{(r)})^+\}, \quad r = 0, 1, \ldots.$$

Here $x^+ \equiv \max(0, x)$, $x_1 \Lambda x_2 \equiv \max(x_1, x_2)$, $A_n^{(r)} = \sum_{j=1}^{r} A_{n-j}$, and L_{n-r} is the waiting time of the $(n-r)$-th customer. We shall proceed next with (c), (d) and (e) accordingly.

(c) Denoting $\tilde{Y}_n = Y_n - x$ and $\tilde{V}_n = A_n \Lambda (L_n + \tilde{Y}_n)$ we can first write (5.2.52) as

$$H_V(x, \boldsymbol{v}) = 1 - \frac{\mathbb{E}_{\boldsymbol{u}} \sum_{n=1}^{\tau} \tilde{V}_n}{\mathbb{E}_{\boldsymbol{u}} \tau}, \quad \boldsymbol{u} \doteq (\boldsymbol{v}, x),$$

and then proceed similarly to (5.2.48) and (5.2.51).

(d) We have

$$H_R(r, \boldsymbol{v}) = P\{R_n \leq r\}$$

$$= \mathbb{E}_{\boldsymbol{v}} I_{(-\infty, 0]}(R_n - r) == \frac{\mathbb{E}_{\boldsymbol{v}} \sum_{n=1}^{\tau} I_n(\tilde{R}_n)}{\mathbb{E}_{\boldsymbol{v}} \tau} \tag{5.2.55}$$

where $0 \leq r \leq n,\ 1 \leq n < \tau$, and

$$\tilde{R}_n = L_{n-r} - \sum_{j=1}^{r} A_{n-j}. \tag{5.2.56}$$

Under the probability measure $g(z)$, we can write $H_R(r, \boldsymbol{v})$ and $\tilde{H}_{RN}(r, \boldsymbol{v})$ as

$$H_R(r, \boldsymbol{v}) = \frac{\mathbb{E}_g \sum_{n=1}^{\tau} I_n \widetilde{W}_n(\boldsymbol{v})}{\mathbb{E}_g \sum_{n=1}^{\tau} \widetilde{W}_n(\boldsymbol{v})} \tag{5.2.57}$$

and

$$\tilde{H}_{RN}(r, v) = \frac{\sum_{i=1}^{N} \sum_{n=1}^{\tau_i} I_{ni} \widetilde{W}_{ni}(v)}{\sum_{i=1}^{N} \sum_{n=1}^{\tau_i} \widetilde{W}_{ni}(v)}, \tag{5.2.58}$$

respectively, where $\widetilde{W}_n = \prod_{j=1}^{n} W_j$, $W_j = f(Z_j, v)/f(Z_j)$, $Z \sim g(z)$ and similarly for $\nabla^k \tilde{H}_{RN}(r, v)$. Note that in this case there is *no need* for the "push out" method, since both I_{ni} and $\nabla^k \widetilde{W}_{ni}(v)$ can be calculated *simultaneously* for different values of r and v, while using the SF method.

Remark 5.2.2 Taking into account that $L_n = \sum_{j=1}^{n}(Y_j - A_j)$, it is readily seen that the events

$$\left\{ L_{n-r} \leq \sum_{j=1}^{r} A_{n-j} \right\} \quad \text{and} \quad \left\{ L_n \leq \sum_{j=1}^{r} Y_{n-j} \right\} \tag{5.2.59}$$

are equivalent. Note that for computational purposes, it is desirable to use in (5.2.57) and (5.2.58) the event $\{L_{n-r} \leq \sum_{j=1}^{r} A_{n-j}\}$ instead of the event $\{L_n \leq \sum_{j=1}^{r} Y_{n-j}\}$.

(e) Consider (5.2.55). In this case, we can again use the standard SF method to estimate $\nabla^k H_M(r, v)$. We, however, omit its derivation.

Remark 5.2.3 Assume that we want to estimate

$$\ell(v) = \mathbb{E}_v M = \frac{\mathbb{E}_v \int_0^{\eta} M_t dt}{\mathbb{E}_v \eta} . \tag{5.2.60}$$

Taking into account that $\mathbb{E}_v \int_0^{\eta} M_t dt = \mathbb{E}_v \sum_{n=1}^{\tau} L_n$ and $\mathbb{E}_v \eta = \mathbb{E}_v \sum_{n=1}^{\tau} A_n$ we can rewrite (5.2.60) (see also Glynn and Whitt (1989)) as

$$\ell(v) = \frac{\mathbb{E}_v \sum_{n=1}^{\tau} L_n}{\mathbb{E}_v \sum_{n=1}^{\tau} A_n} = \frac{\mathbb{E}_g \sum_{n=1}^{\tau} L_n \widetilde{W}_n}{\mathbb{E}_g \sum_{n=1}^{\tau} A_n \widetilde{W}_n}$$

and then estimate $\ell(v)$ and $\nabla^k \ell(v)$ accordingly.

The $GI/D/1$ and $D/G/1$ Queues

We consider the $GI/D/1$ queue only, since the treatment of $D/G/1$ is similar. Let L_n be the steady-state sojourn time process in the $GI/D/1$ queue. Taking into account that for the $GI/D/1$ queue

$$L_n = \sum_{j=1}^{n}(A_j - v_2) = \sum_{j=1}^{n-1}(A_j - v_2) + (A_n - v_2), \quad n = 1, \ldots, \tau,$$

we can write (5.2.46) as

$$H_L(x, v) = \mathbb{E}_u I_{(-\infty,0]}(\tilde{L}_n), \ u = (x, v), \tag{5.2.61}$$

where

$$\tilde{L}_n = \sum_{j=1}^{n-1} U_j + \tilde{U}_n, \ \ U_j = A_j - v_2, \ \ \tilde{U}_n = A_n - v_2 - x, \ \ A_k \sim f(y, v_1),$$

$k = 1, \ldots, n$; $n = 1, \ldots, \tau$ and $v = (v_1, v_2)$. Clearly, with $\tilde{I}$ at hand we can define a POSF estimator of $H_L(x, v)$ similar to (5.2.51) and thus estimate the $H_L(x, v)$ from a *single simulation simultaneously* for different values of x and v. Note, however, that in this case we have to define two auxiliary pdfs, namely $\tilde{f}(y, v)$ and $\breve{f}(y, x, v)$ for the rv's U_j, $j = 1, \ldots, n-1$, and $\tilde{U}_n$, respectively. Note also that for $x = 0$ we have $\tilde{f} = \breve{f}$.

Let us now use the "push out" technique to estimate the expected sojourn time $\ell(v) = \mathbb{E}_v L$, $v = (v_1, v_2)$, in the $GI/D/1$ queue. In this case $\tilde{f} = \breve{f}$ and the POSF estimator can be written as

$$\tilde{\ell}_N(v) = \frac{\sum_{i=1}^{N} \sum_{n=1}^{\tau_i} \tilde{L}_{ni}(\underline{Z}_{ni}) \widetilde{W}_{ni}(\underline{Z}_{ni}, v_1, v_2)}{\sum_{i=1}^{N} \sum_{n=1}^{\tau_i} \widetilde{W}_{ni}(\underline{Z}_{ni}, v_1, v_2)}, \tag{5.2.62}$$

where $\underline{Z}_{ni} = (Z_{1i}, \ldots, Z_{ni})$, $n = 1, \ldots, \tau_i$, $i = 1, \ldots, N$, is a random sample from $g(z)$, $\widetilde{W}_{ni}(v) = \prod_{j=1}^{n} W_{ji}(v)$, $W_j = \tilde{f}(Z_{ji}, v)/g(Z_{ji})$, $Z_{ji} \sim g(z)$, $v = (v_1, v_2)$, and similarly for $\nabla^k \tilde{\ell}_N(v)$.

Note that (5.2.62) coincides with (5.2.51) in the sense that, in order to obtain (5.2.62) from (5.2.51), we have to replace $\tilde{\ell}$ by $\tilde{H}$, v_2 by x, and $\tilde{L}$ by $\tilde{\ell}$ with all other data remaining the same.

(C) Numerical Results

(i) **Static models.** Consider the following simple model:

$$L(Y, v_1) = \min(v_1 + Y_2)(v_1 + Y_3). \tag{5.2.63}$$

Assume that $Y_j \sim \exp(v_j)$, $j = 2, 3$. In this case, the cdf's of $\tilde{Y}_1 = v_1 + Y_2$ and $\tilde{Y}_2 = v_1 + Y_3$ can be written as

$$\tilde{F}_1(y) = P(\tilde{Y}_1 \leq y) = P(Y_2 \leq y - v_1) = F_2(y - v_1)$$

and

$$\tilde{F}_2(y) = P(\tilde{Y}_2 \leq y) = F_3(y - v_1),$$

where F_2 and F_3 are the cdf's of Y_2 and Y_3, respectively. Clearly that

$$\tilde{f}_1(y, v_1, v_2) = \begin{cases} v_2 e^{-v_2(y - v_1)}, & y \geq v_1, \\ 0, & \text{otherwise}, \end{cases} \tag{5.2.64}$$

and

$$\tilde{f}_2(y, v_1, v_3) = \begin{cases} v_3 e^{-v_3(y-v_1)}, & y \geq v_1 \\ 0, & \text{otherwise.} \end{cases} \tag{5.2.65}$$

As a dominating pdf let us choose $g(\boldsymbol{z}) = g_2(z_2)\, g_3(z_3)$, where g_i is an exponintial pdf with parameter v_{0i}, $i = 2, 3$. In this case

$$W(\boldsymbol{v}) = \begin{cases} \frac{v_2 e^{-v_2(Z_1 - v_1)}}{v_{02} e^{-v_{02} Z_1}} \frac{v_3 e^{-v_3(Z_2 - v_1)}}{v_{03} e^{-v_{03} Z_2}}, & \text{if } Z_1 \geq v_1 \text{ and } Z_2 \geq v_2, \\ \\ 0, & \text{otherwise.} \end{cases} \tag{5.2.66}$$

Table 5.2.1 presents the theoretical values $\ell(\boldsymbol{v})$, the point estimators $\tilde{\ell}_N(\boldsymbol{v})$, the associated sample variance of $\tilde{\ell}_N(\boldsymbol{v})$ (denoted $\hat{\sigma}^2\{\tilde{\ell}_N(\boldsymbol{v})\}$), and the 95% confidence intervals for $\ell(\boldsymbol{v})$ as a function of relative perturbation γ, defined as $\gamma = (|v - v_0|)/v_0$. We fix the reference vector $\boldsymbol{v}_0 = (v_{01}, v_{02}, v_{03}) = (1, 2, 3)$ and estimate $\ell(\boldsymbol{v})$ by the POSF estimator $\tilde{\ell}_N(\boldsymbol{v})$ for $\gamma = 0.1,\ 0.3,\ 0.5$ and $N = 10^3$ and 10^4. Notice that for $\gamma = 0.1$ we have $\boldsymbol{v} = (v_1, v_2, v_3) = (0.9,\ 1.8,\ 2.7)$ and similarly for $\gamma =\ 0.3$ and 0.5.

Table 5.2.2 presents similar data with the POSF estimator $\tilde{\nabla}\ell_N(\boldsymbol{v})$. More specifically, it presents theoretical values $\partial\ell(\boldsymbol{v})/\partial v_j$, $j = 1, 2, 3$, the point estimators $\partial\tilde{\ell}(\boldsymbol{v})/\partial v_j$, $j = 1, 2, 3$, and the 95% confidence intervals for $\boldsymbol{v}_0 = \boldsymbol{v} = (v_1, v_2, v_3) = (3.0,\ 0.2,\ 0.3)$ and $N = 10^4,\ 10^5$.

Note that in deriving $\nabla\tilde{\ell}_N(\boldsymbol{v})$ we took into account that

$$\frac{\mathrm{d}}{\mathrm{d}v} \int_{a(v)}^{b(v)} g(v, x)\mathrm{d}x = \int_{a(v)}^{b(v)} \frac{\partial}{\partial v} g(v, x)\mathrm{d}x$$

$$+ \left[\frac{\mathrm{d}}{\mathrm{d}v} b(v)\right] g(v, b(v)) - \left[\frac{\mathrm{d}}{\mathrm{d}v} a(v)\right] g(v, a(v)),$$

which for our model

$$\ell(\boldsymbol{v}) = \mathbb{E}_{\boldsymbol{v}} L(\tilde{Y}_1, \tilde{Y}_2) = \int_{v_1}^{\infty} \int_{v_1}^{\infty} L(y_1, y_2) \tilde{f}_1(y_1, v_1, v_2) \tilde{f}_2(y_2, v_1, v_3) \mathrm{d}y_1 \mathrm{d}y_2$$

after some simple algebra results in

$$\nabla\tilde{\ell}_N(\boldsymbol{v}) = \begin{bmatrix} N^{-1} \sum_{i=1}^{N} [(\tilde{L} - v_1)(v_2 + v_3)] \\ \\ N^{-1} \sum_{i=1}^{N} [\tilde{L}(v_2^{-1} - (\tilde{Y}_{1i} - v_1))] \\ \\ N^{-1} \sum_{i=1}^{N} [\tilde{L}(v_3^{-1} - (\tilde{Y}_{2i} - v_1))] \end{bmatrix}. \tag{5.2.67}$$

Table 5.2.1 Performance of the POSF estimators $\tilde{\ell}_N(\boldsymbol{v})$ for the stochastic PERT model (5.2.63).

N	γ	$\ell(\boldsymbol{v})$	$\tilde{\ell}_N(\boldsymbol{v})$	$\hat{\sigma}^2\{\ell_N(\boldsymbol{v})\}$	Conf. Interval	
10^3	0.1	1.2812	1.2910	1.7422	1.2092	1.3728
10^4	0.1	1.2812	1.2798	1.2299	1.258	1.3015
10^3	0.3	1.4538	1.4822	8.8243	1.2981	1.6663
10^4	0.3	1.4538	1.4655	6.9395	1.4139	1.5172
10^3	0.5	1.6333	1.6331	35.3296	1.2647	2.0015
10^4	0.5	1.6333	1.6477	36.5346	1.5293	1.7662

Table 5.2.2 Performance of the POSF estimators $\nabla\tilde{\ell}_N(\boldsymbol{v})$ for the stochastic PERT model (5.2.63).

N	$\frac{\partial\ell}{\partial v_1}$	$\frac{\partial\ell}{\partial v_2}$	$\frac{\partial\ell}{\partial v_3}$	$\frac{\partial\tilde{\ell}}{\partial v_1}$	$\frac{\partial\tilde{\ell}}{\partial v_2}$	$\frac{\partial\tilde{\ell}}{\partial v_3}$	Conf. Intervals		
10^4	1.0	-4.0	-4.0	1.0	-3.85	-3.76	0.98, 1.02	-3.25, -4.45	-3.29, -4.22
10^5	1.0	-4.0	-4.0	1.0	-3.98	-3.92	0.98, 0.99	-3.78, -4.17	-3.77, -4.07

Table 5.2.3 presents the theoretical values $H_L(x, v)$ and $\nabla H_L(x, v)$, their corresponding POSF estimates $\tilde{H}_{LN}(x, v)$ and $\nabla\tilde{H}_{LN}(x, v)$ (see (5.2.32) and (5.2.34), respectively) and the associated sample variances denoted $\hat{\sigma}^2\{\tilde{H}_{LN}\}$ and $\hat{\sigma}^2\{\nabla\tilde{H}_{LN}\}$, respectively, while simulating $N = 10^5$ customers with the $M/M/\infty$ queue. We chose the dominating density $g(z) = f(z, x_0, v_0)$ as in (5.2.43), took $x_0 = 8,\ v_0 = 0.3$ and estimated the cdf $H_L(x, v)$ and the pdf

$\nabla H_L(x,v) = f_L(x,v)$ from a single simulation run for different values of x and v, assuming $1 \le x \le \infty$ and $0.3 \le v \le 0.6$, respectively.

Table 5.2.3 Performance of the POSF estimators of $H_L(x,v)$ and $\nabla H_L(x,v) = f_L(x,v)$ for the $M/M/\infty$ queue.

v	x	$H_L(x,v)$	$\tilde{H}_{LN}(x,v)$	$\nabla H_L(x,v)$	$\nabla \tilde{H}_{LN}(x,v)$	$\hat{\sigma}^2\{\tilde{H}_{LN}\}$	$\hat{\sigma}^2\{\nabla \tilde{H}_{LN}\}$
	1.0	0.259	0.267	0.222	0.221	2.111	0.492
	2.5	0.528	0.528	0.142	0.141	2.311	0.148
.3	4.0	0.699	0.698	0.090	0.090	1.829	0.081
	5.0	0.777	0.779	0.067	0.067	1.309	0.045
	6.5	0.858	0.862	0.043	0.042	0.412	0.013
	8.0	0.909	0.910	0.027	0.027	0.082	0.007
	1.0	0.330	0.340	0.268	0.267	3.425	0.770
	2.5	0.632	0.631	0.147	0.146	3.226	0.155
.4	4.0	0.798	0.796	0.081	0.080	2.419	0.069
	5.0	0.865	0.867	0.054	0.054	1.663	0.031
	6.5	0.926	0.930	0.030	0.030	0.513	0.006
	8.0	0.959	0.959	0.016	0.016	0.126	0.002
	1.0	0.393	0.406	0.303	0.302	4.899	1.112
	2.5	0.714	0.712	0.143	0.143	4.075	0.150
.5	4.0	0.865	0.862	0.068	0.067	2.954	0.055
	5.0	0.918	0.921	0.041	0.041	2.001	0.020
	6.5	0.961	0.966	0.019	0.019	0.652	0.003
	8.0	0.982	0.981	0.009	0.009	0.221	0.001
	1.0	0.451	0.467	0.329	0.328	6.481	1.483
	2.5	0.777	0.774	0.134	0.133	4.871	0.134
.6	4.0	0.909	0.906	0.054	0.054	3.464	0.040
	5.0	0.950	0.953	0.030	0.030	2.345	0.012
	6.5	0.980	0.984	0.012	0.012	0.824	0.001
	8.0	0.992	0.991	0.005	0.005	0.347	0.000

(ii) **Queueing models.** Table 5.2.4 presents the Crude Monte Carlo (CMC) estimator $\bar{H}_{LN}(x,v)$, the POSF estimator $\tilde{H}_{LN}(x,v)$ (see (5.2.62)) and the associated sample variance $\hat{\sigma}^2\{\bar{H}_{LN}\}$ and $\hat{\sigma}^2\{\tilde{H}_{LN}\}$ for the $M/G/1$ queue with interarrival rate $\lambda = 1$ and the following service time pdf:

$$f(z,k,v) = v\exp(-v(z-k)), \quad z \ge k, \; k = 0.3. \tag{5.2.68}$$

Note that the CMC estimator $\bar{H}_{LN}(x,v)$ was produced by using the standard regenerative ratio estimator, that is while simulating the $M/G/1$ queue separately for each value v (the total of four simulation runs, see table 5.2.4). Note also that (5.2.68) presents a shifted (by a factor k) exponential pdf.

We chose the dominating pdf $g(z)$ as

$$g(z) = f(z, k_0, v_0) = v_0 \exp(-v_0(z - k_0)), \quad z \geq k_0,$$

with $v_0 = 3.0$, $k_0 = 0.3$ and estimated the cdf $H_L(x, v)$ from a single run based on $N = 10^5$ customers for different values of x and v. More precisely, we fixed $v = 4, 5, 6$ and 10, and obtained reliable estimators of the cdf H in the intervals $0.34 \leq x \leq 0.58$ and $0.32 \leq x \leq 0.60$, respectively. It is readily seen that $\mathbb{E}_g(Z) = 0.667$. Hence the reference traffic intensity $\rho_0 = \lambda \mathbb{E}_g(Z) = 0.667$. Note finally that in order to expand the prediction interval of H we need to run an additional simulation with different values of the reference parameters v_0 and k_0 in the dominating pdf $g(z)$.

Table 5.2.4 Performance of the CMC estimator $\bar{H}_{LN}(x, v)$ and the POSF estimator $\tilde{H}_{LN}(x, v)$ for the $M/G/1$ queue with the service time pdf given in (5.2.68).

v	ρ	x	$\bar{H}_{LN}(x,v)$	$\hat{\sigma}^2\{\bar{H}_{LN}\}$	$\tilde{H}_{LN}$	$\hat{\sigma}^2\{\tilde{H}_{LN}(x,v)\}$
		0.340	0.067	0.421E-06	0.067	0.123E-05
		0.380	0.127	0.370E-06	0.126	0.409E-05
4.00	0.550	0.460	0.233	0.263E-06	0.231	0.135E-04
		0.500	0.276	0.203E-06	0.274	0.187E-04
		0.540	0.317	0.137E-06	0.326	0.273E-04
		0.580	0.356	0.701E-0.4	0.362	0.325E-04
		0.340	0.091	0.907E-0.6	0.090	0.183E-05
		0.380	0.171	0.170E-06	0.167	0.581E-05
5.00	0.500	0.460	0.301	0.292E-06	0.294	0.174E-04
		0.500	0.356	0.339E-06	0.345	0.237E-04
		0.540	0.405	0.382E-06	0.405	0.333E-04
		0.580	0.444	0.418E-06	0.445	0.392E-04
		0.340	0.116	0.112E-0.6	0.115	0.215E-05
		0.380	0.212	0.197E-06	0.208	0.639E-05
6.00	0.467	0.460	0.358	0.314E-06	0.357	0.178E-04
		0.500	0.418	0.353E-06	0.413	0.235E-04
		0.540	0.470	0.383E-06	0.477	0.319E-04
		0.580	0.517	0.407E-06	0.520	0.367E-04
		0.360	0.279	0.224E-06	0.290	0.947E-05
		0.400	0.403	0.286E-06	0.416	0.187E-04
10.00	0.400	0.460	0.531	0.321E-06	0.544	0.298E-04
		0.500	0.592	0.326E-06	0.605	0.362E-04
		0.560	0.666	0.610E-06	0.678	0.451E-04
		0.600	0.709	0.115E-0.4	0.737	0.533E-04

Based on the results of table 5.2.4, fig. 5.2.1 presents the empirical cdf's $\bar{H}_{LN}(x,v)$ and $\tilde{H}_{LN}(x,v)$ as functions of x (denoted CMC H_{LN} and POSF H_{LN}, respectively) and the associated 95% confidence intervals (denoted 95%CI) for the POSF estimator $\tilde{H}_{LN}(x,v)$ with $v = 10$ ($\rho = 0.4$).

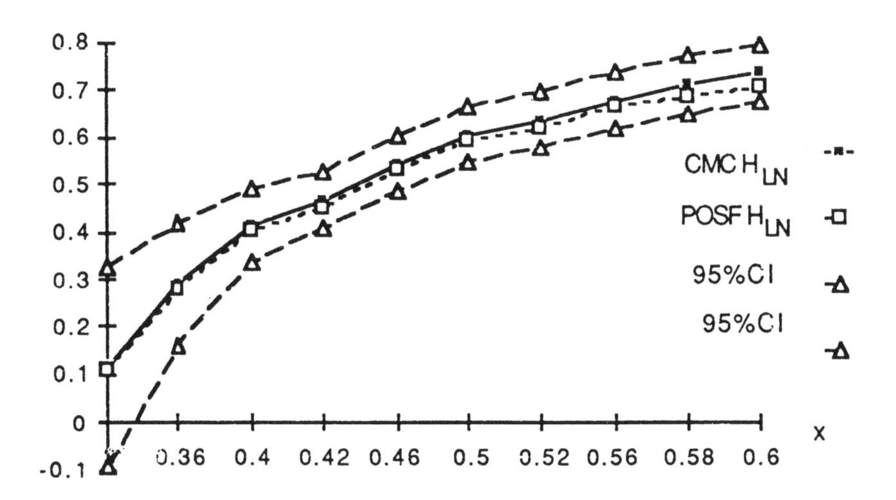

Figure 5.2.1 Performance of the CMC estimator $\bar{H}_{LN}(x,v)$ and the POSF estimator $\tilde{H}_{LN}(x,v)$ for the $M/G/1$ queue with the service time pdf $f(z,k,v) = v\ \exp(-v(z-k)), z > k,\ k = 0.3,\ v = 10$ and $\rho = 0.4$.

Table 5.2.5 presents the theoretical values $\ell(v)$, the corresponding POSF estimators $\tilde{\ell}_N(v)$ (see (5.2.62)) and the associated sample variance $\hat{\sigma}^2\{\tilde{\ell}_N(v)\}$ for the $M/D/1$ queue with interarrival rate $\lambda = 1$ and traffic density $\rho = \lambda v_2 = v_2$ while simulating $N = 500000$ customers. We chose the dominating density $g(z) = f(z, v_{01}, v_{02}) = v_{01}\exp(-v_{01}(v_{02} + z))$ (see (5.2.43)) assuming that $v_{01} = \lambda = 1$, $v_{02} = 0.6$ (the reference traffic intensity $\rho_0 = \lambda v_{02} = 0.9$), and predicted the expected sojourn time for different v_2, in the interval $0.1 \le v_2 \le 0.9$.

Table 5.2.5 Performance of the POSF estimators $\tilde{\ell}_N(\boldsymbol{v})$ for the $M/D/1$ queue with the reference traffic intensity $\rho_0 = 0.9$.

v_2	ρ	$\ell(\boldsymbol{v})$	$\tilde{\ell}_N(\boldsymbol{v})$	$\hat{\sigma}^2\{\tilde{\ell}_N(\boldsymbol{v})\}$
0.100	0.100	5.560E-03	5.600E-03	3.090E-08
0.150	0.150	1.320E-02	1.410E-02	1.430E-07
0.200	0.200	2.500E-02	2.550E-02	4.030E-07
0.250	0.250	4.170E-02	4.110E-02	1.080E-06
0.300	0.300	6.430E-02	6.250E-02	2.570E-06
0.350	0.350	9.420E-02	9.420E-02	1.690E-05
0.400	0.400	1.330E-01	1.330E-01	2.180E-05
0.450	0.450	1.840E-01	1.730E-01	3.100E-05
0.500	0.500	2.500E-01	2.280E-01	4.710E-05
0.550	0.550	3.360E-01	3.140E-01	1.280E-04
0.600	0.600	4.500E-01	4.130E-01	2.280E-04
0.650	0.650	6.040E-01	6.140E-01	8.810E-04
0.700	0.700	8.170E-01	7.380E-01	5.390E-04
0.750	0.750	1.120E+00	1.020E+00	9.980E-03
0.800	0.800	1.600E+00	1.630E+00	2.250E-02
0.850	0.850	2.410E+00	2.300E+00	1.750E-02
0.900	0.900	4.050E+00	4.170E+00	1.000E-02

It follows from the results of our tables 5.2.1–5.2.5 that the POSF estimators perform satisfactorily.

We performed extensive simulation studies using the POSF method for different interarrival and service distributions of the $GI/G/1$ queue and for simple queueing networks. Our simulation results are similar to those in tables 5.2.4 and 5.2.5 and indicate reasonably good performance of the POSF estimators.

5.3 Sensitivity Analysis with Autocorrelated Inputs

5.3.1 Introduction

In this section we extend some of our earlier results on sensitivity analysis of DES from independent input sequences $\{\boldsymbol{Y}_t : t > 0\}$ to autocorrelated ones, which we denote $\{\boldsymbol{X}_t : t > 0\}$ (see (5.1.5)). In particular, we shall show how to evaluate the expected steady-state performance

$$\ell(\boldsymbol{v}) = \mathbb{E}_{\boldsymbol{v}}\{L(\underline{X}_t)\} \tag{5.3.1}$$

and the associated sensitivities $\nabla^k \ell(\boldsymbol{v})$, $k \geq 1$, for queueing models, with special emphasis on autocorrelated TES sequences (see (5.1.8) and (5.1.9)).

If not stated otherwise we assume as before that $\{L_t : t > 0\}$ is the waiting time process in a stable $GI/G/1$ queue.

Before proceding further it is important to understand that with autocorrelated inputs the output process $\{L_t : t > 0\}$ is *not regenerative.* This is so since the regeneration is violated owing to dependence of the waiting time of the *first* customer in the current busy cycle and the waiting time of the *last* customer in the previous busy cycle. Therefore, we cannot use the direct regenerative SF estimators based on formulas (5.2.6) and (5.2.7). We shall use here instead the truncated SF (TSF) estimators, which are suitable for stationary and ergodic processess (see section 3.3 and Melamed and Rubinstein (1992)). Sensitivity analysis of queueing models with autocorrelated inputs, while using *Harris recurrence* (see Asmussen (1987)) to set a regenerative framework, is discussed in Asmussen and Melamed (1992).

The rest of this section is organized as follows. In subsection 5.3.2 we use direct TSF estimators, which are similar to those of subsection 5.2.1 and are based on the representation of $\ell(\boldsymbol{v})$ as

$$\ell(\boldsymbol{v}) = \mathbb{E}_{\boldsymbol{v}_1}\{L_t[x_t(\underline{Y}_t, \boldsymbol{v}_2)]\}. \tag{5.3.2}$$

It is important to note that the subscripts $\boldsymbol{v}_1$ and $\boldsymbol{v}$ in $\mathbb{E}_{\boldsymbol{v}_1}\{L_t[x_t(\underline{Y}_t, \boldsymbol{v}_2)]\}$ and $\mathbb{E}_{\boldsymbol{v}}\{L_t(\underline{X}_t)\}$ (see (5.3.2) and (5.3.1)) indicate that the expectations are taken with respect to the pdf's $f(y, \boldsymbol{v}_1)$ and $\tilde{f}(x, \boldsymbol{v})$ of the steady-state random variables Y and X, respectively. In subsections 5.3.3 and 5.3.4 we apply the "push in" (IPA) and the "push out" techniques discussed in subsections 5.2.3 and 5.2.4, respectively, to estimate $\nabla^k \ell(\boldsymbol{v})$, $k = 0, 1$. Subsection 5.3.5 presents some numerical results while using AR(1) and TES input sequences for the $GI/G/1$ queue.

5.3.2 Direct Estimators

Suppose for a moment that the input sequence $\{X_t : t > 0\}$ is uncorrelated. In this case the truncated version of $\nabla^k \ell(\boldsymbol{v})$ and the associated TSF estimator can be written (see (3.3.12) and 3.3.13)) as

$$\nabla^k \ell^{tr}(\boldsymbol{v}, m) = \mathbb{E}_g\{L_t \nabla^k \widetilde{W}_t^{tr}(\boldsymbol{v}, m)\} \tag{5.3.3}$$

and

$$\nabla^k \overline{\ell}_N^{tr}(\boldsymbol{v}, m) = N^{-1} T^{-1} \sum_{i=1}^{N} \sum_{t=1}^{T} L_{ti} \nabla^k \widetilde{W}_{ti}^{tr}(\boldsymbol{v}, m), \tag{5.3.4}$$

$k = 0, 1, \ldots$, respectively. Here

$$\widetilde{W}_t^{tr}(\boldsymbol{v}, m) = \prod_{j=t-m+1}^{t} W_j,$$

m is the truncation parameter and T is the batch size (see section 1.2). Assume now that the inputs $\{X_t : t > 0\}$ are correlated. Arguing similarly to (5.2.3) and (5.2.4) and taking (5.3.3) and (3.3.11) into account we obtain

$$\nabla^k_{\boldsymbol{v}_1} \ell^{tr}(\boldsymbol{v}, m) = \mathbb{E}_g\{\check{L}_t(\underline{Z}_t, \boldsymbol{v}_2) \nabla^k_{\boldsymbol{v}_1} \widetilde{W}_t^{tr}(\underline{Z}_t, \boldsymbol{v}_1, m)\} \tag{5.3.5}$$

and

$$\nabla^k_{\boldsymbol{v}_2} \ell^{tr}(\boldsymbol{v}, m) = \mathbb{E}_g\{\nabla^k_{\boldsymbol{v}_2} \check{L}_t(\underline{Z}_t, \boldsymbol{v}_2) \widetilde{W}_t^{tr}(\underline{Z}_t, \boldsymbol{v}_1, m)\}, \tag{5.3.6}$$

$k = 1, 2, \ldots$, respectively, where $\check{L}_t(\cdot) = L_t[\boldsymbol{x}_t(\cdot)]$.

Consider formula (5.3.5). It allows us to estimate the performance $\ell(\boldsymbol{v})$ and *all* sensitivities $\nabla^k_{\boldsymbol{v}_1} \ell(\boldsymbol{v})$, $k = 1, 2, \ldots$, from a *single simulation simultaneously* for different values of $\boldsymbol{v}_1$, provided $\boldsymbol{v}_2$ is fixed. The corresponding TSF estimator can be written as

$$\nabla^k_{\boldsymbol{v}_1} \bar{\ell}^{tr}_N(\boldsymbol{v}, m) = N^{-1} T^{-1} \sum_{i=1}^{N} \sum_{t=1}^{T} \check{L}_{ti}(\underline{Z}_{ti}, \boldsymbol{v}_2) \nabla^k_{\boldsymbol{v}_1} \widetilde{W}_{ti}^{tr}(\underline{Z}_{ti}, \boldsymbol{v}_1, m), \tag{5.3.7}$$

and similarly for the estimator of (5.3.6). Here

$$\underline{Z}_{ti} = (Z_{1i}, \ldots, Z_{Ti}),\ Z_{ti},\ \ t = 1, \ldots, T,\ i = 1, \ldots, N,$$

is a random sample from the dominating pdf $g(z)$. Note that in this case, using an appropriate dominating pdf $g(\boldsymbol{z})$), we can estimate $\nabla \ell(\boldsymbol{v})$, $\boldsymbol{v} = (\boldsymbol{v}_1, \boldsymbol{v}_2)$, for an entire region of permissible values of $\boldsymbol{v}$.

The algorithm for the estimation of $\nabla^k_{\boldsymbol{v}_1} \ell(\boldsymbol{v})$ while using the TSF estimators (5.3.6) applied to the TES input sequence can be written as follows:

Algorithm 5.3.1 :

1. *Generate the TES sequence* $\{Z_t : t > 0\}$ *according to (5.1.7) and (5.1.8).*

2. *Generate the output process* $\{L_t(\underline{Z}_t) : t > 0\}$.

3. *Estimate the parameter vector* $\nabla^k_{\boldsymbol{v}_1} \ell(\boldsymbol{v})$, $k = 0, 1$, *according to (5.3.6).*

The reader is asked to write a similar algorithm for estimation of $\nabla^k_{\boldsymbol{v}_2} \ell(\boldsymbol{v})$.

5.3.3 The "Push in" Technique

We shall consider here only the nonregenerative IPA approach for estimation of the gradient $\nabla\ell(\boldsymbol{v})$, $\boldsymbol{v} = (\boldsymbol{v}_1, \boldsymbol{v}_2)$, where $\ell(\boldsymbol{v})$ is the expected steady-state waiting time in a stable $GI/G/1$ queue. To do so we apply again Lindley's recursive equation

$$L_{t+1} = \max\{0, L_t + Z_t\}, \quad t \geq 1, \; L_1 = 0,$$

to the autocorrelated interarrival and service time sequences. Here as before $Z_t = X_{1t} - X_{2(t+1)}$. The corresponding IPA algorithm for estimation of $\nabla\ell(\boldsymbol{v})$, while using, say, the TES sequence coincides with algorithm 5.2.2. We have to replace

$$\hat{Z}_t(\boldsymbol{v}) = F_1^{-1}(U_{1t}, v_1) - F_2^{-1}(U_{2(t+1)}, v_2)$$

in step 2 of algorithm 5.2.2 by

$$\hat{Z}_t(\boldsymbol{v}) = F_1^{-1}[\eta_{1t}^+(U_{1t}, v_{11}), \boldsymbol{v}_{21}] - F_2^{-1}[\eta_{2t}^+(U_{2(t+1}, v_{12}), \boldsymbol{v}_{22}],$$

where $\boldsymbol{v}_1 = (\boldsymbol{v}_{11}, \boldsymbol{v}_{12})$ and $\boldsymbol{v}_2 = (v_{21}, v_{22})$ with all other data remaining the same, and similarly for other autocorrelated sequences.

The reader is asked to write the algorithm explicitly and to prove the unbiasedness of the IPA estimator.

5.3.4 The "Push out" Technique

We now show how the "push out" technique (see subsections 5.2.2 and 5.2.4) can be applied to estimate $\nabla^k\ell(\boldsymbol{v})$, $k = 0, 1$, $\boldsymbol{v} = (\boldsymbol{v}_1, \boldsymbol{v}_2)$. Note first that for an autocorrelated sequence $\{X_t : t > 0\}$ of the form $X_t = h(X_{t-1}, Y_j, \boldsymbol{v}_2)$, $t = 1, 2, \ldots$, we just have $\tilde{L}(\underline{\boldsymbol{x}}_t) \equiv L(\underline{\boldsymbol{x}}_t)$, where $\tilde{L}(\underline{\boldsymbol{x}}_t) = L(\underline{\boldsymbol{y}}_t, \boldsymbol{v}_2)$ (see (5.2.9) and (5.2.10)). Therefore application of the "push out" technique just means finding the joint t-dimensional pdf $\tilde{f}_t(\boldsymbol{x}_1, \ldots, \boldsymbol{x}_t, \boldsymbol{v})$ of the random vector $\underline{X}_t = (X_1, \ldots, X_t)$ in terms of the pdf $f(y, \boldsymbol{v}_2)$ of the rv Y. By the Markovian property of $\{X_t : t > 0\}$ the joint pdf $\tilde{f}_t(\boldsymbol{x}_1, \ldots, \boldsymbol{x}_t, \boldsymbol{v})$ can be written as

$$\tilde{f}_t(\boldsymbol{x}_1, \ldots, \boldsymbol{x}_t, \boldsymbol{v}) = \tilde{f}(\boldsymbol{x}_1, \boldsymbol{v}) \cdot \tilde{f}(\boldsymbol{x}_2|\boldsymbol{x}_1, \boldsymbol{v}) \cdots \tilde{f}(\boldsymbol{x}_t|\boldsymbol{x}_{t-1}, \boldsymbol{v}), \tag{5.3.8}$$

where $\boldsymbol{x}_0 = 0$. We next find the conditional pdf's $\tilde{f}(\boldsymbol{x}_t|\boldsymbol{x}_{t-1}, \boldsymbol{v})$, $t = 1, 2, \ldots$, in terms of the unconditional pdf $f(y, \boldsymbol{v}_1)$ for the AR(1) and TES sequences. As an easy exercise the reader is ask to do so for the Minification/maxification sequence.

Example 5.3.1 AR(1) sequences. Consider the following simple modification of (5.1.6):

$$X_t(v_2) = \begin{cases} X_1, & \text{if } t = 1, \\ v_2 X_{t-1} + (1 - v_2) Y_t, & \text{if } t > 1, \end{cases} \tag{5.3.9}$$

where $0 < v_2 < 1$. In this case it is readily seen that

$$\tilde{f}(\boldsymbol{x}_t | \boldsymbol{x}_{t-1}, \boldsymbol{v}) = f\left(\frac{\boldsymbol{x}_t - v_2 \boldsymbol{x}_{t-1}}{1 - v_2}, \boldsymbol{v}_1\right), \tag{5.3.10}$$

$t = 1, 2, \ldots$. In particular, let $Y \sim \exp(v_1)$. Then

$$f\left(\frac{\boldsymbol{x}_t - v_2 \boldsymbol{x}_{t-1}}{1 - v_2}, \boldsymbol{v}_1\right) = v_1 \exp\left(\frac{-v_1(\boldsymbol{x}_t - v_2 \boldsymbol{x}_{t-1})}{(1 - v_2)}\right) \cdot I_{(-\infty, 0]}(\boldsymbol{x}_t - v_2 \boldsymbol{x}_{t-1}).$$

Example 5.3.2 TES sequences. It is shown in Melamed (1992) that

$$\tilde{f}_t(\boldsymbol{x}_t | \boldsymbol{x}_{t-1}, \boldsymbol{v}) = 1/(\theta_2 - \theta_1) f(\boldsymbol{x}_t, \boldsymbol{v}_1) I\{F(\boldsymbol{x}_t, \boldsymbol{v}_1) \in C[F(\boldsymbol{x}_{t-1}, \boldsymbol{v}_1), \boldsymbol{v}_2]\}, \tag{5.3.11}$$

where $t = 1, 2, \ldots$, $\boldsymbol{v}_2 = (\theta_1, \theta_2)$ and C is a circular interval defined in Jagerman and Melamed (1992a).

With (5.3.10) and (5.3.11) at hand we shall show how to estimate $\nabla^k \ell(\boldsymbol{v})$, $k = 0, 1$, *simultaneously for different values of* $\boldsymbol{v}_1$ and $\boldsymbol{v}_2$ while using the TSF estimators. Indeed, in analogy to (5.3.3), (3.3.11) we obtain

$$\nabla^k \ell^{tr}(\boldsymbol{v}, m) = \mathbb{E}_g\{\tilde{L}_t(\underline{Z}_t) \nabla^k \widehat{W}_t^{tr}(\underline{Z}_t, \boldsymbol{v}, m)\} \tag{5.3.12}$$

and

$$\nabla^k \tilde{\ell}_N^{tr}(\boldsymbol{v}, m) = N^{-1} T^{-1} \sum_{i=1}^{N} \sum_{t=1}^{T} \tilde{L}_{ti}(\underline{Z}_{ti}) \nabla^k \widehat{W}_{ti}^{tr}(\underline{Z}_{ti}, \boldsymbol{v}, m). \tag{5.3.13}$$

Here

$$\widehat{W}_t^{tr}(\underline{Z}_{ti}, \boldsymbol{v}, m) = \prod_{j=t-m+1}^{t} W_j(Z_{ji}, \boldsymbol{v}),$$

m is the truncation parameter, and $Z_{11}, \ldots, Z_{TN}$ is a random sample from the pdf $g(z)$ which for each $t > 0$ dominates the conditional pdf $\tilde{f}_t(\boldsymbol{x}_t | \boldsymbol{x}_{t-1}, \boldsymbol{v})$.

We shall calculate now $\widehat{W}_t^{tr}(\underline{z}_t, \boldsymbol{v}, m)$ for the AR(1) and TES sequences, assuming that $g(\boldsymbol{z}) = \tilde{f}_t(\boldsymbol{x}_t | \boldsymbol{x}_{t-1}, \boldsymbol{v})$ and $\boldsymbol{v}_0 = (\boldsymbol{v}_{01}, \boldsymbol{v}_{02})$. The reader is again asked to do so for the Minification/maxification sequence.

Example 5.3.3 AR(1) sequences. In this case we have

$$\widehat{W}_t^{tr}(\underline{z}_t, \boldsymbol{v}, m) = \prod_{j=t-m+1}^{t} \frac{f(z_j - v_2 z_{j-1}, \boldsymbol{v}_1)(1 - v_{02})}{f(z_j - v_{02} z_{j-1}, \boldsymbol{v}_{01})(1 - v_2)}. \tag{5.3.14}$$

In particular, let $Y \sim \exp(v_1)$. Then

$$\begin{aligned} \widehat{W}_t^{tr}(\underline{z}_t, \boldsymbol{v}, m) &= (v_1/v_{01})^m \\ \times \quad & \exp\left\{ -[(v_1 - v_1 v_2)/(1 - v_2) - (v_{01} - v_{01} v_{02})/(1 - v_{02})] \sum_{j=1}^{m-1} z_j \right\} \\ \times \quad & \exp\{-(v_1/(1 - v_2) - v_{01}/(1 - v_{02}))z_m\} \cdot \prod_{j=t-m+1}^{t} I(z_j - v_2 z_{j-1}) \end{aligned} \tag{5.3.15}$$

where $v_{02} \leq v_2$.

Example 5.3.4 TES sequences. In this case we have

$$\begin{aligned} \widehat{W}_t^{tr}(\underline{z}_t, \boldsymbol{v}, m) &= \left(\frac{\theta_{02} - \theta_{01}}{\theta_2 - \theta_1} \right)^m \prod_{j=t-m+1}^{t} \frac{f(z_j, \boldsymbol{v}_1)}{f(z_j, \boldsymbol{v}_{01})} \\ \times \quad & \prod_{k=t-m+1}^{t} I\{F(z_k, \boldsymbol{v}_1) \in C[F(z_{k-1}, \boldsymbol{v}_1), \boldsymbol{v}_2]\}, \end{aligned} \tag{5.3.16}$$

where $\boldsymbol{v}_2 = (\theta_1, \theta_2)$.

The reader is asked to derive explicitly $\nabla \widehat{W}_t^{tr}(\underline{z}_t, \boldsymbol{v}, m)$ for both AR(1) and TES models assuming that $Y \sim \exp(v_1)$.

Note that care must be taken when differentiating the performance $\ell(\boldsymbol{v})$ containing indicators like (5.3.15) and (5.3.16). The interested reader is referred to Uryas'ev (1992), who established general conditions for differentiability of performance measures of the type

$$\ell(\boldsymbol{v}) = \int_{\Omega(\boldsymbol{v})} Q(\boldsymbol{y}, \boldsymbol{v}) \mathrm{d}\boldsymbol{y},$$

where the integration domain $\Omega(\boldsymbol{v})$ depends on $\boldsymbol{v}$ and is defined by the system of inequality constraints as follows:

$$\Omega(\boldsymbol{v}) = \{q_j(\boldsymbol{y}, \boldsymbol{v}) \leq 0, \ j = 1, \ldots, k\}.$$

The algorithm for estimation of the parameter vector $\nabla^k \ell(\boldsymbol{v})$, $k > 0$, *everywhere* in an entire region of permissible values of $\boldsymbol{v} \in V$, $\boldsymbol{v} = (\boldsymbol{v}_1, \boldsymbol{v}_2)$ using the "push out" method can be written as

Algorithm 5.3.2 :

(1) Generate a sample $Z_{11}, \ldots, Z_{TN}$ *from the dominating pdf* $g(z)$.

(2) Generate the output processes $\tilde{L}_{ti}(\underline{Z}_{ti})$ *and* $\nabla^k \widehat{W}_{ti}^{tr}(\underline{Z}_{ti}, \boldsymbol{v}, m)$.

(3) Calculate $\nabla^k \bar{\ell}_N^{tr}(\boldsymbol{v}, m)$ *according to (5.3.13).*

5.3.5 Numerical Results

We present now simulation results while using the direct TSF estimators for estimating the expected steady-state waiting time in the standard $M/M/1$ queue with autocorrelated inputs. Depending on whether the arrival times, or the service times, or both the interarrival and the service times are autocorrelated we shall denote the corresponding queues as $M_c/M/1$, $M/M_c/1$ and $M_c/M_c/1$, respectively, and similarly for the $GI/G/1$ queue. We assumed that the interarrival times are autocorrelated, while the service times are independent, i.e. we considered an $M_c/M/1$ queue, chose the interarrival rate $\lambda = 1$, the reference parameter $\mu_0 = 1.667$ ($\rho_0 = 0.6$), the truncation parameter $m = 15$, and estimated the parameters $\nabla_{\boldsymbol{v}_1}^k \ell(\boldsymbol{v})$, $k = 0, 1$, for $\mu = 2.5$ ($\rho = 0.4$), where $\ell(\boldsymbol{v})$ is the expected steady-state waiting time. To compare our results with those obtained by the Crude Monte Carlo method we performed additional simulation of the $M/M/1$ queue with $\mu_0 = \mu = 2.5$.

Table 5.3.1 presents the point estimators $\nabla_{v_1}^k \bar{\ell}_N^{tr}(\boldsymbol{v}, \boldsymbol{v}, m)$ and $\nabla_{v_1}^k \bar{\ell}_N^{tr}(\boldsymbol{v}, \boldsymbol{v}_0, m)$, $k = 0, 1$, based on $N = 10^5$ customers, and the corresponding half (in %) with of the relative confidence intervals, denoted $w^k(\boldsymbol{v})$, $k = 0, 1$, and $w^k(\boldsymbol{v}_0)$, $k = 0, 1$, respectively as functions of θ_1 (see (5.1.8)) for the TES model. Here $\boldsymbol{v} = (v_1, \boldsymbol{v}_2)$, $v_1 = \mu$, $\boldsymbol{v}_2 = (\theta_1, \theta_2 = 0)$ and similarly for $\boldsymbol{v}_0$.

Table 5.3.1 Performance of TSF estimators $\nabla_{v_1}^k \bar{\ell}_N^{tr}(\boldsymbol{v}, \boldsymbol{v}, m)$, and $\nabla_{v_1}^k \bar{\ell}_N^{tr}(\boldsymbol{v}, \boldsymbol{v}_0, m)$, $k = 0, 1$, as functions of the autocorrelation parameter $\boldsymbol{v}_2 = (\theta_1, \theta_2)$, $\theta_2 = 0)$ in the TES model.

θ_1	Without the change of measure				With the change of measure			
	$\bar{\ell}_N^{tr}(\boldsymbol{v})$	$w^0(\boldsymbol{v})$	$\nabla_{v_1} \bar{\ell}_N^{tr}(\boldsymbol{v})$	$w^1(\boldsymbol{v})$	$\bar{\ell}_N^{tr}(\boldsymbol{v}_0)$	$w^0(\boldsymbol{v}_0)$	$\nabla v_1 \bar{\ell}_N^{tr}(\boldsymbol{v}_0)$	$w^1(\boldsymbol{v}_0)$
0.1	0.227	6.94	-0.250	17.29	0.248	8.40	-0.286	11.96
0.2	0.202	7.65	-0.239	19.19	0.225	10.15	-0.271	12.25
0.3	0.187	7.85	-0.231	17.77	0.213	11.04	-0.265	15.25

Table 5.3.2 presents similar data with AR(1) autocorrelated sequences.

Table 5.3.2 Performance of TSF estimators $\nabla^k_{v_1} \bar{\ell}^{tr}_N(v, v, m)$, and $\nabla^k_{v_1} \bar{\ell}^{tr}_N(v, v_0, m)$, $k = 0, 1$, as functions of the autocorrelation parameter v_2 in the AR(1) model.

v_2	Without the change of measure				With the change of measure			
	$\bar{\ell}^{tr}_N(v)$	$w^0(v)$	$\nabla_{v_1} \bar{\ell}^{tr}_N(v)$	$w^1(v)$	$\bar{\ell}^{tr}_N(v_0)$	$w^0(v_0)$	$\nabla v_1 \bar{\ell}^{tr}_N(v_0)$	$w^1(v_0)$
0.1	0.408	5.39	-0.389	15.68	0.401	6.18	-0.398	12.65
0.2	0.280	5.04	-0.283	16.42	0.271	6.52	-0.270	14.95
0.3	0.239	5.00	-0.239	17.33	0.227	7.89	-0.246	18.74

The results of these tables are self-explanatory. As expected, increasing autocorrelations give rise to increased waiting times. Furthermore, the effect is more dramatic for TES than for AR(1), since apparently TES autocorrelations decay more slowly.

We also found that TSF estimators (both the direct and the "push out") with autocorrelated inputs perform reasonably well for the $GI_c/G_c/1$ queue and related queueing models. For more details on the performance of TSF and related score function estimators with autocorrelated inputs, see Gilat (1992) and Izxaki (1992).

Chapter summary

In this chapter we treated the extended versions of the model $\ell(v) = \mathbb{E}_v L(\underline{Y}_t)$, namely the model $\ell(v) = \mathbb{E}_{v_1}\{L(\underline{Y}_t, v_2)\}$, $v = (v_1, v_2)$, where both the underlying pdf f and the sample performance L depend on the parameter vectors v_1 and v_2, respectively. Particular emphasis was placed on queueing models with autocorrelated inputs, where v_2 is called the autocorrelation parameter vector. We showed

(a) How to combine it with the standard SF method and in particular for queueing models with autocorrelated inputs. To evaluate (estimate) the performance and the sensitivities $\nabla^k \ell(v) = \{\nabla^k_{v_1} \ell(v), \nabla^k_{v_2} \ell(v)\}$, $k = 0, 1$, we introduced the direct, "push out" and "push in" (IPA) estimators.

(b) That the direct estimators of $\nabla^k_{v_1} \ell(v)$ involve differentiation of the pdf $f(y, v_1)$ only, that is they present the standard SF estimators, while the direct estimators of $\nabla^k_{v_2} \ell(v)$ involve differentiation of the sample function $L(v_2)$, and require that $L(v_2)$ is both available analytically and smooth (differentiable) with respect to v_2, which is typically not the case.

To overcome this difficulty we introduced a new technique, called the "push out" technique. We showed

(a) That the "push out" technique, which is based on the change of variables method, merely replaces the original sample function $L(\boldsymbol{v}_2)$ by an auxiliary one $\tilde{L}$ while "pushing out" the parameter vector $\boldsymbol{v}_2$ from $L(\boldsymbol{v}_2)$ to an auxiliary pdf $\tilde{f}(\boldsymbol{y}, \boldsymbol{v})$.

(b) How both the auxiliary sample performance and the auxiliary pdf can be obtained from their original counterparts, and how combining them allows us to obtain consistent estimators of $\nabla^k \ell(\boldsymbol{v}),\ k = 0, 1$.

We finally showed

(a) That the IPA method introduced by Ho, Eyler and Chien (1979) corresponds to the "push in" technique, which by itself can be viewed as a dual of the "push out" technique.

(b) That the "push in" transformation $\boldsymbol{x} = \boldsymbol{x}(\boldsymbol{y}, \boldsymbol{v}_1)$ and its particular case $x = F(y, \boldsymbol{v}_1)$ typically leads to a nonsmooth sample performance function $\hat{L}(\boldsymbol{u}, \boldsymbol{v})$, depending on both $\boldsymbol{v}_1$ and $\boldsymbol{v}_2$. As a result the required interchangeability conditions of expectation and differentiation often do not hold. This is the main reason why the IPA technique has limited applications. Nevertheless, we presented several examples where the IPA estimators are useful and can outperform (be more accurate than) their straightforward SF counterparts (e.g. when using the standard regenerative SF estimators in queueing models with long regenerative cycles).

Based on our extensive single-run simulation studies with rather general queueing networks for sensitivity analysis (estimation of $\nabla^k \ell(\boldsymbol{v}),\ k = 0, 1$, simultaneously for different values $\boldsymbol{v}$) and their optimization we recommend to:

(1) Use the direct SF estimators, provided (i) the variances (confidence intervals) are reasonable, say when the expected number of waiting customers in a busy cycle is, say less than 10, and (ii) the sample performance function $L(\boldsymbol{v}_2)$ is smooth (differentiable) with respect to $\boldsymbol{v}_2$. (Clearly (ii) holds when L does not depend on $\boldsymbol{v}_2$.)

(2) Use the low variance DSF (decomposable score function) and TSF (truncated score function) estimators, or the CSF (conditional score function) method of McLeish and Rollans (1992), if (1) fails (the variance is large). Our numerical studies clearly indicate that the DSF and TSF estimators are highly efficient when optimizing DES.

(3) Use the "push out" technique, if (2) fails and the required transformation $\boldsymbol{x} = \boldsymbol{x}(\boldsymbol{y}, \boldsymbol{v}_2)$ is available.

We believe that this discussion provides a basic guidance and gives an insight into when and how to use the direct, "push out" and "push in" (IPA) estimators for performance evaluation, sensitivity analysis and optimization of DES.

5.4 Exercises

1. Consider example 5.2.2. Derive

$$\nabla \ell(\boldsymbol{v}) = \{\frac{\partial \ell(\boldsymbol{v})}{\partial v_1}, \frac{\partial \ell(\boldsymbol{v})}{\partial v_2}\}. \tag{5.4.17}$$

   Take the derivatives inside the expected value and obtain the associated direct estimator $\nabla \bar{\ell}_N(\boldsymbol{v})$. Prove its consistency.

2. Consider examples 5.2.5 and 5.2.6. Derive (5.4.17). Take the derivatives inside the expected values and obtain the associated IPA estimator $\nabla \bar{\ell}_N(\boldsymbol{v})$ in terms of $\hat{L}(U_i, \boldsymbol{v})$. Prove its consistency.

3. Consider algorithm 5.2.2. Derive (5.4.17). Take the derivatives inside the expected values and obtain the associated IPA estimator $\nabla \bar{\ell}_N(\boldsymbol{v})$ in terms of $\hat{L}(U_i, \boldsymbol{v})$. Prove its consistency.

4. Consider algorithm 5.3.2. Derive (5.4.17), while using the TES input sequence. Take the derivatives inside the expected values and obtain the associated IPA estimator $\nabla \bar{\ell}_N(\boldsymbol{v})$ in terms of $\hat{L}(U_i, \boldsymbol{v})$. Prove its consistency.

5. Consider examples 5.3.3 and 5.3.4. Based on (5.3.14) and (5.3.16), derive expressions for $\nabla \widehat{W}_t^{tr}(\underline{z}_t, \boldsymbol{v}, m)$ while using AR(1) and TES sequences, assuming that $Y \sim \exp(v_1)$. Derive similar expressions for the Minification/maxification sequence.

Chapter 6

Statistical Inference of Stochastic Optimization Programs

6.1 Introduction

In this chapter we discuss the statistical properties of estimators derived from stochastic counterparts (approximations) of the "true" optimization programs. In the previous chapters such approximations were obtained while replacing expected value functions by their sample mean estimators constructed via simulation. We outline here a general theory sufficiently flexible to deal with this basic approximation scheme as well as with some of its extensions. Some preliminary results in this direction have already been given in appendix A of chapter 2.

This chapter is organized as follows. Later on in this section we formulate some useful results about interchangeability of the expectation and differentiation operators (theorem 6.1.1) and give a uniform version of the Law of Large Numbers (theorem 6.1.2). We also discuss measurability properties of the optimal value and the optimal solutions in stochastic programming. In section 6.2 we show that under mild regularity conditions stochastic programming estimators converge with probability one to their "true" counterparts. By its nature, asymptotic analysis of the considered stochastic programs is closely related to (deterministic) sensitivity analysis of optimization problems. A connection between the deterministic and stochastic parts of the analysis is often accomplished by the delta method, which will be the subject of section 6.3. The delta method will be applied to asymptotic analysis of the optimal value function in section 6.4 Deterministic sensitivity analysis of nonlinear programs has an independent interest and can be employed, for example, in the investigation of sensitivity (robustness) of the optimal solutions with respect to distribution contaminations. This will be the subject of section 6.5. Finally, in section 6.6 we give a general discussion of the asymptotic properties of the optimal solutions in stochastic programming.

Consider the following optimization problem involving equality and inequality constraints:

$$(\mathrm{P}_0) \qquad \begin{array}{lll} \text{minimize} & \ell_0(\boldsymbol{v}), & \boldsymbol{v} \in V, \\ \text{subject to} & \ell_j(\boldsymbol{v}) \leq 0, & j = 1, \dots, k, \\ & \ell_j(\boldsymbol{v}) = 0, & j = k+1, \dots, M, \end{array}$$

where $\ell_j(\boldsymbol{v})$, $j = 0, 1, \dots, M$, are real-valued functions and V is a subset of the n-dimensional vector space $\mathbb{R}^n$. Suppose that some of the functions $\ell_j(\boldsymbol{v})$ are not available analytically and have to be approximated, say by the Monte Carlo (simulation) methods. That is, instead of solving the program (P_0), referred to as the *true* program, we solve a sequence of approximating programs

$$(\bar{\mathrm{P}}_\mathrm{N}) \qquad \begin{array}{lll} \text{minimize} & \psi_{0N}(\boldsymbol{v}), & \boldsymbol{v} \in V, \\ \text{subject to} & \psi_{jN}(\boldsymbol{v}) \leq 0, & j = 1, \dots, k, \\ & \psi_{jN}(\boldsymbol{v}) = 0, & j = k+1, \dots, M. \end{array}$$

Here, for every $j = 0, 1, \dots, M$, $\{\psi_{jN}(\boldsymbol{v})\}$ is a sequence of real-valued (random) functions converging, asymptotically in N, to the corresponding function $\ell_j(\boldsymbol{v})$. The calculated the optimal value $\bar{\varphi}_N$ and an the optimal solution $\bar{\boldsymbol{v}}_N$ of the program $(\bar{\mathrm{P}}_\mathrm{N})$ are viewed as estimators of the optimal value φ^* and the optimal solution $\boldsymbol{v}^*$ of the true program (P_0), respectively.

The basic scheme for the true and approximating programs has already been employed in previous chapters and is constructed as follows. Let $(\Omega, \mathcal{F})$ be a measurable space and P be a probability measure on $(\Omega, \mathcal{F})$. Consider real-valued functions $h_j(\zeta, \boldsymbol{v})$ of two variables $\zeta \in \Omega$ and $\boldsymbol{v} \in \mathbb{R}^n$ and let Z be a random variable taking values in Ω and distributed according to the probability measure P. Suppose that the functions $\ell_j(\boldsymbol{v})$ can be represented in the form of expected values as follows:

$$\ell_j(\boldsymbol{v}) = \mathbb{E}\{h_j(Z, \boldsymbol{v})\} = \int_\Omega h_j(\zeta, \boldsymbol{v}) \mathrm{d}P(\zeta) , \tag{6.1.1}$$

$j = 0, 1, \dots, M$. Now let $Z_1, \dots, Z_N$ be a random sample of independent identically distributed (iid) random variables taking values in Ω with the common probability measure (distribution) P. Then the expected value functions $\ell_j(\boldsymbol{v})$ can be estimated (approximated) by the corresponding sample mean functions

$$\bar{\ell}_{jN}(v) = N^{-1} \sum_{i=1}^{N} h_j(Z_i, v). \tag{6.1.2}$$

The obtained optimization program ($\bar{\text{P}}_{\text{N}}$), with $\psi_{jN}(v) = \bar{\ell}_{jN}(v)$, can be considered as a stochastic counterpart of the true program (P_0).

We shall refer to the above scheme as the *sample mean construction.* The associated minimizers $\bar{v}_N$ of the program ($\bar{\text{P}}_{\text{N}}$) are closely related to the M-estimators of Huber (1964, 1981). Notice that if for some j the function $h_j(\zeta, v) = h_j(v)$ is independent of ζ, then the expected value function $\ell_j(v)$ and the sample mean function $\bar{\ell}_{jN}(v)$ coincide and are equal to the function $h_j(v)$. In this case we call the corresponding constraint (objective) function *deterministic.*

The basic scheme of the sample mean construction can be extended by considering composite functions which are composed from the expected value and the sample mean functions, respectively. For example, the objective function $\ell_0(v)$ of the true program may be given in the form of the ratio $\ell_0(v) = \ell_1(v)/\ell_2(v)$ with $\ell_1(v)$ and $\ell_2(v)$ being the expected value functions representable in the form (5.1.1) . In this case it is natural to estimate $\ell_0(v)$ by the ratio $\psi_{0N}(v) = \bar{\ell}_{1N}(v)/\bar{\ell}_{2N}(v)$ of the corresponding sample mean functions.

While investigating DESS we dealt with functions $h_j(\zeta, v)$ of the form $L_j(z)W\ (z, v)$, where $\zeta = z$ was a finite-dimensional vector. In this case it is sufficient to take $\Omega = \mathbb{R}^m$ equipped with its Borel σ-algebra $\mathcal{F}$, and to consider random vectors Z taking values in $\mathbb{R}^m$. (Recall that the Borel σ-algebra of $\mathbb{R}^m$, or more generally of a metric space, is generated by the family of open sets.) A more sophisticated construction of the probability space $(\Omega,\ \mathcal{F}, P)$ is required in order to handle DEDS considered in chapter 3.

Let $Y_1, Y_2, \ldots,$ be an input sequence of iid m-dimensional random vectors generated from a pdf $f(y, v)$ depending on the parameter vector v. Consider an output process $L_t = L_t(\underline{Y}_t)$, $\underline{Y}_t = (Y_1, \cdots, Y_t)$, driven by the input sequence $\{Y_t\}$. The sequences $\{Y_t\}$ and $\{L_t\}$, $t \in \mathbb{N}$, can be viewed as discrete time stochastic processes. (Here $\mathbb{N}$ denotes the set of positive integers.) For any k-tuple $(t_1, \ldots, t_k)$ of distinct elements of $\mathbb{N}$, the distribution of the random vector $(Y_{t_1}, \ldots, Y_{t_k})$ is called a finite-dimensional distribution of the process $\{Y_t, t \in \mathbb{N}\}$.

Consider the product space $\Omega = \mathbb{R}^m \times \mathbb{R}^m \times \ldots$, the Cartesian product of copies of $\mathbb{R}^m$, one copy for each $t = 1, 2 \ldots$ The product space Ω can be equipped with the σ-algebra $\mathcal{F}$ generated by the so-called finite-dimensional sets, or cylinders. Kolmogorov's existence theorem ensures that there exists a probability measure P defined on $(\Omega,\ \mathcal{F})$ such that the coordinate-variable process $\{\boldsymbol{Y}'_t,\ t \in \mathbb{N}\}$ on $(\Omega,\ \mathcal{F},\ P)$ has the same finite dimensional distributions as $\{\boldsymbol{Y}_t,\ t \in \mathbb{N}\}$ (see, for example, Billingsley (1979, section 36), for details). We identify $\{\boldsymbol{Y}_t,\ t \in \mathbb{N}\}$ with $\{\boldsymbol{Y}'_t,\ t \in \mathbb{N}\}$. Notice that the pdf $f(\boldsymbol{y}, \boldsymbol{v})$ and hence the constructed probability measure $P = P_{\boldsymbol{v}}$, depend on the parameter vector $\boldsymbol{v}$.

Now let τ be a measurable mapping from the probability space $(\Omega,\ \mathcal{F},\ P)$ into $\mathbb{N}$. (The mapping $\tau : \Omega \to \mathbb{N}$ is measurable in the sense that the set $\{\omega : \tau(\omega) = t\}$ lies in $\mathcal{F}$ for any $t \in \mathbb{N}$.) We may look at τ as a discrete random variable defined on $(\Omega,\ \mathcal{F},\ P)$. Assume that the occurrence of the event $\{\tau = t\}$, $t \in \mathbb{N}$, is determined by $\underline{\boldsymbol{Y}}_t = (\boldsymbol{Y}_1, \ldots, \boldsymbol{Y}_t)$ alone and hence τ is a *stopping time* for the process $\{\boldsymbol{Y}_t, t \in \mathbb{N}\}$. Let $g(\boldsymbol{z})$ be a density function on $\mathbb{R}^m$ and let $\boldsymbol{Z}_1, \boldsymbol{Z}_2, \ldots$, be a sequence of iid random vectors generated from $g(\boldsymbol{z})$. Then it was shown in section 3.2 of chapter 3 that the expected value of the random variable $X = \sum_1^{\tau} L_t(\underline{\boldsymbol{Y}}_t)$ can be represented in the form (see equation (3.2.25))

$$\mathbb{E}_{\boldsymbol{v}} X = \mathbb{E}_g \left\{ \sum_{t=1}^{\tau} L_t(\underline{\boldsymbol{Z}}_t) \widetilde{W}_t(\underline{\boldsymbol{Z}}_t, \boldsymbol{v}) \right\} , \tag{6.1.3}$$

where

$$\widetilde{W}_t(\underline{\boldsymbol{z}}_t, \boldsymbol{v}) = f_t(\underline{\boldsymbol{z}}_t, \boldsymbol{v}) g_t(\underline{\boldsymbol{z}}_t)$$

and

$$f_t(\underline{\boldsymbol{z}}_t, \boldsymbol{v}) = \prod_{i=1}^{t} f(\boldsymbol{z}_i, \boldsymbol{v}),\ g_t(\underline{\boldsymbol{z}}_t) = \prod_{i=1}^{t} g(\boldsymbol{z}_i) .$$

Note that the expectation in the left-hand side of (6.1.3) is taken with respect to the probability measure $P_{\boldsymbol{v}}$, while the expectation in the right-hand side of (6.1.3) is taken with respect to the probability measure P_g on $(\Omega,\ \mathcal{F})$ which has the finite dimensional distributions determined by the process $\{\boldsymbol{Z}_t,\ t \in \mathbb{N}\}$.

A certain amount of care should be exercised when choosing the pdf $g(\boldsymbol{z})$. It is assumed that for every $\boldsymbol{v} \in V$ the expectation $\mathbb{E}_{\boldsymbol{v}}(\tau)$ is finite. It follows then that the stopping time τ is finite w.p.1. That is, $\tau(\omega)$ is finite for every ω in Ω except the set of $P_{\boldsymbol{v}}$-measure zero. This allows us to ignore the event $\{\tau = \infty\}$ when calculating the expectation $\mathbb{E}_{\boldsymbol{v}} X$, which is supposed to be finite for all $\boldsymbol{v} \in V$. In order to derive (6.1.3) one has to verify that the expectation $\mathbb{E}_g(\tau)$ is finite as well. This is one of the requirements that the pdf $g(\boldsymbol{z})$ must satisfy (see below and L'Ecuyer (1991)). Also in order for the functions $\widetilde{W}_t$ to be well defined, $g(\boldsymbol{z})$ must dominate the densities $f(\boldsymbol{z}, \boldsymbol{v})$ in the sense that the support of $f(\boldsymbol{z}, \boldsymbol{v})$ lies in the support of $g(\boldsymbol{z})$ for all $\boldsymbol{v} \in V$.

For an element $\zeta = (\boldsymbol{z}_1, \boldsymbol{z}_2, \ldots)$ of Ω denote by $\underline{\boldsymbol{z}}_t$ the vector $(\boldsymbol{z}_1, \ldots, \boldsymbol{z}_t)$.

Define the function

$$h(\zeta, \boldsymbol{v}) = \sum_{t=1}^{\tau} L_t(\underline{\boldsymbol{z}}_t)\widetilde{W}_t(\underline{\boldsymbol{z}}_t, \boldsymbol{v}) \tag{6.1.4}$$

of $\zeta \in \Omega$ and $\boldsymbol{v} \in \mathbb{R}^n$. Since $\tau = \tau(\zeta)$ is finite w.p.1, the function $h(\zeta, \boldsymbol{v})$ is defined by (6.1.4) for almost every ζ with respect to the measure P_g. We can now write expression (6.1.3) in the form

$$\mathbb{E}_{\boldsymbol{v}} X = \int_{\Omega} h(\zeta, \boldsymbol{v}) \mathrm{d}P_g(\zeta) = \mathbb{E}_g\{h(Z, \boldsymbol{v})\} \,. \tag{6.1.5}$$

The function $\ell_1(\boldsymbol{v}) = \mathbb{E}_{\boldsymbol{v}} X$ can be estimated then by the sample mean function corresponding to the expectation given in the right-hand side of (6.1.5) (cf. (3.2.27)). Similar derivations can be applied to estimation of the function $\ell_2(\boldsymbol{v}) = \mathbb{E}_{\boldsymbol{v}} \tau$ as well. This yields the sample mean construction for DEDS.

In the remainder of this section we formulate some useful theorems which will be important for the subsequent analysis. In appendix A of chapter 2 the results were formulated with respect to random vectors and corresponding finite-dimensional distributions. However, this fact was nowhere used in the proofs and the results are still valid for general probability spaces. In particular, lemma A2 can be formulated in the following form. (Unless stated otherwise, all probability statements will be given with respect to the considered probability space $(\Omega, \mathcal{F}, P)$.)

Theorem 6.1.1 *Let U be an open subset of $\mathbb{R}^n$ and suppose that:*

(i) *There exists an integrable function $k : \Omega \to \mathbb{R}$ such that for all $\boldsymbol{u}, \boldsymbol{v} \in U$ and $\zeta \in \Omega$,*

$$|h(\zeta, \boldsymbol{u}) - h(\zeta, \boldsymbol{v})| \leq k(\zeta)||\boldsymbol{u} - \boldsymbol{v}|| \,. \tag{6.1.6}$$

(ii) *For any given $\boldsymbol{v} \in U$ and almost every ζ, the function $h(\zeta, \cdot)$ is differentiable at $\boldsymbol{v}$.*
Then the expected value function $\ell(\boldsymbol{v}) = \mathbb{E}\{h(Z, \boldsymbol{v})\}$ is Lipschitz continuous and differentiable on U and

$$\nabla \ell(\boldsymbol{v}) = \mathbb{E}\{\nabla h(Z, \boldsymbol{v})\} \,. \tag{6.1.7}$$

Usually the set U is taken to be a (convex) neighborhood of a considered point. In situations where $h(\zeta, \cdot)$ is differentiable everywhere, it follows from the Mean Value Theorem that (6.1.6) holds with

$$k(\zeta) = \sup \{||\nabla h(\zeta, \boldsymbol{v})|| : \boldsymbol{v} \in U\} \,.$$

Therefore in this case conditions (i) of theorem 6.1.1 can be replaced by the following condition.

1. (i$'$) *The family* $\{||\nabla h(\zeta, \boldsymbol{v})||,\ \boldsymbol{v} \in U\}$ *is dominated by an integrable function, i.e. there exists an integrable function* $b : \Omega \to \mathbb{R}$ *such that*

$$||\nabla h(\zeta, \boldsymbol{v})|| \leq b(\zeta)$$

for all $\boldsymbol{v} \in U$ *and* $\zeta \in \Omega$.

As an example let us apply the result of theorem 6.1.1 to the expected value function $\mathbb{E}_{\boldsymbol{v}} X$ given in the form (6.1.3) or (6.1.5). Suppose that the pdf $f(\boldsymbol{z}, \boldsymbol{v})$ is everywhere differentiable in $\boldsymbol{v}$. It follows then that the functions $\widetilde{W}_t(\underline{\boldsymbol{z}}_t, \cdot)$ are differentiable and hence condition (ii) holds. In order to verify condition (i$'$) consider a sequence of functions $\{c_t(\underline{\boldsymbol{z}}_t)\}$ such that

$$||L_t(\underline{\boldsymbol{z}}_t) \nabla f_t(\underline{\boldsymbol{z}}_t, \boldsymbol{v})|| \leq c_t(\underline{\boldsymbol{z}}_t) \tag{6.1.8}$$

for all $\boldsymbol{v} \in U$, $\underline{\boldsymbol{z}}_t = (\boldsymbol{z}_1, \ldots, \boldsymbol{z}_t)$ and $t \in \mathrm{N}$. Then

$$||L_t(\underline{\boldsymbol{z}}_t) \nabla \widetilde{W}_t(\underline{\boldsymbol{z}}, \boldsymbol{v})|| \leq b_t(\underline{\boldsymbol{z}}_t)\ , \tag{6.1.9}$$

where $b_t(\underline{\boldsymbol{z}}_t) = c_t(\underline{\boldsymbol{z}}_t)/g_t(\underline{\boldsymbol{z}}_t)$. Consequently the family $\{||\nabla h(\zeta, \boldsymbol{v})||,\ \boldsymbol{v} \in U\}$ is dominated by the function

$$b(\zeta) = \sum_{t=1}^{\tau} b_t(\underline{\boldsymbol{z}}_t)\ .$$

Therefore condition (i$'$) holds if the functions c_t can be chosen in such a way that the expectation

$$\mathbb{E}_g \left\{ \sum_{t=1}^{\tau} b_t(\underline{\boldsymbol{Z}}_t) \right\} = \int_\Omega b(\zeta) dP_g(\zeta)$$

is finite. It follows then that

$$\nabla(\mathbb{E}_{\boldsymbol{v}} X) = \mathbb{E}_g \left\{ \sum_{t=1}^{\tau} L_t(\underline{\boldsymbol{Z}}_t) \nabla \widetilde{W}_t(\underline{\boldsymbol{Z}}_t, \boldsymbol{v}) \right\}\ .$$

Another useful result was given in lemma A1 of appendix A of chapter 2.

Theorem 6.1.2 *Let* $\ell(\boldsymbol{v})$ *be the expected value of* $h(Z, \boldsymbol{v})$ *and let* $\bar{\ell}_N(\boldsymbol{v})$ *be the sample mean estimator of* $\ell(\boldsymbol{v})$. *Suppose that:*

(i) *For almost every* $\zeta \in \Omega$, *the function* $h(\zeta, \cdot)$ *is continuous on* V.

(ii) *The family* $\{|h(\zeta, \boldsymbol{v})|,\ \boldsymbol{v} \in V\}$ *is dominated by an integrable function. Then* $\ell(\boldsymbol{v})$ *is continuous on* V *and if, in addition, the set* V *is compact, then w.p.1* $\bar{\ell}_N(\boldsymbol{v})$ *converges to* $\ell(\boldsymbol{v})$ *uniformly on* V.

In the case of DEDS, continuity of the functions $h(\zeta, \cdot)$, defined in (6.1.4), follows from continuity of $f(z, \cdot)$ The dominance condition (ii) of theorem 6.1.2 can be verified in a way similar to the analysis following theorem 6.1.1.

Finally we mention some results about measurability of the optimal value and the optimal solutions of stochastic programs. A thorough discussion and historical background of the theory involved can be found in Rockafellar (1976) and Castaing and Valadier (1977). Recall that a mapping $g : \Omega \to \mathbb{R}^n$ is said to be *measurable* if for every Borel set $A \subset \mathbb{R}^n$ its inverse image

$$g^{-1}(A) = \{\omega \in \Omega : g(\omega) \in A\}$$

is an $\mathcal{F}$-measurable subset of Ω. Now let us consider a *multifunction* (also called set valued or point-to-set mapping) $G : \Omega \rightrightarrows \mathbb{R}^n$. That is, the mapping G assigns to each $\omega \in \Omega$ a subset $G(\omega)$ of $\mathbb{R}^n$. The set

$$\text{dom}\, G = \{\omega \in \Omega : G(\omega) \neq \emptyset\}$$

is called the domain of G. It is said that G is *closed-valued* if $G(\omega)$ is a closed subset of $\mathbb{R}^n$ for every $\omega \in \Omega$. Such a closed-valued multifunction G is said to be *measurable*, if for every closed set $A \subset \mathbb{R}^n$ one has

$$G^{-1}(A) = \{\omega \in \Omega : G(\omega) \cap A \neq \emptyset\} \in \mathcal{F} .$$

Note that measurability of G implies that dom $G = G^{-1}(\mathbb{R}^n)$ is an $\mathcal{F}$-measurable subset of Ω.

The introduction of measurable multifunctions is justified by the following theorem on measurable selections which seemingly can be attributed to several authors (see Rockafellar (1976, p. 163), Castaing and Valadier (1977, chapter 3)).

Theorem 6.1.3 Measurable selection theorem *If $G : \Omega \rightrightarrows \mathbb{R}^n$ is a closed-valued measurable multifunction, then there exists at least one measurable selection, i.e. a measurable mapping $g :$ dom $G \to \mathbb{R}^n$ such that $g(\omega) \in G(\omega)$ for all $\omega \in$ dom G.*

Consider now an extended real-valued function $f(\omega, v)$. The function $f : \Omega \times \mathbb{R}^n \to \bar{\mathbb{R}}$ is said to be extended in the sense that it takes values in the extended real line $\bar{\mathbb{R}} = \mathbb{R} \cup \{-\infty, +\infty\}$. We associate with $f(\omega, v)$ the multifunction $\Gamma : \omega \rightrightarrows$ epi $f(\omega, \cdot)$, where

$$\text{epi}\, f(\omega, \cdot) = \{(v, \alpha) \in \mathbb{R}^{n+1} : \alpha \geq f(\omega, v)\}$$

is called the epigraph of the function $f(\omega, \cdot)$. It is said that f is a *random lower semicontinuous* (random lsc) function if the corresponding epigraphical multifunction $\Gamma : \Omega \rightrightarrows \mathbb{R}^{n+1}$ is closed-valued and measurable. (Random lsc functions were called *normal integrands* in Rockafellar (1976).) Notice that the multifunction Γ is closed valued if and only if epi $f(\omega, \cdot)$ is a closed subset of $\mathbb{R}^{n+1}$ for every $\omega \in \Omega$. This in turn is equivalent to lower semicontinuity of the function $f(\omega, \cdot)$.

For our purposes, the usefulness of the random lsc functions lies in the following result which is a consequence of the measurable selection theorem (see Rockafellar (1976, p. 178)).

Theorem 6.1.4 *Let $f(\omega, \boldsymbol{v})$ be a random lsc function and let $G : \Omega \rightrightarrows \mathbb{R}^n$ be a closed-valued measurable multifunction. Then the optimal value function*

$$\phi(\omega) = \inf\ f(\omega, \boldsymbol{v}),\ \ \boldsymbol{v} \in G(\omega),$$

and the optimal solution multifunction

$$M(\omega) = \arg\min\ f(\omega, \boldsymbol{v}),\ \ \boldsymbol{v} \in G(\omega)$$

are both measurable.

Notice that it follows from lower semicontinuity of $f(\omega, \cdot)$ that the multifunction $M(\omega)$ is closed valued. By definition the optimal value $\phi(\omega)$ is $+\infty$ if the set $G(\omega)$ is empty. Also note that because of the measurable selection theorem, measurability of $M(\omega)$ implies existence of a measurable optimal solution $\bar{\boldsymbol{v}}(\omega)$ with $\bar{\boldsymbol{v}}(\omega)$ being an element of $M(\omega)$ for all ω such that $M(\omega)$ is nonempty.

A particular example of random lsc functions are Carathéodory functions. It is said that $f : \Omega \times \mathbb{R}^n \rightarrow \mathbb{R}$ is a *Carathéodory* function if $f(\omega, \boldsymbol{v})$ is continuous in $\boldsymbol{v}$ and measurable in ω. Such functions will be especially convenient for our purposes. We have that if the random functions $\psi_{jN}(\boldsymbol{v}) = \psi_{jN}(\omega, \boldsymbol{v})$, $j = 0, \ldots, M$, of the program $(\bar{\mathrm{P}}_{\mathrm{N}})$ are *Carathéodory* functions, then the corresponding optimal value $\bar{\varphi}_N = \bar{\varphi}_N(\omega)$ is measurable and the optimal solution $\bar{\boldsymbol{v}}_N = \bar{\boldsymbol{v}}_N(\omega)$ can be selected to be measurable.

6.2 Consistency of the Stochastic Estimators of the Optimal Value and the Optimal Solution

In this section we show that under mild regularity conditions, the optimal value $\bar{\varphi}_N$ and the optimal solution $\bar{\boldsymbol{v}}_N$ of $(\bar{\mathrm{P}}_{\mathrm{N}})$ converge to their true counterparts φ^* and $\boldsymbol{v}^*$, respectively. Denote by $\boldsymbol{\Phi}^*$ and $\boldsymbol{\Phi}_N$ the sets of feasible solutions of the programs (P_0) and $(\bar{\mathrm{P}}_{\mathrm{N}})$. That is

$$\begin{aligned} \boldsymbol{\Phi}^* = \{\boldsymbol{v} \in V : \ell_j(\boldsymbol{v}) &\leq 0,\ j = 1, \cdots, k; \\ \ell_j(\boldsymbol{v}) &= 0,\ j = k+1, \cdots, M\} \end{aligned} \tag{6.2.1}$$

and

$$\Phi_N = \{v \in V : \psi_{jN}(v) \leq 0,\ j = 1, \cdots, k; \quad \psi_{jN}(v) = 0,\ j = k+1, \cdots, M\}\,. \tag{6.2.2}$$

First we establish the following deterministic result which can be easily translated into statistical language.

Theorem 6.2.1 *Suppose that:*

(i) *The program* (P_0) *has a unique optimal solution* v^*.

(ii) *The set* V *is compact.*

(iii) *The functions* $\psi_{jN}(v)$, $j = 0, \cdots, M$, *are continuous and converge to* $\ell_j(v)$, $j = 0, \cdots, M$ *uniformly on* V, *i.e.*

$$\lim_{N\to\infty} \sup_{v \in V} |\psi_{jN}(v) - \ell_j(v)| = 0,$$

$j = 0, \ldots, M$.

(iv) *The program* $(\bar{\mathrm{P}}_{\mathrm{N}})$ *possesses a feasible point* u_N, $u_N \in \Phi_N$, *such that* $u_N \to v^*$ *as* $N \to \infty$.

Then $\bar{\varphi}_N$ *and* $\bar{v}_N$ *tend to* φ^* *and* v^*, *respectively, as* $N \to \infty$.

Proof Let us observe that it follows from the continuity and the uniform convergence of $\psi_{jN}(v)$ that the functions $\ell_j(v)$ are continuous on V. Condition (iv) means that the set Φ_N is nonempty. Moreover, we have that Φ_N is a closed subset of V and hence is compact. It follows that $\psi_{0N}(v)$ has a minimizer $\bar{v}_N$ (possibly not unique) over the set Φ_N.

We argue now by a contradiction. Suppose that $\bar{v}_N$ does not tend to v^*. Then since V is compact, we can assume, by passing to a subsequence if necessary, that $\bar{v}_N$ converges to a point $v_0 \in V$, $v_0 \neq v^*$. It follows then from the uniform convergence of $\psi_{jN}(v)$ to $\ell_j(v)$ and continuity of $\ell_j(v)$ that $\psi_{jN}(\bar{v}_N)$ tends to $\ell_j(v_0)$, $j = 0, \ldots, M$, as $N \to \infty$. By passing to a limit we obtain that $\ell_j(v_0)$ is less than or equal to zero for $j = 1, \ldots, k$, and that $\ell_j(v_0)$ is zero for $j = k+1, \ldots, M$. Consequently $v_0 \in \Phi^*$ and hence

$$\ell_0(v_0) > \ell_0(v^*)\,. \tag{6.2.3}$$

Now we have that

$$\psi_{0N}(u_N) \geq \psi_{0N}(\bar{v}_N)$$

and since $u_N \to v^*$,

$$\lim_{N\to\infty} \psi_{0N}(u_N) = \ell_0(v^*)\,.$$

It thus follows that

$$\ell_0(\boldsymbol{v}^*) \geq \lim_{N\to\infty} \psi_{0N}(\bar{\boldsymbol{v}}_N) = \ell_0(\boldsymbol{v}_0) ,$$

which contradicts (6.2.3). This proves that $\bar{\boldsymbol{v}}_N$ tends to $\boldsymbol{v}^*$. Th convergence of $\bar{\varphi}_N$ to φ^* then follows. □

The deterministic result of theorem 6.2.1 can be translated into its stochastic analogue by adding in assumptions (iii) and (iv) as well as in the conclusion the words "w.p.1". For the sample mean construction, the uniform convergence w.p.1 of $\psi_{jN}(\boldsymbol{v})$ to $\ell_j(\boldsymbol{v})$ is ensured by the Strong Law of Large Numbers (recall lemma A1 from the appendix A of chapter 2 and theorem 6.1.2). For an extension of the uniform convergence to composite functions see exercise 6.2.

Some remarks about the regularity assumptions of theorem 6.2.1 will now follow. Assumption (i) is needed to ensure convergence of the minimizer $\bar{\boldsymbol{v}}_N$ but is not required for the convergence of $\bar{\varphi}_N$ to φ^*. If the set V^* of the optimal solutions of the program (P_0) contains more than one point, then the above proof can be easily modified to show that the distance $\text{dist}(\bar{\boldsymbol{v}}_N, V^*)$, from $\bar{\boldsymbol{v}}_N$ to V^*, tends to zero (see exercise 6.3). However, we cannot then guarantee that $\bar{\boldsymbol{v}}_N$ converges to a particular point of V^*.

Assumption (ii) can be replaced by various regularity conditions ensuring that the minimizer $\bar{\boldsymbol{v}}_N$ stays (for the stochasic variant stays w.p.1) in a bounded subset of $\mathbb{R}^n$ as N tends to infinity (see exercise 6.4). For a more detailed discussion of this problem see, for example, Huber (1967) and Dupačová and Wets (1988).

In situations where all constraints are deterministic and hence $\Phi_N = \Phi^*$, assumption (iv) holds automatically by taking $\boldsymbol{u}_N = \boldsymbol{v}^*$ for all N. In the general case assumption (iv) is more subtle. It can be guaranteed by various regularity conditions, called constraint qualifications in optimization theory. Suppose that the constraint functions $\ell_j(\boldsymbol{v})$, $j = 1, \ldots, M$, of the program (P_0) are continuously differentiable in a neighborhood of the point $\boldsymbol{v}^*$ and that $\boldsymbol{v}^*$ is an interior point of V. Denote by $J(\boldsymbol{v})$ the index set of inequality constraints active at $\boldsymbol{v}$,

$$J(\boldsymbol{v}) = \{j : \ell_j(\boldsymbol{v}) = 0,\ j = 1, \ldots, k\} . \tag{6.2.4}$$

The following is a constraint qualification of Mangasarian and Fromovitz (1967) applied to the program (P_0) at the point $\boldsymbol{v}^*$ (cf. exercise 6.6).

MF-condition. The gradients $\nabla\ell_j(\boldsymbol{v}^*)$, $j = k+1, \ldots, M$, are linearly independent and there exists a vector $\boldsymbol{b}$ such that

$$\boldsymbol{b}'\nabla\ell_j(\boldsymbol{v}^*) = 0, \quad j = k+1, \ldots, M,$$

$$b'\nabla \ell_j(v^*) < 0, \quad j \in J(v^*).$$

There is no loss of generality in assuming that v^* is an interior point of V. Usually the set V is defined by simple constraints such as that the coordinates of its points must be non-negative, etc. If some of those constraints become active at v^*, we can add them to the list of inequality constraints by writing them explicitly. Note that the MF-condition holds if, but not only if, all the gradients $\nabla \ell_j(v^*)$, $j \in J(v^*) \cup \{k+1, \ldots, M\}$, are linearly independent. We refer to this last condition as the *linear independence condition.*

Under the MF-condition, the system of constraints of the program (P_0) is stable in the following sense. There exists a neighborhood of the point v^* and a positive constant κ such that for every point v in this neighborhood

$$\operatorname{dist}(v, \Phi^*) \le \kappa \left\{ \sum_{j=1}^{k} [\ell_j(v)]^+ + \sum_{j=k+1}^{M} |\ell_j(v)| \right\}, \tag{6.2.5}$$

where $a^+ = \max\{0, a\}$. Of course, if $v \in \Phi^*$, then $[\ell_j(v)]^+$, $j = 1, \ldots, k$, and $|\ell_j(v)|$, $j = k+1, \ldots, M$, are zeros and the inequality (6.2.5) holds. More generally, the right-hand side of (6.2.5) provides a bound on the distance from a point v, sufficiently close to v^*, to the feasible set Φ^*. This bound is especially convenient since it is given in terms of the constraint functions $\ell_j(v)$.

The above stability result can be extended to the programs with constraints sufficiently close, in a certain sense, to those of the program (P_0). Since this result will be important in several ways we state it explicitly.

Theorem 6.2.2 Stability theorem *Suppose that :*

(i) *The point v^* is an interior point of V and the MF - condition, applied to the program (P_0) at v^*, holds.*

(ii) *The functions $\psi_{jN}(v)$, $j = 1, \ldots, M$, are continuously differentiable in a neighborhood of v^*.*

(iii) *The functions $\psi_{jN}(v)$ and their gradients $\nabla\psi_{jN}(v)$, $j = 1, \ldots, M$, converge to $\ell_j(v)$ and $\nabla \ell_j(v)$, respectively, as $N \to \infty$, uniformly in v in a neighborhood of v^*.*

Then there exists a positive constant κ such that for all v in a neighborhood of v^* and all N large enough

$$\operatorname{dist}(v, \Phi_N) \le \kappa \left\{ \sum_{j=1}^{k} [\psi_{jN}(v)]^+ + \sum_{j=k+1}^{M} |\psi_{jN}(v)| \right\}. \tag{6.2.6}$$

Notice that the inequality (6.2.5) follows from (6.2.6) if we take $\psi_{jN}(\boldsymbol{v}) = \ell_j(\boldsymbol{v})$ for all j and N. A prototype of the stability result of theorem 6.2.2, involving equality constraints in Banach spaces, appeared in earlier works of Ljusternik (1934) and Graves (1950). For inequality constraints this result is a consequence of the Robinson (1976)–Ursescu (1975) stability theorem.

It immediately follows from (6.2.4) that, under the assumptions of theorem 6.2.2, the distance from $\boldsymbol{v}^*$ to Φ_N tends to zero as $N \to \infty$ and hence condition (iv) of theorem 6.2.1 holds. Let us summarize our discussion of the consistency of $\bar{\varphi}_N$ and $\bar{\boldsymbol{v}}_N$ in the case of the sample mean construction.

Theorem 6.2.3 *Consider the sample mean construction and suppose that :*

(i) *The program* $(\mathrm{P_0})$ *has a unique optimal solution* $\boldsymbol{v}^*$.

(ii) *The set* V *is compact.*

(iii) *The functions* $h_j(\zeta, \boldsymbol{v}),\ j = 0, \ldots, M$, *are Carathéodory functions.*

(iv) *For almost every* ζ *the functions* $h_j(\zeta, \cdot),\ j = 1, \ldots, M$, *are continuously differentiable in a neighbourhood* U *of* $\boldsymbol{v}^*$.

(v) *The families* $\{|h_j(\zeta, \boldsymbol{v})|,\ \boldsymbol{v} \in V\},\ j = 0, \ldots, M$, *and* $\{\|\nabla h_j(\zeta, \boldsymbol{v})\|,\ \boldsymbol{v} \in U\},\ j = 1, \ldots, M$, *are dominated by integrable functions.*

(vi) *The point* $\boldsymbol{v}^*$ *is an interior point of* V *and the MF-condition, applied to* $(\mathrm{P_0})$ *at* $\boldsymbol{v}^*$, *holds.*

Then $\bar{\varphi}_N$ *and* $\bar{\boldsymbol{v}}_N$ *converge w.p.1 to* φ^* *and* $\boldsymbol{v}^*$, *respectively.*

A few remarks are now in order. By theorem 6.1.4 it follows from assumption (iii) that $\bar{\varphi}_N$ and $\bar{\boldsymbol{v}}_N$ are measurable. (More precisely, there exists a measurable selection of the optimal solution $\bar{\boldsymbol{v}}_N$.) Since the Carathéodory functions $h_j(\zeta, \boldsymbol{v})$ are supposed to be continuous in $\boldsymbol{v}$, the dominance condition of assumption (v) implies that $\ell_j(\boldsymbol{v})$ are continuous and that the sample mean functions $\bar{\ell}_{jN}(\boldsymbol{v})$ tend to $\ell_j(\boldsymbol{v})$ w.p.1 uniformly in $\boldsymbol{v} \in V$ (see theorem 6.1.2). Because of assumptins (iv) and (v) we have that the functions $\ell_j(\boldsymbol{v})$, $j = 1, \ldots, M$, are continuously differentiable (see theorem 6.1.1) and that $\nabla \bar{\ell}_{jN}(\boldsymbol{v})$ converge to $\nabla \ell_j(\boldsymbol{v})$ w.p.1 uniformly in $\boldsymbol{v}$ in a neighborhood of $\boldsymbol{v}^*$. (The uniform convergence of $\nabla \bar{\ell}_{jN}(\boldsymbol{v})$ follows again from theorem 6.1.2 and from the interchangeability of the expected value and differentiation.) This shows that the corresponding assumptions of theorems 6.2.1 and 6.2.2 are satisfied. Note that if all constraints of $(\bar{\mathrm{P}}_{\mathrm{N}})$ are deterministic and hence $\Phi_N = \Phi^*$, then assumptions (iv), (vi) and the second dominance condition of assumption (v) can be omitted.

The consistency result of theorem 6.2.3 can be extended to deal with functions which are compositions of the expected value and the sample mean functions, respectively (see exercise 6.2).

6.3 The Delta Method

We showed in the previous section that under mild regularity conditions one can expect that the optimal value and the optimal solutions of the approximating program ($\bar{\mathrm{P}}_{\mathrm{N}}$) converge to their true counterparts of the program (P_0). However, the convergence alone is not sufficient for $\bar{\varphi}_N$ and $\bar{\boldsymbol{v}}_N$ to be reasonable estimators of φ^* and $\boldsymbol{v}^*$, respectively. The convergence can be so slow that for practical purposes the estimators $\bar{\varphi}_N$ and $\bar{\boldsymbol{v}}_N$ will be useless. Therefore we also would like to estimate the rate of the convergence. In a statistical context this can be achieved by evaluation of the distributions of $\bar{\varphi}_N$ and $\bar{\boldsymbol{v}}_N$.

The exact distributions of $\bar{\varphi}_N$ and $\bar{\boldsymbol{v}}_N$ are too complicated to be presented in a closed form even for relatively simple cases. Nevertheless, the *asymptotic* distributions can be evaluated under quite general conditions. Often the asymptotic analysis is based on the so-called delta method which is the subject of this section.

Let $\{\boldsymbol{X}_N\}$ be a sequence of m-dimensional random vectors converging in probability to a vector $\boldsymbol{\mu}$. Suppose that there exists a sequence $\{\tau_N\}$ of positive numbers tending to infinity such that $\tau_N(\boldsymbol{X}_N - \boldsymbol{\mu})$ converges in distribution to a random vector $\boldsymbol{Y}$. Let $g : \mathbb{R}^m \to \mathbb{R}^n$ be a mapping and suppose that g is differentiable at $\boldsymbol{\mu}$. This means that

$$g(\boldsymbol{x}) - g(\boldsymbol{\mu}) = [\nabla g(\boldsymbol{\mu})]'(\boldsymbol{x} - \boldsymbol{\mu}) + r(\boldsymbol{x}), \tag{6.3.1}$$

where $\nabla g(\boldsymbol{\mu})$ is the $m \times n$ Jacobian matrix (of partial derivatives) of g at $\boldsymbol{\mu}$ and the remainder $r(\boldsymbol{x})$ is of order $o(||\boldsymbol{x} - \boldsymbol{\mu}||)$. That is, $||\boldsymbol{x} - \boldsymbol{\mu}||^{-1}\, r(\boldsymbol{x})$ tends to zero as $\boldsymbol{x} \to \boldsymbol{\mu}$. It follows from (6.3.1) that

$$\tau_N[g(\boldsymbol{X}_N) - g(\boldsymbol{\mu})] = \tau_N[\nabla g(\boldsymbol{\mu})]'(\boldsymbol{X}_N - \boldsymbol{\mu}) + \tau_N r(\boldsymbol{X}_N)\,. \tag{6.3.2}$$

Since $\tau_N(\boldsymbol{X}_N - \boldsymbol{\mu})$ converges in distribution, it is bounded in probability and hence $||\boldsymbol{X}_N - \boldsymbol{\mu}||$ is of stochastic order $O_p(\tau_N^{-1})$. It follows that

$$r(\boldsymbol{X}_N) = o(||\boldsymbol{X}_N - \boldsymbol{\mu}||) = o_p(\tau_N^{-1})\,,$$

and thus the remainder term $\tau_N\, r(\boldsymbol{X}_N)$ in the right-hand side of (6.3.2) tends in probability to zero. Therefore $\tau_N[g(\boldsymbol{X}_N) - g(\boldsymbol{\mu})]$ has the same limiting distribution as the first term in the right-hand side of (6.3.2). Consequently $\tau_N[g(\boldsymbol{X}_N) - g(\boldsymbol{\mu})]$ converges in distribution to $[\nabla g(\boldsymbol{\mu})]'\boldsymbol{Y}$. We write this as follows:

$$\tau_N\,[g(\boldsymbol{X}_N) - g(\boldsymbol{\mu})] \xrightarrow{\mathcal{D}} [\nabla g(\boldsymbol{\mu})]'\,\boldsymbol{Y}\,, \tag{6.3.3}$$

where the superscript $\mathcal{D}$ stands for the convergence in distribution.

Formula (6.3.3) is routinely employed in multivariable statistical analysis and is known as the (finite-dimensional) delta theorem. In applications τ_N is usually equal to $N^{1/2}$ and the limiting distribution of $N^{1/2}(\boldsymbol{X}_N - \boldsymbol{\mu})$ is multivariable normal with a zero mean vector and a covariance matrix $\boldsymbol{\Sigma}$. Then it follows from (6.3.3) that $N^{1/2}[g(\boldsymbol{X}_N) - g(\boldsymbol{\mu})]$ converges in distribution to normal with zero mean vector and the covariance matrix $\boldsymbol{J}'\boldsymbol{\Sigma}J$, where $\boldsymbol{J} = \nabla g(\boldsymbol{\mu})$.

In some situations it is necessary to deal with mappings which although are nondifferentiable may possess directional derivatives. We say that a mapping $g : \mathbb{R}^m \to \mathbb{R}^n$ is *directionally differentiable* at a point $\boldsymbol{\mu}$ if the one-sided limit

$$g'_{\boldsymbol{\mu}}(\boldsymbol{d}) = \lim_{t \to 0^+} \frac{g(\boldsymbol{\mu} + \boldsymbol{d}) - g(\boldsymbol{\mu})}{t} \tag{6.3.4}$$

exists for all $\boldsymbol{d} \in \mathbb{R}^m$. A simple example of a function $g : \mathbb{R} \to \mathbb{R}$ which is directionally differentiable but nondifferentiable (at zero), is $g(x) = |x|$. As we shall see later such mappings arise naturally in optimization problems. Note that it follows from the definition that the directional derivative $g'_{\boldsymbol{\mu}}(\boldsymbol{d})$ is positively homogeneous in $\boldsymbol{d}$, i.e. $g'_{\boldsymbol{\mu}}(t\boldsymbol{d}) = tg'_{\boldsymbol{\mu}}(\boldsymbol{d})$ for any $t \geq 0$ and all $\boldsymbol{d} \in \mathbb{R}^m$.

We can write (6.3.4) in an equivalent form as follows

$$g(\boldsymbol{\mu} + \boldsymbol{d}) - g(\boldsymbol{\mu}) = g'_{\boldsymbol{\mu}}(\boldsymbol{d}) + r(\boldsymbol{d}) \, , \tag{6.3.5}$$

where $g'_{\boldsymbol{\mu}}(\cdot)$ is positively homogeneous and $t^{-1}r(t\boldsymbol{d}) \to \boldsymbol{0}$ as $t \to 0^+$ for every $\boldsymbol{d}$. The remainder term $r(\boldsymbol{d})$ in (6.3.5) is then of a small order in every fixed direction $\boldsymbol{d}$. A stronger notion of directional differentiability is obtained when $r(\boldsymbol{d})$ is supposed to be of order $o(||\boldsymbol{d}||)$, i.e. $||\boldsymbol{d}||^{-1}r(\boldsymbol{d}) \to \boldsymbol{0}$ as $\boldsymbol{d} \to \boldsymbol{0}$. We say that g is *Fréchet directionally differentiable* at $\boldsymbol{\mu}$ if (6.3.5) holds with $g'_{\boldsymbol{\mu}}(\cdot)$ being positively homogeneous and $r(\boldsymbol{d}) = o(||\boldsymbol{d}||)$. In order to distinguish between these two concepts of directional differentiability, we sometimes refer to (6.3.4) as directional differentiability in the *sense of Gâteaux*. Note that (Fréchet or Gâteaux) directional differentiability of g at $\boldsymbol{\mu}$ and *linearity* of $g'_{\boldsymbol{\mu}}(\cdot)$ imply (Fréchet or Gâteaux) differentiability of g at $\boldsymbol{\mu}$.

Similarly to (6.3.1)–(6.3.3) we have that if g is Fréchet directionally differentiable at $\boldsymbol{\mu}$ and the directional derivative $g'_{\boldsymbol{\mu}}(\boldsymbol{d})$ is continuous in $\boldsymbol{d}$, and if $\tau_N(\boldsymbol{X}_N - \boldsymbol{\mu})$ converges in distribution to $\boldsymbol{Y}$, then

$$\tau_N[g(\boldsymbol{X}_N) - g(\boldsymbol{\mu})] \overset{\mathcal{D}}{\to} \to g'_{\boldsymbol{\mu}}(\boldsymbol{Y}) \, . \tag{6.3.6}$$

The distribution of $g'_{\boldsymbol{\mu}}(\boldsymbol{Y})$ can be quite complicated even if $\boldsymbol{Y}$ is multivariable normal. In particular, asymptotic normality of $\boldsymbol{X}_N$ and Fréchet directional differentiability of g do not imply asymptotic normality of $g(\boldsymbol{X}_N)$ unless $g'_{\boldsymbol{\mu}}(\cdot)$ is linear .

Consider now the program $(\bar{\mathrm{P}}_{\mathrm{N}})$. We may look at the functions ψ_{jN} as *random elements* or *random functions* in an appropriate functional space given by a class of real-valued functions defined on the set V. Then $\bar{\varphi}_N$ and $\bar{\boldsymbol{v}}_N$ can be considered as functions of these random elements. This is somewhat similar to an idea employed in appendix A of chapter 2, where $\bar{\boldsymbol{v}}_N$ was considered as a function of the corresponding distribution (probability measure). Typically functional spaces where the random functions ψ_{jN} live, are infinite-dimensional normed spaces. In order to deal with such spaces we need an extension of the delta method. Such an extension is possible due to the theory of weak convergence in normed (metric) spaces. An exposition of the required theory can be found in Billingsley (1968), which we shall use as a reference book.

Let B_1 and B_2 be two Banach spaces, $g : B_1 \to B_2$ and $\boldsymbol{\mu} \in B_1$. The concepts of Gâteaux and Fréchet directional differentiability can be extended from (6.3.5) to the considered mapping word by word. That is, g is said to be Gâteaux directionally differentiable at $\boldsymbol{\mu}$ if (6.3.5) holds with $g'_{\boldsymbol{\mu}}(\cdot)$ being positively homogeneous and $t^{-1}r(t\boldsymbol{d}) \to \mathbf{0}$ as $t \to 0^+$ for every $\boldsymbol{d} \in B_1$. Fréchet directional differentiability follows if $||\boldsymbol{d}||^{-1}r(\boldsymbol{d}) \to \mathbf{0}$ as $\boldsymbol{d} \to \mathbf{0}$. We also will need an intermediate notion of Hadamard differentiability. It is said that g is *Hadamard directionally differentiable* at $\boldsymbol{\mu}$ if for any sequence $\{\boldsymbol{d}_n\}$ converging to a vector $\boldsymbol{d} \in B_1$ and every sequence $\{t_n\}$ of positive numbers converging to zero, the limit

$$g'_{\boldsymbol{\mu}}(\boldsymbol{d}) = \lim_{n\to\infty} \frac{g(\boldsymbol{\mu} + t_n\boldsymbol{d}_n) - g(\boldsymbol{\mu})}{t_n} \tag{6.3.7}$$

exists (in the norm topology of B_2).

The limit (6.3.7) can be written in the form

$$g'_{\boldsymbol{\mu}}(\boldsymbol{d}) = \lim_{\substack{t\to 0^+ \\ \boldsymbol{d}'\to\boldsymbol{d}}} \frac{g(\boldsymbol{\mu} + t\boldsymbol{d}') - g(\boldsymbol{\mu})}{t}$$

and therefore Hadamard directional differentiability implies Gâteaux directional differentiability. Some elementary properties of the above concepts of directional differentiability are given in exercise 6.8. Note that the concept of Hadamard directional differentiability is suitable for deriving the chain rule for composite mappings (see exercise 6.9).

Now let B_1 and B_2 be equipped with their Borel σ-algebras $\mathcal{B}_1$ and $\mathcal{B}_2$, respectively. (the Borel σ-algebra of a normed (metric) space is the σ-algebra generated by the family of open sets.) This makes $(B_1, \mathcal{B}_1)$ and $(B_2, \mathcal{B}_2)$ measurable spaces. A measurable mapping from a probability space into B_1 is called a *random element* of B_1 (Billingsley (1968, section 4)). Consider a sequence $\{\boldsymbol{Y}_N\}$ of random elements of B_1. It is said that $\{\boldsymbol{Y}_N\}$ converges in *distribution* (weakly) to a random element $\boldsymbol{Y}$ of B_1 if the expected values $\mathrm{E}\{f(\boldsymbol{Y}_N)\}$ converge to $\mathbb{E}\{f(\boldsymbol{Y})\}$, as $N \to \infty$, for every bounded and continuous function $f : B_1 \to \mathbb{R}$ (see Billingsley (1968, section 4), for details). We then write $\boldsymbol{Y}_N \xrightarrow{\mathcal{D}} \boldsymbol{Y}$.

Consider a sequence of random elements in B_1 of the form $\boldsymbol{Y}_N = \tau_N(\boldsymbol{X}_N - \boldsymbol{\mu})$, where $\{\tau_N\}$ is a sequence of positive numbers tending to infinity and $\boldsymbol{\mu} \in B_1$. Suppose that $\boldsymbol{Y}_N \xrightarrow{\mathcal{D}} \boldsymbol{Y}$ and that the mapping $g : B_1 \to B_2$ in some sense is directionally differentiable at $\boldsymbol{\mu}$. Then we would like to have the delta theorem in the form given in (6.3.6). Notice that in order for the formula (6.3.6) to make sense, the mapping g as well as the directional derivative $g'_{\boldsymbol{\mu}}$ should be measurable. These mapping are measurable, in particular, if they are continuous.

The notion of Gâteaux differentiability is too weak to guarantee formula (6.3.6). On the other hand, the stronger notion of Fréchet differentiability often is not applicable in many interesting situations. It was suggested by Reeds (1976) that the intermediate concept of Hadamard differentiability is most suitable for a formulation of the delta method. Reeds' approach was described and further developed in Fernholz (1983). We give now a first version of the delta theorem where we follow Gill (1989), Grübel (1988), and King (1989).

Theorem 6.3.1 Delta theorem *Let B_1 and B_2 be Banach spaces, $g : B_1 \to B_2$ and let $\boldsymbol{Y}_N = \tau_N(\boldsymbol{X}_N - \boldsymbol{\mu})$ be a sequence of random elements of B_1 with $\tau_N \to +\infty$. Suppose that the space B_1 is separable, the mapping g is measurable and Hadamard directionally differntiable at $\boldsymbol{\mu}$ and that $\{\boldsymbol{Y}_N\}$ converges in distribution to a random element $\boldsymbol{Y}$. Then*

$$\tau_N \left[g(\boldsymbol{X}_N) - g(\boldsymbol{\mu})\right] \xrightarrow{\mathcal{D}} g'_{\boldsymbol{\mu}}(\boldsymbol{Y}) \tag{6.3.8}$$

and

$$\tau_N \left[g(\boldsymbol{X}_N) - g(\boldsymbol{\mu})\right] = g'_{\boldsymbol{\mu}}(\boldsymbol{Y}_N) + o_p(1)\,. \tag{6.3.9}$$

Recall that a normed space is said to be separable if it has a countable, dense subset. Let us also note that Hadamard directional differentiability of g implies that the directional derivative $g'_{\boldsymbol{\mu}}(\cdot)$ is continuous (exercise 6.8(c)) and hence is measurable. The following proof of theorem 6.3.1 is taken from Gill (1989) and is based on the Skorohod-Dudley-Wichura almost sure representation theorem (e.g. Pollard (1984, p. 71)).

Representation theorem *Suppose that $\boldsymbol{Y}_N \xrightarrow{\mathcal{D}} \boldsymbol{Y}$ in a normed space B, where $\boldsymbol{Y}$ takes values in a separable subset of B. Then there exists a sequence $\boldsymbol{Y}'_N$, $\boldsymbol{Y}'$, defined on a single probability space such that $\boldsymbol{Y}'_N \overset{\mathcal{D}}{=} \boldsymbol{Y}_N$, for all N, $\boldsymbol{Y}' \overset{\mathcal{D}}{=} \boldsymbol{Y}$, and $\boldsymbol{Y}'_N \to \boldsymbol{Y}'$ w.p.1.*

Here $\boldsymbol{Y}' \stackrel{\mathcal{D}}{=} \boldsymbol{Y}$ means that the probability measures induced by $\boldsymbol{Y}'$ and $\boldsymbol{Y}$ coincide.

Proof of theorem 6.3.1 Consider the sequence $\boldsymbol{Y}_N = \tau_N(\boldsymbol{X}_N - \boldsymbol{\mu})$ of random elements of B_1. By the Representation theorem, there exist $\boldsymbol{Y}'_N$, $\boldsymbol{Y}'$, on a single probability space with $\boldsymbol{Y}'_N \stackrel{\mathcal{D}}{=} \boldsymbol{Y}_N$, $\boldsymbol{Y}' \stackrel{\mathcal{D}}{=} \boldsymbol{Y}$, and $\boldsymbol{Y}'_N \to \boldsymbol{Y}'$ w.p.1. Now for $\boldsymbol{X}'_N = \boldsymbol{\mu} + \tau_N^{-1}\boldsymbol{Y}'_N$ we have $\boldsymbol{X}'_N \stackrel{\mathcal{D}}{=} \boldsymbol{X}_N$ for all N. It follows from Hadamard directional differentiability of g that

$$\tau_N \left[g(\boldsymbol{X}'_N) - g(\boldsymbol{\mu})\right] \to g'_{\boldsymbol{\mu}}(\boldsymbol{Y}') \text{ w.p.1.} \tag{6.3.10}$$

Since convergence w.p.1 implies convergence in distribution and the terms in (6.3.10) have the same distributions as the corresponding terms in (6.3.8), the asymptotic result (6.3.8) follows.

Now since $g'_{\boldsymbol{\mu}}(\cdot)$ is continuous, we have that

$$g'_{\boldsymbol{\mu}}(\boldsymbol{Y}'_N) \to g'_{\boldsymbol{\mu}}(\boldsymbol{Y}) \quad \text{w.p.1.}$$

Together with (5.3.10) this implies that the difference between $g'_{\boldsymbol{\mu}}(\boldsymbol{Y}'_N)$ and the left-hand side of (6.3.10) tends with probability one and hence in probability, to zero. We obtain that

$$\tau_N \left[g(\boldsymbol{X}'_N) - g(\boldsymbol{\mu})\right] = g'_{\boldsymbol{\mu}} \left[\tau_N(\boldsymbol{X}'_N - \boldsymbol{\mu})\right] + o_p(1),$$

which implies (6.3.9). □

In some situations we shall need a notion of Hadamard directional differentiability tangentially to a set. Let K be a subset of the Banach space B_1. We say that the mapping $g : B_1 \to B_2$ is Hadamard directionally differentiable at $\boldsymbol{\mu}$ *tangentially* to K if the limit (6.3.7) exists for all sequences $\{\boldsymbol{d}_n\}$ converging to $\boldsymbol{d}$, of the form $\boldsymbol{d}_n = t_n^{-1}(\boldsymbol{x}_n - \boldsymbol{\mu})$, where $\boldsymbol{x}_n \in K$ and $t_n \to 0^+$. Such vectors $\boldsymbol{d}$ form a cone, called the *contingent* (Bouligand) cone to K at $\boldsymbol{\mu}$ and denoted $T_{\boldsymbol{\mu}}(K)$. The contingent cone is always closed and convex if the set K is convex.

We formulate now a second version of the delta theorem. Its proof is similar to the one of theorem 6.3.1 (see Shapiro (1991a)).

Theorem 6.3.2 *Let K be a subset of B_1 and suppose that :*

(i) *The mapping g is measurable and Hadamard directionally differentiable at $\boldsymbol{\mu}$ tangentially to K.*

(ii) *There is a sequence $\tau_N \to +\infty$ of positive numbers such that $\tau_N(\boldsymbol{X}_N - \boldsymbol{\mu})$ converges in distribution to a random element $\boldsymbol{Y}$, where the distribution of $\boldsymbol{Y}$ is concentrated on a separable subset of B_1.*

(iii) *For all N large enough,*

$$\mathrm{P}\{\boldsymbol{X}_N \in K\} = 1 .$$

Then

$$\tau_N\,[g(\boldsymbol{X}_N) - g(\boldsymbol{\mu})] \xrightarrow{\mathcal{D}} g'_{\boldsymbol{\mu}}(\boldsymbol{Y}) .$$

Note that it follows from (ii) and (iii) that the distribution of $\boldsymbol{Y}$ is concentrated on the contingent cone $T_{\boldsymbol{\mu}}(K)$. (This can be proved quite easily by employing again the Representation theorem.) Therefore the distribution of $g'_{\boldsymbol{\mu}}(\boldsymbol{Y})$ is well defined.

6.4 Asymptotic Analysis of the Optimal Value

In this section we investigate the aymptotic behavior of the optimal value $\bar{\varphi}_N$ of the program $(\bar{\mathrm{P}}_\mathrm{N})$. It appears that for this purpose the delta method of theorem 6.3.1 is especially convenient. We shall work here directly with functions of the programs (P_0) and $(\bar{\mathrm{P}}_\mathrm{N})$, which are viewed as elements of an appropriate functional space, rather than with the corresponding (empirical) measures and von Mises statistical functionals. In the multivariable cases this allows us to avoid considerable technical difficulties. The idea of the asymptotic analysis of stochastic programs via functional spaces is due to King (1986).

Let us start by considering the situation where all constraints of the program $(\bar{\mathrm{P}}_\mathrm{N})$ are deterministic. That is, the constraint functions of the program $(\bar{\mathrm{P}}_\mathrm{N})$ coincide identically with the corresponding constraint functions of the program (P_0). In this case the feasible sets Φ^* and Φ_N, defined in (6.2.1) and (6.2.2), respectively, are the same. We denote this feasible set by S, i.e. $S = \Phi^* = \Phi_N$. Then

$$\varphi^* = \inf\{\ell_0(\boldsymbol{v}) : \boldsymbol{v} \in S\} \tag{6.4.1}$$

and

$$\bar{\varphi}_N = \inf\{\psi_{0N}(\boldsymbol{v}) : \boldsymbol{v} \in S\} . \tag{6.4.2}$$

It will be assumed that the set S is *nonempty* and *compact* and that the functions $\ell_0(\boldsymbol{v})$ and $\psi_{0N}(\boldsymbol{v})$ are *continuous* on S. This means that ℓ_0 and ψ_{0N} belong to the linear (vector) space $C(S)$ of continuous functions $\xi : S \to \mathbb{R}$. The space $C(S)$ endowed with the sup-norm

$$||\xi|| = \sup\{|\xi(\boldsymbol{v})| : \boldsymbol{v} \in S\} ,$$

becomes a separable Banach space (see, for example, Dunford and Schwartz (1958)). Consider the optimal value function $\phi : C(S) \to \mathbb{R}$ defined as

$$\phi(\xi) = \inf \{\xi(\boldsymbol{v}) : \boldsymbol{v} \in S\}, \tag{6.4.3}$$

and denote by $S^*(\xi)$ the corresponding set of minimizers of $\xi(\boldsymbol{v})$ over S,

$$S^*(\xi) = \arg\min \{\xi(\boldsymbol{v}) : \boldsymbol{v} \in S\}. \tag{6.4.4}$$

Since S is compact, the set $S^*(\xi)$ is nonempty and compact for every $\xi \in C(S)$ and $\phi(\xi)$ is finite. We have, of course, that $\varphi^* = \phi(\ell_0)$ and $\bar{\varphi}_N = \phi(\psi_{0N})$.

Let us observe that the function ϕ is continuous and hence is measurable with respect to the Borel σ-algebra of $C(S)$. Indeed, for any $\xi,\ \zeta \in C(S)$,

$$|\phi(\xi) - \phi(\zeta)| \leq \sup_{\boldsymbol{v} \in S} |\xi(\boldsymbol{v}) - \zeta(\boldsymbol{v})| = ||\xi - \zeta||\,.$$

Therefore, in fact, ϕ is Lipschitz continuous on $C(S)$. It also follows from the definition that ϕ is positively homogeneous and that

$$\phi(\xi + \zeta) \geq \phi(\xi) + \phi(\zeta)$$

for all $\xi,\ \zeta \in C(S)$. This implies that the function ϕ is concave.

In order to proceed with the delta method we need to calculate directional derivatives of ϕ. The following result is related to work of Danskin (1967) and also can be derived by methods of convex analysis (e.g. Ioffe and Tihomirov (1979, section 4.5.2)).

Theorem 6.4.1 *Suppose that the set S is compact and that $\phi : C(S) \to \mathbb{R}$ is the optimal value function defined in (6.4.3). Then for any $\mu \in C(S)$, ϕ is Hadamard directionally differentiable at μ and*

$$\phi'_\mu(\delta) = \min_{\boldsymbol{v} \in S^*(\mu)} \delta(\boldsymbol{v})\,. \tag{6.4.5}$$

Proof Sinse ϕ is Lipschitz continuous, it will be sufficient to establish formula (6.4.5) with respect to directional differentiability in the sense of Gâteaux (see Exercise 6.8(f)). For every $t \geq 0$ consider the function $\mu + t\delta$ and its minimizer $\bar{\boldsymbol{v}}(t)$ over the set S, i.e. $\bar{\boldsymbol{v}}(t) \in S^*(\mu + t\delta)$. We have that for all $\boldsymbol{v} \in S^*(\mu)$,

$$\phi(\mu) = \mu(\boldsymbol{v})\,,$$

$$\phi(\mu + t\delta) \leq \mu(\boldsymbol{v}) + t\delta(\boldsymbol{v})\,,$$

and hence

$$\phi(\mu + t\delta) - \phi(\mu) \leq t\delta(\boldsymbol{v})\,.$$

Consequently,

$$\limsup_{t\to 0^+} \frac{\phi(\mu + t\delta) - \phi(\mu)}{t} \leq \min_{\boldsymbol{v}\in S^*(\mu)} \delta(\boldsymbol{v}) . \qquad (6.4.6)$$

In a similar way we have that

$$\phi(\mu + t\delta) - \phi(\mu) \geq t\delta(\bar{\boldsymbol{v}}(t)) . \qquad (6.4.7)$$

Note that every accumulation point of $\bar{\boldsymbol{v}}(t)$ as $t \to 0^+$ belongs to the set $S^*(\mu)$ (see theorem 6.2.1 and exercise 6.3). Therefore, by the arguments of compactness, it follows from (6.4.7) that there exists a point $\boldsymbol{v}^* \in S^*(\mu)$ such that

$$\liminf_{t\to 0^+} \frac{\phi(\mu + t\delta) - \phi(\mu)}{t} \geq \delta(\boldsymbol{v}^*) .$$

This last inequality together with (6.4.6) imply that

$$\lim_{t\to 0^+} \frac{\phi(\mu + t\delta) - \phi(\mu)}{t} = \min_{\boldsymbol{v}\in S^*(\mu)} \delta(\boldsymbol{v})$$

and hence the proof is complete. □

Formula (6.4.5) shows that the directional derivative $\phi'_\mu(\delta)$ is linear in δ if and only if the optimal set $S^*(\mu)$ is a *singleton*, i.e. contains only one point, $S^*(\mu) = \{\boldsymbol{v}^*\}$, say. Therefore the optimal value function ϕ is Hadamard differentiable at μ if and only if $S^*(\mu) = \{\boldsymbol{v}^*\}$ is a singleton. In the last case

$$\phi'_\mu(\delta) = \delta(\boldsymbol{v}^*) . \qquad (6.4.8)$$

It is interesting to note that the assumptions of theorem 6.4.1 do not guarantee Fréchet differentiability of ϕ at μ even if $S^*(\mu)$ is a singleton.

In order to derive the asymptotic distribution of the estimator $\bar{\varphi}_N$ we can apply the delta method (theorem 6.3.1) to the optimal value function ϕ. The following result is then a consequence of formula (6.4.5) of theorem 6.4.1.

Theorem 6.4.2 *Let $\{\psi_{0N}\}$ be a sequence of random elements in $C(S)$, $\ell_0 \in C(S)$, and let φ^* and $\bar{\varphi}_N$ be defined as in (6.4.1) and (6.4.2), respectively. Suppose that $N^{1/2}(\psi_{0N} - \ell_0)$ converges in distribution to a random element Y of $C(S)$. Then*

$$N^{1/2}(\bar{\varphi}_N - \varphi^*) \xrightarrow{\mathcal{D}} \min_{\boldsymbol{v}\in S^*(\ell_0)} Y(\boldsymbol{v}) . \qquad (6.4.9)$$

In particular, if $\ell_0(\boldsymbol{v})$ attains its minimum over S at a unique point $\boldsymbol{v}^*$, then $N^{1/2}(\bar{\varphi}_N - \varphi^*)$ converges in distribution to $Y(\boldsymbol{v}^*)$.

Notice that the random element Y takes values in $C(S)$ and can be viewed as a random function from S to $\mathbb{R}$. For any finite subset $\{\boldsymbol{v}_1, \ldots, \boldsymbol{v}_m\}$ of S, with Y is associated the finite-dimensional random vector $(Y(\boldsymbol{v}_1), \ldots, Y(\boldsymbol{v}_m))$. The distribution of this random vector is called the *finite dimensional distribution* of Y. Formula (6.4.9) shows that the asymptotic distribution of $\bar{\varphi}_N$ has the form of the distribution of the minimum of, generally interdependent, variables. It may be impossible to calculate such a distribution in a closed form even if the finite-dimensional distributions of Y are known, say normal. The situation is simplified considerably if the true program (P_0) has only one optimal solution $\boldsymbol{v}^*$. In this case the asymptotic distribution of $N^{1/2}(\bar{\varphi}_N - \varphi^*)$ is given by the distribution of the (real-valued) random variable $Y(\boldsymbol{v}^*)$, which in most applications will be normal.

Theorem 6.4.2 involves some "stochastic" assumptions on which we elaborate now. It was assumed that the functions ψ_{0N} can be considered as random elements of the Banach space $C(S)$. This means that ψ_{0N} are given by measurable mappings from a probability space $(\Omega, \mathcal{F}, P)$ into the space $C(S)$. That is, the mapping $\omega \to \psi_{0N}(\omega, \cdot)$, from Ω to $C(S)$, is measurable with respect to the σ-algebra $\mathcal{F}$ and the Borel σ-algebra of $C(S)$. This mapping is measurable, in particular, if $\psi_{0N}(\omega, \boldsymbol{v})$ is a Carathéodory function. That is, for every $\boldsymbol{v} \in S$ the function $\psi_{0N}(\cdot, \boldsymbol{v}) : \Omega \to \mathbb{R}$ is measurable and for all $\omega \in \Omega$ the function $\psi_{0N}(\omega, \cdot)$ is continuous on S.

It was also assumed in theorem 6.4.2 that the sequence $\{N^{1/2}(\psi_{0N} - \ell_0)\}$ converges in distribution in $C(S)$. Let us consider now the sample mean construction with $\ell_0(\boldsymbol{v})$ given as the expected value of $h_0(Z, \boldsymbol{v})$ and $\psi_{0N}(\boldsymbol{v}) = \bar{\ell}_{0N}(\boldsymbol{v})$.

Assumption A1

(i) For every $\boldsymbol{v} \in S$, the function $h_0(\cdot, \boldsymbol{v})$ is measurable.

(ii) For some point $\boldsymbol{v}_0 \in S$, the expectation $\mathbb{E}\{h_0(Z, \boldsymbol{v}_0)^2\}$ is finite.

(iii) There exists a function $b : \Omega \to \mathbb{R}$ such that $\mathbb{E}\{b(Z)^2\}$ is finite and that

$$|h_0(\zeta, \boldsymbol{u}) - h_0(\zeta, \boldsymbol{v})| \leq b(\zeta)\|\boldsymbol{u} - \boldsymbol{v}\|$$

for all $\boldsymbol{u}, \boldsymbol{v} \in S$ and $\zeta \in \Omega$.

It follows from condition (iii) of the above assumption that for almost every ζ, the function $h_0(\zeta, \cdot)$ is Lipschitz continuous on S. Moreover, since S is compact and hence is bounded, it also follows from (iii) that for every $\boldsymbol{v} \in S$ the function $|h_0(\cdot, \boldsymbol{v})|$ is dominated by the function $h_0(\cdot, \boldsymbol{v}_0) + cb(\cdot)$, where c is a sufficiently large positive number. Because of conditions (ii) and (iii), this last function is integrable and hence, by the Lebesgue dominated convergence theorem, the expected value function $\ell_0(\boldsymbol{v})$ is continuous on S. That is $\ell_0 \in C(S)$. We also have that, because of (i) and (iii), the function $h_0(\zeta, \boldsymbol{v})$ and hence the corresponding sample mean functions, are Carathéodory functions. Consequently we can consider $\psi_{0N} = \bar{\ell}_{0N}$ as random elements of $C(S)$.

The conditions of assumption A1 are sufficient for the corresponding Central Limit theorem in $C(S)$ to hold. That is, the sequence $Y_N = N^{1/2}(\bar{\ell}_{0N} - \ell_0)$ of random elements of $C(S)$ converges in distribution to a random element Y. (See Araujo and Giné (1980) for details. This result is also discussed in King (1986, Proposition A5).) Note that the finite-dimensional distributions of the random element (random function, random process) Y are multivariable normal. In particular, for any $\boldsymbol{v}^* \in S$ the distribution of $Y(\boldsymbol{v}^*)$ is normal with zero mean and variance

$$\sigma^2 = \mathbb{E}\{h_0(Z, \boldsymbol{v}^*)^2\} - \ell_0(\boldsymbol{v}^*)^2 \,. \tag{6.4.10}$$

We sumarize our discussion in the following theorem.

Theorem 6.4.3 *Consider the sample mean construction. Suppose that assumption A1 holds and that $\ell_0(\boldsymbol{v})$ attains its minimum over S at a unique point $\boldsymbol{v}^*$. Then $N^{1/2}(\bar{\varphi}_N - \varphi^*)$ converges in distribution to a normal with zero mean and variance σ^2 given in (6.4.10).*

If $Z_1, \ldots, Z_N$ is a random sample from $(\Omega, \mathcal{F}, P)$, then the asymptotic variance σ^2, given in (6.4.10), can be consistently estimated by

$$\hat{\sigma}^2 = N^{-1} \sum_{i=1}^{N} h_0(Z_i, \bar{\boldsymbol{v}}_N)^2 - \bar{\ell}_{0N}(\bar{\boldsymbol{v}}_N)^2 \,. \tag{6.4.11}$$

In the case of DEDS the objective functions of the programs (P_0) and $(\bar{\mathrm{P}}_\mathrm{N})$ are ratios of the expected value and the sample mean functions, respectively. We leave this as an exercise to derive the corresponding asymptotic variance and its consistent estimator (see exercise 6.11).

In situations where constraints of the program $(\bar{\mathrm{P}}_\mathrm{N})$ can be nondeterministic, the asymptotic analysis of $\bar{\varphi}_N$ is more delicate. We investigate first the convex case. That is, suppose that the programs (P_0) and $(\bar{\mathrm{P}}_\mathrm{N})$ involve the inequality constraints only, the set V is convex and compact and all functions $\ell_0(\boldsymbol{v}), \ldots, \ell_k(\boldsymbol{v})$ and $\psi_{0N}(\boldsymbol{v}), \ldots, \psi_{kN}(\boldsymbol{v})$ are convex. It follows then that the associated sets of feasible solutions,

$$\Phi^* = \{\boldsymbol{v} \in V : \ell_j(\boldsymbol{v}) \le 0, \quad j = 1, \ldots, k\}$$

and

$$\Phi_N = \{\boldsymbol{v} \in V : \psi_{jN}(\boldsymbol{v}) \le 0, \quad j = 1, \ldots, k\},$$

are also convex.

Consider the space $B = C(V)\times\ldots\times C(V)$ which is given by the Cartesian product of $k+1$ replications of the space $C(V)$. The space B can be equipped, for example, with the max-norm

$$||\boldsymbol{\xi}|| = \max\{||\xi_i|| : \ 0 \le i \le k\} ,$$

where $\boldsymbol{\xi} = (\xi_0, \ldots, \xi_k) \in B$ and $||\xi_i||$, $i = 0, \ldots, k$, are the sup-norms of the functions $\xi_i \in C(V)$. For a given $\boldsymbol{\xi} \in B$, denote by $S^*(\boldsymbol{\xi})$ the set of minimizers of the function $\xi_0(v)$ over V subject to the constraints $\xi_j(v) \le 0$, $j = 1, \ldots, k$, and by $\phi(\boldsymbol{\xi})$ the corresponding optimal value

$$\phi(\boldsymbol{\xi}) = \inf\{\xi_0(\boldsymbol{v}) : \boldsymbol{v} \in V, \ \xi_j(\boldsymbol{v}) \le 0, \ j = 1, \ldots, k\} . \tag{6.4.12}$$

In particular, for $\boldsymbol{\mu} = (\ell_0, \ldots, \ell_k)$ and $\boldsymbol{X}_N = (\psi_{0N}, \ldots, \psi_{kN})$ we have that $S^*(\boldsymbol{\mu})$ and $S^*(\boldsymbol{X}_N)$ are the sets of the optimal solutions of the programs $(\mathrm{P_0})$ and $(\bar{\mathrm{P}}_\mathrm{N})$, respectively, and $\varphi^* = \phi(\boldsymbol{\mu})$ and $\bar{\varphi}_N = \phi(\boldsymbol{X}_N)$.

With every $\boldsymbol{\xi} \in B$ and $\boldsymbol{\lambda} = (\lambda_1, \ldots, \lambda_k)$ we associate the following *Lagrangian function*:

$$\mathcal{L}(\boldsymbol{v}, \boldsymbol{\lambda}, \boldsymbol{\xi}) = \xi_0(\boldsymbol{v}) + \sum_{j=1}^{k} \lambda_j \xi_j(\boldsymbol{v}) . \tag{6.4.13}$$

Suppose that the program $(\mathrm{P_0})$ is convex and that the *Slater condition* for the program $(\mathrm{P_0})$ holds, i.e. there exists a point $\boldsymbol{v} \in V$ such that $\ell_j(\boldsymbol{v}) < 0$ for all $j = 1, \ldots, k$. Then to every optimal point $\bar{\boldsymbol{v}} \in S^*(\boldsymbol{\mu})$ corresponds a vector $\bar{\boldsymbol{\lambda}} = (\bar{\lambda}_1, \ldots, \bar{\lambda}_k)$ of Lagrange multipliers such that

$$\mathcal{L}(\bar{\boldsymbol{v}}, \bar{\boldsymbol{\lambda}}, \boldsymbol{\mu}) = \min_{\boldsymbol{v} \in V} \mathcal{L}(\boldsymbol{v}, \bar{\boldsymbol{\lambda}}, \boldsymbol{\mu}) , \tag{6.4.14}$$

$$\bar{\lambda}_j \ge 0 \text{ and } \bar{\lambda}_j \ell_j(\bar{\boldsymbol{v}}) = 0, \ \ j = 1, \ldots, k, \tag{6.4.15}$$

(see, for example, Ioffe and Tihomirov (1979, p. 68)). Condition (6.4.14) means that the Lagrangian function $\mathcal{L}(\cdot, \bar{\boldsymbol{\lambda}}, \boldsymbol{\mu})$ attains its minimum over V at the point $\bar{\boldsymbol{v}}$. The above optimality conditions can also be formulated in subdifferential form. In particular, if $\mathcal{L}(\cdot, \bar{\boldsymbol{\lambda}}, \boldsymbol{\mu})$ is differentiable at $\bar{\boldsymbol{v}}$ and $\bar{\boldsymbol{v}}$ is an interior point of V, then (6.4.14) is equivalent to a more familiar condition that the gradient $\nabla \mathcal{L}(\bar{\boldsymbol{v}}, \bar{\boldsymbol{\lambda}}, \boldsymbol{\mu})$ is zero. (As before all gradients here are taken with respect to $\boldsymbol{v}$.)

Denote by $\Lambda(\boldsymbol{\mu})$ the set of vectors $\bar{\boldsymbol{\lambda}}$ of Lagrange multipliers satisfying the optimality conditions (6.4.14) and (6.4.15). It follows from the duality theory of convex programming that under the Slater condition, the set $\Lambda(\boldsymbol{\mu})$ is *nonempty* and *bounded* and is the same for all points $\bar{\boldsymbol{v}} \in S(\boldsymbol{\mu})$. For a discussion of the duality theory in convex programming we refer to Rockafellar (1970) and Golshtein (1972).

Let us consider a subset K of the space B formed by vectors $\boldsymbol{\xi}$ such that each component ξ_i, $i = 0, \ldots, k$, of $\boldsymbol{\xi}$ is a convex function on an open neighborhood of V. Convexity of ($\mathrm{P_0}$) and ($\bar{\mathrm{P}}_\mathrm{N}$) implies, of course, that $\boldsymbol{\mu} = (\ell_0, \ldots, \ell_k)$ and $\boldsymbol{X}_N = (\psi_{0N}, \ldots, \psi_{kN})$ belong to the set K. The following theorem is an extension of a result due to Golshtein (1972).

Theorem 6.4.4 *Suppose that the program ($\mathrm{P_0}$) is convex and that the Slater condition for the program ($\mathrm{P_0}$) holds. Then the optimal value function $\phi(\boldsymbol{\xi})$ is Hadamard directionally differentiable at $\boldsymbol{\mu} = (\ell_0, \ldots, \ell_k)$ tangentially to the set K and for $\boldsymbol{\delta} \in T_{\boldsymbol{\mu}}(K)$,*

$$\phi'_{\boldsymbol{\mu}}(\boldsymbol{\delta}) = \min_{\boldsymbol{v} \in S^*(\boldsymbol{\mu})} \max_{\boldsymbol{\lambda} \in \Lambda(\boldsymbol{\mu})} \mathcal{L}(\boldsymbol{v}, \boldsymbol{\lambda}, \boldsymbol{\delta}) . \tag{6.4.16}$$

Proof Let $\boldsymbol{u}$ be a point in V such that $\ell_j(\boldsymbol{u}) < -\epsilon$, $j = 1, \ldots, k$, for some $\epsilon > 0$. The existence of such a point is guaranteed by the Slater condition. It also follows from the Slater condition and the compactness of V that the sets $S^*(\boldsymbol{\mu})$ and $\Lambda(\boldsymbol{\mu})$ are nonempty and compact. Consider a sequence $\boldsymbol{\xi}_n = (\xi_{on}, \ldots, \xi_{kn})$ in B, converging to $\boldsymbol{\mu}$ and consider a point $\boldsymbol{v}^* \in S^*(\boldsymbol{\mu})$. Since V is convex, the segment joining $\boldsymbol{v}^*$ and $\boldsymbol{u}$ is contained in V. It also follows from convexity of the functions $\ell_j(\boldsymbol{v})$ that

$$\ell_j[\boldsymbol{v}^* + \alpha(\boldsymbol{u} - \boldsymbol{v}^*)] \le (1 - \alpha)\ell_j(\boldsymbol{v}^*) + \alpha \ell_j(\boldsymbol{u}) \le -\alpha\epsilon$$

for all $\alpha \in [0, 1]$. Since the functions $\xi_{jn}(\boldsymbol{v})$ converge to $\ell_j(\boldsymbol{v})$ uniformly on V, we obtain that there exists a sequence $\alpha_n \to 0^+$ such that for all n large enough, $\xi_{jn}(\boldsymbol{u}_n) \le 0$, $j = 1, \ldots, k$, where $\boldsymbol{u}_n = \boldsymbol{v}^* + \alpha_n(\boldsymbol{u} - \boldsymbol{v}^*)$. It follows that for sufficiently large n the optimal set $S^*(\boldsymbol{\xi}_n)$ is nonempty and if $\boldsymbol{v}_n \in S^*(\boldsymbol{\xi}_n)$, then the distance from $\boldsymbol{v}_n$ to $S^*(\boldsymbol{\mu})$ tends to zero (see theorem 6.2.1 and exercise 6.3).

Now consider a sequence $\{\boldsymbol{\xi}_n\} \subset K$ of the form $\boldsymbol{\xi}_n = \boldsymbol{\mu} + t_n \boldsymbol{\delta}_n$ such that $t_n \to 0^+$ and $\{\boldsymbol{\delta}_n\}$ converges to a vector $\boldsymbol{\delta} \in B$. Let $\bar{\boldsymbol{\lambda}} \in \Lambda(\boldsymbol{\mu})$ and $\boldsymbol{v}_n \in S^*(\boldsymbol{\xi}_n)$. Then

$$\phi(\boldsymbol{\xi}_n) \ge \mathcal{L}(\boldsymbol{v}_n, \bar{\boldsymbol{\lambda}}, \boldsymbol{\xi}_n)$$

and for any $\bar{\boldsymbol{v}} \in S^*(\boldsymbol{\mu})$,

$$\phi(\boldsymbol{\mu}) = \mathcal{L}(\bar{\boldsymbol{v}}, \bar{\boldsymbol{\lambda}}, \boldsymbol{\mu}) .$$

It follows from the optimality condition (6.4.14) that

$$\mathcal{L}(\bar{\boldsymbol{v}}, \bar{\boldsymbol{\lambda}}, \boldsymbol{\mu}) \le \mathcal{L}(\boldsymbol{v}_n, \bar{\boldsymbol{\lambda}}, \boldsymbol{\mu}) .$$

Consequently,

$$\phi(\boldsymbol{\xi}_n) - \phi(\boldsymbol{\mu}) \ge \mathcal{L}(\boldsymbol{v}_n, \bar{\boldsymbol{\lambda}}, \boldsymbol{\xi}_n) - \mathcal{L}(\boldsymbol{v}_n, \bar{\boldsymbol{\lambda}}, \boldsymbol{\mu}) = t_n \mathcal{L}(\boldsymbol{v}_n, \bar{\boldsymbol{\lambda}}, \boldsymbol{\delta}_n) . \tag{6.4.17}$$

Suppose now that $\boldsymbol{v}_n$ tends to a point $\boldsymbol{v}^*$. Then $\boldsymbol{v}^* \in S^*(\boldsymbol{\mu})$ and $\mathcal{L}(\boldsymbol{v}_n, \bar{\boldsymbol{\lambda}}, \boldsymbol{\delta}_n)$ tends to $\mathcal{L}(\boldsymbol{v}^*, \bar{\boldsymbol{\lambda}}, \boldsymbol{\delta})$. Since $\bar{\boldsymbol{\lambda}}$ was an arbitrary element of $\Lambda(\boldsymbol{\mu})$, it follows then from (6.4.17) that

$$\liminf_{n\to\infty} \frac{\phi(\boldsymbol{\xi}_n) - \phi(\boldsymbol{\mu})}{t_n} \geq \min_{\boldsymbol{v}\in S^*(\boldsymbol{\mu})} \max_{\boldsymbol{\lambda}\in\Lambda(\boldsymbol{\mu})} \mathcal{L}(\boldsymbol{v}, \boldsymbol{\lambda}, \boldsymbol{\delta}) .$$

On the other hand consider a point $\bar{\boldsymbol{v}} \in S^*(\boldsymbol{\mu})$ and let $\boldsymbol{\lambda}_n \in \Lambda(\boldsymbol{\xi}_n)$ and $\boldsymbol{v}_n \in S^*(\boldsymbol{\xi}_n)$. Notice that by the arguments of duality, the distance from $\boldsymbol{\lambda}_n$ to $\Lambda(\boldsymbol{\mu})$ tends to zero as well as the distance from $\boldsymbol{v}_n$ to $S^*(\boldsymbol{\mu})$. We have that

$$\phi(\boldsymbol{\xi}_n) = \mathcal{L}(\boldsymbol{v}_n, \boldsymbol{\lambda}_n, \boldsymbol{\xi}_n)$$

and by the assumption of convexity and optimality of $\boldsymbol{v}_n$,

$$\mathcal{L}(\boldsymbol{v}_n, \boldsymbol{\lambda}_n, \boldsymbol{\xi}_n) \leq \mathcal{L}(\bar{\boldsymbol{v}}, \boldsymbol{\lambda}_n, \boldsymbol{\xi}_n) .$$

Also

$$\phi(\boldsymbol{\mu}) \geq \mathcal{L}(\bar{\boldsymbol{v}}, \boldsymbol{\lambda}_n, \boldsymbol{\mu})$$

and hence

$$\phi(\boldsymbol{\xi}_n) - \phi(\boldsymbol{\mu}) \leq \mathcal{L}(\bar{\boldsymbol{v}}, \boldsymbol{\lambda}_n, \boldsymbol{\xi}_n) - \mathcal{L}(\bar{\boldsymbol{v}}, \boldsymbol{\lambda}_n, \boldsymbol{\mu}) = t_n \mathcal{L}(\bar{\boldsymbol{v}}, \boldsymbol{\lambda}_n, \boldsymbol{\delta}_n) .$$

Since the point $\bar{\boldsymbol{v}}$ was an arbitrary point of $S^*(\boldsymbol{\mu})$ and by continuity of the Lagrangian function, it follows that

$$\limsup_{n\to\infty} \frac{\phi(\boldsymbol{\xi}_n) - \phi(\boldsymbol{\mu})}{t_n} \leq \min_{\boldsymbol{v}\in S^*(\boldsymbol{\mu})} \max_{\boldsymbol{\lambda}\in\Lambda(\boldsymbol{\mu})} \mathcal{L}(\boldsymbol{v}, \boldsymbol{\lambda}, \boldsymbol{\delta})$$

and hence the proof is complete. □

Formula (6.4.16) shows that, in the convex case, the directional derivative $\phi'_{\boldsymbol{\mu}}(\boldsymbol{\delta})$ is linear in $\boldsymbol{\delta}$ if *both* sets $S^*(\boldsymbol{\mu})$ and $\Lambda(\boldsymbol{\mu})$ are singletons, say $S^*(\boldsymbol{\mu}) = \{\boldsymbol{v}^*\}$ and $\Lambda(\boldsymbol{\mu}) = \{\boldsymbol{\lambda}^*\}$. Then

$$\phi'_{\boldsymbol{\mu}}(\boldsymbol{\delta}) = \mathcal{L}(\boldsymbol{v}^*, \boldsymbol{\lambda}^*, \boldsymbol{\delta}) = \delta_0(\boldsymbol{v}^*) + \sum_{j=1}^{k} \lambda_j^* \delta_j(\boldsymbol{v}^*) . \tag{6.4.18}$$

We discuss later conditions which guarantee the uniqueness of the Lagrange multipliers.

By definition we take $\phi(\boldsymbol{\xi}) = +\infty$ if the set of points $\boldsymbol{v} \in V$ satisfying the constraints $\boldsymbol{\xi}_j(\boldsymbol{v}) \leq 0$, $j = 1, \ldots, k$, is empty. Note that $\phi(\boldsymbol{\xi})$ is always greater than $-\infty$, since it was assumed that V is compact. It follows from the result of theorem 6.1.4 that $\phi(\boldsymbol{\xi})$ is measurable. (It is also not difficult to show that $\phi(\boldsymbol{\xi})$ is lower semi-continuous; see exercise 6.12.)

Now theorems 6.3.2 and 6.4.4 imply the following result.

Theorem 6.4.5 *Let $\boldsymbol{X}_N = (\psi_{0N}, \ldots, \psi_{kN})$ be a sequence of random elements in the space $B = C(V) \times \ldots \times C(V)$. Suppose that :*

(i) *The program (P_0) is convex.*

(ii) *For all N large enough, the program $(\bar{\mathrm{P}}_\mathrm{N})$ is convex w.p.1.*

(iii) *The Slater condition for the program (P_0) holds.*

(iv) *The random elements $N^{1/2}(\boldsymbol{X}_N - \boldsymbol{\mu})$ converge in distribution to a random element $\boldsymbol{Y}$.*

Then

$$N^{1/2}(\bar{\varphi}_N - \varphi^*) \xrightarrow{\mathcal{D}} \min_{\boldsymbol{v} \in S^*(\boldsymbol{\mu})} \max_{\boldsymbol{\lambda} \in \Lambda(\boldsymbol{\mu})} \mathcal{L}(\boldsymbol{v}, \boldsymbol{\lambda}, \boldsymbol{Y}) . \tag{6.4.19}$$

In particular, if $S^*(\boldsymbol{\mu}) = \{\boldsymbol{v}^*\}$ and $\Lambda(\boldsymbol{\mu}) = \{\boldsymbol{\lambda}^*\}$ are singletons, then

$$N^{1/2}(\bar{\varphi}_N - \varphi^*) \xrightarrow{\mathcal{D}} Y_0(\boldsymbol{v}^*) + \sum_{j=1}^{k} \lambda_j^* Y_j(\boldsymbol{v}^*) . \tag{6.4.20}$$

It may be noted that if all constraints of the program $(\bar{P}_N)$ are deterministic, then the random elements Y_j, $j = 1, \ldots, k$, are identically zero. In this case formula (6.4.19) coincides with the corresponding formula (6.4.9) of theorem 6.4.2.

Now let us consider the sample mean construction, with $\ell_j(\boldsymbol{v})$ given by the expected values of $h_j(Z, \boldsymbol{v})$ and $\psi_{jN} = \bar{\ell}_{jN}$, $j = 0, \ldots, k$, and make the following assumptions.

Assumtion A2 The set V is convex and compact and there exists an open convex neighborhood U of V such that for every $\zeta \in \Omega$, the functions $h_j(\zeta, \cdot)$, $j = 0, \ldots, k$, are convex on U.

Assumption A3

(i) For every $\boldsymbol{v} \in V$, the functions $h_j(\cdot, \boldsymbol{v})$, $j = 0, \ldots, k$, are measurable.

(ii) For some point $\boldsymbol{v}_0 \in V$, the expectations $\mathbb{E}\{h_j(Z, \boldsymbol{v}_0)^2\}$, $j = 0, \ldots, k$, are finite.

(iii) There exist functions $b_j : \Omega \to \mathbb{R}$, $j = 0, \ldots, k$, such that $\mathbb{E}\{b_j(Z)^2\}$ are finite and

$$|h_j(\zeta, \boldsymbol{u}) - h_j(\zeta, \boldsymbol{v})|| \leq b_j(\zeta)||\boldsymbol{u} - \boldsymbol{v}||$$

for all $\boldsymbol{u}, \boldsymbol{v} \in V$ and $\zeta \in \Omega$.

It follows from assumption A2 that the sample mean functions $\bar{\ell}_{jN}$ and the expected value functions ℓ_j, $j = 0, \dots, k$, are convex. Consequently, ($\mathrm{P_0}$) and ($\mathrm{\bar{P}_N}$) are convex programs. Assumption A3 involves the same conditions as assumption A1 applied to every function $h_j(\zeta, \boldsymbol{v})$. It follows that $N^{1/2}(\boldsymbol{X}_N - \boldsymbol{\mu})$ converges in distribution to a random element $\boldsymbol{Y}$ of B. For any given $\boldsymbol{v} \in V$ the corresponding random vector $(Y_0(\boldsymbol{v}), \dots, Y_k(\boldsymbol{v}))$ is distributed multivariable normal with zero mean vector and covariances

$$\mathbb{E}\{Y_i(\boldsymbol{v})Y_j(\boldsymbol{v})\} = \mathbb{E}\{h_i(Z, \boldsymbol{v})h_j(Z, \boldsymbol{v})\} - \ell_i(\boldsymbol{v})\ell_j(\boldsymbol{v}),\ i, j = 0, \dots, k\,.$$

The following result is now a consequence of theorem 6.4.5.

Theorem 6.4.6 *Suppose that assumptions A2 and A3 are satisfied, that the Slater condition for the program* ($\mathrm{P_0}$) *holds and that the program* ($\mathrm{P_0}$) *possesses a unique optimal solution* $\boldsymbol{v}^*$ *and a unique vector* $\boldsymbol{\lambda}^*$ *of Lagrange multipliers. Then* $N^{1/2}(\bar{\varphi}_N - \varphi^*)$ *converges in distribution to a normal with zero mean and variance*

$$\sigma^2 = \mathbb{E}\left\{\left[h_0(Z, \boldsymbol{v}^*) + \sum_{j=1}^{k} \lambda_j^* h_j(Z, \boldsymbol{v}^*)\right]^2\right\} - \ell_0(\boldsymbol{v}^*)^2\,. \tag{6.4.21}$$

Notice that by the optimality conditions (6.4.15) we have that $\lambda_j^* \ell_j(\boldsymbol{v}^*) = 0$, $j = 1, \dots, k$. Consequently the expected value of $\lambda_j^* h_j(Z, \boldsymbol{v}^*)$ is zero for $j = 1, \dots, k$, and $\ell_0(\boldsymbol{v}^*)$ is equal to $\mathcal{L}(\boldsymbol{v}^*, \boldsymbol{\lambda}^*, \boldsymbol{\mu})$.

It is remarkable that in the two cases considered cases involving deterministic constraints and convex programs, the asymptotic analysis of the optimal value does not require differentiability of the functions $\ell_j(\boldsymbol{v})$ or $\psi_{jN}(\boldsymbol{v})$. In situations where convexity is not presented it should be replaced by differentiability assumptions. Consider the programs ($\mathrm{P_0}$) and ($\mathrm{\bar{P}_N}$), with equality and inequality constraints. Suppose that the set V is compact and that all functions $\ell_j(\boldsymbol{v})$ and $\psi_{jN}(\boldsymbol{v})$, $j = 0, \dots, M$, belong to the space $C^1(V)$ of real-valued, continuously differentiable on an open neighborhood of V functions. The space $C^1(V)$, equipped with the norm

$$||\xi|| = \sup_{\boldsymbol{v} \in V} |\xi(\boldsymbol{v})| + \sup_{\boldsymbol{v} \in V} ||\nabla \xi(\boldsymbol{v})||,$$

becomes a separable Banach space.

We work now in the Banach space $B = C^1(V) \times \dots \times C^1(V)$, which is given by the Cartesian product of $M + 1$ replications of the space $C^1(V)$. With every $\boldsymbol{\xi} \in B$ and $\boldsymbol{\lambda} = (\lambda_1, \dots, \lambda_M)$ we associate the following Lagrangian function

$$\mathcal{L}(\boldsymbol{v}, \boldsymbol{\lambda}, \boldsymbol{\xi}) = \xi_0(\boldsymbol{v}) + \sum_{j=1}^{M} \lambda_j \xi_j(\boldsymbol{v})\,. \tag{6.4.22}$$

In particular, for $\boldsymbol{\mu} = (\ell_0, \dots, \ell_M)$ we have

$$\mathcal{L}(\boldsymbol{v}, \boldsymbol{\lambda}, \boldsymbol{\mu}) = \ell_0(\boldsymbol{v}) + \sum_{j=1}^{M} \lambda_j \ell_j(\boldsymbol{v}) .$$

Let $\boldsymbol{v}^*$ be the optimal solution of the program (P_0). Suppose that $\boldsymbol{v}^*$ is an interior point of V and that a constraint qualification, for example the Mangasarian-Fromovitz (MF) condition, holds at $\boldsymbol{v}^*$. Then the standard first-order (Kuhn-Tucker) necessary optimality conditions follow. That is, there exists a vector $\boldsymbol{\lambda}^* = (\lambda_1^*, \dots, \lambda_M^*)$ of Lagrange multipliers such that

$$\nabla \mathcal{L}(\boldsymbol{v}^*, \boldsymbol{\lambda}^*, \boldsymbol{\mu}) = \mathbf{0} , \tag{6.4.23}$$

$$\lambda_j^* \geq 0 \quad \text{and} \quad \lambda_j^* \ell_j(\boldsymbol{v}^*) = 0, \;\; j = 1, \dots, k . \tag{6.4.24}$$

It may happen that more than one vector $\boldsymbol{\lambda}^*$ satisfies the necessary conditions (6.4.23) and (6.4.24). Uniqueness of $\boldsymbol{\lambda}^*$ is guaranteed by the linear independence condition which requires for the gradient vectors $\nabla \ell_j(\boldsymbol{v}^*)$, $j \in J(\boldsymbol{v}^*) \cup \{k+1, \dots, M\}$, to be linearly independent. Here, as before, $J(\boldsymbol{v}^*)$ denotes the index set of inequality constraints active at $\boldsymbol{v}^*$. Although the linear independence condition is sufficient for uniqueness of the Lagrange multipliers, it is not necessary in general. We give now necessary and sufficient conditions for the uniqueness of $\boldsymbol{\lambda}^*$. For a point $\boldsymbol{v}^*$ and a corresponding vector $\boldsymbol{\lambda}$ of Lagrange multipliers denote

$$J_+(\boldsymbol{v}^*, \boldsymbol{\lambda}) = \{j \in J(\boldsymbol{v}^*) : \lambda_j > 0\} \tag{6.4.25}$$

and

$$J_0(\boldsymbol{v}^*, \boldsymbol{\lambda}) = \{j \in J(\boldsymbol{v}^*) : \lambda_j = 0\} . \tag{6.4.26}$$

Lagrange multipliers uniqueness (LMU) condition. The gradients $\nabla \ell_j(\boldsymbol{v}^*)$, $j \in J_+(\boldsymbol{v}^*, \boldsymbol{\lambda}^*) \cup \{k+1, \dots, M\}$, are linearly independent and there exists a vector $\boldsymbol{b}$ such that

$$\boldsymbol{b}' \nabla \ell_j(\boldsymbol{v}^*) = 0, \quad j \in J_+(\boldsymbol{v}^*, \boldsymbol{\lambda}^*) \cup \{k+1, \dots, M\} ,$$

$$\boldsymbol{b}' \nabla \ell_j(\boldsymbol{v}^*) < 0, \quad j \in J_0(\boldsymbol{v}^*, \boldsymbol{\lambda}^*) .$$

It is not difficult to show that the following two statements hold (see, for example, Kyparisis (1985)).

(i) If there exists a unique vector $\boldsymbol{\lambda}^*$ of Lagrange multipliers satisfying necessary conditions (6.4.23) and (6.4.24), then the associated LMU-condition holds.

(ii) Conversely, if the LMU-condition holds for some vector $\boldsymbol{\lambda}^*$ satisfying (6.4.23) and (6.4.24), then the Lagrange multipliers vector $\boldsymbol{\lambda}^*$ is unique.

The LMU-condition is somewhere between the linear independence and the MF-conditions. It is weaker than the linear independence condition, but is stronger than the MF-condition. In particular, if $J_+(\boldsymbol{v}^*, \boldsymbol{\lambda}^*) = J(\boldsymbol{v}^*)$, i.e. all Lagrange multipliers corresponding to the active inequality constraints are strictly positive, then the LMU-condition coincides with the linear independence condition. If on the other hand $J_+(\boldsymbol{v}^*, \boldsymbol{\lambda}^*)$ is empty and hence $J_0(\boldsymbol{v}^*, \boldsymbol{\lambda}^*) = J(\boldsymbol{v}^*)$, then the LMU-condition becomes the same as the MF-condition.

We describe now directional derivatives of the optimal value function

$$\phi(\boldsymbol{\xi}) = \inf \{\xi_0(\boldsymbol{v}) : \boldsymbol{v} \in \Phi(\boldsymbol{\xi})\} , \tag{6.4.27}$$

where $\xi = (\xi_0, \ldots, \xi_M) \in B$ and

$$\Phi(\boldsymbol{\xi}) = \{\boldsymbol{v} \in V : \xi_j(\boldsymbol{v}) \leq 0,\ j = 1, \ldots, k;\ \xi_j(\boldsymbol{v}) = 0,\ j = k+1, \ldots, M\}$$

is the corresponding set of feasible solutions. By $S^*(\boldsymbol{\xi})$ we denote the associated set of the optimal solutions

$$S^*(\boldsymbol{\xi}) = \arg\ \min\{\boldsymbol{\xi}_0(\boldsymbol{v}) : \boldsymbol{v} \in \Phi(\boldsymbol{\xi})\} . \tag{6.4.28}$$

The following theorem is closely related to the results of Gauvin and Dubeau (1982), Lempio and Maurer (1980) and Levitin (1974).

Theorem 6.4.7 *Suppose that the set V is compact and $\Phi^* = \Phi(\boldsymbol{\mu})$ is nonempty, that all points of the optimal set $S^*(\boldsymbol{\mu})$ of the program* (P_0) *are interior points of V and that to every $\boldsymbol{v} \in S^*(\boldsymbol{\mu})$ corresponds a unique vector $\bar{\boldsymbol{\lambda}}(\boldsymbol{v})$ of Lagrange multipliers satisfying the first-order (Kuhn-Tucker) necessary conditions. Then the optimal value function $\phi(\boldsymbol{\xi})$ is Hadamard directionally differentiable at $\boldsymbol{\mu}$ and*

$$\phi'_{\boldsymbol{\mu}}(\boldsymbol{\delta}) = \min_{\boldsymbol{v} \in S^*(\boldsymbol{\mu})} \mathcal{L}(\boldsymbol{v}, \bar{\boldsymbol{\lambda}}(\boldsymbol{v}), \boldsymbol{\delta}) . \tag{6.4.29}$$

Proof First we observe that since V is compact and the feasible set Φ^* of the program (P_0) is nonempty, the optimal set $S^*(\boldsymbol{\mu})$ is nonempty and compact.

Consider a point $\boldsymbol{v}^* \in S(\boldsymbol{\mu})$ and the corresponding vector $\boldsymbol{\lambda}^* = \bar{\boldsymbol{\lambda}}(\boldsymbol{v}^*)$ of Lagrange multipliers. Since it is supposed that $\boldsymbol{\lambda}^*$ is unique, the LMU-condition follows. Note that the LMU-condition becomes the MF-condition if the inequality constraints corresponding to the positive Lagrange multipliers are treated as equality constraints. Therefore we can apply the stability result of theorem 6.2.2. It follows from (6.2.6) that there exists a positive constant c such that for every $\boldsymbol{\xi} = (\xi_0, \ldots, \xi_M)$ in a neighborhood of $\boldsymbol{\mu}$ there is a point $\bar{\boldsymbol{u}} = \bar{\boldsymbol{u}}(\boldsymbol{\xi})$ satisfying the constraints

$$\begin{aligned} \xi_j(\bar{\boldsymbol{u}}) &= 0, \qquad j \in J_+(\boldsymbol{v}^*, \boldsymbol{\lambda}^*) \cup \{k+1, \ldots, M\}, \\ \xi_j(\bar{\boldsymbol{u}}) &\leq 0, \qquad j \in J_0(\boldsymbol{v}^*, \boldsymbol{\lambda}^*), \end{aligned} \tag{6.4.30}$$

and such that

$$\|\boldsymbol{v}^* - \bar{\boldsymbol{u}}(\boldsymbol{\xi})\| \leq c\|\boldsymbol{\xi} - \boldsymbol{\mu}\|. \tag{6.4.31}$$

Notice that the stability result of theorem 6.2.2 is applicable here since convergence of a sequence $\{\boldsymbol{\xi}_n\} \subset B$ to $\boldsymbol{\mu}$ in the considered norm means that $\{\xi_{jn}(\boldsymbol{v})\}$ and $\{\nabla \xi_{jn}(\boldsymbol{v})\}$ converge to $\boldsymbol{\ell}(\boldsymbol{v})$ and $\nabla \ell_j(\boldsymbol{v})$, respectively, uniformly on V.

Consider a sequence $\boldsymbol{\xi}_n = \boldsymbol{\mu} + t_n \boldsymbol{\delta}_n$, with $\boldsymbol{\delta}_n \to \boldsymbol{\delta}$ and $t_n \to 0^+$, and let $\boldsymbol{u}_n = \bar{\boldsymbol{u}}(\boldsymbol{\xi}_n)$. Clearly, $\boldsymbol{u}_n \in \Phi(\boldsymbol{\xi}_n)$ and it follows from (6.4.31) that $\boldsymbol{u}_n \to \boldsymbol{v}^*$. Therefore, for n large enough, the optimal set $S^*(\boldsymbol{\xi}_n)$ is nonempty and if $\boldsymbol{v}_n \in S^*(\boldsymbol{\xi}_n)$, then the distance from $\boldsymbol{v}_n$ to $S^*(\boldsymbol{\mu})$ tends to zero (see theorem 6.2.1 and exercise 6.3). Now since $\boldsymbol{u}_n \in \Phi(\boldsymbol{\xi}_n)$ we have that $\phi(\boldsymbol{\xi}_n)$ is less than or equal to $\xi_{0n}(\boldsymbol{u}_n)$. Moreover, because of the constraints (6.4.30), $\lambda_j^* \xi_{jn}(\boldsymbol{u}_n)$ is zero for all $j = 1, \ldots, M$, and hence $\xi_{0n}(\boldsymbol{u}_n)$ is equal to $\mathcal{L}(\boldsymbol{u}_n, \boldsymbol{\lambda}^*, \xi_n)$. Consequently

$$\phi(\boldsymbol{\xi}_n) \leq \mathcal{L}(\boldsymbol{u}_n, \boldsymbol{\lambda}^*, \xi_n) = \mathcal{L}(\boldsymbol{u}_n, \boldsymbol{\lambda}^*, \boldsymbol{\mu}) + t_n \mathcal{L}(\boldsymbol{u}_n, \boldsymbol{\lambda}^*, \boldsymbol{\delta}_n).$$

Since by the first-order necessary conditions the gradient $\nabla \mathcal{L}(\boldsymbol{v}^*, \boldsymbol{\lambda}^*, \boldsymbol{\mu})$ is zero and because of (6.4.31), we have that

$$\mathcal{L}(\boldsymbol{u}_n, \boldsymbol{\lambda}^*, \boldsymbol{\mu}) = \mathcal{L}(\boldsymbol{v}^*, \boldsymbol{\lambda}^*, \boldsymbol{\mu}) + o(t_n).$$

Furthermore, $\phi(\boldsymbol{\mu})$ is equal to $\mathcal{L}(\boldsymbol{v}^*, \boldsymbol{\lambda}^*, \boldsymbol{\mu})$ and hence

$$\phi(\boldsymbol{\xi}_n) - \phi(\boldsymbol{\mu}) \leq t_n \mathcal{L}(\boldsymbol{u}_n, \boldsymbol{\lambda}^*, \boldsymbol{\delta}_n) + o(t_n).$$

Since $\boldsymbol{v}^*$ was an arbitrary point of $S^*(\boldsymbol{\mu})$ and because of the continuity of the Lagrangian we obtain then that

$$\limsup_{n \to \infty} \frac{\phi(\boldsymbol{\xi}_n) - \phi(\boldsymbol{\mu})}{t_n} \leq \min_{\boldsymbol{v} \in S^*(\boldsymbol{\mu})} \mathcal{L}(\boldsymbol{v}, \bar{\boldsymbol{\lambda}}(\boldsymbol{v}), \boldsymbol{\delta}).$$

On the other hand consider a sequence $\boldsymbol{v}_n \in S^*(\boldsymbol{\xi}_n)$ converging to a point $\boldsymbol{v}^*$. Then $\boldsymbol{v}^* \in S^*(\boldsymbol{\mu})$ and we can again apply the stability result of theorem 6.2.2 (at the point $\boldsymbol{v}^*$). It follows that for n large enough, there exists a point $\boldsymbol{w}_n$ such that $\|\boldsymbol{v}_n - \boldsymbol{w}_n\|$ is of order $O(\|\boldsymbol{\xi}_n - \boldsymbol{\mu}\|) = O(t_n)$ and

$$\begin{aligned} \ell_j(\boldsymbol{w}_n) &= 0, \qquad j \in J_+(\boldsymbol{v}^*, \boldsymbol{\lambda}^*) \cup \{k+1, \ldots, M\}, \\ \ell_j(\boldsymbol{w}_n) &\leq 0, \qquad j \in J_0(\boldsymbol{v}^*, \boldsymbol{\lambda}^*), \end{aligned}$$

where $\boldsymbol{\lambda}^* = \bar{\boldsymbol{\lambda}}(\boldsymbol{v}^*)$. Since $\boldsymbol{w}_n \in \Phi(\boldsymbol{\mu})$, we have then that $\phi(\boldsymbol{\mu})$ is less than or equal to $\ell_0(\boldsymbol{w}_n)$. It also follows that $\ell_0(\boldsymbol{w}_n)$ is equal to $\mathcal{L}(\boldsymbol{w}_n, \boldsymbol{\lambda}^*, \boldsymbol{\mu})$ and hence

$$\phi(\boldsymbol{\mu}) \leq \mathcal{L}(\boldsymbol{w}_n, \boldsymbol{\lambda}^*, \boldsymbol{\mu}) .$$

Moreover,

$$\phi(\boldsymbol{\xi}_n) = \xi_{0n}(\boldsymbol{v}_n) \geq \mathcal{L}(\boldsymbol{v}_n, \boldsymbol{\lambda}^*, \boldsymbol{\xi}_n) = \mathcal{L}(\boldsymbol{v}_n, \boldsymbol{\lambda}^*, \boldsymbol{\mu}) + t_n \mathcal{L}(\boldsymbol{v}_n, \boldsymbol{\lambda}^*, \boldsymbol{\delta}_n) .$$

Now by the Mean Value Theorem

$$|\mathcal{L}(\boldsymbol{v}_n, \boldsymbol{\lambda}^*, \boldsymbol{\mu}) - \mathcal{L}(\boldsymbol{w}_n, \boldsymbol{\lambda}^*, \boldsymbol{\mu})| \leq ||\boldsymbol{v}_n - \boldsymbol{w}_n|| \; ||\nabla \mathcal{L}(\boldsymbol{x}_n, \boldsymbol{\lambda}^*, \boldsymbol{\mu})|| ,$$

where $\boldsymbol{x}_n$ is a point on the segment joining $\boldsymbol{v}_n$ and $\boldsymbol{w}_n$. Since $\boldsymbol{x}_n \to \boldsymbol{v}^*$ and because $\nabla \mathcal{L}(\boldsymbol{v}^*, \boldsymbol{\lambda}^*, \boldsymbol{\mu})$ is zero, we have that $\nabla \mathcal{L}(\boldsymbol{x}_n, \boldsymbol{\lambda}^*, \boldsymbol{\mu}) \to \mathbf{0}$. Also $||\boldsymbol{v}_n - \boldsymbol{w}_n||$ is of order $O(t_n)$ and hence

$$\mathcal{L}(\boldsymbol{v}_n, \boldsymbol{\lambda}^*, \boldsymbol{\mu}) - \mathcal{L}(\boldsymbol{w}_n, \boldsymbol{\lambda}^*, \boldsymbol{\mu}) = o(t_n) .$$

It follows that

$$\phi(\boldsymbol{\xi}_n) - \phi(\boldsymbol{\mu}) \geq t_n \mathcal{L}(\boldsymbol{v}_n, \boldsymbol{\lambda}^*, \boldsymbol{\delta}_n) + o(t_n) .$$

We obtain that

$$\liminf_{n \to \infty} \frac{\phi(\boldsymbol{\xi}_n) - \phi(\boldsymbol{\mu})}{t_n} \geq \min_{\boldsymbol{v} \in S^*(\boldsymbol{\mu})} \mathcal{L}(\boldsymbol{v}, \bar{\boldsymbol{\lambda}}(\boldsymbol{v}), \boldsymbol{\delta})$$

and the proof is complete. □

It was assumed in theorem 6.4.7 that to every point $\boldsymbol{v} \in S^*(\boldsymbol{\mu})$ corresponds a *unique* vector of Lagrange multipliers. If this uniqueness condition does not hold, then one may think of an analogue of the min-max formula (6.4.16). However, in general (nonconvex) situations such a formula may be incorrect (see Gauvin and Dubeau (1982) for a counterexample). Then, surprisingly, investigation of directional differentiability of the optimal value function involves second-order derivatives of the considered functions (Gauvin and Janin (1988) and Bonnans, Joffe and Shapiro (1992)).

It follows from (6.4.29) that under the assumptions of theorem 6.4.7, the directional derivative $\phi'_{\boldsymbol{\mu}}(\boldsymbol{\delta})$ is linear in $\boldsymbol{\delta}$ if $S^*(\boldsymbol{\mu}) = \{\boldsymbol{v}^*\}$ is a singleton. In the last case

$$\phi'_{\boldsymbol{\mu}}(\boldsymbol{\delta}) = \mathcal{L}(\boldsymbol{v}^*, \boldsymbol{\lambda}^*, \boldsymbol{\delta}) = \delta_0(\boldsymbol{v}^*) + \sum_{j=1}^{M} \lambda_j^* \delta_j(\boldsymbol{v}^*) , \tag{6.4.32}$$

where $\boldsymbol{\lambda}^* = \bar{\boldsymbol{\lambda}}(\boldsymbol{v}^*)$.

Theorem 6.4.7 and the delta method imply the following result.

Theorem 6.4.8 *Let $\boldsymbol{x}_N = (\psi_{0N}, \ldots, \psi_{MN})$ be a sequence of random elements of the space $B = C^1(V) \times \ldots \times C^1(V)$ such that $N^{1/2}(\boldsymbol{X}_N - \boldsymbol{\mu})$ converges in distribution to a random element $\boldsymbol{Y}$. Suppose that the assumptions of theorem 6.4.7, with respect to the program* (P_0), *hold. Then*

$$N^{1/2}(\bar{\varphi}_N - \varphi^*) \xrightarrow{\mathcal{D}} \min_{\boldsymbol{v} \in S^*(\boldsymbol{\mu})} \mathcal{L}(\boldsymbol{v}, \bar{\boldsymbol{\lambda}}(\boldsymbol{v}), \boldsymbol{Y})\,. \tag{6.4.33}$$

In particular, if $S^*(\boldsymbol{\mu}) = \{\boldsymbol{v}^*\}$ is a singleton, then

$$N^{1/2}(\bar{\varphi}_N - \varphi^*) \xrightarrow{\mathcal{D}} Y_0(\boldsymbol{v}^*) + \sum_{j=1}^{M} \lambda_j^* Y_j(\boldsymbol{v}^*)\,, \tag{6.4.34}$$

where $\boldsymbol{\lambda}^* = \bar{\boldsymbol{\lambda}}(\boldsymbol{v}^*)$ is the corresponding vector of Lagrange multipliers.

Formulas (6.4.33) and (6.4.34) are quite similar to those of theorem 6.4.5. However, it should be understood that here we work in the spaces of continuously differentiable functions. Therefore regularity conditions, which are required to ensure the "stochastic" assumptions of theorem 6.4.8, will be more restrictive. Let us consider now the sample mean construction.

Assumption A4

(i) For every $\boldsymbol{v} \in V$ the functions $h_j(\cdot, \boldsymbol{v})$, $j = 0, \ldots, M$, are measurable and for almost every ζ the functions $h_j(\zeta, \cdot)$, $j = 0, \ldots, M$, are continuously differentiable on an open neighborhood of V.

(ii) For some point $\boldsymbol{v}_0 \in V$, the expectations $\mathbb{E}\{h_j(Z, \boldsymbol{v}_0)^2\}$ and $\mathbb{E}\{||\nabla h_j(Z, \boldsymbol{v}_0)||^2\}$, $j = 0, \ldots, M$, are finite.

(iii) There exist functions $b_j : \Omega \to \mathbb{R}$, $j = 0, \ldots, M$, such that $\mathbb{E}\{b_j(Z)^2\}$ are finite and

$$||\nabla h_j(\zeta, \boldsymbol{u}) - \nabla h_j(\zeta, \boldsymbol{v})|| \leq b_j(\zeta)||\boldsymbol{u} - \boldsymbol{v}||$$

for all $\boldsymbol{u}, \boldsymbol{v} \in V$ and $\zeta \in \Omega$.

Since a pointwise limit of a measurable function is measurable, it follows from the measurability of $h_j(\cdot, \boldsymbol{v})$ that $\nabla h_j(\cdot, \boldsymbol{v})$ are also measurable. Consequently $\boldsymbol{X}_N = (\bar{\ell}_{0N}, \ldots, \bar{\ell}_{MN})$ can be considered as random elements of the space B. Conditions (ii) and (iii) of assumption A4 imply that the families $\{||\nabla h_j(\zeta, \boldsymbol{v})||,\ \boldsymbol{v} \in V\}$, $j = 0, \ldots, M$, are dominated by integrable functions. Therefore we have here that the expected value functions $\ell_j(\boldsymbol{v})$, $j = 0, \ldots, M$, are continuously differentiable and the first-order partial derivatives of $\ell_j(\boldsymbol{v})$ can be taken inside the expected value. Consequently $\boldsymbol{\mu} = (\ell_0, \ldots, \ell_M) \in B$.

Assumption A4 implies that the sequence $\{N^{1/2}(\boldsymbol{X}_N - \boldsymbol{\mu})\}$ converges in distribution to a random element $\boldsymbol{Y}$ of the space B. We obtain then from theorems 6.4.7 and 6.4.8 the following result.

Theorem 6.4.9 *Suppose that the set V is compact, that assumption A4 holds and that the program (P_0) possesses a unique optimal solution $\boldsymbol{v}^*$ and a unique vector $\boldsymbol{\lambda}^*$ of Lagrange multipliers with $\boldsymbol{v}^*$ being an interior point of V. Then $N^{1/2}(\bar{\varphi}_N - \varphi^*)$ converges in distribution to a normal with zero mean and variance*

$$\sigma^2 = \mathbb{E}\left\{\left[h_0(Z, \boldsymbol{v}^*) + \sum_{j=1}^{M} \lambda_j^* h_j(Z, \boldsymbol{v}^*)\right]^2\right\} - \ell_0(\boldsymbol{v}^*)^2 . \tag{6.4.35}$$

We leave as an exercise to extend the result of theorem 6.4.9 to situations where the programs (P_0) and $(\bar{\mathrm{P}}_\mathrm{N})$ involve functions that are representable as compositions of the expected value and the sample mean functions, respectively (compare with exercise 6.11).

Notice that if for some j, $1 \le j \le M$, the function $h_j(\zeta, \boldsymbol{v})$ is independent of ζ, i.e. $h_j(\zeta, \boldsymbol{v}) = h_j(\boldsymbol{v})$, then $\lambda_j^* h_j(\boldsymbol{v}^*)$ is zero by the first-order necessary conditions. Consequently this term can be omitted from the expression for the asymptotic variance σ^2 given in (6.4.35). In particular, if all constraints are deterministic, then formula (6.4.35) coincides with (6.4.10).

It was assumed in theorems 6.4.8 and 6.4.9 that the functions ψ_{jN} of the program $(\bar{\mathrm{P}}_\mathrm{N})$ are continuously differentiable although the corresponding derivatives did not appear in any way in the derived asymptotics of the optimal value $\bar{\varphi}_N$. In section 6.6 we present a somewhat different approach which allows us, to a certain extent, to relax the differentiability assumptions about the functions ψ_{jN}.

6.5 Sensitivity and Robustness Analysis of the Optimal Solutions

In this section we consider the following parametric mathematical programming problem

$$(\mathrm{P}_{\boldsymbol{\xi}}) \qquad \begin{array}{lll} \text{minimize} & g_0(\boldsymbol{v}, \boldsymbol{\xi}), & \boldsymbol{v} \in V, \\ \text{subject to} & g_j(\boldsymbol{v}, \boldsymbol{\xi}) \le 0\ , & j = 1, \ldots, k, \\ & g_j(\boldsymbol{v}, \boldsymbol{\xi}) = 0\ , & j = k+1, \ldots, M. \end{array}$$

Here $\boldsymbol{\xi}$ is viewed as a vector of parameters giving perturbations of the program $(\mathrm{P}_{\boldsymbol{\xi}})$. It will be assumed that $\boldsymbol{\xi} \in \mathcal{Z}$, where $\mathcal{Z}$ is a linear normed space.

We study the differentiability properties of the optimal solution $\bar{\boldsymbol{v}}(\boldsymbol{\xi})$ of $(\mathrm{P}_{\boldsymbol{\xi}})$, considered as a function of the parameter vector $\boldsymbol{\xi}$, in a vicinity of a given point $\boldsymbol{\xi}_0$. It will be assumed that $g_j(\cdot, \boldsymbol{\xi}_0) = \ell_j(\cdot)$, $j = 0, \ldots, M$, and hence the "unperturbed" program $(\mathrm{P}_{\boldsymbol{\xi}_0})$ coincides with the "true" program (P_0) introduced in section 6.1. The parameter space $\mathcal{Z}$ can be finite-dimensional or it can be one of the functional spaces considered in section 6.4.

It will be useful in several respects to have a general result describing the differential behavior of $\bar{\boldsymbol{v}}(\boldsymbol{\xi})$. In chapter 2 we already used such results to investigate the asymptotic properties of the optimal solutions in the unconstrained case. In this section we discuss the general case of the parametric program $(\mathrm{P}_{\boldsymbol{\xi}})$ involving equality and inequality constraints. For the unperturbed (or true) program (P_0) we use the same notation as in the previous sections. In particular, the index sets $J(\boldsymbol{v})$, $J_+(\boldsymbol{v}^*, \boldsymbol{\lambda})$ and $J_0(\boldsymbol{v}^*, \boldsymbol{\lambda})$ are defined in (6.2.4), (6.4.25) and (6.4.26), respectively.

In order to get an idea about the differentiability properties of $\bar{\boldsymbol{v}}(\boldsymbol{\xi})$ let us consider the following, relatively simple, situation. Suppose that the space $\mathcal{Z}$ is finite-dimensional, $\mathcal{Z} = \mathbb{R}^m$ say, that the functions $g_j : \mathbb{R}^n \times \mathbb{R}^m \to \mathbb{R}$, $j = 0, \ldots, M$, are twice continuously differentiable and that $\bar{\boldsymbol{v}}(\boldsymbol{\xi})$ is an interior point of V. Then, under a constraint qualification, $\bar{\boldsymbol{v}}(\boldsymbol{\xi})$ must satisfy the first-order (Kuhn-Tucker) necessary conditions. Suppose further that the unperturbed program (P_0) has a unique optimal solution $\boldsymbol{v}^*$, that $\bar{\boldsymbol{v}}(\boldsymbol{\xi})$ tends to $\boldsymbol{v}^*$ as $\boldsymbol{\xi} \to \boldsymbol{\xi}_0$ and that the linear independence condition holds at the point $\boldsymbol{v}^*$. This last condition implies that there exists a unique vector $\boldsymbol{\lambda}^*$ of Lagrange multipliers corresponding to the optimal solution $\boldsymbol{v}^*$. Suppose also that the following condition holds.

Strict complementarity condition All Lagrange multipliers λ_j^*, $j \in J(\boldsymbol{v}^*)$ of the program (P_0) corresponding to the active at $\boldsymbol{v}^*$ inequality constraints, are strictly positive.

Then, by the arguments of continuity, the linear independence condition holds and the corresponding Lagrange multipliers $\bar{\lambda}_j(\boldsymbol{\xi})$, $j \in J(\boldsymbol{v}^*)$, remain strictly positive for all $\boldsymbol{\xi}$ sufficiently close to $\boldsymbol{\xi}_0$. This implies that the index set of inequality constraints that are active at $\bar{\boldsymbol{v}}(\boldsymbol{\xi})$ is the same for all $\boldsymbol{\xi}$ in a neighborhood of $\boldsymbol{\xi}_0$. Consequently we can write the first-order necessary conditions in the form of equations. For the sake of notational convenience let us consider only the inequality constraints that are *active* at $\boldsymbol{v}^*$, i.e. suppose that $J(\boldsymbol{v}^*) = \{1, \ldots, k\}$.

We have that for all $\boldsymbol{\xi}$ sufficiently close to $\boldsymbol{\xi}_0$, the optimal solution $\bar{\boldsymbol{v}}(\boldsymbol{\xi})$ and the corresponding vector $\bar{\boldsymbol{\lambda}}(\boldsymbol{\xi})$ of Lagrange multipliers are solutions of the equations

$$\nabla_{\boldsymbol{v}} L(\boldsymbol{v}, \boldsymbol{\lambda}, \boldsymbol{\xi}) = \mathbf{0} , \tag{6.5.1}$$

$$G(\boldsymbol{v}, \boldsymbol{\xi}) = \mathbf{0}\ , \tag{6.5.2}$$

where

$$L(\boldsymbol{v}, \boldsymbol{\lambda}, \boldsymbol{\xi}) = g_0(\boldsymbol{v}, \boldsymbol{\xi}) + \sum_{j=1}^{M} \lambda_j g_j(\boldsymbol{v}, \boldsymbol{\xi}) \tag{6.5.3}$$

is the Lagrangian function of the program $(\mathrm{P}_{\boldsymbol{\xi}})$ and

$$G(\boldsymbol{v}, \boldsymbol{\xi}) = [g_1(\boldsymbol{v}, \boldsymbol{\xi}), \ldots, g_M(\boldsymbol{v}, \boldsymbol{\xi})]\ . \tag{6.5.4}$$

Note that without the strict complementarity condition some of the inequality constraints can become inactive for other values of $\boldsymbol{\xi}$ than $\boldsymbol{\xi}_0$. In this case we cannot write the corresponding optimality conditions in the form of parametric equations.

The necessary conditions (6.5.1) and (6.5.2) represent $n + M$ equations in $n + M$ unknowns given by the components of the vector $\boldsymbol{z} = (\boldsymbol{v}, \boldsymbol{\lambda})$. We can write them in compact form as follows:

$$\nabla_{\boldsymbol{z}} L(\boldsymbol{z}, \boldsymbol{\xi}) = \mathbf{0}\ . \tag{6.5.5}$$

The solution $\bar{\boldsymbol{z}}(\boldsymbol{\xi}) = (\bar{\boldsymbol{v}}(\boldsymbol{\xi})\ ,\ \bar{\boldsymbol{\lambda}}(\boldsymbol{\xi}))$ of (6.5.5) depends on the parameter vector $\boldsymbol{\xi}$. In order to calculate derivatives of $\bar{\boldsymbol{z}}(\boldsymbol{\xi})$ we can apply the Implicit Function theorem to equations (6.5.5) at the point $(\boldsymbol{z}^*, \boldsymbol{\xi}_0) = (\boldsymbol{v}^*, \boldsymbol{\lambda}^*, \boldsymbol{\xi}_0)$. Suppose that the $(n + M) \times (n + M)$ Hessian matrix

$$\boldsymbol{H} = \nabla^2_{\boldsymbol{z}\boldsymbol{z}} L(\boldsymbol{z}^*, \boldsymbol{\xi}_0) = \begin{bmatrix} \nabla^2_{\boldsymbol{v}\boldsymbol{v}} L(\boldsymbol{v}^*, \boldsymbol{\lambda}^*, \boldsymbol{\xi}_0) & \vdots & \nabla_{\boldsymbol{v}} G(\boldsymbol{v}^*, \boldsymbol{\xi}_0) \\ \cdots\cdots\cdots\cdots & \vdots & \cdots\cdots\cdots \\ \nabla_{\boldsymbol{v}} G(\boldsymbol{v}^*, \boldsymbol{\xi}_0)' & \vdots & \mathbf{0} \end{bmatrix} \tag{6.5.6}$$

is nonsingular. (This is equivalent to a second-order sufficient condition for the program (P_0); see exercise 6.13.) Then the mapping $\bar{\boldsymbol{z}}(\boldsymbol{\xi})$ is differentiable at $\boldsymbol{\xi} = \boldsymbol{\xi}_0$ and the corresponding $(n + M) \times m$ Jacobian matrix $\nabla \bar{\boldsymbol{z}}(\boldsymbol{\xi}_0)$ can be calculated according to the formula

$$\nabla \bar{\boldsymbol{z}}(\boldsymbol{\xi}_0) = -\boldsymbol{H}^{-1} \nabla^2_{\boldsymbol{z}\boldsymbol{\xi}} L(\boldsymbol{z}^*, \boldsymbol{\xi}_0)\ , \tag{6.5.7}$$

where

$$\nabla^2_{\boldsymbol{z}\boldsymbol{\xi}} L(\boldsymbol{z}^*, \boldsymbol{\xi}_0) = \begin{bmatrix} \nabla^2_{\boldsymbol{v}\boldsymbol{\xi}} L(\boldsymbol{v}^*, \boldsymbol{\lambda}^*, \boldsymbol{\xi}_0) \\ \nabla_{\boldsymbol{\xi}} G(\boldsymbol{v}^*, \boldsymbol{\xi}_0)' \end{bmatrix}.$$

For a detailed discussion of this approach, based on the classical Implicit Function theorem, see Fiacco (1983).

In the following analysis we remove some of the above regularity assumptions, in particular the strict complementarity condition. We assume subsequently that the unperturbed program (P_0) has a *unique* optimal solution $\boldsymbol{v}^*$ and a *unique* vector $\boldsymbol{\lambda}^*$ of the corresponding Lagrange multipliers. Unless stated otherwise all considered derivatives (gradients, Hessians) will be taken with respect to the components of the vector $\boldsymbol{v}$.

Let us consider the following quadratic programming problem depending on the parameter vector $\boldsymbol{\xi}$:

$$(Q_{\boldsymbol{\xi}}) \quad \begin{array}{ll} \underset{\boldsymbol{u} \in \mathbb{R}^n}{\text{minimize}} & \boldsymbol{u}'\nabla L(\boldsymbol{v}^*, \boldsymbol{\lambda}^*, \boldsymbol{\xi}) + \frac{1}{2}\boldsymbol{u}'\nabla^2 L(\boldsymbol{v}^*, \boldsymbol{\lambda}^*, \boldsymbol{\xi}_0)\boldsymbol{u} \\ \text{subject to} & \boldsymbol{u}'\nabla g_j(\boldsymbol{v}^*, \boldsymbol{\xi}_0) + g_j(\boldsymbol{v}^*, \boldsymbol{\xi}) \leq 0, \qquad j \in J_0(\boldsymbol{v}^*, \boldsymbol{\lambda}^*), \\ & \boldsymbol{u}'\nabla g_j(\boldsymbol{v}^*, \boldsymbol{\xi}_0) + g_j(\boldsymbol{v}^*, \boldsymbol{\xi}) = 0, \\ & j \in J_+(\boldsymbol{v}^*, \boldsymbol{\lambda}^*) \cup \{k+1, \ldots, M\}. \end{array}$$

We show that under certain regularity conditions, the optimal solution $\bar{\boldsymbol{u}}(\boldsymbol{\xi})$ of the program ($Q_{\boldsymbol{\xi}}$) gives a first-order approximation of $\bar{\boldsymbol{v}}(\boldsymbol{\xi})$ at $\boldsymbol{\xi}_0$

Before proceeding further we need to discuss the second-order optimality conditions for the program (P_0) . Let us consider the so-called *critical cone* $C(\boldsymbol{v}^*)$ associated with the point $\boldsymbol{v}^*$,

$$\begin{aligned} C(\boldsymbol{v}^*) = \{: \boldsymbol{u}'\nabla \ell_j(\boldsymbol{v}^*) &\leq 0, \; j \in J(\boldsymbol{v}^*) \cup \{0\}, \\ \boldsymbol{u}'\nabla \ell_j(\boldsymbol{v}^*) &= 0, j = k+1, \ldots, M\}\,. \end{aligned} \tag{6.5.8}$$

Under a constraint qualification (e.g. the MF-condition), the vectors $\boldsymbol{u}$ satisfying the linear constraints $\boldsymbol{u}'\nabla \ell_j(\boldsymbol{v}^*) \leq 0$, $j \in J(\boldsymbol{v}^*)$, $\boldsymbol{u}'\nabla \ell_j(\boldsymbol{v}^*) = 0$, $j = k+1, \ldots, M$, form a cone of feasible directions at the point $\boldsymbol{v}^*$. The first-order necessary conditions ensure that the linearization $\boldsymbol{u}'\nabla \ell_0(\boldsymbol{v}^*)$ of the objective function is non-negative for every feasible direction $\boldsymbol{u}$. Therefore we actually can replace the inequality $\boldsymbol{u}'\nabla \ell_0(\boldsymbol{v}^*) \leq 0$ in the definition of $C(\boldsymbol{v}^*)$ by the corresponding equality $\boldsymbol{u}'\nabla \ell_0(\boldsymbol{v}^*) = 0$. Intuitively the critical cone $C(\boldsymbol{v}^*)$ represents those feasible directions where the first-order Taylor expansion does not give any information about the local behavior of $\ell_0(\boldsymbol{v})$. This is compensated by the following second-order conditions.

Second-order sufficient conditions. There exists a vector $\boldsymbol{\lambda}^*$ of Lagrange multipliers, satisfying the first-order necessary conditions, such that for every nonzero vector $\boldsymbol{u} \in C(\boldsymbol{v}^*)$ the inequality

$$\boldsymbol{u}'\nabla^2 L(\boldsymbol{v}^*, \boldsymbol{\lambda}^*, \boldsymbol{\xi}_0)\boldsymbol{u} > 0 \tag{6.5.9}$$

holds.

Under the MF-condition, the above second-order sufficient conditions guarantee that $\boldsymbol{v}^*$ is a locally optimal solution of (P_0). In the case when the Lagrange multipliers vector $\boldsymbol{\lambda}^*$ is unique, these second-order sufficient conditions are complete in the sense that the corresponding second-order necessary conditions are obtained if the strict inequality sign in (6.5.9) is replaced by the sign "$\geq$". We leave this as an exercise to show that if $\boldsymbol{\lambda}$ is a vector of Lagrange multipliers, then the critical cone can be written in the alternative form as follows:

$$C(\boldsymbol{v}^*) = \{\boldsymbol{u} : \begin{array}{ll} \boldsymbol{u}'\nabla\ell_j(\boldsymbol{v}^*) \leq 0, & j \in J_0(\boldsymbol{v}^*, \boldsymbol{\lambda}) , \\ \boldsymbol{u}'\nabla\ell_j(\boldsymbol{v}^*) = 0, & j \in \{k+1, \ldots, M\} \cup J_+(\boldsymbol{v}^*, \boldsymbol{\lambda})\} . \end{array} \tag{6.5.10}$$

For a discussion of the second-order optimality conditions see, for example, Ioffe (1979) and Ben-Tal (1980).

For $\boldsymbol{\xi} = \boldsymbol{\xi}_0$ the quadratic programming problem $(Q_{\boldsymbol{\xi}_0})$ is reduced to minimization of the function $\boldsymbol{u}'\nabla^2 L(\boldsymbol{v}^*, \boldsymbol{\lambda}^*, \boldsymbol{\xi}_0)\boldsymbol{u}$ over the critical cone $C(\boldsymbol{v}^*)$. It follows then from the second-order sufficient conditions that the zero vector is the only optimal solution of $(Q_{\boldsymbol{\xi}_0})$, i.e. $\bar{\boldsymbol{u}}(\boldsymbol{\xi}_0) = \mathbf{0}$. However, the considered second-order sufficient conditions do not guarantee that $(Q_{\boldsymbol{\xi}})$ has a unique optimal solution for $\boldsymbol{\xi} \neq \boldsymbol{\xi}_0$. In order to ensure uniqueness of $\bar{\boldsymbol{u}}(\boldsymbol{\xi})$ it is required to strengthen these conditions. Consider the linear space

$$S(\boldsymbol{v}^*) = \{\boldsymbol{u} : \boldsymbol{u}'\nabla\ell_j(\boldsymbol{v}^*) = 0 , \quad j \in \{k+1, \ldots, M\} \cup J_+(\boldsymbol{v}^*, \boldsymbol{\lambda}^*)\} . \tag{6.5.11}$$

Note that the critical cone $C(\boldsymbol{v}^*)$ is contained in the linear space $S(\boldsymbol{v}^*)$.

Strong second-order sufficient conditions. For every nonzero vector $\boldsymbol{u} \in S(\boldsymbol{v}^*)$, inequality (6.5.9) holds.

We can formulate now the main result of this section.

Theorem 6.5.1 *Suppose that :*

(i) *For every $\boldsymbol{\xi}$ in a neighborhood of $\boldsymbol{\xi}_0$, the functions $g_j(\cdot, \boldsymbol{\xi})$, $j = 0, \ldots, M$, are twice differentiable and the corresponding derivatives are continuous jointly in $\boldsymbol{v}$ and $\boldsymbol{\xi}$.*

(ii) *The program (P_0) has a unique optimal solution $\boldsymbol{v}^*$ and a unique vector $\boldsymbol{\lambda}^*$ of the corresponding Lagrange multipliers.*

(iii) *The set V is compact and $\boldsymbol{v}^*$ is an interior point of V.*

(iv) *There exists a constant c such that*

$$|g_j(\boldsymbol{v}^*, \boldsymbol{\xi}) - g_j(\boldsymbol{v}^*, \boldsymbol{\xi}_0)| \leq c\|\boldsymbol{\xi} - \boldsymbol{\xi}_0\| ,$$

$$\|\nabla g_j(\boldsymbol{v}^*, \boldsymbol{\xi}) - \nabla g_j(\boldsymbol{v}^*, \boldsymbol{\xi}_0)\| \leq c\|\boldsymbol{\xi} - \boldsymbol{\xi}_0\| ,$$

$j = 0, \ldots, M$, for all $\boldsymbol{\xi}$ in a neighborhood of $\boldsymbol{\xi}_0$.

(v) *The strong second-order sufficient conditions hold.*

Then

$$\bar{\boldsymbol{v}}(\boldsymbol{\xi}) = \boldsymbol{v}^* + \bar{\mathbf{u}}(\boldsymbol{\xi}) + o(||\boldsymbol{\xi} - \boldsymbol{\xi}_0||)\,, \tag{6.5.12}$$

where $\bar{\mathbf{u}}(\boldsymbol{\xi})$ *is the optimal solution of* $(\mathrm{Q}_{\boldsymbol{\xi}})$.

We prove theorem 6.5.1 in several steps. Let us first recall that the assumption about the existence and uniqueness of $\boldsymbol{\lambda}^*$ is in itself a constraint qualification (the LMU-condition) which implies the MF-condition. It follows then by the stability theorem that for all $\boldsymbol{\xi}$ sufficiently close to $\boldsymbol{\xi}_0$, the program $(\mathrm{P}_{\boldsymbol{\xi}})$ has a feasible solution converging to $\boldsymbol{v}^*$ as $\boldsymbol{\xi} \to \boldsymbol{\xi}_0$. This and the compactness of V imply that for all $\boldsymbol{\xi}$ near $\boldsymbol{\xi}_0$, $(\mathrm{P}_{\boldsymbol{\xi}})$ has an optimal solution $\bar{\boldsymbol{v}}(\boldsymbol{\xi})$ and that $\bar{\boldsymbol{v}}(\boldsymbol{\xi})$ tends to $\boldsymbol{v}^*$ as $\boldsymbol{\xi} \to \boldsymbol{\xi}_0$. Moreover, we have that there exists a vector $\bar{\boldsymbol{\lambda}}(\boldsymbol{\xi})$ of Lagrange multipliers (possibly not unique) corresponding to the optimal solution $\bar{\boldsymbol{v}}(\boldsymbol{\xi})$ and that $\bar{\boldsymbol{\lambda}}(\boldsymbol{\xi}) \to \boldsymbol{\lambda}^*$ as $\boldsymbol{\xi} \to \boldsymbol{\xi}_0$. By the arguments of continuity we obtain then that the Lagrange multipliers $\bar{\lambda}_j(\boldsymbol{\xi})$, $j \in J_+(\boldsymbol{v}^*, \boldsymbol{\lambda}^*)$, remain positive and hence the corresponding inequality constraints are active at $\bar{\boldsymbol{v}}(\boldsymbol{\xi})$ for all $\boldsymbol{\xi}$ in a neighborhood of $\boldsymbol{\xi}_0$. It follows that for all $\boldsymbol{\xi}$ sufficiently close to $\boldsymbol{\xi}_0$, program $(\mathrm{P}_{\boldsymbol{\xi}})$ and the program

$$(\mathrm{P'}_{\boldsymbol{\xi}}) \qquad \begin{array}{lll} \text{minimize} & L(\boldsymbol{v}, \boldsymbol{\lambda}^*, \boldsymbol{\xi}), & \boldsymbol{v} \in V, \\ \text{subject to} & g_j(\boldsymbol{v}, \boldsymbol{\xi}) \le 0, & j \in J_0(\boldsymbol{v}^*, \boldsymbol{\lambda}^*), \\ & g_j(\boldsymbol{v}, \boldsymbol{\xi}) = 0, & j \in J_+(\boldsymbol{v}^*, \boldsymbol{\lambda}^*) \cup \{k+1, \ldots, M\}, \end{array}$$

have the same optimal solutions in a neighborhood of the point $\boldsymbol{v}^*$. We shall work subsequently with the program $(\mathrm{P'}_{\boldsymbol{\xi}})$ rather than $(\mathrm{P}_{\boldsymbol{\xi}})$.

Lemma 6.5.1 *Suppose that assumptions (i)–(iv) of theorem 6.5.1 hold and that the second-order sufficient conditions are satisfied. Then*

$$\bar{\boldsymbol{v}}(\boldsymbol{\xi}) - \boldsymbol{v}^* = O(||\boldsymbol{\xi} - \boldsymbol{\xi}_0||)\,. \tag{6.5.13}$$

Proof We argue by a contradiction. Suppose that (6.5.13) is false. Then there is a sequence $\boldsymbol{\xi}_n \to \boldsymbol{\xi}_0$ and $\boldsymbol{v}_n = \bar{\boldsymbol{v}}(\boldsymbol{\xi}_n) \to \boldsymbol{v}^*$ such that

$$\lim_{n\to\infty} t_n^{-1} ||\boldsymbol{\xi}_n - \boldsymbol{\xi}_0|| = 0\,, \tag{6.5.14}$$

where $t_n = ||\boldsymbol{v}_n - \boldsymbol{v}^*||$. We also can assume, by passing to a subsequence if necessary, that $t_n^{-1}||\boldsymbol{v}_n - \boldsymbol{v}^*||$ converges to a vector $\boldsymbol{d}$, $||\boldsymbol{d}|| = 1$. We have that for all n large enough,

$$g_j(\boldsymbol{v}_n, \boldsymbol{\xi}_n) \le 0\,, \quad j \in J_0(\boldsymbol{v}^*, \boldsymbol{\lambda}^*), \tag{6.5.15}$$

$$g_j(\boldsymbol{v}_n, \boldsymbol{\xi}_n) = 0\,, \quad j \in J_+(\boldsymbol{v}^*, \boldsymbol{\lambda}^*) \cup \{k+1, \ldots, M\}\,. \tag{6.5.16}$$

By the Mean Value Theorem,

$$g_j(\boldsymbol{v}_n, \boldsymbol{\xi}_n) = (\boldsymbol{v}_n - \boldsymbol{v}^*)'\nabla g_j(\tilde{\boldsymbol{v}}_n, \boldsymbol{\xi}_n) + g_j(\boldsymbol{v}^*, \boldsymbol{\xi}_n),$$

where $\tilde{\boldsymbol{v}}_n$ is a point on the segment joining $\boldsymbol{v}_n$ and $\boldsymbol{v}^*$. Also because of assumption (iv) and (6.5.14),

$$g_j(\boldsymbol{v}^*, \boldsymbol{\xi}_n) - g_j(\boldsymbol{v}^*, \boldsymbol{\xi}_0) = O(||\boldsymbol{\xi}_n - \boldsymbol{\xi}_0||) = o(t_n) .$$

Then by passing to the limit in (6.5.15) and (6.5.16) we obtain

$$\boldsymbol{d}'\nabla g_j(\boldsymbol{v}^*, \boldsymbol{\xi}_0) \leq 0 , \quad j \in J_0(\boldsymbol{v}^*, \boldsymbol{\lambda}^*) ,$$

$$\boldsymbol{d}'\nabla g_j(\boldsymbol{v}^*, \boldsymbol{\xi}_0) = 0 , \quad j \in J_+(\boldsymbol{v}^*, \boldsymbol{\lambda}^*) \cup \{k+1, \ldots, M\} ,$$

and hence $\boldsymbol{d} \in C(\boldsymbol{v}^*)$.

We have that for n large enough the optimal value $\phi(\boldsymbol{\xi}_n)$ is equal to $L(\boldsymbol{v}_n, \boldsymbol{\lambda}^*, \boldsymbol{\xi}_n)$. Employing the second-order Taylor expansion we obtain

$$\begin{aligned} L(\boldsymbol{v}_n, \boldsymbol{\lambda}^*, \boldsymbol{\xi}_n) - L(\boldsymbol{v}^*, \boldsymbol{\lambda}^*, \boldsymbol{\xi}_n) &= (\boldsymbol{v}_n - \boldsymbol{v}^*)'\nabla L(\boldsymbol{v}*, \boldsymbol{\lambda}^*, \boldsymbol{\xi}_n) \\ + \quad 1/2(\boldsymbol{v}_n - \boldsymbol{v}^*)'\nabla^2 L(\bar{\boldsymbol{v}}_n, \boldsymbol{\lambda}^*, \boldsymbol{\xi}_n)(\boldsymbol{v}_n - \boldsymbol{v}^*) , & \qquad (6.5.17) \end{aligned}$$

where $\bar{\boldsymbol{v}}_n$ is a point on the segment joining $\boldsymbol{v}_n$ and $\boldsymbol{v}^*$. Again because of assumption (iv) and (6.5.14) and since $\nabla L(\boldsymbol{v}^*, \boldsymbol{\lambda}^*, \boldsymbol{\xi}_0) = \boldsymbol{0}$ we have

$$||\nabla L(\boldsymbol{v}^*, \boldsymbol{\lambda}^*, \boldsymbol{\xi}_n)|| = O(||\boldsymbol{\xi}_n - \boldsymbol{\xi}_0||) = o(t_n) . \qquad (6.5.18)$$

Moreover, because of the continuity of the second-order derivatives it follows then from (6.5.17) that

$$\phi(\boldsymbol{\xi}_n) - L(\boldsymbol{v}^*, \boldsymbol{\lambda}^*, \boldsymbol{\xi}_n) = \tfrac{1}{2}t_n^2 \boldsymbol{d}'\nabla^2 L(\boldsymbol{v}^*, \boldsymbol{\lambda}^*, \boldsymbol{\xi}_0)\boldsymbol{d} + o(t_n^2) . \qquad (6.5.19)$$

Note that by the second-order sufficient conditions the number $\boldsymbol{d}'\nabla^2 L(\boldsymbol{v}^*, \boldsymbol{\lambda}^*, \boldsymbol{\xi}_0)\boldsymbol{d}$ is greater than zero.

On the other hand because of the stability theorem and assumption (iv), there exists a sequence of points $\boldsymbol{u}_n$ satisfying the contraints of $(\mathrm{P}'_{\boldsymbol{\xi}_n})$ such that

$$||\boldsymbol{u}_n - \boldsymbol{v}^*|| = O(||\boldsymbol{\xi}_n - \boldsymbol{\xi}_0||) . \qquad (6.5.20)$$

We have that

$$\begin{aligned} \phi(\boldsymbol{\xi}_n) \leq L(\boldsymbol{u}_n, \boldsymbol{\lambda}^*, \boldsymbol{\xi}_n) &= L(\boldsymbol{v}^*, \boldsymbol{\lambda}^*, \boldsymbol{\xi}_n) \\ + \quad (\boldsymbol{u}_n - \boldsymbol{v}^*)'\nabla L(\boldsymbol{v}^*, \boldsymbol{\lambda}^*, \boldsymbol{\xi}_n) &+ O(||\boldsymbol{u}_n - \boldsymbol{v}^*||^2) . \end{aligned}$$

It follows then from (6.5.18), (6.5.20) and (6.5.14) that

$$\phi(\boldsymbol{\xi}_n) \leq L(\boldsymbol{v}^*, \boldsymbol{\lambda}^*, \boldsymbol{\xi}_n) + o(t_n^2) .$$

This last inequality contradicts (6.5.19) and hence the proof is complete. □

Before proceeding further, we derive an estimate of the distance between the optimal solutions of the two optimization problems

$$\text{minimize} \quad f(\boldsymbol{x}), \ \boldsymbol{x} \in S, \tag{6.5.21}$$

and

$$\text{minimize} \quad g(\boldsymbol{x}), \ \boldsymbol{x} \in T. \tag{6.5.22}$$

Here S and T are subsets of a normed vector space and programs (6.5.21) and (6.5.22) are supposed to be, in some sense, close to each other.

We say that the *second-order growth condition*, for the program (6.5.21) at a point $\boldsymbol{x}_0 \in S$, holds if there exists a positive number α and a neighborhood W of $\boldsymbol{x}_0$ such that

$$f(\boldsymbol{x}) \geq f(\boldsymbol{x}_0) + \alpha||\boldsymbol{x} - \boldsymbol{x}_0||^2 \tag{6.5.23}$$

for all $\boldsymbol{x} \in S \cap W$. The second-order growth condition implies, of course, that $\boldsymbol{x}_0$ is a local minimizer of $f(\boldsymbol{x})$ over S.

Consider the function $h(\boldsymbol{x}) = g(\boldsymbol{x}) - f(\boldsymbol{x})$ and the constants

$$\kappa = \sup\left\{\frac{|h(\boldsymbol{x}) - h(\boldsymbol{x}_0)|}{||\boldsymbol{x} - \boldsymbol{x}_0||} : \boldsymbol{x} \in W \ , \ \boldsymbol{x} \neq \boldsymbol{x}_0\right\} , \tag{6.5.24}$$

$$\delta_1 = \sup\{\text{dist}\,(\boldsymbol{x}, S') : \boldsymbol{x} \in T'\} , \tag{6.5.25}$$

$$\delta_2 = \text{dist}\,(\boldsymbol{x}_0, T'), \tag{6.5.26}$$

where $S' = S \cap W$ and $T' = T \cap W$.

Lemma 6.5.2 *Suppose that the second-order growth condition holds, that $f(\boldsymbol{x})$ and $g(\boldsymbol{x})$ are Lipschitz continuous on W modulus k_1 and k_2, respectively, and let $\bar{\boldsymbol{x}}$ be an optimal solution of (6.5.22) such that $\bar{\boldsymbol{x}} \in W$. Then*

$$||\bar{\boldsymbol{x}} - \boldsymbol{x}_0|| \leq \alpha^{-1}\kappa + 2\delta_1 + \alpha^{-1/2}(k_1\delta_1 + k_2\delta_2)^{1/2} . \tag{6.5.27}$$

Proof Let $\boldsymbol{x}$ be an element of T'. Then we have that $g(\bar{\boldsymbol{x}}) \leq g(\boldsymbol{x})$ and hence

$$\begin{aligned} f(\bar{\boldsymbol{x}}) - f(\boldsymbol{x}_0) &= h(\boldsymbol{x}_0) - h(\bar{\boldsymbol{x}}) + g(\bar{\boldsymbol{x}}) - g(\boldsymbol{x}) + g(\boldsymbol{x}) - g(\boldsymbol{x}_0) \\ &\leq \kappa||\bar{\boldsymbol{x}} - \boldsymbol{x}_0|| + k_2||\boldsymbol{x} - \boldsymbol{x}_0||. \end{aligned}$$

Since $\boldsymbol{x}$ was an arbitrary point of T', it follows that

$$f(\bar{\boldsymbol{x}}) - f(\boldsymbol{x}_0) \leq \kappa||\bar{\boldsymbol{x}} - \boldsymbol{x}_0|| + k_2\delta_2 . \tag{6.5.28}$$

Now let $\boldsymbol{y}$ be an element of S'. Then

$$f(\bar{\boldsymbol{x}}) - f(\boldsymbol{x}_0) \geq f(\boldsymbol{y}) - f(\boldsymbol{x}_0) - |f(\bar{\boldsymbol{x}}) - f(\boldsymbol{y})| \geq f(\boldsymbol{y}) - f(\boldsymbol{x}_0) - k_1 ||\bar{\boldsymbol{x}} - \boldsymbol{y}|| .$$

By the second-order growth condition it follows that

$$\begin{aligned} f(\bar{\boldsymbol{x}}) - f(\boldsymbol{x}_0) &\geq \alpha ||\boldsymbol{y} - \boldsymbol{x}_0||^2 - k_1 ||\bar{\boldsymbol{x}} - \boldsymbol{y}|| \\ &\geq \alpha(||\boldsymbol{y} - \bar{\boldsymbol{x}}|| - ||\bar{\boldsymbol{x}} - \boldsymbol{x}_0||)^2 - k_1 ||\bar{\boldsymbol{x}} - \boldsymbol{y}|| \\ &\geq \alpha ||\bar{\boldsymbol{x}} - \boldsymbol{x}_0||^2 - 2\alpha ||\bar{\boldsymbol{x}} - \boldsymbol{y}|| \, ||\bar{\boldsymbol{x}} - \boldsymbol{x}_0|| - k_1 ||\bar{\boldsymbol{x}} - \boldsymbol{y}|| . \end{aligned}$$

Since $\boldsymbol{y}$ was an arbitrary element of S' we obtain that

$$f(\bar{\boldsymbol{x}}) - f(\boldsymbol{x}_0) \geq \alpha ||\bar{\boldsymbol{x}} - \boldsymbol{x}_0||^2 - 2\alpha\beta ||\bar{\boldsymbol{x}} - \boldsymbol{x}_0|| - k_1 \beta ,$$

where $\beta = \text{dist}\,(\bar{\boldsymbol{x}}, S')$. Since $\beta \leq \delta_1$, it follows that

$$f(\bar{\boldsymbol{x}}) - f(\boldsymbol{x}_0) \geq \alpha ||\bar{\boldsymbol{x}} - \boldsymbol{x}_0||^2 - 2\alpha\delta_1 ||\bar{\boldsymbol{x}} - \boldsymbol{x}_0|| - k_1 \delta_1 . \tag{6.5.29}$$

Putting (6.5.28) and (6.5.29) together we obtain

$$\alpha ||\bar{\boldsymbol{x}} - \boldsymbol{x}_0||^2 - (2\alpha\delta_1 + \kappa) ||\bar{\boldsymbol{x}} - \boldsymbol{x}_0|| - (k_1\delta_1 + k_2\delta_2) \leq 0 .$$

This quadratic inequality yields

$$||\bar{\boldsymbol{x}} - \boldsymbol{x}_0|| \leq \frac{2\alpha\delta_1 + \kappa + [(2\alpha\delta_1 + \kappa)^2 + 4\alpha(k_1\delta_1 + k_2\delta_2)]^{1/2}}{2\alpha},$$

which implies (6.5.27). □

Note that if the function $h(\boldsymbol{x})$ is differentiable on W and the neighborhood W is convex, then it follows from the Mean Value Theorem that the constant κ given in (6.5.24) is less than or equal to

$$\bar{\kappa} = \sup\{||\nabla h(\boldsymbol{x})|| : \boldsymbol{x} \in W\} . \tag{6.5.30}$$

The second-order growth condition can be ensured by various forms of second-order sufficient conditions (see, for example, Robinson (1982)). In particular, in the case when the set S is convex and $f(\boldsymbol{x})$ is differentiable, we have that if $\boldsymbol{x}_0$ is a locally optimal solution of (6.5.21), then by the first-order necessary conditions $(\boldsymbol{x} - \boldsymbol{x}_0)'\nabla f(\boldsymbol{x}_0) \geq 0$ for all $\boldsymbol{x} \in S$. The second-order growth condition will follow if $f(\boldsymbol{x})$ has the second-order Taylor expansion

$$f(\boldsymbol{x}) = f(\boldsymbol{x}_0) + (\boldsymbol{x} - \boldsymbol{x}_0)'\nabla f(\boldsymbol{x}_0) + \tfrac{1}{2}(\boldsymbol{x} - \boldsymbol{x}_0)'\nabla^2 f(\boldsymbol{x}_0)(\boldsymbol{x} - \boldsymbol{x}_0) + o(||\boldsymbol{x} - \boldsymbol{x}_0||^2)$$

with $\nabla^2 f(\boldsymbol{x}_0)$ being positive definite on a linear space containing all vectors of the form $\boldsymbol{x} - \boldsymbol{x}_0, \;\; \boldsymbol{x} \in S$.

Proof of theorem 6.5.1 The idea of the proof is to show that the programs $(\mathrm{Q}_{\boldsymbol{\xi}})$ and $(\mathrm{P}'_{\boldsymbol{\xi}})$ are sufficiently close to each other in terms of the constants appearing in the right-hand side of (6.5.27), and that the second-order growth condition holds.

The strong second-order sufficient conditions mean that the quadratic function $\boldsymbol{u}'\nabla^2 L(\boldsymbol{v}^*, \boldsymbol{\lambda}^*, \boldsymbol{\xi}_0)\boldsymbol{u}$ is positive definite on the linear space $S(\boldsymbol{v}^*)$. This implies that there exists a positive number α such that

$$\boldsymbol{u}'\nabla^2 L(\boldsymbol{v}^*, \boldsymbol{\lambda}^*, \boldsymbol{\xi}_0)\boldsymbol{u} \geq \alpha||\boldsymbol{u}||^2$$

for all $\boldsymbol{u} \in S(v^*)$. Since the objective function of $(\mathrm{Q}_{\boldsymbol{\xi}})$ is quadratic and the constraints are linear, it follows then that the second-order growth condition, for $(\mathrm{Q}_{\boldsymbol{\xi}})$ at $\bar{\boldsymbol{u}}(\boldsymbol{\xi})$, holds with the corresponding constant α independent of $\boldsymbol{\xi}$ and $W = \mathbb{R}^n$.

Consider now

$$r(\boldsymbol{\xi}) = \max\{||\bar{\boldsymbol{v}}(\boldsymbol{\xi}) - \boldsymbol{v}^*||, \quad ||\bar{\boldsymbol{u}}(\boldsymbol{\xi})||\} ,$$

the neighborhood

$$W(\boldsymbol{\xi}) = \{\boldsymbol{v} : ||\boldsymbol{v} - \boldsymbol{v}^*|| \leq r(\boldsymbol{\xi})\}$$

of $\boldsymbol{v}^*$ and the function

$$h(\boldsymbol{u}, \boldsymbol{\xi}) = L(\boldsymbol{v}^* + \boldsymbol{u}, \boldsymbol{\lambda}^*, \boldsymbol{\xi}) - \boldsymbol{u}'\nabla L(\boldsymbol{v}^*, \boldsymbol{\lambda}^*, \boldsymbol{\xi}) - \tfrac{1}{2}\boldsymbol{u}'\nabla^2 L(\boldsymbol{v}^*, \boldsymbol{\lambda}^*, \boldsymbol{\xi}_0)\boldsymbol{u}$$

representing the difference between objective functions of the programs $(\mathrm{P}'_{\boldsymbol{\xi}})$ and $(\mathrm{Q}_{\boldsymbol{\xi}})$. Note that $\bar{\boldsymbol{v}}(\boldsymbol{\xi})$ and $\boldsymbol{v}^* + \bar{\boldsymbol{u}}(\boldsymbol{\xi})$ lie in $W(\boldsymbol{\xi})$ and that, by lemma 6.5.1, $r(\boldsymbol{\xi})$ is of order $O(||\boldsymbol{\xi} - \boldsymbol{\xi}_0||)$. We have to estimate the constant

$$\bar{\kappa}(\boldsymbol{\xi}) = \sup\{||\nabla h(\boldsymbol{u}, \boldsymbol{\xi})|| : ||\boldsymbol{u}|| \leq r(\boldsymbol{\xi})\} .$$

Since

$$\nabla h(\boldsymbol{u}, \boldsymbol{\xi}) = \nabla L(\boldsymbol{v}^* + \boldsymbol{u}, \boldsymbol{\lambda}^*, \boldsymbol{\xi}) - \nabla L(\boldsymbol{v}^*, \boldsymbol{\lambda}^*, \boldsymbol{\xi}) - \nabla^2 L(\boldsymbol{v}^*, \boldsymbol{\lambda}^*, \boldsymbol{\xi}_0)\boldsymbol{u} ,$$

it follows by the Mean Value Theorem that

$$||\nabla h(\boldsymbol{u}, \boldsymbol{\xi})|| \leq ||\boldsymbol{u}|| \max_{t\in[0,1]} ||\nabla^2 L(\boldsymbol{v}^* + t\boldsymbol{u}, \boldsymbol{\lambda}^*, \boldsymbol{\xi}) - \nabla^2 L(\boldsymbol{v}^*, \boldsymbol{\lambda}^*, \boldsymbol{\xi}_0)|| .$$

Since $r(\boldsymbol{\xi}) = O(||\boldsymbol{\xi} - \boldsymbol{\xi}_0||)$ and because of the continuity of $\nabla^2 L(\boldsymbol{v}, \boldsymbol{\lambda}^*\boldsymbol{\xi})$ we obtain that $\bar{\kappa}(\boldsymbol{\xi})$ is of order $o(||\boldsymbol{\xi} - \boldsymbol{\xi}_0||)$.

Furthermore, because of the Stability Theorem, the constants $\delta_1(\boldsymbol{\xi})$ and $\delta_2(\boldsymbol{\xi})$ corresponding to the distances between the feasible sets of the programs $(\mathrm{P}'_{\boldsymbol{\xi}})$ and $(\mathrm{Q}_{\boldsymbol{\xi}})$ are of order

$$\sup\{|g_j(\boldsymbol{v}^* + \boldsymbol{u}, \boldsymbol{\xi}) - g_j(\boldsymbol{v}^*, \boldsymbol{\xi}) - \boldsymbol{u}'\nabla g_j(\boldsymbol{v}^*, \boldsymbol{\xi}_0)| : ||\boldsymbol{u}|| \leq r(\boldsymbol{\xi})\} .$$

Again it follows by the Mean Value Theorem that these constants are of order $o(||\boldsymbol{\xi} - \boldsymbol{\xi}_0||)$.

Finally, since $\nabla L(\boldsymbol{v}^*, \boldsymbol{\lambda}^*, \boldsymbol{\xi}_0) = \mathbf{0}$ and because of assumption (iv), we have that the Lipschitz constant $k_2(\boldsymbol{\xi})$ of $L(\cdot, \boldsymbol{\lambda}^*, \boldsymbol{\xi})$ in the neighborhood $W(\boldsymbol{\xi})$ of $\boldsymbol{v}^*$ is of order $O(||\boldsymbol{\xi} - \boldsymbol{\xi}_0||)$. Similarly for the objective function of the program $(\mathrm{Q}_{\boldsymbol{\xi}})$. Consequently $k_1(\boldsymbol{\xi})\delta_1(\boldsymbol{\xi})$ and $k_2(\boldsymbol{\xi})\delta_2(\boldsymbol{\xi})$ are of order $o(||\boldsymbol{\xi} - \boldsymbol{\xi}_0||^2)$. It follows then from the estimate (6.5.27) of lemma 6.5.2 that $||\bar{\boldsymbol{v}}(\boldsymbol{\xi}) - \boldsymbol{v}^* - \bar{\boldsymbol{u}}(\boldsymbol{\xi})||$ is of order $o(||\boldsymbol{\xi} - \boldsymbol{\xi}_0||)$ and hence the proof is complete. □

The result (6.5.12) of theorem 6.5.1 gives us a first-order expansion of $\bar{\boldsymbol{v}}(\boldsymbol{\xi})$ in terms of the optimal solution of the quadratic programming problem $(\mathrm{Q}_{\boldsymbol{\xi}})$. Suppose now that $\nabla L(\boldsymbol{v}^*, \boldsymbol{\lambda}^*, \cdot)$ and $g_j(\boldsymbol{v}^*, \cdot)$ are (Fréchet) differentiable at $\boldsymbol{\xi}_0$, i.e.

$$\nabla L(\boldsymbol{v}^*, \boldsymbol{\lambda}^*, \boldsymbol{\xi}) = \nabla^2_{\boldsymbol{v}\boldsymbol{\xi}} L(\boldsymbol{v}^*, \boldsymbol{\lambda}^*, \boldsymbol{\xi}_0)(\boldsymbol{\xi} - \boldsymbol{\xi}_0) + o(||\boldsymbol{\xi} - \boldsymbol{\xi}_0||)\,,$$

$$g_j(\boldsymbol{v}^*, \boldsymbol{\xi}) = \nabla_{\boldsymbol{\xi}} g_j(\boldsymbol{v}^*, \boldsymbol{\xi}_0)(\boldsymbol{\xi} - \boldsymbol{\xi}_0) + o(||\boldsymbol{\xi} - \boldsymbol{\xi}_0||)\,,$$

$j \in J(\boldsymbol{v}^*) \cup \{k+1, \ldots, M\}$. Then program $(\mathrm{Q}_{\boldsymbol{\xi}})$ can be approximated by the program

$$(\mathrm{Q}^*_{\boldsymbol{\delta}}) \quad \begin{array}{ll} \underset{\boldsymbol{u} \in \mathbb{R}^n}{\text{minimize}} & \boldsymbol{u}'\nabla^2_{\boldsymbol{v}\boldsymbol{\xi}} L(\boldsymbol{v}^*, \boldsymbol{\lambda}^*, \boldsymbol{\xi}_0)\boldsymbol{\delta} + \tfrac{1}{2}\boldsymbol{u}'\nabla^2_{\boldsymbol{u}\boldsymbol{u}} L(\boldsymbol{v}^*, \boldsymbol{\lambda}^*, \boldsymbol{\xi}_0)\boldsymbol{u} \\ \text{subject to} & \boldsymbol{u}'\nabla_{\boldsymbol{v}} g_j(\boldsymbol{v}^*, \boldsymbol{\xi}_0) + \nabla_{\boldsymbol{\xi}} g_j(\boldsymbol{v}, \boldsymbol{\xi}_0)\boldsymbol{\delta} \le 0, \quad j \in J_0(\boldsymbol{v}^*, \boldsymbol{\lambda}^*)\,, \\ & \boldsymbol{u}'\nabla_{\boldsymbol{v}} g_j(\boldsymbol{v}^*, \boldsymbol{\xi}_0) + \nabla_{\boldsymbol{\xi}} g_j(\boldsymbol{v}, \boldsymbol{\xi}_0)\boldsymbol{\delta} = 0\,, \\ & j \in J_+(\boldsymbol{v}^*, \boldsymbol{\lambda}^*) \cup \{k+1, \ldots, M\} \end{array}$$

with $\boldsymbol{\delta} = \boldsymbol{\xi} - \boldsymbol{\xi}_0$. It can be shown in a similar way that under the assumptions of theorem 6.5.1 the optimal solution $\boldsymbol{u}^*(\boldsymbol{\delta})$ of $(\mathrm{Q}^*_{\boldsymbol{\delta}})$ also gives a first-order approximation of $\bar{\boldsymbol{v}}(\boldsymbol{\xi})$, that is

$$\bar{\boldsymbol{v}}(\boldsymbol{\xi}) = \boldsymbol{v}^* + \boldsymbol{u}^*(\boldsymbol{\xi} - \boldsymbol{\xi}_0) + o(||\boldsymbol{\xi} - \boldsymbol{\xi}_0||)\,. \tag{6.5.31}$$

If the strict complementarity condition holds, i.e. the index set $J_0(\boldsymbol{v}^*, \boldsymbol{\lambda}^*)$ is empty, then the program $(\mathrm{Q}^*_{\boldsymbol{\delta}})$ involves (linear) equality constraints only. In this case the optimal solution $\boldsymbol{u}^*(\boldsymbol{\delta})$ can be written in a closed form and it is linear in $\boldsymbol{\delta}$. Consequently $\bar{\boldsymbol{v}}(\boldsymbol{\xi})$ is (Fréchet) differentiable at $\boldsymbol{\xi}_0$ and the corresponding derivatives can be obtained as solutions of the linear equations representing first-order optimality conditions for $(\mathrm{Q}^*_{\boldsymbol{\delta}})$ (see 6.5.7)).

In any case $\boldsymbol{u}^*(\boldsymbol{\delta})$ is positively homogeneous and hence, by (6.5.31), $\boldsymbol{u}^*(\boldsymbol{\delta})$ gives the directional derivatives (in the sense of Fréchet) of $\bar{\boldsymbol{v}}(\boldsymbol{\xi})$ at $\boldsymbol{\xi}_0$. It is also possible to derive a second-order expansion of the optimal value function $\phi(\boldsymbol{\xi})$ of the program $(\mathrm{P}_{\boldsymbol{\xi}})$ in the terms the optimal value of the program $(\mathrm{Q}_{\boldsymbol{\xi}})$

or (Q^*_δ) (see exercise 6.14). For various extensions and generalizations of sensitivity analysis in mathematical programming see Auslender and Cominetti (1990), Bonnans (1992), Bonnans, Joffe and Shapiro (1992), Gauvin and Janin (1988), King and Rockafellar (1992a) and Shapiro (1985, 1988, 1992).

The sensitivity analysis result of theorem 6.5.1 can be applied in a straightforward manner to the investigation of robustness of the optimal solutions with respect to contamination of the corresponding pdf $f(\boldsymbol{y}, \boldsymbol{v})$ (see section 3.6). That is, consider the contaminated pdf $\tilde{f}_t(\boldsymbol{y}, \boldsymbol{v})$ in the form of section 3.6. The associated expected performances

$$q_j(\boldsymbol{v}, t) = \mathbb{E}_{\tilde{f}_t}\{L_j(\boldsymbol{y}, \boldsymbol{v})\}$$

can be written as

$$q_j(\boldsymbol{v}, t) = \ell_j(\boldsymbol{v}) + t[\eta_j(\boldsymbol{v}) - \ell_j(\boldsymbol{v})] ,$$

where

$$\eta_j(\boldsymbol{v}) = \int L_j(\boldsymbol{y}, \boldsymbol{v}) h(\boldsymbol{y}, \boldsymbol{v}) \mathrm{d}\boldsymbol{y} .$$

Let $\tilde{\boldsymbol{v}}(t)$ be the optimal solution of the corresponding parametric program

$$(\mathrm{P}_t) \quad \begin{array}{lll} \text{minimize} & q_0(\boldsymbol{v}, t) , & \boldsymbol{v} \in V , \\ \text{subject to} & q_j(\boldsymbol{v}, t) \le 0, & j = 1, \dots, k, \\ & q_j(\boldsymbol{v}, t) = 0, & j = k+1, \dots, M . \end{array}$$

This optimal solution depends on the parameter $t \in [0, 1]$, and its derivative at $t = 0$ represents sensitivity of the optimal value of the parameter vector with respect to the considered contamination. We have that under suitable regularity conditions the derivative $\mathrm{d}\tilde{\boldsymbol{v}}(t)/\mathrm{d}t|_{t=0}$ is given by the optimal solution $\boldsymbol{u}^*$ of the quadratic programming problem

$$(\mathrm{Q}^*) \quad \begin{array}{ll} \text{minimize} & \boldsymbol{u}' \left[\nabla \eta_0(\boldsymbol{v}^*) + \sum_{j=1}^M \lambda_j^* \nabla \eta_j(\boldsymbol{v}^*)\right] + \frac{1}{2} \boldsymbol{u}' \nabla^2 L(\boldsymbol{v}^*, \boldsymbol{\lambda}^*) \boldsymbol{u} \\ \text{subject to} & \boldsymbol{u}^* \nabla \ell_j(\boldsymbol{v}^*) + \eta_j(\boldsymbol{v}^*) \le 0, \quad j \in J_0(\boldsymbol{v}^*, \boldsymbol{\lambda}^*) , \\ & \boldsymbol{u}' \nabla \ell_j(\boldsymbol{v}^*) + \eta_j(\boldsymbol{v}^*) = 0, \\ & j \in J_+(\boldsymbol{v}^*, \boldsymbol{\lambda}^*) \cup \{k+1, \dots, M\}, \end{array}$$

where

$$L(\boldsymbol{v}, \boldsymbol{\lambda}) = \ell_0(\boldsymbol{v}) + \sum_{j=1}^M \lambda_j \ell_j(\boldsymbol{v}) .$$

One can also apply the result of theorem 6.5.1 in order to evaluate the sensitivity of the optimal solution $\boldsymbol{v}^*(\boldsymbol{r})$ of the program

$$(\mathrm{P}_{\boldsymbol{r}}) \quad \begin{array}{lll} \text{minimize} & \ell_0(\boldsymbol{v}, \boldsymbol{r}) , & \boldsymbol{v} \in V, \\ \text{subject to} & \ell_j(\boldsymbol{v}, \boldsymbol{r}) \le 0, & j = 1, \dots, k, \\ & \ell_j(\boldsymbol{v}, \boldsymbol{r}) = 0, & j = k+1, \dots, M. \end{array}$$

Here $\ell_j(\boldsymbol{v}, \boldsymbol{r})$ are expected performances depending on the vectors $\boldsymbol{v}$ and $\boldsymbol{r}$. For example, $\boldsymbol{r}$ can represent the interarrival rates while $\boldsymbol{v}^*(\boldsymbol{r})$ gives the corresponding optimal service rates (see section 1.1). Suppose that for $\boldsymbol{r} = \boldsymbol{r}_0$ the optimal solution $\boldsymbol{v}^* = \boldsymbol{v}^*(\boldsymbol{r}_0)$ is known. We have then that for $\boldsymbol{r}$ close to $\boldsymbol{r}_0$ the optimal solution $\boldsymbol{v}^*(\boldsymbol{r})$ is approximated by $\boldsymbol{v}^* + \boldsymbol{u}^*(\Delta\boldsymbol{r})$, where $\Delta\boldsymbol{r} = \boldsymbol{r} - \boldsymbol{r}_0$ and $\boldsymbol{u}^*(\Delta\boldsymbol{r})$ is the optimal solution of the associated quadratic programming problem $(\mathrm{Q}^*_{\Delta\boldsymbol{r}})$. The vector $\boldsymbol{u}^* = \boldsymbol{u}^*(\Delta\boldsymbol{r})$ represents the sensitivity of $\boldsymbol{v}^*(\boldsymbol{r})$ with respect to small perturbations $\Delta\boldsymbol{r}$ of the parameter vector $\boldsymbol{r}$.

6.6 Asymptotic Analysis of the Optimal Solutions

In this section we study the asymptotic behavior of the optimal solutions $\bar{\boldsymbol{v}}_N$ of the stochastic program $(\bar{\mathrm{P}}_\mathrm{N})$. One possible approach to the asymptotic analysis of $\bar{\boldsymbol{v}}_N$ is to apply the delta method (theorem 6.3.1) together with calculated directional derivatives of the optimal solution mapping in an appropriate functional space. Following this approach King (1986) and King and Rockafellar, (1992,a,b) developed a theory that analyzes the central limit properties of the solutions using a uniform CLT and concepts of the generalized derivatives for multivalued solution mappings. We take a different approach, however. It will allow us to relax the asumptions about the differentiability properties of the random functions $\psi_{jN}(\boldsymbol{v})$. First we show that under mild regularity conditions, the program $(\bar{\mathrm{P}}_\mathrm{N})$ is asymptotically equivalent to a stochastic programming problem depending on a single random vector. We then derive the asymptotic distribution of $\bar{\boldsymbol{v}}_N$ by applying the finite-dimensional delta theorem.

Consider the Lagrangian functions

$$L(\boldsymbol{v}, \boldsymbol{\lambda}) = \ell_0(\boldsymbol{v}) + \sum_{j=1}^{M} \lambda_j \ell_j(\boldsymbol{v}) \,, \tag{6.6.1}$$

$$\mathcal{L}_N(\boldsymbol{v}, \boldsymbol{\lambda}) = \psi_{0N}(\boldsymbol{v}) + \sum_{j=1}^{M} \lambda_j \psi_{jN}(\boldsymbol{v}) \,, \tag{6.6.2}$$

of the programs (P_0) and $(\bar{\mathrm{P}}_\mathrm{N})$, respectively. We assume that the program (P_0) has a *unique* optimal solution $\boldsymbol{v}^*$ and a *unique* vector $\boldsymbol{\lambda}^*$ of the corresponding Lagrange multipliers satisfying the first-order (Kuhn-Tucker) necessary conditions. We also assume that $\boldsymbol{v}^*$ is an interior point of V and that $\bar{\boldsymbol{v}}_N$ converges w.p.1 (w.p.1) to $\boldsymbol{v}^*$ (see section 6.2 for a discussion of the consistency of $\bar{\boldsymbol{v}}_N$).

Suppose that the random functions $\psi_{jN}(\boldsymbol{v})$ are continuously differentiable. Then, under a constraint qualification, the optimal solutions $\bar{\boldsymbol{v}}_N$ and the associated vectors $\bar{\boldsymbol{\lambda}}_N$ of the Lagrange multipliers must satisfy the first-order necessary conditions. Suppose further that $\nabla\psi_{jN}(\boldsymbol{v})$, $j = 0,\ldots,M$, converge to $\nabla\ell_j(\boldsymbol{v})$ w.p.1 uniformly in a neighborhood of $\boldsymbol{v}^*$. It follows then that $\bar{\boldsymbol{\lambda}}_N \rightarrow \boldsymbol{\lambda}^*$ w.p.1. If we assume, in addition, that the strict complementarity condition holds, i.e. the set $J_0(\boldsymbol{v}^*, \boldsymbol{\lambda}^*)$ is empty, we obtain that w.p.1 the Lagrange multipliers $\bar{\lambda}_{jN}$, $j \in J(\boldsymbol{v}^*)$, are positive and hence the corresponding inequality constraints are active at $\bar{\boldsymbol{v}}_N$ for all N large enough. Without the strict complementarity conditions the set of active at $\bar{\boldsymbol{v}}_N$ inequality constraints can vary with $\bar{\boldsymbol{v}}_N$ and it is no longer possible to write the first-order optimality conditions in the form of equations. In order to overcome this difficulty we show how these optimality conditions can be represented in the form of so-called generalized equations in the terminology of Robinson (1979).

Before proceeding further we shall need a few simple facts from convex analysis. Consider a convex cone K, $K \subset \mathbb{R}^m$. Its polar (negative dual) cone is defined by

$$K^- = \{\boldsymbol{y} : \boldsymbol{y}'\boldsymbol{x} \leq 0 \quad \text{for all} \quad \boldsymbol{x} \in K\}\,.$$

In a sense the polar cone K^- can be viewed as a cone which is orthogonal to the cone K. In particular, if K is a linear space, then K^- represents the orthogonal complement of K. That is, K^- is formed by vectors $\boldsymbol{y}$ such that $\boldsymbol{y}'\boldsymbol{x} = 0$ for all $\boldsymbol{x} \in K$. If K is a closed convex cone, the duality relation $(K^-)^- = K$ follows. Now let S be a closed convex subset of $\mathbb{R}^m$. The normal cone to S at $\boldsymbol{x}$ is defined by

$$N_S(\boldsymbol{x}) = \{\boldsymbol{y} : \boldsymbol{y}'(\boldsymbol{z} - \boldsymbol{x}) \leq 0 \quad \text{for all } \boldsymbol{z} \in S\}$$

when $\boldsymbol{x} \in S$ and $N_S(\boldsymbol{x}) = \emptyset$ if $\boldsymbol{x} \notin S$. In the case when the set $S = K$ is a closed convex cone it follows that

$$N_K(\boldsymbol{x}) = \{\boldsymbol{y} \in K^- : \boldsymbol{y}'\boldsymbol{x} = 0\}.$$

An important property of normal cones considered as a function of $\boldsymbol{x}$, is monotonicity. That is, for any two points $\boldsymbol{x}_1$, $\boldsymbol{x}_2$ and $\boldsymbol{y}_1 \in N_S(\boldsymbol{x}_1)$, $\boldsymbol{y}_2 \in N_S(\boldsymbol{x}_2)$ we have

$$(\boldsymbol{y}_1 - \boldsymbol{y}_2)'(\boldsymbol{x}_1 - \boldsymbol{x}_2) \geq 0\,.$$

Let us consider now the cone $K = \mathbb{R}^k_+ \times \mathbb{R}^{M-k}$, $K \subset \mathbb{R}^M$. It is not difficult to see that $K^- = \mathbb{R}^k_- \times \{\mathbf{0}\}$. We have that the Lagrange multipliers vector must lie in K and the constraints of the program (P_0) can be written in the compact form $A(\boldsymbol{v}) \in K^-$, where

$$A(\boldsymbol{v}) = [\ell_1(\boldsymbol{v}), \ldots, \ell_M(\boldsymbol{v})]\,.$$

Moreover, because of the necessary condition $\boldsymbol{\lambda}^{*'} A(\boldsymbol{v}^*) = \mathbf{0}$ we actually have that $A(\boldsymbol{v}^*) \in N_K(\boldsymbol{\lambda})$. Putting all of this together we can write the first-order necessary conditions for the program (P$_0$) in the form

$$\mathbf{0} \in G(\boldsymbol{z}) + T(\boldsymbol{z}) \,, \tag{6.6.3}$$

where $\boldsymbol{z} = (\boldsymbol{v}, \boldsymbol{\lambda})$,

$$G(\boldsymbol{z}) = (\nabla L(\boldsymbol{v}, \boldsymbol{\lambda}), -A(\boldsymbol{v})) \tag{6.6.4}$$

and $T(\boldsymbol{z})$ is the multifunction given by

$$T(\boldsymbol{z}) = N_{\mathbf{R}^n \times K}(\boldsymbol{z}) = \{\mathbf{0}\} \times N_K(\boldsymbol{\lambda}) \,. \tag{6.6.5}$$

We refer to (6.6.3), with $G(\boldsymbol{z})$ being a mapping and $T(\boldsymbol{z})$ being a multifunction, as *generalized equations* (cf. Robinson (1979, 1982)). In particular, if $\boldsymbol{\lambda}^*$ is an interior point of K, i.e. the strict complementarity condition holds, then $N_K(\boldsymbol{\lambda}) = \{\mathbf{0}\}$ for all $\boldsymbol{\lambda}$ sufficiently close to $\boldsymbol{\lambda}^*$. In this case the generalized equations (6.6.3) become (locally) equations in the ordinary sense.

The stochastic counterpart of the generalized equations (6.6.3) can be written in the form

$$\mathbf{0} \in \Gamma_N(\boldsymbol{z}) + \boldsymbol{\Upsilon}_N + T(\boldsymbol{z}) \,, \tag{6.6.6}$$

where

$$\Gamma_N(\boldsymbol{z}) = (\nabla \mathcal{L}_N(\boldsymbol{v}, \boldsymbol{\lambda}), -\psi_N(\boldsymbol{v})) \tag{6.6.7}$$

and

$$\psi_N(\boldsymbol{v}) = [\psi_{1N}(\boldsymbol{v}), \ldots, \psi_{MN}(\boldsymbol{v})] \,.$$

The generalized equations (6.6.6) represent the first-order optimality conditions for ($\bar{\text{P}}_\text{N}$). The term $\boldsymbol{\Upsilon}_N = (\boldsymbol{\Upsilon}_{1N}, \boldsymbol{\Upsilon}_{2N})$ appears in order to compensate for a possible error in the solution of the stochastic program ($\bar{\text{P}}_\text{N}$). It will be assumed that this error term $\boldsymbol{\Upsilon}_N$ tends in probability to zero as $N \to \infty$ at a rate which will be specified later.

We study the asymptotic behavior of solutions $\bar{\boldsymbol{z}}_N = (\bar{\boldsymbol{v}}_N, \bar{\boldsymbol{\lambda}}_N)$ of (6.6.6) by employing the particular structure of the generalized equations (6.6.3) and (6.6.6). It will be assumed that the functions $\psi_{jN}(\boldsymbol{v})$, $j = 0, \ldots, M$, are *locally Lipschitz continuous.* (By Rademacher's theorem the set of points where a locally Lipschitz function fails to be differentiable has Lebesgue measure zero.) At points where $\mathcal{L}_N(\cdot, \boldsymbol{\lambda})$ is not differentiable, the corresponding gradient in the definition of $\Gamma_N(\boldsymbol{z})$ can be replaced by an element of the so-called generalized gradient in the sense of Clarke (1983). At such points the mapping $\Gamma_N(\boldsymbol{z})$ can be discontinuous.

Let us discuss the following "nondifferentiable" example. Consider the function $h(z, v) = |z - v|$, $z, v \in \mathbb{R}$, and let Z be a random variable distributed according to a cdf $F(z)$. The associated expected value function

$\ell(v) = \mathbb{E}\{h(Z,v)\}$ is convex and continuous provided it is finite valued for all $v \in \mathbb{R}$. A minimizer v^* of $\ell(v)$ over $V = \mathbb{R}$ is determined by the condition that $F(z) \le 1/2$ for all $z \le v^*$ and $F(z) \ge 1/2$ for all $z \ge v^*$. Suppose that such a number v^* is unique and that $F(z)$ is continuous at $z = v^*$. Then $F(v^*) = 1/2$ and v^* is called the median of the cdf $F(z)$.

The function $h(z,v)$ is not differentiable at those points where $v = z$. Nevertheless, the expected value function $\ell(v)$ is continuously differentiable provided the cdf $F(z)$ is continuous (see exercise 6.1). Let $Z_1, \ldots, Z_N$ be a random sample generated from $F(z)$ and let

$$\psi_N(v) = N^{-1} \sum_{i=1}^{N} |Z_i - v| \tag{6.6.8}$$

be the corresponding sample mean estimator of $\ell(v)$. Note that $\psi_N(v)$ is not differentiable at points $v = Z_i$, $i = 1, \ldots, N$. A minimizer $\bar{v}_N$ of $\psi_N(v)$ over $V = \mathbb{R}$ is called the sample median and can be calculated as follows. Suppose that the sample is arranged in the increasing order $Z_{(1)} \le Z_{(2)} \le \cdots \le Z_{(N)}$, i.e. $Z_{(1)}$ is the smallest among the numbers Z_i, $i = 1, \ldots, N$, and so on. Then $\bar{v}_N = Z_{(k)}$ with $k = (N+1)/2$ for odd N, and $\bar{v}_N$ can be any number from the interval $[Z_{(k)}, Z_{(k+1)}]$, $k = N/2$, for even N. We see that for even N the minimizer $\bar{v}_N$ is not defined uniquely and for odd N the function $\psi_N(v)$ is not differentiable and its derivative $\psi_N'(v)$ is discontinuous at $\bar{v}_N$. We shall return to this example later on.

The generalized equations (6.6.3) and (6.6.6) have a specific structure which will be essential for our analysis. First let us observe that the multifunction $T : \mathbb{R}^{n+M} \to \mathbb{R}^{n+M}$ is *monotone*, i.e. for all $\boldsymbol{z}_1$, $\boldsymbol{z}_2$ and $\boldsymbol{y}_1 \in T(\boldsymbol{z}_1), \boldsymbol{y}_2 \in T(\boldsymbol{z}_2)$, we have

$$(\boldsymbol{y}_1 - \boldsymbol{y}_2)'(\boldsymbol{z}_1 - \boldsymbol{z}_2) \ge 0 . \tag{6.6.9}$$

This follows from the monotonicity property of normal cone operators.

Consider now the mapping $G(\boldsymbol{z})$ defined in (6.6.4). Suppose that the functions $\ell_j(v)$, $j = 0, \ldots, M$, are twice continuously differentiable. Then the $(n+M) \times (n+M)$ Jacobian matrix of $G(\boldsymbol{z})$ can be written in the form

$$\nabla G(\boldsymbol{z}) = \begin{bmatrix} \nabla^2 L(\boldsymbol{v}, \boldsymbol{\lambda}) & \nabla A(\boldsymbol{v}) \\ -\nabla A(\boldsymbol{v})' & \boldsymbol{0} \end{bmatrix} .$$

It follows that for any vector $\boldsymbol{y} = (\boldsymbol{u}, \boldsymbol{\mu}) \in \mathbb{R}^{n+M}$ we have $\boldsymbol{y}'\nabla G(\boldsymbol{z})\boldsymbol{y} = \boldsymbol{u}'\nabla^2 L(\boldsymbol{v}, \boldsymbol{\lambda})\boldsymbol{u}$. Also for $\boldsymbol{z}_1 = (\boldsymbol{v}_1, \boldsymbol{\lambda}_1)$ and $\boldsymbol{z}_2 = (\boldsymbol{v}_2, \boldsymbol{\lambda}_2)$ we have

$$G(\boldsymbol{z}_1) - G(\boldsymbol{z}_2) = \int_0^1 \nabla G(t\boldsymbol{z}_1 + (1-t)\boldsymbol{z}_2)(\boldsymbol{z}_1 - \boldsymbol{z}_2)dt$$

and hence

$$(\boldsymbol{z}_1 - \boldsymbol{z}_2)'[G(\boldsymbol{z}_1) - G(\boldsymbol{z}_2)] = \int_0^1 (\boldsymbol{z}_1 - \boldsymbol{z}_2)'\nabla G(t\boldsymbol{z}_1 + (1-t)\boldsymbol{z}_2)(\boldsymbol{z}_1 - \boldsymbol{z}_2)dt$$

$$= \int_0^1 (\boldsymbol{v}_1 - \boldsymbol{v}_2)'\nabla^2 L(t\boldsymbol{z}_1 + (1-t)\boldsymbol{z}_2)(\boldsymbol{v}_1 - \boldsymbol{v}_2)dt \,. \tag{6.6.10}$$

Suppose now that the Hessian matrix $\nabla^2 L(\boldsymbol{v}^*, \boldsymbol{\lambda}^*)$ is positive definite. It follows then from the integral representation (6.6.10) and continuity of $\nabla^2 L(\boldsymbol{v}, \boldsymbol{\lambda})$ that there exist a positive constant α and a neighborhood W of $\boldsymbol{z}^* = (\boldsymbol{v}^*, \boldsymbol{\lambda}^*)$ such that

$$(\boldsymbol{z}_1 - \boldsymbol{z}_2)'[G(\boldsymbol{z}_1) - G(\boldsymbol{z}_2)] \geq \alpha||\boldsymbol{v}_1 - \boldsymbol{v}_2||^2 \tag{6.6.11}$$

for all $\boldsymbol{z}_1$, $\boldsymbol{z}_2 \in W$.

This motivates us to make the following assumptions related to the generalized equations (6.6.3).

Assumption A1 The multifunction $T(\boldsymbol{z})$ is monotone.

Assumption A2 The generalized equation (6.6.3) has unique solution $\boldsymbol{z}^* = (\boldsymbol{v}^*, \boldsymbol{\lambda}^*)$.

Assumption A3 There exist a constant $\alpha > 0$ and a neighborhood W of $\boldsymbol{z}^*$ such that the inequality (6.6.11) holds for all $\boldsymbol{z}_1$, $\boldsymbol{z}_2 \in W$.

Consider now the parametric family of the generalized equations

$$\boldsymbol{0} \in G(\boldsymbol{z}) + \boldsymbol{\delta} + T(\boldsymbol{z}) \tag{6.6.12}$$

depending on the parameter vector $\boldsymbol{\delta} = (\boldsymbol{\delta}_1, \boldsymbol{\delta}_2)$, $\boldsymbol{\delta}_1 \in \mathbb{R}^n$, $\boldsymbol{\delta}_2 \in \mathbb{R}^M$. We make the following assumption about the generalized equation (6.6.12).

Assumption A4 For every $\boldsymbol{\delta}$ in a neighborhood of zero, the generalized equations (6.6.12) have a solution $\tilde{\boldsymbol{z}}(\boldsymbol{\delta})$ such that

$$||\tilde{\boldsymbol{z}}(\boldsymbol{\delta}) - \boldsymbol{z}^*|| = O(||\boldsymbol{\delta}||) \,. \tag{6.6.13}$$

In the case when $G(\boldsymbol{z})$ and $T(\boldsymbol{z})$ are given in (6.6.4) and (6.6.5), respectively, the generalized equations (6.6.12) represent the first-order necessary conditions corresponding to the program

$$(\mathrm{P}^*_{\boldsymbol{\delta}}) \qquad \begin{array}{ll} \text{minimize} & \ell_0(\boldsymbol{v}) + \boldsymbol{\delta}_1'\boldsymbol{v} \\ \text{subject to} & \ell(\boldsymbol{v}) - \delta_{2j} \leq 0 \,, \quad j = 1, \ldots, k \\ & \ell_j(\boldsymbol{v}) - \delta_{2j} = 0, \quad j = k+1, \ldots, M. \end{array}$$

In this case the existence of $\tilde{\boldsymbol{z}}(\boldsymbol{\delta}) = (\tilde{\boldsymbol{v}}(\boldsymbol{\delta}), \tilde{\boldsymbol{\lambda}}(\boldsymbol{\delta}))$ is implied by standard conditions ensuring the existence of locally optimal solutions of $(\mathrm{P}^*_{\boldsymbol{\delta}})$ and by verification of the corresponding first-order necessary conditions. The Lipschitzian stability condition (6.6.13) follows by results in the previous section (see lemma 6.5.1) essentially under the second-order sufficient conditions for the program (P_0).

We introduce now some required "stochastic" assumptions. Consider random mappings

$$\Delta_N(z) = \Gamma_N(z) - G(z)\,. \tag{6.6.14}$$

Assumption B1

$$||\Delta_N(z^*)|| = O_p(N^{-1/2})\,. \tag{6.6.15}$$

Assumption B2 There is a neighborhood W of z^* such that

$$\sup_{z \in W} \frac{||\Delta_N(z) - \Delta_N(z^*)||}{N^{-1/2} + ||z - z^*||} = o_p(1)\,. \tag{6.6.16}$$

Condition (6.6.15) of assumption B1 means that the random vectors $N^{1/2}[\Gamma_N(z^*) - G(z^*)]$ are bounded in probability. In the case of the sample mean construction we have by the Central Limit theorem that, under standard regularity conditions, these random vectors converge in distribution and hence (6.6.15) follows.

Assumption B2 is more delicate. Consider the sample mean construction. Suppose that the corresponding functions $h_j(\zeta, \cdot)$, $j = 0, \ldots, M$, are twice continuously differentiable in a neighborhood U of v^* and that the families $\{||\nabla h_j(\zeta, v)||,\ v \in U\}$ and $\{||\nabla^2 h(\zeta, v)||,\ v \in U\}$, $j = 0, \ldots, M$, are dominated by integrable functions. By theorem 6.1.2 we have then that $\nabla \bar{\ell}_{jN}(v)$ and $\nabla^2 \bar{\ell}_{jN}(v)$ tend w.p.1 to $\nabla \ell_j(v)$ and $\nabla^2 \ell_j(v)$, respectively, uniformly on U (provided U is compact). By the Mean Value Theorem this imples that

$$\sup_{\substack{v \in U \\ v \neq v^*}} \frac{|\delta_{jN}(v) - \delta_{jN}(v^*)|}{||v - v^*||} \longrightarrow 0 \quad \text{w.p.1},$$

where $\delta_{jN}(v) = \bar{\ell}_{jN}(v) - \ell_j(v)$, $j = 0, \ldots, M$, and similarly for the gradients $\nabla \delta_{jN}(v)$. It follows that there is a neighborhood W of z^* such that

$$\sup_{\substack{z \in W \\ z \neq z^*}} \frac{||\Delta_N(z) - \Delta_N(z^*)||}{||z - z^*||} \longrightarrow 0 \quad \text{w.p.1}$$

Therefore in smooth (differentiable) cases assumption B2 holds and the term $N^{-1/2}$ in the denominator of the ratio in (6.6.16) can be omitted.

In nondifferentiable situations assumption B2 is more involved. Consider the "nondifferentiable" example discussed earlier in this section with

$h(\zeta, v) = |\zeta - v|$. We have that the difference function $\delta_N(v) = \psi_N(v) - \ell(v)$ is differentiable at v and its derivative is given by (see exercise 6.1)

$$\delta'_N(v) = 2F_N(v) - 2F(v)$$

provided the cdf $F(z)$ and the empirical cdf $F_N(z)$ are continuous at $z = v$. Suppose that $F(z) = z$, $0 \leq z \leq 1$, i.e. the considered distribution is uniform $(0,1)$. Then $v^* = 1/2$ and since (w.p.1) F_N is constant in a sufficiently small neighborhood of v^*, we have that

$$\frac{|\delta'_N(v) - \delta'_N(v^*)|}{|v - v^*|} = 2$$

for all v sufficiently close to v^*. Therefore the supremum of $|\delta'_N(v) - \delta'_N(v^*)|/|v - v^*|$ over any neighborhood of v^* is greater or equal to 2. This demonstrates that the term $N^{-1/2}$ in the denominator of the associated ratio is essential here.

Conditions of type (6.6.16) for the sample mean constructions, were studied by Huber (1967) in his work on the statistical theory of the M-estimators. He showed that such conditions hold under mild regularity assumptions even in nondifferentiable cases. An interesting discussion of the ideas involved can be found in Pollard (1989).

Finally we shall need the following assumption.

Assumption B3

$$||\bar{\boldsymbol{\lambda}}_N - \boldsymbol{\lambda}^*|| = O_p(||\bar{\boldsymbol{v}}_N - \boldsymbol{v}^*|| + N^{-1/2}) . \tag{6.6.17}$$

This assumption requires an explanation. We show in a moment that if $G(\boldsymbol{z}), T(\boldsymbol{z})$ and $\Gamma_N(\boldsymbol{z})$ are given as in (6.6.4), (6.6.5) and (6.6.7), respectively, then there exists a positive constant β, depending on $\ell_j(\boldsymbol{v}^*)$ and $\nabla \ell_j(\boldsymbol{v}^*)$, $j = 1, \ldots, M$, only, such that

$$\begin{aligned} ||\bar{\boldsymbol{\lambda}}_N - \boldsymbol{\lambda}^*|| &\leq \beta(1 + ||\bar{\boldsymbol{\lambda}}_N||)\{||\nabla\psi_{0N}(\bar{\boldsymbol{v}}_N) - \nabla\ell_0(\boldsymbol{v}^*)|| + ||\nabla\psi_N(\bar{\boldsymbol{v}}_N) \\ &- \nabla A(\boldsymbol{v}^*)|| + ||\psi_N(\bar{\boldsymbol{v}}_N) - A(\boldsymbol{v}^*)|| + ||\boldsymbol{\Upsilon}_N||\} . \end{aligned} \tag{6.6.18}$$

Suppose now that $\bar{\boldsymbol{v}}_N$ converges w.p.1 to $\boldsymbol{v}^*$, that $\nabla \ell_j(\boldsymbol{v})$ are Lipschitz continuous near $\boldsymbol{v}^*$, that $||\boldsymbol{\Upsilon}_N|| = O_p(N^{-1/2})$ and that the following analogues of assumptions B1 and B2 hold for the random mappings $\Delta_{1N}(\boldsymbol{v}) = \nabla\psi_{0N}(\boldsymbol{v}) - \nabla\ell_0(\boldsymbol{v})$, $\Delta_{2N}(\boldsymbol{v}) = \psi_N(\boldsymbol{v}) - A(\boldsymbol{v})$ and $\Delta_{3N}(\boldsymbol{v}) = \nabla\psi_N(\boldsymbol{v}) - \nabla A(\boldsymbol{v})$.

Assumption B1′

$$||\Delta_{iN}(\boldsymbol{v}^*)|| = O_p(N^{-1/2}), \quad i = 1, 2, 3. \tag{6.6.19}$$

Assumption B2′ There is a neighborhood U of $\boldsymbol{v}^*$ such that

$$\sup_{\boldsymbol{v}\in U} \frac{\|\Delta_{iN}(\boldsymbol{v}) - \Delta_{iN}(\boldsymbol{v}^*)\|}{N^{-1/2} + \|\boldsymbol{v} - \boldsymbol{v}^*\|} = o_p(1), \quad i = 1, 2, 3. \tag{6.6.20}$$

We obtain that the Lagrange multipliers vectors $\bar{\boldsymbol{\lambda}}_N$ converge w.p.1 to $\boldsymbol{\lambda}^*$ and hence are bounded in probability. It is not difficult to see then that, under the above assumptions, (6.6.17) follows from (6.6.18). Note also that assumptions B1′ and B2′ imply assumptions B1 and B2.

In order to prove the upper bound (6.6.18) we use the following geometrical result. Consider an $m \times n$ matrix $\boldsymbol{G}$, the corresponding linear transformation $g : \mathbb{R}^n \to \mathbb{R}^m$, $g(\boldsymbol{x}) = \boldsymbol{G}\boldsymbol{x}$, and let K be a closed convex polyhedral subset of $\mathbb{R}^n$ (i.e. K is defined by a finite number of linear constraints). Denote by Haus(A, B) the Hausdorff distance between two nonempty subsets A, B of $\mathbb{R}^n$,

$$\text{Haus}(A, B) = \max\{\sup_{\boldsymbol{x}\in B} \text{dist}\,(\boldsymbol{x}, A), \quad \sup_{\boldsymbol{x}\in A} \text{dist}(\boldsymbol{x}, B)\}.$$

It follows then that the multifunction $S : \mathbb{R}^m \to \mathbb{R}^n$, defined by

$$S(\boldsymbol{y}) = \mathrm{g}^{-1}(\boldsymbol{y}) \cap K = \{\boldsymbol{x} \in K : \; \boldsymbol{G}\boldsymbol{x} = \boldsymbol{y}\}$$

is Lipschitz continuous in the Hausdorff metric (Walkup and Wets (1969). That is, there exists a constant γ, depending only on g and K, such that for any $\boldsymbol{y}_1$, $\boldsymbol{y}_2 \in \mathbb{R}^m$ and if the sets $S(\boldsymbol{y}_1) = g^{-1}(\boldsymbol{y}_1) \cap K$ and $S(\boldsymbol{y}_2) = g^{-1}(\boldsymbol{y}_2) \cap K$ are nonempty, then

$$\text{Haus}(S(\boldsymbol{y}_1), S(\boldsymbol{y}_2)) \leq \gamma \|\boldsymbol{y}_1 - \boldsymbol{y}_2\| \,. \tag{6.6.21}$$

Consider now the programs (P_0) and $(\bar{\mathrm{P}}_\mathrm{N})$. The corresponding Lagrange multipliers vectors must lie in the (convex, polyhedral) set $K = \mathbb{R}^k_+ \times \mathbb{R}^{M-k}$ and satisfy the linear equations

$$\boldsymbol{b} + \boldsymbol{B}\boldsymbol{\lambda} = \boldsymbol{0}, \; \boldsymbol{c}'\boldsymbol{\lambda} = \epsilon \,, \tag{6.6.22}$$

with $\epsilon = 0$ and $\boldsymbol{b}, \boldsymbol{B}$ and $\boldsymbol{c}$ are given by $\nabla \ell_0(\boldsymbol{v}^*)$, $\nabla A(\boldsymbol{v}^*)$ and $A(\boldsymbol{v}^*)$, respectively, for the program (P_0) and by $\nabla \psi_{0N}(\bar{\boldsymbol{v}}_N) + \boldsymbol{\Upsilon}_{1N}$ and $\nabla \psi_N(\bar{\boldsymbol{v}}_N)$, $\psi_N(\bar{\boldsymbol{v}}_N) - \boldsymbol{\Upsilon}_{2N}$, for the program $(\bar{\mathrm{P}}_\mathrm{N})$. By the above geometrical result we have that the set of solutions of equations (6.6.22) subject to $\boldsymbol{\lambda} \in K$, depends Lipschitz continuously on $\boldsymbol{b}$ and ϵ with respect to the Hausdorff metric. That is, there exists a constant β, depending on $\boldsymbol{B} = \nabla A(\boldsymbol{v}^*)$ and $\boldsymbol{c} = A(\boldsymbol{v}^*)$, such that if $\boldsymbol{\lambda}^* \in K$ is the unique solution of equations (6.6.22) with $\epsilon = 0$, and $\bar{\boldsymbol{\lambda}} \in K$ is a solution of the equations

$$\bar{\boldsymbol{b}} + \bar{\boldsymbol{B}}\boldsymbol{\lambda} = \boldsymbol{0} \,, \quad \bar{\boldsymbol{c}}'\boldsymbol{\lambda} = 0 \,,$$

then

$$||\bar{\boldsymbol{\lambda}} - \boldsymbol{\lambda}^*|| \leq \beta[\, ||\bar{\boldsymbol{b}} - \boldsymbol{b} + (\bar{\boldsymbol{B}} - \boldsymbol{B})\bar{\boldsymbol{\lambda}}|| + |(\bar{\boldsymbol{c}} - \boldsymbol{c})'\bar{\boldsymbol{\lambda}}|\,]$$

and hence

$$||\bar{\boldsymbol{\lambda}} - \boldsymbol{\lambda}^*|| \leq \beta(1 + ||\bar{\boldsymbol{\lambda}}||)(||\bar{\boldsymbol{b}} - \boldsymbol{b}|| + ||\bar{\boldsymbol{B}} - \boldsymbol{B}|| + ||\bar{\boldsymbol{c}} - \boldsymbol{c}||)\,.$$

The last inequality implies (6.6.18).

We can formulate now the main result of this section. Consider the generalized equations (6.6.12) with $\boldsymbol{\delta} = \boldsymbol{\delta}_N = \Delta_N(\boldsymbol{z}^*)$,

$$\mathbf{0} \in G(\boldsymbol{z}) + \Delta_N(\boldsymbol{z}^*) + T(\boldsymbol{z})\,. \tag{6.6.23}$$

By assumption A4, the generalized equations (6.6.23) have a solution $\tilde{\boldsymbol{z}}_N = (\tilde{\boldsymbol{v}}_N, \tilde{\boldsymbol{\lambda}}_N)$ for all $\boldsymbol{\delta}_N = \Delta_N(\boldsymbol{z}^*)$ sufficiently close to zero.

Theorem 6.6.1 *Suppose that assumptions A1-A4 and B1-B3 hold,* $\boldsymbol{\Upsilon}_N = o_p(N^{-1/2})$ *and the component* $\bar{\boldsymbol{v}}_N$ *of the solution* $\bar{\boldsymbol{z}}_N$ *of the generalized equations (6.6.6) converges in probability to* $\boldsymbol{v}^*$*. Then*

$$\bar{\boldsymbol{v}}_N = \tilde{\boldsymbol{v}}_N + o_p(N^{-1/2}), \tag{6.6.24}$$

where $\tilde{\boldsymbol{v}}_N$ *is the component of the solution* $\tilde{\boldsymbol{z}}_N$ *of the generalized equations (6.6.23).*

Proof Let us show first that

$$||\bar{\boldsymbol{z}}_N - \boldsymbol{z}^*|| = O_p(N^{-1/2})\,. \tag{6.6.25}$$

We have that

$$-G(\boldsymbol{z}^*) \in T(\boldsymbol{z}^*)$$

and

$$-G(\bar{\boldsymbol{z}}_N) - \Delta_N(\bar{\boldsymbol{z}}_N) - \boldsymbol{\Upsilon}_N \in T(\bar{\boldsymbol{z}}_N)\,.$$

Since the multifunction $T(\boldsymbol{z})$ is monotone, it follows that

$$[G(\bar{\boldsymbol{z}}_N) + \Delta_N(\bar{\boldsymbol{z}}_N) + \boldsymbol{\Upsilon}_N - G(\boldsymbol{z}^*)]'(\bar{\boldsymbol{z}}_N - \boldsymbol{z}^*) \leq 0$$

and hence

$$[G(\bar{\boldsymbol{z}}_N) - G(\boldsymbol{z}^*)]'(\bar{\boldsymbol{z}}_N - \boldsymbol{z}^*) \leq -[\Delta_N(\bar{\boldsymbol{z}}_N) + \boldsymbol{\Upsilon}_N]'(\bar{\boldsymbol{z}}_N - \boldsymbol{z}^*)\,. \tag{6.6.26}$$

Since $\bar{\boldsymbol{v}}_N$ tends in probability to $\boldsymbol{v}^*$ and because of assumption B3 we have that $||\bar{\boldsymbol{z}}_N - \boldsymbol{z}^*|| = o_p(1)$. Therefore with probability tending to one as $N \to \infty$, $\bar{\boldsymbol{z}}_N \in W$ and hence by assumption A3,

$$[G(\bar{\boldsymbol{z}}_N) - G(\boldsymbol{z}^*)]'(\bar{\boldsymbol{z}}_N - \boldsymbol{z}^*) \geq \alpha||\bar{\boldsymbol{v}}_N - \boldsymbol{v}^*||^2\,. \tag{6.6.27}$$

It follows from (6.6.26) and (6.6.27) that

$$\alpha||\bar{\boldsymbol{v}}_N - \boldsymbol{v}^*||^2 \leq ||\Delta_N(\bar{\boldsymbol{z}}_N) + \boldsymbol{\Upsilon}_N||\ ||\bar{\boldsymbol{z}}_N - \boldsymbol{z}^*||$$

and hence

$$\alpha||\bar{\boldsymbol{v}}_N - \boldsymbol{v}^*||^2 \leq (||\Delta_N(\boldsymbol{z}^*)|| + ||\Delta_N(\bar{\boldsymbol{z}}_N) - \Delta(\boldsymbol{z}^*)|| + ||\boldsymbol{\Upsilon}_N||)\ ||\bar{\boldsymbol{z}}_N - \boldsymbol{z}^*||\,.$$

Together with assumptions B1, B2 and since $\boldsymbol{\Upsilon}_N = o_p(N^{-1/2})$, this implies

$$\alpha||\bar{\boldsymbol{v}}_N - \boldsymbol{v}^*||^2 \leq [O_p(N^{-1/2}) + o_p(||\bar{\boldsymbol{z}}_N - \boldsymbol{z}^*||)]||\bar{\boldsymbol{z}}_N - \boldsymbol{z}^*||.$$

Moreover, because of assumption B3 we have then that

$$\alpha||\bar{\boldsymbol{v}}_N - \boldsymbol{v}^*||^2 \leq [O_p(N^{-1/2}) + o_p(||\bar{\boldsymbol{v}}_N - \boldsymbol{v}^*||)]O_p(||\bar{\boldsymbol{v}}_N - \boldsymbol{v}^*|| + N^{-1/2})\,.$$

Consequently

$$[\alpha + o_p(1)]||\bar{\boldsymbol{v}}_N - \boldsymbol{v}^*||^2 \leq O_p(N^{-1/2})||\bar{\boldsymbol{v}}_N - \boldsymbol{v}^*|| + O_p(N^{-1})$$

which implies that

$$||\bar{\boldsymbol{v}}_N - \boldsymbol{v}^*|| = O_p(N^{-1/2})\,.$$

By assumption B3, (6.6.25) then follows.

Now again because of monotonicity of $T(\boldsymbol{z})$ we have

$$[G(\bar{\boldsymbol{z}}_N) + \Delta_N(\bar{\boldsymbol{z}}_N) + \boldsymbol{\Upsilon}_N - G(\tilde{\boldsymbol{z}}_N) - \Delta_N(\boldsymbol{z}^*)]'(\bar{\boldsymbol{z}}_N - \tilde{\boldsymbol{z}}_N) \leq 0$$

and hence

$$[G(\bar{\boldsymbol{z}}_N) - G(\tilde{\boldsymbol{z}}_N)]'(\bar{\boldsymbol{z}}_N - \tilde{\boldsymbol{z}}_N) \leq -[\Delta_N(\bar{\boldsymbol{z}}_N) + \boldsymbol{\Upsilon}_N - \Delta_N(\boldsymbol{z}^*)]'(\bar{\boldsymbol{z}}_N - \tilde{\boldsymbol{z}}_N)\,.$$

Together with assumption A3 this implies

$$\alpha||\bar{\boldsymbol{v}}_N - \tilde{\boldsymbol{v}}_N||^2 \leq ||\Delta_N(\bar{\boldsymbol{z}}_N) + \boldsymbol{\Upsilon}_N - \Delta_N(\boldsymbol{z}^*)||\ ||\bar{\boldsymbol{z}}_N - \tilde{\boldsymbol{z}}_N||\,. \qquad (6.6.28)$$

Notice that it follows from condition (6.6.13) of assumption A4 and from (6.6.15) that $||\tilde{\boldsymbol{z}}_N - \boldsymbol{z}^*||$ is of order $O_p(N^{-1/2})$ and hence, because of (6.6.25),

$$||\bar{\boldsymbol{z}}_N - \tilde{\boldsymbol{z}}_N|| = O_p(N^{-1/2})\,. \qquad (6.6.29)$$

It follows from (6.6.28), (6.6.29), assumption B2 and $\boldsymbol{\Upsilon}_N = o_p(N^{-1/2})$ that

$$\alpha||\bar{\boldsymbol{v}}_N - \tilde{\boldsymbol{v}}_N||^2 \leq [o_p(N^{-1/2}) + o_p(||\bar{\boldsymbol{z}}_N - \boldsymbol{z}^*||)]O_p(N^{-1/2})$$

which together with (6.6.25) imply (6.6.24). □

Theorem 6.6.1 shows that the components $\bar{\boldsymbol{v}}_N$ and $\tilde{\boldsymbol{v}}_N$ of the respective solutions of the generalized equations (6.6.6) and (6.6.23) are asymptotically equivalent. It can be shown (see exercise 6.15) that the component $\tilde{\boldsymbol{v}}_N$ is determined uniquely for all solutions $\tilde{\boldsymbol{z}}_N$ sufficiently close to $\boldsymbol{z}^*$. This is not so for the component $\tilde{\boldsymbol{\lambda}}_N$ of $\tilde{\boldsymbol{z}}_N$. The specified assumptions do not imply uniqueness of $\tilde{\boldsymbol{\lambda}}_N$ for $\tilde{\boldsymbol{z}}_N \neq \boldsymbol{z}^*$. This is because the LMU-condition (unlike the linear independence or MF-conditions) is not preserved under small perturbations. Uniqueness of $\tilde{\boldsymbol{\lambda}}_N$ will follow if we assume that the linear independence condition holds. It is left as an exercise to show that, under this additional assumption of linear independence, $\bar{\boldsymbol{\lambda}}_N$ and $\tilde{\boldsymbol{\lambda}}_N$ are also asymptotically equivalent, i.e. $\bar{\boldsymbol{\lambda}}_N - \tilde{\boldsymbol{\lambda}}_N$ is of stochastic order $o_p(N^{-12})$.

The generalized equations (6.6.23) depend on the single random vector $\Delta_N(\boldsymbol{z}^*)$. We have that if $N^{1/2}\Delta_N(\boldsymbol{z}^*) \xrightarrow{\mathcal{D}} \boldsymbol{Y}$ and the component $\tilde{\boldsymbol{v}}(\boldsymbol{\delta})$ of the solution $\tilde{\boldsymbol{z}}(\boldsymbol{\delta})$ of the generalized equations (6.6.12) is Hadamard directionally differentiable at $\boldsymbol{\delta} = \boldsymbol{0}$, with the corresponding directional derivative $\tilde{\boldsymbol{v}}_0'(\cdot)$, then by the (finite-dimensional) delta theorem

$$N^{1/2}(\tilde{\boldsymbol{v}}_N - \boldsymbol{v}^*) \xrightarrow{\mathcal{D}} \tilde{\boldsymbol{v}}_0{}'(\boldsymbol{Y})\,.$$

Together with (6.6.24) this implies that

$$N^{1/2}(\bar{\boldsymbol{v}}_N - \boldsymbol{v}^*) \xrightarrow{\mathcal{D}} \tilde{\boldsymbol{v}}_0{}'(\boldsymbol{Y})\,.$$

In order to calculate the required directional derivatives of $\tilde{\boldsymbol{v}}(\boldsymbol{\delta})$ we can employ results from section 6.5. Recall that if $G(\boldsymbol{z})$ and $T(\boldsymbol{z})$ are given as in (6.6.4) and (6.6.5), respectively, then $\tilde{\boldsymbol{v}}(\boldsymbol{\delta})$ can be represented as an optimal solution of the parametric program ($\mathrm{P}^*_{\boldsymbol{\delta}}$). Suppose that the functions $\ell_j(\boldsymbol{v})$, $j = 0, \ldots, M$, are twice continuously differentiable, the program (P_0) has a unique optimal solution $\boldsymbol{v}^*$ and a unique vector $\boldsymbol{\lambda}^*$ of Lagrange multipliers, $\boldsymbol{v}^*$ is an interior point of V and the strong second-order sufficient conditions hold. Then, by theorem 6.5.1, we have that $\tilde{\boldsymbol{v}}(\boldsymbol{\delta})$ is Hadamard (Fréchet) directionally differentiable at $\boldsymbol{\delta} = \boldsymbol{0}$ and the corresponding directional derivative $\tilde{\boldsymbol{v}}_0{}'(\boldsymbol{d})$ in the direction $\boldsymbol{d} = (\boldsymbol{d}_1, \boldsymbol{d}_2)$ is given by the optimal solution $\bar{\boldsymbol{u}}(\boldsymbol{d})$ of the quadratic programming problem

$$\begin{array}{lll} \text{minimize} & \boldsymbol{u}'\boldsymbol{d}_1 + \frac{1}{2}\boldsymbol{u}'\nabla^2 L(\boldsymbol{v}^*, \boldsymbol{\lambda}^*)\boldsymbol{u} & \\ \text{subject to} & \boldsymbol{u}'\nabla\ell_j(\boldsymbol{v}^*) - d_{2j} \leq 0, & j \in J_0(\boldsymbol{v}^*, \boldsymbol{\lambda}^*), \\ & \boldsymbol{u}'\nabla\ell_j(\boldsymbol{v}^*) - d_{2j} = 0, & j \in J_+(\boldsymbol{v}^*, \boldsymbol{\lambda}^*) \cup \{k+1, \ldots, M\}. \end{array} \tag{6.6.30}$$

We obtain the following result (cf. Shapiro (1989, 1990b)). For the sake of notational convenience only the inequality constraints which are *active* at the point $\boldsymbol{v}^*$ will be considered. That is, we suppose that $J(\boldsymbol{v}^*) = \{1, \ldots, k\}$.

Theorem 6.6.2 *Suppose that :*

(i) *The functions* $\ell_j(\boldsymbol{v})$, $j = 0, \ldots, M$, *are twice continuously differentiable.*

(ii) *The true program* ($\mathrm{P_0}$) *has a unique optimal solution* $\boldsymbol{v}^*$ *and a unique vector* $\boldsymbol{\lambda}^*$ *of the Lagrange multipliers with* $\boldsymbol{v}^*$ *being an interior point of* V.

(iii) *The Hessian matrix* $\nabla^2 L(\boldsymbol{v}^*, \boldsymbol{\lambda}^*)$ *is positive definite.*

(iv) *The random functions* $\psi_{jN}(\boldsymbol{v})$, $j = 0, \ldots, M$, *are Lipschitz continuous in a neighborhood of* $\boldsymbol{v}^*$ *and differentiable at* $\boldsymbol{v}^*$ *w.p.1.*

(v) *Assumptions* B1′ *and* B2 ′ *hold.*

(vi) *Random vectors* $N^{1/2}(\nabla \mathcal{L}_N(\boldsymbol{v}^*, \boldsymbol{\lambda}^*), \psi_N(\boldsymbol{v}^*))$ *converge in distribution as* $N \to \infty$ *to a random vector* $\boldsymbol{Y} = (\boldsymbol{Y}_1, \boldsymbol{Y}_2)$.

Let $\bar{\boldsymbol{v}}_N$ *be an optimal solution of* ($\bar{\mathrm{P}}_\mathrm{N}$) *converging in probability as* $N \to \infty$ *to* $\boldsymbol{v}^*$. *Then*

$$N^{1/2}(\bar{\boldsymbol{v}}_N - \boldsymbol{v}^*) \ \xrightarrow{\mathcal{D}}\ \bar{\boldsymbol{u}}(\boldsymbol{Y}) , \tag{6.6.31}$$

where $\bar{\boldsymbol{u}}(\boldsymbol{Y})$ *is the optimal solution of the quadratic programming problem*

$$\begin{array}{lll} \text{minimize} & \boldsymbol{u}'\boldsymbol{Y}_1 + \frac{1}{2}\boldsymbol{u}'\nabla^2 L(\boldsymbol{v}^*, \boldsymbol{\lambda}^*)\boldsymbol{u} & \\ \text{subject to} & \boldsymbol{u}'\nabla \ell_j(\boldsymbol{v}^*) + Y_{2j} \le 0, & j \in J_0(\boldsymbol{v}^*, \boldsymbol{\lambda}^*), \\ & \boldsymbol{u}'\nabla \ell_j(\boldsymbol{v}^*) + Y_{2j} = 0, & j \in J_+(\boldsymbol{v}^*, \boldsymbol{\lambda}^*) \cup \{k+1, \ldots, M\}. \end{array} \tag{6.6.32}$$

Note that by the first-order optimality conditions $\nabla L(\boldsymbol{v}^*, \boldsymbol{\lambda}^*) = \boldsymbol{0}$ and therefore

$$\Delta_N(\boldsymbol{z}^*) = \Gamma_N(\boldsymbol{z}^*) = (\nabla \mathcal{L}_N(\boldsymbol{v}^*, \boldsymbol{\lambda}^*), -\psi_N(\boldsymbol{v}^*)) \ \xrightarrow{\mathcal{D}} (\boldsymbol{Y}_1, -\boldsymbol{Y}_2) .$$

The strong second-order sufficient conditions for the program ($\mathrm{P_0}$) are implied by the assumption of positive definiteness of the matrix $\nabla^2 L(\boldsymbol{v}^*, \boldsymbol{\lambda}^*)$ (assumption (iii)).

In the case of the sample mean construction convergence in distribution of the random vectors $N^{1/2}(\nabla \mathcal{L}_N(\boldsymbol{v}^*, \boldsymbol{\lambda}^*), \psi_N(\boldsymbol{v}^*))$ is ensured by the Central Limit theorem under standard regularity conditions. The limiting distribution of $N^{1/2}(\nabla \mathcal{L}_N(\boldsymbol{v}^*, \boldsymbol{\lambda}^*),\ \psi_N(\boldsymbol{v}^*))$ is then multivariable normal $N(\boldsymbol{0}, \boldsymbol{\Sigma})$ with zero mean vector and covariance matrix

$$\boldsymbol{\Sigma} = \begin{bmatrix} \boldsymbol{\Sigma}_{11} & \boldsymbol{\Sigma}_{12} \\ \boldsymbol{\Sigma}_{21} & \boldsymbol{\Sigma}_{22} \end{bmatrix}$$

given by

$$\boldsymbol{\Sigma}_{11} = \mathbb{E}\left\{[\nabla h_0(\boldsymbol{Z},\boldsymbol{v}^*) + \sum_{j=1}^{M} \lambda_j^* \nabla h_j(\boldsymbol{Z},\boldsymbol{v}^*)][\nabla h_0(\boldsymbol{Z},\boldsymbol{v}^*) + \sum_{j=1}^{M} \lambda_j^* \nabla h_j(\boldsymbol{Z},\boldsymbol{v}^*)]'\right\} . \tag{6.6.33}$$

$$\boldsymbol{\Sigma}_{12} = \mathbb{E}\left\{\left[\nabla h_0(\boldsymbol{Z},\boldsymbol{v}^*) + \sum_{j=1}^{M} \lambda_j^* \nabla h_j(\boldsymbol{Z},\boldsymbol{v}^*)\right] H(\boldsymbol{Z},\boldsymbol{v}^*)'\right\} , \tag{6.6.34}$$

$$\boldsymbol{\Sigma}_{22} = \mathbb{E}\{H(\boldsymbol{Z},\boldsymbol{v}^*)H(\boldsymbol{Z},\boldsymbol{v}^*)'\} , \tag{6.6.35}$$

where $H(\boldsymbol{Z},\boldsymbol{v}) = [h_1(\boldsymbol{Z},\boldsymbol{v}),\ldots,h_M(\boldsymbol{Z},\boldsymbol{v})]'$.

The optimal solution $\bar{\boldsymbol{u}}(\boldsymbol{d})$ of the parametric program (6.6.30) is linear in $\boldsymbol{d}$ if and only if the set $J_0(\boldsymbol{v}^*,\boldsymbol{\lambda}^*)$ is empty, i.e. the strict complementarity condition is satisfied. Therefore we have that if

$$N^{1/2}(\nabla\mathcal{L}_N(\boldsymbol{v}^*,\boldsymbol{\lambda}^*),\ \psi_N(\boldsymbol{v}^*)) \xrightarrow{\mathcal{D}} N(\boldsymbol{0},\boldsymbol{\Sigma}) ,$$

then $\bar{\boldsymbol{v}}_N$ is asymptotically normal if and only if the strict complementarity condition holds. In the last case $N^{1/2}(\bar{\boldsymbol{v}}_N - \boldsymbol{v}^*, \bar{\boldsymbol{\lambda}}_N - \boldsymbol{\lambda}^*)$ converges in distribution to multivariable normal with zero mean vector and the covariance matrix $\boldsymbol{B}^{-1}\boldsymbol{\Sigma}\boldsymbol{B}^{-1}$, where

$$\boldsymbol{B} = \begin{bmatrix} \nabla^2 L(\boldsymbol{v}^*,\boldsymbol{\lambda}^*) & \nabla A(\boldsymbol{v}^*) \\ \nabla A(\boldsymbol{v}^*)' & \boldsymbol{0} \end{bmatrix} . \tag{6.6.36}$$

If some of the constraints of $(\bar{\mathrm{P}}_{\mathrm{N}})$ are deterministic, then the corresponding coordinate variables of the random vector $\boldsymbol{Y}_2$ are identically zero. In particular, if all the constraints of $(\bar{\mathrm{P}}_{\mathrm{N}})$ are deterministic, then $Y_{2j} = 0$, $j = 1,\ldots,M$, and the feasible set of the program (6.6.32) becomes the critical cone $C(\boldsymbol{v}^*)$. In this case the limiting distribution of $N^{1/2}(\bar{\boldsymbol{v}}_N - \boldsymbol{v}^*)$ is given by the distribution of the minimizer $\bar{\boldsymbol{u}}(\boldsymbol{Y}_1)$ of the quadratic function $\boldsymbol{u}'\boldsymbol{Y}_1 + \frac{1}{2}\boldsymbol{u}'\nabla^2 L(\boldsymbol{v}^*,\boldsymbol{\lambda}^*)\boldsymbol{u}$ over the critical cone $C(\boldsymbol{v}^*)$.

Let us discuss now the question of uniqueness of the optimal solutions $\bar{\boldsymbol{v}}_N$ of $(\bar{\mathrm{P}}_{\mathrm{N}})$ for sufficiently large N. It was demonstrated in the "median" example for the nondifferentiable cases that the uniqueness of $\bar{\boldsymbol{v}}_N$ is not guaranteed even in the unconstrained situations. Therefore we shall need here certain differentiability assumptions. Suppose that the random functions $\psi_{jN}(\boldsymbol{v})$ are *twice continuously differentiable* and that the Hessian matrices $\nabla^2\psi_{jN}(\boldsymbol{v})$, $j = 0,\ldots,M$, converge w.p.1 to $\nabla^2\ell_j(\boldsymbol{v})$ *uniformly* in a neighborhood of $\boldsymbol{v}^*$. We know that $\bar{\boldsymbol{v}}_N$ can be represented as the component of a solution $\bar{z}_N = (\bar{v}_N, \bar{\lambda}_N)$ of the generalized equations (6.6.6) with $\Upsilon_N = 0$. We also have that for $z_1 = (v_1, \lambda_1)$ and $z_2 = (v_2, \lambda_2)$,

$$(z_1 - z_2)'[\Gamma_N(z_1) - \Gamma_N(z_2)] = \int_0^1 (v_1 - v_2)'\nabla^2 \mathcal{L}_N(tz_1 + (1-t)z_2)(v_1 - v_2)dt \, .$$

Suppose that the matrix $\nabla^2 L(v^*, \lambda^*)$ is *positive definite* and let $\bar{z}_N = (\bar{v}_N, \bar{\lambda}_N)$ and $\widehat{z}_N = (\widehat{v}_N, \widehat{\lambda}_N)$ be two solutions of the generalized equations (6.6.6) (with $\Upsilon_N = 0$) converging w.p.1 to $z^* = (v^*, \lambda^*)$. It follows then from the above integral representation that there exists a positive constant α such that w.p.1 for all sufficiently large N,

$$(\widehat{z}_N - \bar{z}_N)'[\Gamma_N(\widehat{z}_N) - \Gamma_N(\bar{z}_N)] \geq \alpha \|\widehat{v}_N - \bar{v}_N\|^2 \, . \tag{6.6.37}$$

Moreover, we have by monotonicity of $T(z)$ that

$$(\widehat{z}_N - \bar{z}_N)'[\Gamma_N(\widehat{z}_N) - \Gamma_N(\bar{z}_N)] \leq 0 \, . \tag{6.6.38}$$

The inequalities (6.6.37) and (6.6.38) imply that $\widehat{v}_N = \bar{v}_N$. We obtain that under the above differentiability assumptions and if $\nabla^2 L(v^*, \lambda^*)$ is positive definite, then strongly consistent optimal solutions $\bar{v}_N$ of $(\bar{\mathrm{P}}_\mathrm{N})$ are unique w.p.1 for all sufficiently large N.

6.7 Exercises

1. Consider the function $h(z, v) = |z - v|$, $z, v \in \mathbb{R}$, and let Z be a random variable distributed according to a cdf $F(z)$. Show that the expected value function $\ell(v) = \mathbb{E}\{h(Z, v)\}$ is differentiable at a point v if and only if $F(z)$ is continuous at $z = v$ and in the last case

$$d\ell(v)/dv = \mathbb{E}\{\partial h(Z, v)/\partial v\} = 2F(v) - 1 \, .$$

   Compare this example with the result of theorem 6.1.1.

2. Consider the real-valued continuous functions $\ell_j(v)$, $j = 1, \ldots, k$, defined on a compact set $V \subset \mathbb{R}^n$, and let $g(u)$ be a real-valued continuous function defined on an open set $U \subset \mathbb{R}^k$ containing all vectors of the form $u = (\ell_1(v), \ldots, \ell_k(v))$, $v \in V$. Let $\{\psi_{jN}(v)\}$, $j = 1, \ldots, k$, be sequences of real-valued functions converging to $\ell_j(v)$, as $N \to \infty$, uniformly in $v \in V$.

(a) Show that the sequence of composite functions

$$\{g(\psi_{1N}(\boldsymbol{v}), \ldots, \psi_{kN}(\boldsymbol{v}))\}$$

converges to $g(\ell_1(\boldsymbol{v}), \ldots, \ell_k(\boldsymbol{v}))$, as $N \to \infty$, uniformly in $\boldsymbol{v} \in V$.

(b) As an example, apply the result of (a) to a particular case with $k = 2$ and $g(\boldsymbol{u}) = u_1/u_2$ to show that under the above assumptions and if $\ell_2(\boldsymbol{v}) \neq 0$ for all $\boldsymbol{v} \in V$, then $\psi_{1N}(\boldsymbol{v})/\psi_{2N}(\boldsymbol{v})$ converges to $\ell_1(\boldsymbol{v})/\ell_2(\boldsymbol{v})$ uniformly in $\boldsymbol{v} \in V$.

(c) Formulate an extension of theorem 6.2.3 to include situations where the objective and the constraint functions of the program (P_0) are representable as compositions of the expected value functions in the form (6.1.1).

3. Consider theorem 6.2.1. Suppose that assumption (i) is omitted, so that the set V^* of optimal solutions of the program (P_0) may contain more than one point. Also modify assumption (iv) to :
(iv′) *The program* $(\bar{\mathrm{P}}_{\mathrm{N}})$ *possesses a feasible point* $\boldsymbol{u}_N$, $\boldsymbol{u}_N \in \Phi_N$, *such that the distance* $dist(\boldsymbol{u}_N, V^*)$ *tends to zero as* $N \to \infty$.
Show then that $\mathrm{dist}(\bar{\boldsymbol{v}}_N, V^*) \to 0$ and $\bar{\varphi}_N \to \varphi^*$.

4. Show that the conclusions of theorem 6.2.1 still hold if the compactness assumption (ii) is replaced by the following so-called inf-compactness, condition :
(ii′) *There exists a number* α, $\alpha > \varphi^*$, *and a compact subset* V_0 *of* V *such that for all* N *large enough,*

$$\{\boldsymbol{v} \in \Phi_N : \psi_{0N}(\boldsymbol{v}) \leq \alpha\} \subset V_0 \,.$$

In this case it is enough to suppose in assumption (iii) that the uniform convergence occurs on the subset V_0 of V.

5. Let $\ell(\boldsymbol{v})$ be the expected value of $h(Z, \boldsymbol{v})$, $\boldsymbol{v} \in V$, and let $\bar{\ell}_N(\boldsymbol{v})$ be its sample mean estimator. Consider

$$b_\delta(\zeta, \boldsymbol{v}) = \sup_{\|\boldsymbol{u}-\boldsymbol{v}\| \leq \delta} |h(\zeta, \boldsymbol{u}) - h(\zeta, \boldsymbol{v})| \,.$$

Suppose that for every $\boldsymbol{v} \in V$, the expected value $\ell(\boldsymbol{v})$ is finite and that

$$\lim_{\delta \to 0^+} \mathbb{E}\{b_\delta(Z, \boldsymbol{v})\} = 0 \,.$$

(a) Show that $\ell(\boldsymbol{v})$ is continuous on V.

(b) If, in addition, V is compact show that with probability one $\bar{\ell}_N(\boldsymbol{v})$ converges to $\ell(\boldsymbol{v})$ uniformly on V.

(c) Formulate an extension of theorem 6.2.3 to deal with situations where $h_0(\zeta, \boldsymbol{v})$ is not necessarily continuous in $\boldsymbol{v}$.

6. Show that the Mangasarian-Fromovitz constraint qualification, applied to the program (P_0) at the point $\boldsymbol{v}^*$, is equivalent to the following condition. The system

$$\sum_{j=1}^{M} \lambda_j \nabla \ell_j(\boldsymbol{v}^*) = \mathbf{0} \,,$$

$$\lambda_j \geq 0 \ \text{ and } \lambda_j \ell_j(\boldsymbol{v}^*) = 0, \ j = 1, \ldots, k;$$

has only one solution, namely $\lambda_j = 0, \ j = 1, \ldots, M$.

7. Let $\boldsymbol{\mu} = (\mu_1, \mu_2)$, $\mu_2 \neq 0$, and let $\boldsymbol{X}_N = (X_{1N}, X_{2N})$ be a sequence of two-dimensional random vectors such that $N^{1/2}(\boldsymbol{X}_N - \boldsymbol{\mu})$ converges in distribution to normal with zero mean vector and a covariance matrix $\boldsymbol{\Sigma} = (\sigma_{ij})$. Use the delta theorem to show that $N^{1/2}(X_{1N}/X_{2N} - \mu_1/\mu_2)$ converges in distribution to normal with zero mean and variance

$$\mu_2^{-2}\sigma_{11} - 2\mu_1\mu_2^{-3}\sigma_{12} + \mu_1^2\mu_2^{-4}\sigma_{22} \,.$$

8. Let B_1 and B_2 be Banach spaces, $g : B_1 \to B_2$ and $\boldsymbol{\mu} \in B_1$. Show that:

(a) If g is Fréchet directionally differentiable at $\boldsymbol{\mu}$, then g is continuous at $\boldsymbol{\mu}$.

(b) If g is Fréchet directionally differentiable at $\boldsymbol{\mu}$ and is continuous in a neighborhood of $\boldsymbol{\mu}$, then $g'_{\boldsymbol{\mu}}(\cdot)$ is continuous on B_1.

(c) If g is Hadamard directionally differentiable at $\boldsymbol{\mu}$, then $g'_{\boldsymbol{\mu}}(\cdot)$ is continuous on B_1.

(d) If g is Fréchet directionally differentiable at $\boldsymbol{\mu}$ and $g'_{\boldsymbol{\mu}}(\cdot)$ is continuous, then g is Hadamard directionally differentiable at $\boldsymbol{\mu}$.

(e) If g is Hadamard directionally differentiable at $\boldsymbol{\mu}$ and the space B_1 is finite-dimensional, then g is Fréchet directionally differentiable at $\boldsymbol{\mu}$.

(f) If g is Gâteaux directionally differentiable at $\boldsymbol{\mu}$ and is Lipschitz continuous in a neighborhood of $\boldsymbol{\mu}$, then g is Hadamard directionally differentiable at $\boldsymbol{\mu}$. (Recall that g is said to be Lipschitz continuous in a neighborhood of $\boldsymbol{\mu}$ if there exists a constant k such that for any two points $\boldsymbol{x}$ and $\mathbf{y}$ in this neighborhood,

$$\|g(\boldsymbol{x}) - g(\boldsymbol{y})\| \leq k\|\boldsymbol{x} - \boldsymbol{y}\| \text{ .)}$$

9. Let B_1, B_2 and B_3 be Banach spaces, $f : B_1 \to B_2$, $g : B_2 \to B_3$ and $\boldsymbol{\eta} = f(\boldsymbol{\mu})$. Show that if f is Hadamard (Gâteaux) directionally differentiable at $\boldsymbol{\mu}$ and g is Hadamard directionally differentiable at $\boldsymbol{\eta}$, then the composite mapping $g \circ f : B_1 \to B_3$ is Hadamard (Gâteaux) directionally differentiable at $\boldsymbol{\mu}$ and the chain rule

$$(g \circ f)'_{\boldsymbol{\mu}} = g'_{\boldsymbol{\eta}} \circ f'_{\boldsymbol{\mu}}$$

holds. (For detailed discussions of the concepts of (directional) differentiability in normed (topological) vector spaces, see Averbukh and Smolyanov (1968), Nashed (1971) and Shapiro (1990a).)

10. Prove the following extention of theorem 6.1.1. Let $\ell(\boldsymbol{v})$ be the expected value function of $h(\boldsymbol{Z}, \boldsymbol{v})$ and let U be an open neighborhood of a point $\boldsymbol{v}_0$. Suppose that:

    (a) There exists an integrable function $k : \Omega \to \mathbb{R}$ such that for all $\boldsymbol{u}, \boldsymbol{v} \in U$ and $\zeta \in \Omega$,

    $$|h(\zeta, \boldsymbol{u}) - h(\zeta, \boldsymbol{v})| \leq k(\zeta)\|\boldsymbol{u} - \boldsymbol{v}\| \text{ .}$$

    (b) For almost every ζ, the function $h(\zeta, \cdot)$ is Gâteaux directionally differentiable at $\boldsymbol{v}_0$.

    Then $\ell(\boldsymbol{v})$ is Fréchet directionlly differentiable at $\boldsymbol{v}_0$ and

    $$\ell'_{\boldsymbol{v}_0}(\boldsymbol{d}) = \mathbb{E}\{h'_{\boldsymbol{v}_0}(\boldsymbol{Z}, \boldsymbol{d})\} \text{ ,}$$

    where $h'_{\boldsymbol{v}_0}(\zeta, \boldsymbol{d})$ denotes the directional derivative of $h(\zeta, \cdot)$ at $\boldsymbol{v}_0$ in the direction $\boldsymbol{d}$.

    Hints. Use the Lebesgue dominated convergence theorem to show that the directional derivatives can be taken inside the expected value. Employ the result of exercise 6.8 (e,f) to show that Fréchet directional differentiability follows.

11. Let S be a compact subset of $\mathbb{R}^n$. Consider the Banach space $B = C(S) \times \ldots \times C(S)$, given by the Cartesian product of k replications of the space $C(S)$, and a point $\boldsymbol{\mu} = (\mu_1, \ldots, \mu_k) \in B$. Let $g : \mathbb{R}^k \to \mathbb{R}$ and consider the corresponding mapping $G : B \to C(S)$ defined by

$$G(\xi_1, \ldots, \xi_k)(\boldsymbol{v}) = g(\xi_1(\boldsymbol{v}), \ldots, \xi_k(\boldsymbol{v})) \text{ .}$$

(a) Show that if g is continuous on an open set $U \subset \mathbb{R}^k$ containing all points of the form $\boldsymbol{u} = (\mu_1(\boldsymbol{v}), \ldots, \mu_k(\boldsymbol{v}))$, $\boldsymbol{v} \in S$, then G is continuous in a neighborhood of the point $\boldsymbol{\mu}$ (compare with exercise 6.2(a)).

(b) Show that if g is continuously differentiable on the set U, then the mapping G is Fréchet differentiable at $\boldsymbol{\mu}$ and

$$G'_{\boldsymbol{\mu}}(\boldsymbol{\delta})(\boldsymbol{v}) = (\delta_1(\boldsymbol{v}), \ldots, \delta_k(\boldsymbol{v}))' \nabla g(\mu_1(\boldsymbol{v}), \ldots, \mu_k(\boldsymbol{v})) ,$$

for any $\boldsymbol{\delta} = (\delta_1, \ldots, \delta_k) \in B$.

(c) Let $\boldsymbol{X}_N = (\psi_{1N}, \ldots, \psi_{kN})$ be a sequence of random elements of B such that $N^{1/2}(\boldsymbol{X}_N - \boldsymbol{\mu})$ converges in distribution to a random element $\boldsymbol{Y}$. Use the delta method to calculate the asymptotic distribution of $N^{1/2}[G(\boldsymbol{X}_N) - G(\boldsymbol{\mu})]$.

(d) Let $\ell_j(\boldsymbol{v})$ be the expected value functions, corresponding to $h_j(Z, \boldsymbol{v})$, $j = 1, \ldots, k$, and let $\bar{\ell}_{jN}(\boldsymbol{v})$ be their sample mean estimators. Extend the asymptotic result of theorem 6.4.3 to deal with situations where the programs ($\mathrm{P_0}$) and ($\bar{\mathrm{P}}_\mathrm{N}$) involve minimization of the functions $g(\ell_1(\boldsymbol{v}), \ldots, \ell_k(\boldsymbol{v}))$ and $g(\bar{\ell}_{1N}(\boldsymbol{v}), \ldots, \bar{\ell}_{kN}(\boldsymbol{v}))$, respectively, over S. Show that if $\boldsymbol{v}^*$ is the unique optimal solution of the true program and certain regularity conditions hold, then $N^{1/2}(\bar{\varphi}_N - \varphi^*)$ is asymptotically normal and the corresponding the asymptotic variance is given by

$$\sigma^2 = [\nabla g(\boldsymbol{u}^*)]' \, \boldsymbol{\Psi} \, [\nabla g(\boldsymbol{u}^*)] ,$$

where $\boldsymbol{u}^* = (\ell_1(\boldsymbol{v}^*), \ldots, \ell_k(\boldsymbol{v}^*))$ and $\boldsymbol{\Psi}$ is the $k \times k$ covariance matrix with typical elements

$$[\boldsymbol{\Psi}]_{ij} = \mathbb{E}\{h_i(Z, \boldsymbol{v}^*) h_j(Z, \boldsymbol{v}^*)\} - \ell_i(\boldsymbol{v}^*)\ell_j(\boldsymbol{v}^*) \, , \; i, j = 1, \ldots, k \, .$$

(e) Given a random sample $Z_1, \ldots, Z_N$ from $(\Omega, \mathcal{F}, P)$, derive an estimator $\hat{\sigma}^2$ of the asymptotic variance σ^2 described in (d). Specify regularity conditions under which $\hat{\sigma}^2$ is a consistent estimator of σ^2.

12. Consider a compact set $V \subset \mathbb{R}^n$ and the corresponding Banach space B given by the Cartesian product of $M + 1$ replications of the space $C(V)$. Consider the optimal value function $\phi : B \to \mathbb{R} \cup \{+\infty\}$ defined in (6.4.26).

    (a) Show that the function ϕ is lower semi-continuous.

(b) Show that ϕ is also lower semi-continuous if the space B is given by the Cartesian product of the spaces $C^1(V)$.

13. Consider a symmetric $(n+M) \times (n+M)$ matrix $\mathbf{H}$ of the form

$$\mathbf{H} = \begin{bmatrix} \mathbf{B} & J \\ J' & \mathbf{0} \end{bmatrix}.$$

Assume that $\boldsymbol{u}'\mathbf{B}\boldsymbol{u} \geq 0$ for any vector $\boldsymbol{u}$ such that $J'\boldsymbol{u} = \mathbf{0}$. (If $\mathbf{H}$ is the Hessian matrix given in (6.5.6) with $\mathbf{B} = \nabla^2_{\boldsymbol{vv}} L(\boldsymbol{v}^*, \boldsymbol{\lambda}^*, \boldsymbol{\xi}_0)$ and $J = \nabla_{\boldsymbol{v}} \mathbf{G}(\boldsymbol{v}^*, \boldsymbol{\xi}_0)$, then this last condition corresponds to the second-order necessary conditions for the program (P_0) at the point $\boldsymbol{v}^*$.) Show that $\mathbf{H}$ is nonsingular if and only if any one of the following two conditions holds.

(a) The matrix J is of full column rank M and there exists a positive number α such that the matrix $\mathbf{B} + \alpha J J'$ is positive definite.

(b) The matrix $\mathbf{J}$ is of full column rank M and if $\boldsymbol{u}$ is a nonzero vector such that $J'\boldsymbol{u} = \mathbf{0}$, then $\boldsymbol{u}'\mathbf{B}\boldsymbol{u} > 0$.

14. Consider the optimal value function $\phi(\boldsymbol{\xi})$ corresponding to the program $(\mathrm{P}_{\boldsymbol{\xi}})$ of section 6.5.

(a) Show that if assumptions (i)–(iv) of theorem 6.5.1 are satisfied and the second-order sufficient conditions hold, then

$$\phi(\boldsymbol{\xi}) = L(\boldsymbol{v}^*, \boldsymbol{\lambda}^*, \boldsymbol{\xi}) + \kappa(\boldsymbol{\xi}) + o(\|\boldsymbol{\xi} - \boldsymbol{\xi}_0\|^2),$$

where $\kappa(\boldsymbol{\xi})$ is the optimal value of the program $(\mathrm{Q}_{\boldsymbol{\xi}})$.

(b) Show that if, in addition, $g_j(\boldsymbol{v}^*, \cdot)$, $j = 0, \ldots, M$, have the second-order Taylor expansion at $\boldsymbol{\xi}_0$, then the second-order directional derivative

$$\lim_{t \to 0+} \frac{\phi(\boldsymbol{\xi}_0 + t\boldsymbol{\delta}) - \phi(\boldsymbol{\xi}_0) - t\nabla_{\boldsymbol{\xi}} L(\boldsymbol{v}^*, \boldsymbol{\lambda}^*, \boldsymbol{\xi}_0)\boldsymbol{\delta}}{t^2}$$

exists and is equal to the optimal value of the program $(\mathrm{Q}'_{\boldsymbol{\delta}})$ plus the term $\frac{1}{2}\boldsymbol{\delta}'\nabla^2_{\boldsymbol{\xi\xi}} L(\boldsymbol{v}^*, \boldsymbol{\lambda}^*, \boldsymbol{\xi}_0)\boldsymbol{\delta}$.

15. Suppose that assumptions A1 and A2 of section 6.6 hold. Show that if $\tilde{\boldsymbol{z}}_1(\boldsymbol{\delta}) = (\tilde{\boldsymbol{v}}_1(\boldsymbol{\delta}), \tilde{\boldsymbol{\lambda}}_1(\boldsymbol{\delta}))$ and $\tilde{\boldsymbol{z}}_2(\boldsymbol{\delta}) = (\tilde{\boldsymbol{v}}_2(\boldsymbol{\delta}), \tilde{\boldsymbol{\lambda}}_2(\boldsymbol{\delta}))$ are solutions of the generalized equations (6.6.12) and $\tilde{\boldsymbol{z}}_1(\boldsymbol{\delta}), \tilde{\boldsymbol{z}}_2(\boldsymbol{\delta}) \in W$, then $\tilde{\boldsymbol{v}}_1(\boldsymbol{\delta}) = \tilde{\boldsymbol{v}}_2(\boldsymbol{\delta})$.

16. Derive the asymptotic result of theorem 6.6.2 by using the delta method (theorem 6.3.1) and sensitivity analysis of section 6.5. What regularity conditions will then be required?

Bibliography

[1] Abate, J. and Whitt, W. (1987a). "Transient behavior of regulated Brownian motion, I: starting at the origin", *Adv. Appl. Prob.*, 19, 560–598.

[2] Abate, J. and Whitt, W. (1987b). "Transient behavior of regulated Brownian motion, II: non-zero initial conditions", *Adv. Appl. Prob.*, 19, 599–631.

[3] Abate, J. and Whitt, W. (1987c). "Transient behavior of the M/M/1 queue: starting at the origin", *Queueing Systems*, 2, 41–65.

[4] Abate, J. and Whitt, W. (1988). "The correlation functions of RBM and M/M/1", *Stochastic Models*, 4, 315–359.

[5] Aleksandrov, V.M., Sysoyev, V.I. and Shemeneva, V.V. (1968). "Stochastic Optimization", *Engineering Cybernetics*, 5, 11–16.

[6] Anantharam, V., Heidelberger, P. and Tsoucas, P. (1990). "Analysis of rare events in continuous time Markov chains via time reversal and fluid approximation", Research Report, RC 16280 (#71858) 10/11/90, IBM Research Division, T.J. Watson Research center, Yorktown Heights, NY 10598.

[7] Araujo, A. and Giné, E. (1980). *The Central Limit Theorem for Real and Banach Valued Random Variables*, Wiley, New York.

[8] Arsham, H. (1988). "Sensitivity and optimization of computer simulation models", Manuscript, University of Baltimore.

[9] Arsham, H., Feuerverger, A., McLeish, D.L., Kreimer, J. and Rubinstein, R.Y. (1989). "Sensitivity analysis and the 'What if' problem in simulation analysis", *Mathematical and Computational Modeling*, 12,193–219.

[10] Arsham, H. (1991). "Perturbation analysis in discrete-event simulation", *International Journal of Modelling and Simulation*, 11 (1), 21–28.

[11] Asmussen, S. (1982). "Conditioned limit theorems relating a random walk to its associate, with applications to risk reserve processes and the $GI/G/1$ queue", *Adv. Appl. Probab.*,14, 143-170.

[12] Asmussen, S. (1985). "Conjugate processes and the simulation of ruin problems", *Stoch. Proc. Appl.*, 20, 213–229.

[13] Asmussen, S. (1987). *Applied Probability and Queues*, Wiley, New York.

[14] Asmussen, S. (1989). "Validating the heavy traffic performance of regenerative simulation", *Stochastic Models*, 5 (4), 617–628.

[15] Asmussen, S. (1990). "Exponential families and regression in the Monte Carlo study of queues and random walks", *The Annals of Statistics*, 18 (4), 1851–1867.

[16] Asmussen, S. (1992). "Queueing simulation in heavy traffic", *Math. Opns. Res.*, 17, 84-111.

[17] Asmussen, S. and Melamed, B. (1992). "Regeneration and likelihood ratios in TES processes", Manuscript, Institute of Electronic Systems, Aalborg University, Denmark.

[18] Asmussen, S. and Rubinstein, R.Y. (1992a). "The efficiency and heavy traffic properties of the score function method in sensitivity analysis of queueing models", *Advances of Applied Probability*, 24(1), 172-201.

[19] Asmussen, S. and Rubinstein, R.Y. (1992b). "Performance evaluation and sensitivity analysis via the score function method: from M/M/1 queues to bottleneck networks", Manuscript, Institute of Electronic Systems, Aalborg University, Denmark.

[20] Asmussen, S., Rubinstein, R.Y. and Ch. Wang (1992). "Estimating rare events via likelihood ratios: from M/M/1 queues to bottleneck networks" Manuscript, Institute of Electronic Systems, Aalborg University, Denmark.

[21] Auslender, A. and Cominetti, R. (1990). "First and second-order sensitivity analysis of nonlinear programs under directional constraint qualification conditions", *Optimization*, 21, 351–363.

[22] Averbukh, V.I. and Smolyanov, O.G. (1968). "The various definitions of the derivative in linear topological spaces", *Rusian Mathematical Surveys*, 23, 67-113.

[23] Avriel, M. (1976). *Nonlinear Programming Analysis and Methods*, Prentice-Hall, Englewood Cliffs, N.J.

[24] Barlow, R. and Proschan, F. (1975). *Statistical Theory of Reliability and Life Testing Probability Models*, Holt, Reinhart and Winston, New York.

[25] Beckman, R.J. and McKay, M.D. (1987). "Monte Carlo estimation under different distributions using the same simulation", *Technometrics*, 29 (2),153–160.

[26] Ben-Tal, A. (1980). "Second order and related extremality conditions in nonlinear programnning", *J. Optim. Theory Appl.*, 31, 143–165.

[27] Bertsekas , A. and Gallager, R. (1987). *Data Networks*, Prentice-Hall, Englewood Cliffs, N.J.

[28] Biles, E.B. and Ozmen, H.T. (1987). "Optimization of simulation responses in a multicomputing environment", Proceedings of the 1987 Winter Simulation Conference, Atlanta, Georgia, U.S.A.

[29] Billingsley, P. (1968). *Convergence of Probability Measures*, Wiley, New York.

[30] Billingsley, P. (1979). *Probability and Measure*, Wiley, New York.

[31] Bonnans, J.F. (1992). "Directional derivatives of optimal solutions in smooth nonlinear programming", *Journal of Optimization Theory and Application,* to appear.

[32] Bonnans, J.F., Ioffe, A.D. and Shapiro, A. (1992). "Expansion of exact and approximate solutions in nonlinear programming", in: D. Pallaschke (ed.), *Proceedings of Franco-German Conference on Optimization, Lecture Notes in Economics and Mathematical Systems*, Springer-Verlag, Berlin, pp. 103-117.

[33] Borovkov, A.A. (1986). "Limit theorems for queueing networks I", *Theor. Probab. Appl.*, 31 (3), 413–427.

[34] Brémaud, P. (1990). "On computing derivatives with respect to the rate of a Poisson process trajectorywise: the phantom method", to appear *QUESTA*.

[35] Brémaud, P. (1991). "Maximal coupling and rare perturbation analysis", Manuscript, Laboratoire des Signaux et Systèmes, CNRS-ESE, Plateau de Moulon, 91190 Gif-sur-Yvette, France.

[36] Brémaud, P. and Vázquez-Abad, F.J. (1991). "On the pathwise computation of derivatives with respect to the rate of a point process: the phantom RPA method", Manuscript, Laboratoire des Signaux et Systèmes, CNRS-ESE, Plateau de Moulon, 91190 Gif-sur-Yvette, France.

[37] Bucklew, J.A., Ney, P. and Sadowsky, J.S. (1990). "Monte Carlo simulation and large deviations theory for uniformly recurrent Markov chains", *J. Appl. Prob.*, 27, 44–59.

[38] Castaing, C. and Valadier, M. (1977). *Convex Analysis and Measurable Multifunctions*, Lecture Notes in Mathematics, 580, Springer-Verlag, Berlin. 26: 71–75.

[39] Cheng, R.C.H. and Feast, G.M. (1980). "Control variables with known mean and variance", *J. Operational Res. Soc.*, 31, 51–56.

[40] Clarke, F.H. (1983). *Optimization and Nonsmooth Analysis*, Wiley, New York.

[41] Cochran, W.G. (1977). *Sampling Techniques* (third edition), Wiley, New York.

[42] Conway, A.E. and Goyal, A., "Monte Carlo simulation of computer system availability/reliability models", Manuscript, McGill University, Canada and IBM Research Center, NY.

[43] Cottrell, M., Fort, J.C. and Malgouyres, C. (1983). "Large deviations and rare events in the study of stochastic algorithms", *IEEE Transactions on Automatic Control*, AC-28, 907–920.

[44] Crane, M.A. and Iglehart, D.L. (1974a). "Simulating stable stochastic systems, I: General multiserver queues", *J. Assoc. Comput. Mach.*, 21, 103–113.

[45] Crane, M.A. and Iglehart, D.L. (1974b). "Simulating stable stochastic systems, II: Markov chains", *J. Assoc. Comput. Mach.*, 21, 114–123.

[46] Daduna, H. (1988). Note on the ergodicity of closed queueing networks, preprint Nr. 88-3, Hamburg University.

[47] Danskin, J.M. (1967). *The Theory of Max-min and its Applications to Weapons Allocation Problems*, Springer-Verlag, New York.

[48] Devroye, L. (1986). *Non-Uniform Random Variate Generation*, Springer-Verlag, Berlin.

[49] Dunford, N. and Schwartz, J.T. (1958). *Linear Operators*, Vol. I, Wiley, New York.

[50] Dupačová, J. and Wets, R. (1988). "Asymptotic behavior of statistical estimators and of optimal solutions of stochastic optimization problems", *The Annals of Statistics*, 16, 1517–1549.

[51] Elishakoff, I. (1983a). *Probability Methods in the Theory of Structures*, John Wiley , New York.

[52] Elishakoff, I. (1983b). "How to introduce the imperfection sensitivity concept in design", in: J.M.T. Thompson and G.B. Hunt (eds.) *Collapse: The Buckling of Structures in Theory and Practice*, Cambridge Univ. Press, Cambridge, pp. 345–357.

[53] Ermakov, J.M. (1976). *Monte Carlo Method and Related Questions*, Nauka, Moscow (in Russian).

[54] Ermoliev, M. (1969). "On the method of generalized stochastic gradients and quasi-Fejer sequences", *Cybernetics*, 5 (2), 208–220.

[55] Ermoliev, Y.M. and Wets, R.J.-B. (1988). *Numerical Techniques for Stochastic Optimization*, Springer-Verlag, Berlin.

[56] Ermoliev, Y.M. and Gaivoronski, A.A. (1992). "Stochastic programming techniques for optimization of discrete event systems", *Annals of Operations Research*, to appear.

[57] Fernholz, L.T. (1983). *Von Mises Calculus for Statistical Functionals*, Lecture Notes in Statistics, 19, Springer-Verlag, New York.

[58] Feuerverger, A., McLeish, D.L. and Rubinstein, R.Y. (1986). "A cross-spectral method for sensitivity analysis of computer simulation models", *Comptes Rendus: Mathematical Reports of the Academy of Sciences, Royal Society of Canada*, VIII, (5), 335–339.

[59] Fiacco, A.V. (1983). *Introduction to Sensitivity and Stability Analysis in Nonlinear Programming*, Academic Press, New York.

[60] Fishman, G.S. (1978). *Principles of Discrete Event Simulation*, Wiley, New York.

[61] Fox, B.L. (1987). *Computing the Gradient of Expected Reward up to Absorption*, Technical Report No. 755, Cornell University, New York.

[62] Fox, B.L. and Glasserman, P. (1991). Estimating derivatives via Poisson's equation, *Probability in the Engineering and Informational Sciences*, 5, 415-428.

[63] Frater, M.R., Lennon, T.M. and Anderson, B.D.O. (1989), "Optimally efficient estimation of the statistics of rare events in queueing networks", Manuscript, Australian National University.

[64] Gadrich, T (1989). "Application of Radon-Nikodym derivatives to sensitivity analysis of stochastic models", M.Sc. Thesis, Technion—Israel Institute of Technology, Haifa, Israel (in Hebrew).

[65] Gaivoronski, A.A. (1982). "Approximation methods of solution of stochastic programming problems", *Kibernetika*, 2 (in Russian); English translation in: *Cybernetics*, 18 (2), 241–249).

[66] Gaivoronski, A.A. (1990). "Optimization of stochastic discrete event dynamic systems. A survey of some recent results", in: *Proceedings of the Conference on Computationally Intensive Methods in Simulation and Optimization*, Vienna, 1990, Springer-Verlag, Berlin.

[67] Gaivoronski, A.A. (1991). "Perturbation analysis and optimization of general discrete event dynamic systems using augmented IPA estimates", Part I, Manuscript, V. Glushkov Institute of Cybernetics, Kiev, U.S.S.R.

[68] Gauvin, J. and Dubeau, F. (1982). "Differential properties of the marginal function in mathematical programming", *Math. Programming Study*, 19, 101–119.

[69] Gauvin, J. and Janin, R. (1988). "Directional behavior of optimal solutions in nonlinear mathematical programming", *Mathematics of Operations Research*, 13, 629–649.

[70] Gilat, D. (1992). "The efficiency of the score function method for sensitivity analysis of dicrete event systems with autocorrelated inputs", M.Sc. Thesis, Technion—Israel Institute of Technology, Haifa, Israel (in Hebrew).

[71] Gill, R.D. (1989). "Non-and semiparametric maximum likelihood estimators and the von Mises method (Part I)", *Scandinavian Journal of Statistcs*, 16, 97–124.

[72] Glasserman, P. (1988a). "Infinitesimal perturbation analysis of a birth and death process", *Operations Research Letters*, 7, 43–49.

[73] Glasserman, P. (1988b). "Performance continuity and differentiability in Monte Carlo optimization", *Proceedings of the Winter Simulation Conference,* pp. 518–524.

[74] Glasserman, P. (1990a). Stochastic monotonicity, total positivity, and conditional Monte Carlo for likelihood ratios, At&T Bell Laboratories, Holmdel, NJ.

[75] Glasserman, P. (1990b). "The limiting value of derivative estimators based on perturbation analysis", *Stochastic Models*, 6, 229–258.

[76] Glasserman, P. (1990c). "Discrete time 'inversion' and derivative estimation for Markov chains", *Operations Research Letters*, 9, 305–313.

[77] Glasserman, P. (1991a). "Structural conditions for perturbation analysis derivative estimation: finite time performance indices, *Operations Research*, 39, (5), 724-738.

[78] Glasserman, P. (1991b). "Structural conditions for perturbation analysis of queueing systems", *Journal of the ACM*, 38 (4), 1005-1025.

[79] Glasserman, P. (1991c). *Gradient Estimation via Perturbation Analysis,* Kluwer , Norwell, Mass.

[80] Glasserman, P. (1992a). "Derivative estimates from simulation of continuous time Markov chains", *Operations Research*, to appear.

[81] Glasserman, P. (1992b). "Smoothing complements and randomized score functions", *Annals of O.R.*, special issue on sensitivity analysis and optimization of discrete-event systems, to appear.

[82] Glasserman, P. and Gong, W.B. (1989). "Derivative estimates from discontinuous realizations: smoothing techniques", *Proceedings of the Winter Simulation Conference*, pp. 381–389.

[83] Glasserman, P. and Gong, W.B. (1990). "Smoothed perturbation analysis for a class of discrete event systems", *IEEE Transactions on Automatic Control*, AC-35 (11),1218–1230.

[84] Glasserman, P. and Ho, Y.C. (1989). "Aggregation approximations for sensitivity analysis of multi-class queueing networks", *Performance Evaluation*, 10, 295–308.

[85] Glasserman, P., Hu, J.Q. and Strickland, S.G. (1991). "Strongly consistent steady-state derivative estimates", *Probability in the Engineering and Informational Sciences,*, 5, 391-413.

[86] Glynn, P.W. (1990). "Likelihood ratio gradient estimation for stochastic systems", *Communications of the ACM*, , 33 (10), 75–84.

[87] Glynn, P.W. (1992). "Importance sampling for Markov chains: asymptotics for the variance", Manuscript, Stanford University.

[88] Glynn, P.W. and Iglehart, D.L. (1988). "Simulation method for queues: an overview", *Queueing Systems*, 3, 221–256.

[89] Glynn, P.W. and Iglehart, D.L. (1989). "Importance sampling for stochastic simulations", *Management Sci.*, 35 (11), 1367–1392.

[90] Glynn, P.W. and Iglehart, D.L. (1990). "The theory of standardized time series", *Math. Operations Res.*, 15, 1–16.

[91] Glynn, P.W., L'Ecuyer, P. and Adés, M. (1991). "Gradient estimation for ratios", *Proceedings of the 1991 Winter Simulation Conference*, IEEE Press, pp. 986-993

[92] Glynn, P.W. and Whitt, W. (1987). "Sufficient conditions for functional-limit-theorem versions of $L = \lambda W$", *Queueing Systems*, 1, 279–287.

[93] Glynn, P.W. and Whitt, W. (1989). "Indirect estimation via $L = \lambda w$, *Operations Research*, 37, 82–103.

[94] Glynn, P.W. and Whitt, W. (1991). "Estimating the asymptotic variance with batch means", *Operations Research Letters*, 10, 431–435.

[95] Golshtein, E.G. (1972). *Theory of Convex Programming.* Trans. Math. Monographs 26, American Mathematical Society, Providence, RI.

[96] Gong, W.B. and Ho, Y.C. (1987). "Smoothed (conditional) perturbation analysis of discrete event dynamic systems", *IEEE Transactions on Automatic Control*, AC-32 (10) 858–866.

[97] Goyal, A., Shahabuddin, P., Heidelberger, P., Nicola, V.F. and Glynn, P.W. (1992). "A unified framework for simulating Markovian models of highly dependable systems", *IEEE Trans. on Computers*, Vol. 41,, 1, 36-51.

[98] Graves, L.M. (1950). "Some mapping theorems", *Duke Math. J.*, 17, 111–114.

[99] Gross, D. and Harris, C. (1985). *Fundamentals of Queueing Theory*, John Wiley , New York.

[100] Grübel, R. (1988). "The length of the short", *Ann. Statist.*, 16, 619–628.

[101] Hammersley, J.M. and Handscomb, D.C. (1964). *Monte Carlo Methods*, John Wiley , New York.

[102] Heidelberger, P., Cao, X.R., Suri, R. and Zazanis, M.A. (1988). "Convergence properties of infinitesimal perturbution analysis estimates", *Management Science*, 34,(11), 1281–1302.

[103] Heidelberger, P. and Towsley, D. (1989). "Sensitivity analysis from sample paths using likelihoods", *Management Science*, 35 (12), 1475–1488.

[104] Heidelberger, P. and Welch. P.D. (1981b). "A spectral method for confidence interval generation and run length control in simulations", *Commun. Assoc. Comput. Mach.*, 24, 233–245.

[105] Ho, Y.C. and Cao, X.R. (1991). *Discrete Event Dynamic Systems and Perturbation Analysis*, Kluwer, Norwell, Mass.

[106] Ho, Y.C., Eyler, M.A. and Chien, T.T. (1979). "A gradient technique for general buffer strorage design in a serial production line", *International Journal on Production Research*, 17 (6), 557–580.

[107] Ho, Y.C. and Strickland, S. (1990). "A taxonomy of perturbation analysis techniques", Manuscript, Harvard University.

[108] Huber, P.J. (1964). "Robust estimation of a location parameter", *Ann. Math. Statist.*, 35, 73–101.

[109] Huber, P.J. (1967). "The behavior of maximum likelihood estimates under nonconstrained conditions", *Proceedings of the 5th Berkeley Symposium on Math. Stat. and Probability*, Vol. 1, pp. 221–233.

[110] Huber, P.J. (1981). *Robust Statistics*, Wiley , New York.

[111] Hunkel, V. and Bucklew, J.A. (1988). "Fast simulation for functionals of Markov chains", *Proc. 22nd Ann. Conf. Inform. Sci. & Sys.*, Princeton University, pp. 330–335.

[112] Iglehart, D.L. (1975). "Simulating stochastic systems, V: comparison of ratio estimators", *Nav. Res. Logist. Quart.*, 22, 553–565.

[113] Iglehart, D.L. (1976). "Simulating stable stochastic systems, VI: quantile estimation", *J. Assoc. Comp. Mach.*, 23, 347–360.

[114] Ioffe, A.D. (1979). "Necessary and sufficient conditions for a local minimum; second order conditions and augmented duality", *SIAM Journal, Control and Optimization*,17, 266–288.

[115] Ioffe, A.D. and Tihomirov, V.M. (1979). *Theory of Extremal Problems*, North-Holland , Amsterdam.

[116] Ioffe, M.O. and Katknovnik, V.Y. (1990). "Pointwise and uniform convergence rate with probability 1 of nonparametrical estimators of regression and its derivatives", *Automatic Remote Control*, 51 (1), 23–30.

[117] Iscoe, I., Ney, P. and Nummelin, E. (1985). "Large deviations of uniformly recurrent Markov additive processes", *Adv. Appl. Math.*, 6, 373–411.

[118] Izxaki, J. (1992). "Stochastic optimization of open queueing networks by the score function method". Master's Thesis, Technion, Haifa, Israel.

[119] Jacobson, S.H. and Schruben, L.W. (1989). "Techniques for simulation response optimization", *Operations Research Letters*, 8, 1–9.

[120] Jagerman, D.L. and Melamed, B. (1992a). "The transformation and autocorrelation structure of TES processes. Part 1: general theory", NEC Research Institute, Princeton, New Jersey (to appear in *Stochastic Models*.

[121] Jagerman, D.L. and Melamed, B. (1992b). "The transformation and autocorrelation structure of TES processes. Part 2: special cases", NEC Research Institute, Princeton, New Jersey (to appear in *Stochastic Models*.

[122] Jennrich, R.I. (1969). "Asymptotic properties of nonlinear least squares estimators", *Ann. Math. Statist.*, 40, 633–643.

[123] Kahn, H. and Marshall, A.W. (1953). "Methods of reducing sample size in Monte Carlo computations", *Journal of Operations Research Society of America*, 1 (5), 263–278.

[124] Kahn, H. (1956). "Use of different Monte Carlo sampling techniques",in: *Symposium on Monte Carlo Methods*, H.Meyer (ed.) Weleg, New York, pp. 146–190.

[125] Kalos, M.H. and Whitlock, P.A. (1986). *Monte Carlo Methods*, John Wiley , New York.

[126] Karlin, S. and Taylor, H.J. (1975). *A First Course in Stochastic Processes*, Academic Press, New York.

[127] Kaspi, H. and Mandelbaum, A. (1991). "Regenerative closed queueing networks", Manuscript, Technion, Haifa, Israel.

[128] Katkovnik, V.Y. (1976). *Linear Estimation and Stochastic Optimization Problems (Parametric Smoothing Operators Method)*, Nauka, Moscow, 487 pp. (in Russian).

[129] Katkovnik, V.Y. (1985). *Nonparametric Identification and Smoothing of Data (Local Approximation Method)*, Nauka, Moscow, 386 pp. (in Russian).

[130] Katkovnik, V.Y. and Antonov, G.E. (1972). "Generalization of the concept of statistical gradient", *Autom. Remote Control*, 33 (6), 26–33.

[131] Katkovnik, V.Y. (1992) "Sensitivity analysis of discrete event systems by the local approximation methods", Manuscript, Technion, Haifa, Israel.

[132] King, A.J. (1986). "Asymptotic behavior of solutions in stochastic optimization: nonsmooth analysis and the derivation of non-normal limit distributions", Dissertation, University of Washington.

[133] King, A.J. (1989). "Generalized delta theorems for multivalued mappings and measurable selections", *Mathematics of Operations Research*, 14, 720–736.

[134] King, A.J. and Rockafellar, R.T. (1992a). "Sensitivity analysis for nonsmooth generalized equations", *Mathematical Programming*, to appear.

[135] King, A.J. and Rockafellar, R.T. (1992b). "Asymptotic theory for solutions in statistical estimation and stochastic programming", *Mathematics of Operations Research*, to appear.

[136] Kleijnen, J.P.C. (1974). *Statistical Techniques in Simulation*, Part 1, Marcel Dekker, New York.

[137] Kleijnen, J.P.C. (1986). *Statistical Tools For Simulation Practitioners*, Marcel Dekker, New York.

[138] Kleijnen, J.P.C. (1991). "Sensitivity analysis of simulation experiments: tutorial on regression analysis and statistical design", Tilburg University, Tilburg, The Netherlands

[139] Kleinrock, L. (1975). *Queueing Systems*, Vols. I and II, Wiley , New York.

[140] Kreimer, J. (1984). "Stochastic optimization—an adaptive approach", D.Sc. Thesis, Technion, Haifa, Israel.

[141] Kreimer, J. and Rubinstein, R. (1988). "Smoothed functionals and constrained optimization", *SIAM Journal of Numerical Analysis*, 25, 3.

[142] Kushner, H.J. and Clark, D.S. (1978). *Stochastic Approximation Methods for Constrained and Unconstrained Systems*, Springer-Verlag, Applied Math. Sciences, Vol. 26.

[143] Kyparisis, J. (1985). "On uniqueness of Kuhn-Tucker multipliers in nonlinear programming", *Mathemathical Programming*, 32, 242–246.

[144] Lavenberg, S.S. and Welch, P.D. (1981). "A perspective on the use of control variables to increase the efficiency of Monte Carlo simulation", *Management Science*, 27 (3), 322–335.

[145] Lavenberg, S.S., Moeller, T.L. and Welch, P.D. (1982). "Statistical results on control variables with application to queueing network simulation", *Operations Research*, 30, 182–202.

[146] Law, A.M. and Kelton, W.D. (1991). *Simulation Modeling and Analysis*, McGraw-Hill, New York.

[147] LeCam, L. (1953). "On some asymptotic properties of maximum likelihood estimates and related Bayes' estimates", *University of California Statistics Publication*, 1, 277–329.

[148] L'Ecuyer, P.L (1990a). "A unified version of the IPA, SF, and LR gradient estimation techniques", *Management Science*, 36 (11), 1364–1383.

[149] L'Ecuyer, P.L. (1990b). "Random numbers for simulation", *Communications of the ACM*, 33 (10), 85–98.

[150] L'Ecuyer, P.L (1991a). "An overview of derivative estimation", *Proceedings of the 1991 Winter Simulation Conference*, IEEE Press, pp. 207–217.

[151] L'Ecuyer, P.L. (1991b). "On the interchange of derivative and expectation, and finite moments, for likelihood ratio derivative estimators", Manuscript, Départment d'IRO, Université de Montréal, C.P. Submitted for publication.

[152] L'Ecuyer, P.L. (1992). "Convergence rates for steady-state derivative estimators", To appear in the *Annals of Operations Research.*

[153] L'Ecuyer, P.L. and Côté, S. (1990). "Implementing a random number package with splitting facilities", *ACM Trans. on Math. Softw.* To be published.

[154] L'Ecuyer, P.L., Giroux, N. and Glynn, P.W. (1991). "Stochastic optimization by simulation: convergence proofs and experimental results for the GI/G/1 queue", Report No. G-91-49, GERAD, Université de Montréal.

[155] L'Ecuyer, P.L and Perron, G. (1992). "On the convergence rates of IPA and FDC derivative estimators for finite-horizon stochastic simulations, to appear in *Operations Research.*

[156] L'Ecuyer, P.L. and Tezuka, S. (1991). "Structural properties for two classes of combined generators", *Mathematics of Computation*, 57 (196), 735–746.

[157] Lehmann, E.L. (1983). *Theory of Point Estimation*, Wiley, New York.

[158] Lempio, F, and Maurer, H. (1980). "Differential stability in infinite-dimensional nonlinear programming", *Appl. Math. Optimization*, 6, 139–152.

[159] Levitin, E.S. (1974). "On differential properties of the optimum value of parametric problems of mathematical programming", *Soviet Math. Dokl.*, 15, 603–608.

[160] Lewis, E.E. and Bohm, F. (1984). "Monte Carlo simulation of Markov unreliability models", *Nuclear Engineering and Design*, 77 (1), 49–62.

[161] Lewis, P.A.W. and McKenzie, E. (1991). "Minification processes and their transformations", *J. Appl. Prob.*, 28, 45–57.

[162] Linnik, I. Yu (1972). *Improving the Convergence of the Monte Carlo Method in Certain Queueing Problems.* Translated from Russian into English in 1975, Published by Plenum Publishing Corp., New York, pp. 879–883.

[163] Lirov, Y. and Melamed, B. (1992). "Distributed expert systems for queueing networks capacity planning", *Annals of Operations Research*, to appear.

[164] Ljusternik, L.A. (1934). "Conditional extrema of functionals", *Matem. Sbornik*, 41, 390–401.

[165] Loeve, M. (1977). *Probability Theory I*, Springer-Verlag, New York.

[166] Luke, Y. K. (1969). *The Special Functions and Their Approximations I, II*, Academic Press, New York.

[167] Mangasarian, O.L. and Fromovitz, S. (1967). "The Fritz John necessary optimality conditions in the presence of equality and inequality contraints", *J. Math. Anal. Appl.*, 7, 37–47.

[168] Mann, H.B. and Wald, A. (1943). "On stochastic limit and order relationships", *Annals of Mathematical Statistics*, 14, 217–226.

[169] Marshall, A.W and Olkin, I. (1979). *Inequalities: Theory of Majorization and its Applications*, Academic Press, New York.

[170] McLeish, D.L. (1992). Private communications.

[171] McLeish, D.L. and Rollans, S. (1992). "Conditioning for variance reduction in estimating the sensitivity of simulations", To appear in *Annals of Operations Research.*

[172] Meketon, M.S. (1987). "Optimization in simulation: a survey of recent results", *Proceedings of WSC'87*, The Society for Computer Simulation, pp. 58–67.

[173] Melamed, B. (1991). "A class of methods for generating autocorrelated uniform variates", *ORSA J.Comput.,* 3 (4), 317–329.

[174] Melamed, B. (1992). Private communications.

[175] Melamed, B. and Rubinstein, R.Y. (1992) "Another look at sensitivity analysis of discrete event systems", Manuscript, NEC USA, Inc. C&C Research Laboratories, 4 Independent Way, Princeton, NJ 08540.

[176] Mikhailov, G.A. (1967). "Calculation of system parameter derivatives of functionals of the solutions to the transport equations", *Zh. Vychisyl. Mat. Mat. Fiz.*, 7, 915.

[177] Mikhailov, G.A. (1974). *Some Problems in the Theory of the Monte Carlo Method*, Nauka, Novosibirsk, U.S.S.R. (in Russian).

[178] Miller, L.B. (1967). "Monte Carlo analysis of reactivity coefficients in fast reactors: general theory and applications", ANL-7307 (TID-4500), Argonne National Laboratory, IL.

[179] Nashed, M.Z. (1971). "Differentiability and related properties of nonlinear operators: some aspects of the role of differentials in nonlinear functional analysis", in: L.B. Rall (ed), *Nonlinear Functional Analysis and Applications*, Academic Press, New York.

[180] Nelson, B.L. (1987). "A perspective on variance reduction in dynamic simulation experiments", *Commun. Statist.*, B16, 385–426.

[181] Nemirovskii, A. and R.Y. Rubinstein (1992). "Robust stochastic approximation with application to discrete event systems", Manuscript, Technion, Haifa, Israel.

[182] Parekh, S. and Walrand, J. (1988). "Quick simulation of excessive backlogs in networks of queues", in: W. Fleming and P.L. Lions (eds.), *Stochastic Differential Systems, Stochastic Control Theory and Applications*, IMA Vol. 10, Springer-Verlag, New York, pp. 439–472.

[183] Perez-Luna, A. (1990), "Sensitivity analysys and optimization of queueing networks by the score function method'. Master's Thesis, Technion, Haifa, Israel.

[184] Pflug, G. Ch. (1988). "Sensitivity analysis of semi-Markovian processes", Manuscript University of Giessen, F.R.G.

[185] Pflug, G. Ch. (1989). "Sampling derivatives of probabilities", *Computing*, 42, 315–328.

[186] Pflug, G. Ch. (1990). "On-line optimization of simulated Markovian processes, *Mathematics of Operations Research*, 15 (3), 381–395.

[187] Pflug, G. Ch. (1991). "Simulation and optimization: the interface", in preparation.

[188] Pflug, G. Ch. (1992). "Optimization of simulated discrete events processes", *Annals of Operations Research*, to appear.

[189] Pollard, D. (1984). *Convergence of Stochastic Processes*, Springer-Verlag, New York.

[190] Pollard, D. (1989). "Asymptotics via empirical processes", *Statistaical Sciences*, 4, 341–366.

[191] Polyak, D.G. (1970a). "Precision of statistical simulation of queueing systems", *Eng. Cybern.* (NY), $N_0$1, 72–80.

[192] Polyak, D.G. (1970b). "Increasing the accuracy of simulation queueing systems", *Eng. Cybern.* (NY), $N_0$4, 687–695.

[193] Powell, M.I.D. and Swann, I. (1966). "Weighted uniform sampling—a Monte Carlo technique for reducing variance", *J. Inst. Math. App.*, 2, 228–238.

[194] Raz, I. (1991). "Efficiency of the score function method for sensitivity analysis and optimization of queueing networks", Master's Thesis, Technion, Haifa, Israel.

[195] Reeds, J.A. (1976). "On the definition of von Mises functionals", Ph.D. thesis, Harvard University.

[196] Reiman, M.I. and Weiss, A. (1989a). "Sensitivity analysis for simulations via likelihood rations", *Operations Research*, 37 (5), 830–844.

[197] Reiman, M.I. and Weiss, A. (1989b). "Light traffic derivatives via likelihood ratios", *IEEE Transactions on Information Theory*, 35 (3), 648–654.

[198] Reynolds, J.F. (1975). "The covariance structure of queues and related stochastic processes—a survey of recent work", *Adv. Appl. Prof.*, 7, 383–415.

[199] Rief. H. (1984). "Generalized Monte Carlo perturbation algorithms for correlated sampling and a second order Taylor series approach, *Ann. Nucl. Energy*, 11, 455.

[200] Rief, H. (1988). "Monte Carlo uncertainty analysis", in: Y. Ronen (ed.), *CRC-handbook on Uncertainty Analysis*, CRC Press, Boca Raton, Florida.

[201] Rief, H. (1991). "A synopsis of Monte Carlo perturbation algorithms", Manuscript.

[202] Robinson, S.M. (1976). "Stability theory for systems of inequalities, Part II: Differentiable nonlinear system", *SIAM J. Numer. Anal.*, 13, 497–513.

[203] Robinson, S.M. (1979). "Generalized equations and their solutions, Part I: basic theory", *Mathematical Programming Study*, 10, 128–141.

[204] Robinson, S.M. (1982). "Generalized equations and their solutions, Part II: applications to nonlinear programming", *Mathematical Programming Study*, 19, 200–221.

[205] Rockafellar, R.T. (1970). *Convex Analysis*, Princeton University Press, Princeton, New Jersey.

[206] Rockafellar, R.T. (1976). "Integral functionals, normal integrands and measurable selections", in *Nonlinear Operators and the Calculus of Variations*, Lecture Notes in Mathematics, 543, Springer-Verlag, Berlin, pp. 157-207.

[207] Ross, S.M. (1989). *Introduction to Probability Models*, 4th ed., Academic Press, San Diego.

[208] Rubinstein, R.Y. (1969). "Some problems in Monte Carlo Optimization", PhD Thesis, Riga, Latvia.

[209] Rubinstein, R.Y. (1976). "A Monte Carlo method for estimating the gradient in a stochastic network", Manuscript, Technion, Haifa, Israel.

[210] Rubinstein, R.Y. (1981). *Simulation and the Monte Carlo Method*, John Wiley , New York.

[211] Rubinstein, R.Y. (1986a). *Monte Carlo Optimization Simulation and Sensitivity of Queueing Network*, John Wiley, New York.

[212] Rubinstein, R.Y. (1986b). "The score function approach for sensitivity analysis of computer simulation models", *Mathematics and Computers in Simulation*, 28, 1–29.

[213] Rubinstein, R.Y. (1989). "Sensitivity analysis of computer simulation models via the efficient score", *Operations Research*, 37 (1), 72–81.

[214] Rubinstein, R. (1991). "How to optimize discrete-event systems from a single sample path by the score function method", *Annals of Operations Research*, 27, 175–212.

[215] Rubinstein, R.Y. (1992a). "The 'push out' method for sensitivity analysis of discrete event systems", *Annals of Operations Research*, to appear.

[216] Rubinstein, R.Y. (1992b). "Decomposable score fFunction estimators for sensitivity analysis and optimization of queueing networks", *Annals of Operations Research*, to appear.

[217] Rubinstein, R.Y. (1992c). "Case studies with the score function method for sensitivity analysis and stochastic optimization of queueing networks", in preparation.

[218] Rubinstein, R.Y. (1992d). "Monte Carlo methods for performance evaluation, sensitivity analysis and optimization of stochastic systems", in: A. Kent and J.G. Williams (eds.), *Encyclopedia of Com-*

puter Science and Technology, Vol. 25, Marcel Dekker, New York, pp. 211–233.

[219] Rubinstein, R.Y. and Marcus, R. (1985). "Efficiency of multivariate control variates in Monte Carlo Simulation", *Opererations Research* 33 (3), 661–677.

[220] Rubinstein, R.Y. and Shalmon, M. (1990). "The variance of regenerative estimators with special reference to the sensitivity analysis of queueing systems with Poisson arrivals", First International Conference on Operations Research in Telecommunications, Boca Raton, Florida.

[221] Rubinstein, R.Y. and Shapiro, A. (1990). "Optimization of simulation models by the score function method", *Mathematics and Computers in Simulation*, 32, 373–392.

[222] Rubinstein, R.Y. and Shapiro, A. (1992). "On the optimal choice of reference parameters in the score function method". Manuscript, School of Industrial and Systems Engineering, Georgia Institute of Technology, Atlanta, Georgia 30332-0205, U.S.A.

[223] Schmeiser, B.W. (1980). "Random variate generation: a survey", *Proceedings of 1980 Winter Simulation Conference*, Orlando, Florida, pp. 79–104.

[224] Schmeiser, B.W. (1982). "Batch size effects in the analysis of simulation output", *Operations Research*, 30, 556–568.

[225] Schruben, L.W. (1982). "Detecting initialization bias in simulation output", *Operations Research*, 30, 569–590.

[226] Schruben, L.W. (1983). "Confidence interval estimation using standardized time series", *Operations Research*, 31, 1090–1108.

[227] Serfling, R.J. (1980). *Approximation Theorems of Mathematical Statistics*, Wiley , New York.

[228] Sernik, E.L. and Marcus, S.I. (1990). "Likelihood ratio estimates applied to a Markovian replacement problem", Manuscript, University of Texas at Austin, Texas 78712, U.S.A.

[229] Shaked, M. and Shanthikumar, J.G. (1988). "Stochastic convexity and its applications", *Advances of Applied Probability*, 20 (2), 427–447.

[230] Shalmon, M. (1991). "The moments of the regenerative estimators for the M/G/1 queue and their interpretation as functionals of an associated branching process", presented at the 20th Conference on Stochastic Processes and Their Applications, Israel.

[231] Shalmon M. and Rubinstein, R.Y. (1990). "The variance of regenerative estimators with special reference to the sensitivity analysis of queueing systems with Poisson arrivals", First International Conference on Operations Research in Telecommunications, Boca Raton, Florida.

[232] Shapiro, A. (1985). "Second order sensitivity analysis and asymptotic theory of parameterized nonlinear programs", *Mathematical Programming*, 33, 280–299.

[233] Shapiro, A. (1988). "Sensitivity analysis of nonlinear programs and differentiability properties of metric projections", *SIAM J. Control and Optimization*, 26, 628–645.

[234] Shapiro, A. (1989). "Asymptotic properties of statistical estimators in stochasic programming", *Annals of Statistics*, 17, 841–858.

[235] Shapiro, A. (1990a). "On concepts of directional differentiability", *Journal of Optimization Theory and Applications*, 66, 477–487.

[236] Shapiro, A. (1990b). "On differential stability in stochastic programming", *Mathematical Programming*, 47, 107–116.

[237] Shapiro, A. (1991b). "Asymptotic analysis of stochastic programs", *Annals of Operations Research*, 30, 169–186.

[238] Shapiro, A. (1991b). "Asymptotic behavior of optimal solutions in stochastic programming", *Mathematics of Operations Research*, to appear.

[239] Shapiro, A. (1992). "Perturbation analysis of optimization problems in Banach spaces", *Numer. Funct. Anal. and Optimiz.*, 13, 97–116.

[240] Shwartz A. and Weiss, A. (1991). "Induced rare events: analysis via large deviations and time reversal", Manuscript, Electrical Engineering, Technion, Israel and AT&T Laboratories, Murray Hill, NJ.

[241] Siegmund, D. (1976). "Importance sampling in Monte Carlo study of sequential tests", *Annals of Statistics*, 4, 673–684.

[242] Sigman, K. (1990). "The stability of open queueing networks", *Stoch. Proc. and Appl.*, 35, 11–25.

[243] Sterental, J. (1989). 'Performance evaluation, sensitivity analysis and optimization of computer simulation models". M.Sc. Thesis, Technion—Israel Institute of Technology, Haifa, Israel.

[244] Stroock, D. (1984). *An Introduction to the Theory of Large Deviations*, Springer-Verlag, New York.

[245] Suri, R. (1983). "Implementation of sensitivity calculation on a Monte Carlo experiment", *Journal of Optimization Theory and Applications*, 40 (4), 625–630.

[246] Suri, R. (1987). "Infinitesimal perturbation analysis of general discrete event dynamic systems", *Journal of the ACM*, 34 (3), 686–717.

[247] Suri, R. (1989). "Perturbation analysis: the state of the art and research issues explained via the GI/G/1 queue", *Proceedings of the IEEE*, 77(10, 114–137.

[248] Suri, R. and Zazanis, M.A. (1988). "Perturbation analysis gives strongly consistent sensitivity estimates for the M/G/L queue", *Management Science.* 34 (1), 39-64.

[249] Szidarovsky, F. and Yakowitz, S. (1978). *Principles and Procedures of Numerical Analysis*, Plenum , New York.

[250] Topia, R.A. and Thompson , J.R.(1978) *Nonparametric Density Estimation*, Johns Hopkins University Press, Baltimore.

[251] Ursescu, C. (1975). "Multifunctions with convex closed graph", *Czechoslovak Math. J.*, 25, 438–441.

[252] Uryas'ev, S. (1992). "Derivatives of probability functions and integrals over sets given by inequalities", Working Paper WP-91-38, International Institute for Applied Systems Analysis, A-2361 Laxenburg, Austia .

[253] Vázquez-Abad, F.J. and Kushner, H.J. (1992). "Estimation of the derivative of a stationary measure with respect to a control parameter", to appear in *J.Appl. Prob.*

[254] Vázquez-Abad, F.J. and Kushner, H.J. (1991). "A surrogate estimation approach for adaptive routing in communication networks", Submitted for publication.

[255] Von Mises, R. (1947). "On the asymptotic distribution of differentiable statistical functions", *Ann. Math. Statist.*, 18, 309–348.

[256] Wald, A. (1949). "Note on the consistency of the maximum likelihood estimate", *Ann. Math. Statist.*, 20, 595–601.

[257] Walkup, D.W. and Wets, R.J.B. (1969). "A Lipschitzian characterization of convex polyhedra", *Proceedings of the American Mathematical Society*, 23, 167–173.

[258] Walrand, J. (1987). "Quick simulation of queueing networks: an introduction", *2nd International Workshop on Applied Mathematics and Performance/Reliability Models of Computer/Communication Systems*, University of Rome II, pp. 275–286.

[259] Wardi, Y. (1990) "Stochastic Algorithms with Armijo Step Sizes for Minimization of Functions," *Journal of Optimization Theory and Applications*, Vol. 64, pp. 399-417.

[260] Wardi, Y. and Lee, K. (1991) "Application of Descent Algorithms with Armijo Stepsizes to Simulation-Based Optimization of Queueing Networks," in *Proc. 30th Conf. on Decision and Control*, pp. 110-115.

[261] Wardi, Y., Kellmans, M.H., Cassandras, C.G. and Gong, W.-B. (1992). "Smoothed perturbation analysis algorithms for estimating the derivatives of occupancy-related functions in serial queueing networks", *Annals of OR* (to appear).

[262] Weisman, I. (1992). Private correspondence.

[263] Whitt, W. (1983). "The queueing network analyzer", *BSTJ*, 62 (9), 2779–2813.

[264] Whitt, W. (1989a). "Simulation run length planning", *Proceedings of the 1989 Winter Simulation Conference*, pp. 106-112.

[265] Whitt, W. (1989b). "Planning queueing simulations", *Management Science*, 35 (11), 1341–1366.

[266] Whitt, W. (1991). "The efficiency of one long run versus independent replications in steady-state simulation", *Management Science*, 37 (6) 645–666.

[267] Whitt, W. (1992). "Asymptotic formulas for Markov processes with applications to simulation", *Operations Research*, 40,

[268] Wilson, J.R. (1984). "Variance reduction techniques for digital simulation", *Am. J. Math. Sci.*, 4, 277–312.

[269] Wilson, J.R. and Pritsker, A.A.B. (1984a). "Variance reduction in queueing simulation using generalized concomitant variables", *J. Statist. Comput. Simul.*, 19, 129–153.

[270] Wilson, J.R. and Pritsker, A.A.B. (1984b). "Experimental evaluation of variance reduction techniques for queueing simulation using generalized concomitant variables", *Management Science*, 30, 1459–1472.

[271] Zhang, B. and Ho, Y.C. (1989). "Performance gradient estimation for very large Markov chains", Manuscript, Harvard University.

[272] Zhuguo, T. and Lewis, E.E. (1985). "Component dependency models in Markov Monte Carlo simulation", *Reliability Engineering*, (13), 45–61.

Index